T0336188

UNIT EQUATIONS IN DIOPHANTINE NUMBER THEORY

Diophantine number theory is an active area that has seen tremendous growth over the past century, and in this theory unit equations play a central role. This comprehensive treatment is the first volume devoted to these equations. The authors gather together all the most important results and look at many different aspects, including effective results on unit equations over number fields, estimates on the number of solutions, analogues for function fields, and effective results for unit equations over finitely generated domains. They also present a variety of applications. Introductory chapters provide the necessary background in algebraic number theory and function field theory, as well as an account of the required tools from Diophantine approximation and transcendence theory. This makes the book suitable for young researchers as well as for experts who are looking for an up-to-date overview of the field.

Jan-Hendrik Evertse works at the Mathematical Institute of Leiden University. His research concentrates on Diophantine approximation and applications to Diophantine problems. In this area he has obtained some influential results, in particular on estimates for the numbers of solutions of Diophantine equations and inequalities. He has written more than 75 research papers and co-authored one book with Bas Edixhoven entitled *Diophantine Approximation and Abelian Varieties*.

Kálmán Győry is Professor Emeritus at the University of Debrecen, a member of the Hungarian Academy of Sciences and a well-known researcher in Diophantine number theory. Over his career he has obtained several significant and pioneering results, among others on unit equations, decomposable form equations, and their various applications. His results have been published in one book and 160 research papers. Győry is also the founder and leader of the number theory research group in Debrecen, which consists of his former students and their students.

CAMBRIDGE STUDIES IN ADVANCED MATHEMATICS

Editorial Board:
B. Bollobás, W. Fulton, A. Katok, F. Kirwan, P. Sarnak, B. Simon, B. Totaro

All the titles listed below can be obtained from good booksellers or from Cambridge University Press.
For a complete series listing visit www.cambridge.org/mathematics.

Unit Equations in Diophantine Number Theory

JAN-HENDRIK EVERTSE
Universiteit Leiden

KÁLMÁN GYŐRY
Debreceni Egyetem, Hungary

CAMBRIDGE
UNIVERSITY PRESS

CAMBRIDGE
UNIVERSITY PRESS

University Printing House, Cambridge CB2 8BS, United Kingdom

One Liberty Plaza, 20th Floor, New York, NY 10006, USA

477 Williamstown Road, Port Melbourne, VIC 3207, Australia

314-321, 3rd Floor, Plot 3, Splendor Forum, Jasola District Centre, New Delhi - 110025, India

79 Anson Road, #06-04/06, Singapore 079906

Cambridge University Press is part of the University of Cambridge.

It furthers the University's mission by disseminating knowledge in the pursuit of education, learning and research at the highest international levels of excellence.

www.cambridge.org
Information on this title: www.cambridge.org/9781107097605

© Jan-Hendrik Evertse and Kálmán Győry 2015

First published 2015

A catalogue record for this publication is available from the British Library

ISBN 978-1-107-09760-5 Hardback

Contents

Preface

Diophantine number theory (the study of Diophantine equations, Diophantine inequalities and their applications) is a very active area in number theory with a long history. This book is about *unit equations*, a class of Diophantine equations of central importance in Diophantine number theory, and their applications. Unit equations are equations of the form

$$a_1 x_1 + \cdots + a_n x_n = 1$$

to be solved in elements $x_1, \ldots, x_n$ from a finitely generated multiplicative group Γ, contained in a field K, where $a_1, \ldots, a_n$ are non-zero elements of K. Such equations were studied originally in the cases where the number of unknowns $n = 2$, K is a number field and Γ is the group of units of the ring of integers of K, or more generally, where Γ is the group of S-units in K. Unit equations have a great variety of applications, among others to other classes of Diophantine equations, to algebraic number theory and to Diophantine geometry.

Certain results concerning unit equations and their applications covered in our book were already presented, mostly in special or weaker form, in the books of Lang (1962, 1978, 1983), Győry (1980b), Sprindžuk (1982, 1993), Evertse (1983), Mason (1984), Shorey and Tijdeman (1986), de Weger (1989), Schmidt (1991), Smart (1998), Bombieri and Gubler (2006), Baker and Wüstholz (2007) and Zannier (2009), and in the survey papers of Evertse, Győry, Stewart and Tijdeman (1988b), Győry (1992a, 1996, 2002a, 2010) and Bérczes, Evertse and Győry (2007b).

In 1988, we wrote, together with Stewart and Tijdeman, the survey Evertse, Győry, Stewart and Tijdeman (1988b) on unit equations and their applications giving the state of the art of the subject at that time. Since then, the theory of unit equations has been greatly expanded. In the present book we have tried to give a comprehensive and up-to-date treatment of unit equations and their applications. We prove effective finiteness results for unit equations in two unknowns, describe practical algorithms to solve such equations, give explicit upper bounds for the number of solutions, discuss analogues of unit equations over function fields and over finitely generated domains, and present various applications. The proofs of the results concerning unit equations are mostly based on the very powerful Thue–Siegel–Roth–Schmidt theory from Diophantine approximation and Baker's theory from transcendence theory. We note that there are other important methods and applications, some discovered very recently, that deserve a detailed discussion, but to which we could pay only little or no attention due to lack of time and space.

The present book is the first in a series of two. The second book, titled *Discriminant Equations in Diophantine Number Theory*, also published by Cambridge University Press, is about polynomials and binary forms of given discriminant, with applications to algebraic number theory, Diophantine approximation and Diophantine geometry. There, we will apply the results from the present book. The contents of these two books are an outgrowth of research, done by the two authors since the 1970s.

The present book is aimed at anybody (graduate students and experts) with basic knowledge of algebra (groups, commutative rings, fields, Galois theory) and elementary algebraic number theory. For convenience of the reader, in part I of the book we have provided some necessary background.

Acknowledgments

We are very grateful to Yann Bugeaud, Andrej Dujella, István Gaál, Rafael von Känel, Attila Pethő, Michael Pohst, Andrzej Schinzel and two anonymous referees for carefully reading and critically commenting on some chapters of our book, to Csaba Rakaczki for his careful typing of a considerable part of this book, and to Cambridge University Press, in particular David Tranah, Sam Harrison and Clare Dennison, for their suggestions for and assistance with the final preparation of the manuscript.

The research of the second named author was supported in part by Grants 100339 and 104208 from the Hungarian National Foundation for Scientific Research (OTKA).

Summary

We start with a brief historical overview and then outline the contents of our book. Thue (1909) proved that if $F \in \mathbb{Z}[X, Y]$ is a binary form (i.e., a homogeneous polynomial) of degree at least 3 which is irreducible over $\mathbb{Q}$ and if δ is a non-zero integer, then the equation

$$F(x, y) = \delta \text{ in } x, y \in \mathbb{Z}$$

(nowadays called a *Thue equation*) has only finitely many solutions. To this end, Thue developed a very original Diophantine approximation method concerning the approximation of algebraic numbers by rationals, which was extended later by Siegel, Dyson, Gelfond and Roth.

Thue's result was generalized by Siegel (1921) as follows. Let K be an algebraic number field of degree d with ring of integers O_K, let $F \in O_K[X, Y]$ be a binary form of degree $n > 4d^2 - 2d$ such that $F(1, 0) \neq 0$ and $F(X, 1)$ has no multiple zeros, and let δ be a non-zero element of O_K. Then the equation

$$F(x, y) = \delta \text{ in } x, y \in O_K$$

has only finitely many solutions. This has the following interesting consequence, which was not stated explicitly by Siegel, but which was implicitly proved by him. Denote by O_K^* the group of units of O_K. Let a_1, a_2 be non-zero elements of the number field K. Then the equation

$$a_1 x_1 + a_2 x_2 = 1 \tag{1}$$

has only finitely many solutions in $x_1, x_2 \in O_K^*$. To prove this, choose an integer $n > 4d^2 - 2d$. By Dirichlet's Unit Theorem, the group O_K^* is finitely generated, and thus, any solution $x_1, x_2 \in O_K^*$ of (1) can be written as $x_i = \beta_i \varepsilon_i^n$ for $i = 1, 2$ with $\beta_i, \varepsilon_i \in O_K^*$, such that β_i may assume only finitely many values. Thus, we get a finite number of Thue equations

$$a_1 \beta_1 \varepsilon_1^n + a_2 \beta_2 \varepsilon_2^n = 1,$$

each of which has only finitely many solutions in $\varepsilon_1, \varepsilon_2$.

Mahler (1933a) proved another generalization of Thue's theorem. Let $F \in \mathbb{Z}[X, Y]$ be a binary form of degree $n \geq 3$ such that $F(1, 0) \neq 0$ and $F(X, 1)$ has no multiple zeros, and let $p_1, \ldots, p_t$ be distinct primes. Then the equation

$$F(x, y) = \pm p_1^{z_1} \cdots p_t^{z_t}$$

(today called a *Thue–Mahler equation*) has only finitely many solutions in integers $x, y, z_1, \ldots, z_t$ with $\gcd(x, y) = 1$. A consequence of this result, proved by Mahler in a slightly different formulation, is as follows. Let a_1, a_2 be non-zero rational numbers and let Γ be the multiplicative group generated by -1, $p_1, \ldots, p_t$. Then (1) has only finitely many solutions in $x_1, x_2 \in \Gamma$. The argument is similar to that above. By extending the set of primes $p_1, \ldots, p_t$, we may assume that the numerators and denominators of a_1, a_2 are composed of primes from $p_1 \cdots p_t$. Then, by clearing denominators, we can rewrite (1) as

$$u + v = w,$$

where u, v, w are integers, composed of primes from $p_1, \ldots, p_t$, with $\gcd(u, v, w) = 1$. Choose $n \geq 3$. Then we may write u as ax^n and v as by^n, where a, b, x, y are integers composed of primes from $p_1, \ldots, p_t$ and a, b are from a finite set independent of x_1, x_2. Thus, equation (1) can be reduced to a finite number of Thue–Mahler equations as above with $F = aX^n + bY^n$ which all have only finitely many solutions.

Lang (1960) considered equation (1) with unknowns x_1, x_2 taken from a finitely generated multiplicative group, and was the first to realize the central importance of this equation. He proved the general result that if a_1, a_2 are non-zero elements from an arbitrary field K of characteristic 0 and Γ is an arbitrary finitely generated multiplicative subgroup of K^*, then (1) has only finitely many solutions in elements $x_1, x_2 \in \Gamma$. Inspired by Siegel's original result, equations of type (1) with unknowns from a finitely generated multiplicative group are called *unit equations* (in two unknowns), although the group Γ need not be the unit group of a ring. The proofs of all results mentioned above are based on extensions of Thue's method, which are ineffective in the sense that they do not provide a method to determine the solutions of the equations considered above.

In the 1960s, A. Baker developed a new method in transcendence theory, giving non-trivial effective lower bounds for linear forms in logarithms of algebraic numbers. This turned out to be a very powerful tool to prove *effective* finiteness results for Diophantine equations, that enable one to determine all solutions of the equation, at least in principle. With this method, and extensions thereof, it became possible to give explicit upper bounds for the heights of the solutions of Thue equations and Thue–Mahler equations, and also for the heights of the solutions of equations (1) in units of the ring of integers of a number field or more generally, in *S-units*, these are elements in the number field in whose prime ideal factorizations only prime ideals from a prescribed, finite set *S* occur. Baker (1968b) obtained explicit upper bounds for the solutions of Thue equations. His result was extended by Coates (1969) to Thue–Mahler equations. For explicit upper bounds for the heights of the solutions of unit equations and *S*-unit equations in two unknowns, see Győry (1972, 1973, 1974, 1979), and the many subsequent improvements discussed in Chapter 4. The bounds enabled one to determine, at least in principle, all solutions. Since the 1980s, practical algorithms have been developed, combining Baker's theory with the Lenstra–Lenstra–Lovász (LLL) lattice basis reduction algorithm and enumeration techniques, which allow one to solve in practice concrete Thue equations, Thue–Mahler equations and (*S*-) unit equations, see for instance de Weger (1989), Wildanger (1997) and Smart (1998).

In the 1960s and early 1970s, Schmidt developed his higher dimensional generalization of the Thue–Siegel–Roth method, leading to his *Subspace Theorem* in Schmidt (1972). Schlickewei (1977b) proved an extension of the Subspace Theorem, involving both archimedean and non-archimedean absolute values. Using this so-called *p*-adic Subspace Theorem, several authors obtained finiteness results for the number of soultions of unit equations in an arbitrary number of unknowns, i.e., for linear equations

$$a_1 x_1 + \cdots + a_n x_n = 1 \text{ in } x_1, \ldots, x_n \in \Gamma, \tag{2}$$

where $a_1, \ldots, a_n$ are non-zero elements, and Γ is a finitely generated multiplicative group in a field K of characteristic 0, see Dubois and Rhin (1976), Schlickewei (1977a), Evertse (1984b), Evertse and Győry (1988b) and van der Poorten and Schlickewei (1982, 1991). We mention that the *p*-adic Subspace Theorem is ineffective, and so its consequences for equation (2) are ineffective. It is still open to solve unit equations of the form (2) in more than two unknowns effectively.

In part I of the book, consisting of the first three chapters, we have collected some basic tools. Chapter 1 gives a collection of the results from elementary algebraic number theory that we need throughout the book. In Chapter 2 we recall some basic facts about algebraic function fields. These are used in Chapters 7 and 8. In Chapter 3 we have stated without proof some fundamental results from Diophantine approximation and transcendence theory. We have included some versions of the Subspace Theorem, due to Schmidt, Schlickewei and Evertse, and estimates of Matveev (2000) and Yu (2007) concerning linear forms in logarithms, which are used in Chapters 4, 5 and 6.

Part II, consisting of the other chapters, is the main body of our book. Chapter 4 provides a survey of effective results concerning unit equations in two unknowns over number fields. We derive among others the best effective upper bounds to date, established in Győry and Yu (2006), for the solutions of equation (1) in S-units of a number field. For applications, we give the bounds in completely explicit form. The main tools in the proofs are the results on linear forms in logarithms mentioned above.

In Chapter 5 we address the problem of practically solving concrete equations of the form (1) in units and S-units. Here, we combine estimates for linear forms in logarithms as mentioned in Chapter 3 with the LLL lattice basis reduction algorithm and an enumeration process.

In Chapter 6, we give an overview of the ineffective theory of unit equations in several unknowns. Among other things, we sketch a proof of the theorem of Evertse, Schlickewei and Schmidt (2002), giving an explicit upper bound for the number of those solutions of (2) for which the left side in (2) has no vanishing subsum. The bound depends only on the number n of unknowns and the rank of Γ. We also include a proof of the theorem of Beukers and Schlickewei (1996) which gives a similar, but sharper, result for equations in two unknowns. Further, we discuss some results giving lower bounds for the number of solutions of unit equations.

In Chapter 7, we deal with analogues over function fields of characteristic 0 of some of the effective and ineffective results discussed in Chapters 4 and 6. In particular, we present the Stothers–Mason abc-theorem due to Stothers (1981) and Mason (1984) for algebraic functions, and a result of Evertse and Zannier (2008) on the number of solutions of unit equations in two unknowns over function fields, analogous to the result of Beukers and Schlickewei mentioned above. Further, we give a brief overview of recent results on unit equations over function fields of positive characteristic.

In Chapter 8, the effective results of Chapters 4 and 7 on S-unit equations in two unknowns over number fields and over function fields are combined with some effective specialization argument to prove a general effective finiteness theorem, due to Evertse and Győry (2013), on the solutions of equation (1) in units x_1, x_2 of an arbitrary, effectively given finitely generated integral domain A over $\mathbb{Z}$.

Chapter 9 deals with applications of unit equations to decomposable form equations, which are higher dimensional generalizations of Thue and Thue–Mahler equations. It is proved that unit equations in an arbitrary number of unknowns are in a certain sense equivalent to decomposable form equations, and in particular unit equations in two unknowns are equivalent to Thue equations. Further, a complete description of the set of solutions of decomposable form equations is presented. We give explicit upper bounds for the number of solutions when this number is finite. The bounds do not depend on the coefficients of the decomposable forms involved. We also discuss effective results for some important classes of decomposable form equations, including Thue equations, discriminant form equations, and certain norm form equations. The presented results have many applications, especially to algebraic number theory.

The results on unit equations have many further applications to other Diophantine problems. In Chapter 10 we have made a small selection. We give among other things applications to prime factors of sums of integers, additive unit representations in integral domains, dynamics of polynomial maps, arithmetic graphs, irreducible polynomials, equations and inequalities involving discriminants and resultants, power integral bases in number fields, Diophantine geometry, exponential-polynomial equations, and transcendence theory.

As was mentioned in the Preface, a number of applications of the results of the present book are given in our second book *Discriminant Equations in Diophantine Number Theory*.

At the end of several chapters there are Notes in which some historical remarks are made and further related results, generalizations and applications are mentioned.

PART I

Preliminaries

1

Basic algebraic number theory

We have collected some basic facts about algebraic number fields (finite field extensions of $\mathbb{Q}$), p-adic numbers, and related topics. For further details and proofs, we refer to Lang (1970), chapters I–V, Neukirch (1992), Kapitel I–III and Koblitz (1984).

In the present book, a *ring* is by default a commutative ring with unit element, and an *integral domain* is a commutative ring with unit element and without divisors of 0. Given a ring A, we denote by A^+ its underlying additive group, and by A^* its unit group (multiplicative group of invertible elements).

The ring of integers of an algebraic number field K, that is the integral closure of $\mathbb{Z}$ in K, is denoted by O_K.

1.1 Characteristic polynomial, trace, norm, discriminant

For the moment, let K be any field of characteristic 0. Choose an algebraic closure $\overline{K}$ of K. For every $\alpha \in \overline{K}$, there is a unique, monic, irreducible polynomial $f_\alpha \in K[X]$, such that $f_\alpha(\alpha) = 0$, and f_α divides g for every polynomial $g \in K[X]$ with $g(\alpha) = 0$. We call f_α the *monic minimal polynomial* of α.

Let $f \in K[X]$ be a non-zero polynomial. Then $f = a(X - \alpha_1) \ldots (X - \alpha_r)$ with $a \in K^*, \alpha_1, \ldots, \alpha_r \in \overline{K}$. We call $K(\alpha_1, \ldots, \alpha_r)$ the splitting field of f over K.

Let L be a finite extension of K of degree n. Then there are precisely n distinct K-isomorphic embeddings $L \hookrightarrow \overline{K}, \sigma_1, \ldots, \sigma_n$, say. The composition of the fields $\sigma_1(L), \ldots, \sigma_n(L)$ is called the normal closure of L over K. We define the *characteristic polynomial* of $\alpha \in L$ with respect to L/K by

$$\chi_{L/K,\alpha} := \prod_{i=1}^{n} (X - \sigma_i(\alpha)).$$

In fact, we have $\chi_{L/K,\alpha} = f_\alpha^{[L:K(\alpha)]}$ and $\chi_{L/K,\alpha}$ is the characteristic polynomial of the K-linear map $x \mapsto \alpha x$ from L to L. So $\chi_{L/K,\alpha} \in K[X]$.

3

We define the *trace* and *norm* of $\alpha \in L$ over K by

$$\text{Tr}_{L/K}(\alpha) := \sum_{i=1}^{n} \sigma_i(\alpha), \quad N_{L/K}(\alpha) := \prod_{i=1}^{n} \sigma_i(\alpha).$$

These are up to sign coefficients of $\chi_{L/K,\alpha}$. So we have

$$\text{Tr}_{L/K}(\alpha), \quad N_{L/K}(\alpha) \in K \quad \text{for } \alpha \in L. \tag{1.1.1}$$

Notice that $\text{Tr}_{L/K}$ is a K-linear map $L \to K$ and that $N_{L/K}$ is a multiplicative map $L \to K$. Further, the trace and norm are transitive in towers: let $M \supset L \supset K$ be a tower of finite extension fields; then

$$\left.\begin{array}{l} \text{Tr}_{M/K}(\alpha) = \text{Tr}_{L/K}(\text{Tr}_{M/L}(\alpha)), \\ N_{M/K}(\alpha) = N_{L/K}(N_{M/L}(\alpha)) \end{array}\right\} \quad \text{for } \alpha \in M.$$

Let again L be a finite extension of K of degree n. Take a K-basis $\{\omega_1, \ldots, \omega_n\}$ of L. Then the *discriminant* of this basis is given by

$$D_{L/K}(\omega_1, \ldots, \omega_n) := \det(\text{Tr}_{L/K}(\omega_i \omega_j))_{i,j=1,\ldots,n}.$$

By (1.1.1) we have $D_{L/K}(\omega_1, \ldots, \omega_n) \in K$. The discriminant can be expressed otherwise as

$$D_{L/K}(\omega_1, \ldots, \omega_n) = (\det(\sigma_i(\omega_j))_{i,j=1,\ldots,n})^2,$$

where $\sigma_1, \ldots, \sigma_n$ are the K-isomorphic embeddings of L in $\overline{K}$. For instance, if $L = K(\theta)$, then $\{1, \theta, \ldots, \theta^{n-1}\}$ is a K-basis of L and by Vandermonde's identity,

$$D_{L/K}(1, \theta, \ldots, \theta^{n-1}) = \prod_{1 \le i < j \le n} (\sigma_i(\theta) - \sigma_j(\theta))^2 \ne 0. \tag{1.1.2}$$

Let $\{\theta_1, \ldots, \theta_n\}, \{\omega_1, \ldots, \omega_n\}$ be any two K-bases of L. Then

$$\omega_i = \sum_{j=1}^{n} a_{ij}\theta_j \quad \text{for } i = 1, \ldots, n$$

with $a_{ij} \in K$ and $\det(a_{ij}) \ne 0$. By a straightforward computation we have

$$D_{L/K}(\omega_1, \ldots, \omega_n) = (\det(a_{ij})_{i,j=1,\ldots,n})^2 D_{L/K}(\theta_1, \ldots, \theta_n). \tag{1.1.3}$$

By applying this relation with $\{1, \theta, \ldots, \theta^{n-1}\}$ for $\{\theta_1, \ldots, \theta_n\}$, and using (1.1.2), we deduce that if $\{\omega_1, \ldots, \omega_n\}$ is any K-basis of L, then

$$D_{L/K}(\omega_1, \ldots, \omega_n) \ne 0.$$

We give an application to linear algebra. Let again K be a field of characteristic 0, and let G be a Galois extension of K. For a vector $\mathbf{x} = (x_1, \ldots, x_g) \in G^g$

and for σ in the Galois group $\mathrm{Gal}(G/K)$ of G over K, we define $\sigma(\mathbf{x}) := (\sigma(x_1), \ldots, \sigma(x_g))$.

Lemma 1.1.1 *Let $g \geq 1$, and let V be a G-linear subspace of G^g such that*

$$\sigma(\mathbf{x}) \in V \quad for \ \mathbf{x} \in V, \sigma \in \mathrm{Gal}(G/K).$$

Then V has a basis consisting of vectors from K^g.

Proof. Pick a non-zero vector $\mathbf{b} \in V$. Let $L \subseteq G$ be the smallest Galois extension of K containing the coefficients of $\mathbf{b}$ and choose a K-basis $\{\omega_1, \ldots, \omega_n\}$ of L. Let $\mathrm{Gal}(L/K) = \{\sigma_1, \ldots, \sigma_n\}$. We have $\mathbf{b} = \sum_{j=1}^{n} \omega_j \mathbf{y}_j$ with $\mathbf{y}_j \in K^g$ for $j = 1, \ldots, n$. Then also $\sigma_i(\mathbf{b}) = \sum_{j=1}^{n} \sigma_i(\omega_j) \mathbf{y}_j$ for $i = 1, \ldots, n$. The matrix $(\sigma_i(\omega_j))_{i,j=1,\ldots,n}$ is invertible (the square of its determinant being the discriminant of $\omega_1, \ldots, \omega_n$), hence $\mathbf{y}_1, \ldots, \mathbf{y}_n$ are L-linear combinations of $\sigma_i(\mathbf{b})$ $(i = 1, \ldots, n)$. Now our assumption on V implies that $\mathbf{y}_1, \ldots, \mathbf{y}_n \in V$. It follows that V is generated by vectors from K^g, hence it has a basis from K^g. $\square$

Now let K be an algebraic number field and L a finite extension of K. Then for $\alpha \in L$ we have

$$\alpha \in O_L \iff \chi_{L/K,\alpha} \in O_K[X].$$

As a consequence,

$$\mathrm{Tr}_{L/K}(\alpha), N_{L/K}(\alpha) \in O_K \quad for \ \alpha \in O_L,$$

and

$$D_{L/K}(\omega_1, \ldots, \omega_n) \in O_K$$

for every K-basis $\{\omega_1, \ldots, \omega_n\}$ of L with $\omega_1, \ldots, \omega_n \in O_L$.

1.2 Ideal theory for algebraic number fields

We start with some general notation. Let A be an integral domain with quotient field K. For $\alpha \in K$ and a subset $\mathcal{F}$ of K, we define $\alpha\mathcal{F} := \{\alpha x : x \in \mathcal{F}\}$. A *fractional ideal* of A is a subset $\mathfrak{a}$ of K such that $\mathfrak{a} \neq \{0\}$ and there is $\alpha \in A \setminus \{0\}$ such that $\alpha\mathfrak{a}$ is an ideal of A. In particular, for $\alpha \in K^*$, the set αA is a fractional ideal, which we denote by (α) when it is clear from the context what the underlying domain A is. More generally, given a subset $\mathcal{S} \neq \{0\}$ of K such that there is $\alpha \in A \setminus \{0\}$ with $\alpha\mathcal{S} \subset A$, the set of all finite A-linear combinations with elements from $\mathcal{S}$ is a fractional ideal of A, denoted by $\mathcal{S}A$, called the fractional ideal generated by $\mathcal{S}$.

Let K be an algebraic number field. Recall that its ring of integers O_K is a *Dedekind domain*, that is, O_K is integrally closed, every ideal of O_K is finitely generated, and every non-zero prime ideal of O_K is a maximal ideal (see Lang (1970), chapter 1, sections 2, 3). Henceforth, when we are dealing with prime ideals of O_K, we always exclude (0).

Let $\mathfrak{a}$, $\mathfrak{b}$ be two fractional ideals of O_K. We define their greatest common divisor or sum, lowest common multiple and product by

$$\gcd(\mathfrak{a}, \mathfrak{b}) = \mathfrak{a} + \mathfrak{b} := \{\alpha + \beta : \alpha \in \mathfrak{a}, \beta \in \mathfrak{b}\},$$

$$\mathrm{lcm}(\mathfrak{a}, \mathfrak{b}) := \mathfrak{a} \cap \mathfrak{b},$$

$$\mathfrak{a}\mathfrak{b} := O_K\text{-module generated by all products } \alpha\beta \text{ with } \alpha \in \mathfrak{a} \text{ and } \beta \in \mathfrak{b},$$

respectively. Further, the inverse of a fractional ideal $\mathfrak{a}$ of O_K is defined by

$$\mathfrak{a}^{-1} := \{\alpha \in K : \alpha\mathfrak{a} \subseteq O_K\}.$$

The gcd, lcm and product of two fractional ideals of O_K, and the inverse of a fractional ideal of O_K are again fractional ideals of O_K.

We denote by $\mathcal{P}(O_K)$ the collection of non-zero prime ideals of O_K. The following result comprises the ideal theory for O_K.

Theorem 1.2.1

(i) *The fractional ideals of O_K form an abelian group with product and inverse as defined above, and with unit element $O_K = (1)$.*

(ii) *Every fractional ideal $\mathfrak{a}$ of O_K can be decomposed uniquely as a product of powers of prime ideals*

$$\mathfrak{a} = \prod_{\mathfrak{p} \in \mathcal{P}(O_K)} \mathfrak{p}^{\mathrm{ord}_\mathfrak{p}(\mathfrak{a})},$$

where the exponents $\mathrm{ord}_\mathfrak{p}(\mathfrak{a})$ are rational integers, at most finitely many of which are non-zero.

(iii) *A fractional ideal $\mathfrak{a}$ of O_K is contained in O_K if and only if $\mathrm{ord}_\mathfrak{p}(\mathfrak{a}) \geq 0$ for every $\mathfrak{p} \in \mathcal{P}(O_K)$.*

Proof. See Lang (1970), chapter 1, section 6.   □

The group of fractional ideals of O_K is denoted by $I(O_K)$.

The following consequences are obvious.

Corollary 1.2.2 *Let $\mathfrak{a}$, $\mathfrak{b}$ be two fractional ideals of O_K. Then*

$$\mathfrak{a} \subseteq \mathfrak{b} \iff \mathrm{ord}_\mathfrak{p}(\mathfrak{a}) \geq \mathrm{ord}_\mathfrak{p}(\mathfrak{b}) \text{ for every } \mathfrak{p} \in \mathcal{P}(O_K).$$

Further, we have for every $\mathfrak{p} \in \mathcal{P}(O_K)$,

$$\operatorname{ord}_\mathfrak{p}(\mathfrak{a} \cdot \mathfrak{b}) = \operatorname{ord}_\mathfrak{p}(\mathfrak{a}) + \operatorname{ord}_\mathfrak{p}(\mathfrak{b}),$$
$$\operatorname{ord}_\mathfrak{p}(\mathfrak{a} + \mathfrak{b}) = \min(\operatorname{ord}_\mathfrak{p}(\mathfrak{a}), \operatorname{ord}_\mathfrak{p}(\mathfrak{b})),$$
$$\operatorname{ord}_\mathfrak{p}(\mathfrak{a} \cap \mathfrak{b}) = \max(\operatorname{ord}_\mathfrak{p}(\mathfrak{a}), \operatorname{ord}_\mathfrak{p}(\mathfrak{b})).$$

For $\mathfrak{p} \in \mathcal{P}(O_K)$ we define

$$\operatorname{ord}_\mathfrak{p}(x) := \operatorname{ord}_\mathfrak{p}((x)) \quad \text{if } x \in K^*, \quad \operatorname{ord}_\mathfrak{p}(0) := \infty.$$

Corollary 1.2.2 implies that for every $\mathfrak{p} \in \mathcal{P}(O_K)$, $\operatorname{ord}_\mathfrak{p}$ defines a *discrete valuation* on K, i.e., $\operatorname{ord}_\mathfrak{p}$ is a surjective map from K to $\mathbb{Z} \cup \{\infty\}$ such that for $x, y \in K$ we have

$$\operatorname{ord}_\mathfrak{p}(xy) = \operatorname{ord}_\mathfrak{p}(x) + \operatorname{ord}_\mathfrak{p}(y);$$
$$\operatorname{ord}_\mathfrak{p}(x + y) \geq \min(\operatorname{ord}_\mathfrak{p}(x), \operatorname{ord}_\mathfrak{p}(y)),$$
$$\operatorname{ord}_\mathfrak{p}(x) = \infty \Longleftrightarrow x = 0.$$

The next corollary, whose proof is straightforward, gives some other consequences.

Corollary 1.2.3

(i) *Let* $\mathfrak{a}$ *be a fractional ideal of* O_K. *Then*

$$x \in \mathfrak{a} \Longleftrightarrow \operatorname{ord}_\mathfrak{p}(x) \geq \operatorname{ord}_\mathfrak{p}(\mathfrak{a}) \text{ for all } \mathfrak{p} \in \mathcal{P}(O_K).$$

In particular,

$$x \in O_K \Longleftrightarrow \operatorname{ord}_\mathfrak{p}(x) \geq 0 \text{ for all } \mathfrak{p} \in \mathcal{P}(O_K).$$

(ii) *Let* $\mathfrak{a}$ *be the fractional ideal of* O_K *generated by a set* $\mathcal{S}$. *Then*

$$\operatorname{ord}_\mathfrak{p}(\mathfrak{a}) = \min\{\operatorname{ord}_\mathfrak{p}(\alpha) : \alpha \in \mathcal{S}\} \quad \text{for } \mathfrak{p} \in \mathcal{P}(O_K).$$

1.3 Extension of ideals; norm of ideals

Let K be an algebraic number field and L a finite extension of K of degree n. Every fractional ideal $\mathfrak{a}$ of O_K can be extended to a fractional ideal of O_L,

$$\mathfrak{a}O_L := \{\alpha y : \alpha \in \mathfrak{a}, y \in O_L\},$$

and the map $\mathfrak{a} \mapsto \mathfrak{a}O_L$ gives an injective group homomorphism from the group of fractional ideals of O_K to the group of fractional ideals of O_L. The extension of a prime ideal $\mathfrak{p}$ of O_K can be decomposed in a unique way as a product of powers of prime ideals of O_L, that is,

$$\mathfrak{p}O_L = \prod_{i=1}^{g} \mathfrak{P}_i^{e_i},$$

where $\mathfrak{P}_1, \ldots, \mathfrak{P}_g$ are distinct prime ideals of O_L and $e_1, \ldots, e_g$ are positive integers. We call $\mathfrak{P}_1, \ldots, \mathfrak{P}_g$ the prime ideals of O_L lying above $\mathfrak{p}$. The exponent e_i, henceforth denoted by $e(\mathfrak{P}_i|\mathfrak{p})$, is called the *ramification index* of $\mathfrak{P}_i$ over $\mathfrak{p}$. The residue class ring $O_L/\mathfrak{P}_i$ is a finite field extension of $O_K/\mathfrak{p}$. The degree $[O_L/\mathfrak{P}_i : O_K/\mathfrak{p}]$ of this extension, called the *residue class degree* of $\mathfrak{P}_i$ over $\mathfrak{p}$, is denoted by $f(\mathfrak{P}_i|\mathfrak{p})$. The next proposition gives some properties of ramification indices and residue class degrees.

Proposition 1.3.1 *Let L, $\mathfrak{p}$, $\mathfrak{P}_1, \ldots, \mathfrak{P}_g$ be as above.*

(i) We have $\sum_{i=1}^{g} e(\mathfrak{P}_i|\mathfrak{p})f(\mathfrak{P}_i|\mathfrak{p}) = [L : K]$.

(ii) Assume that L/K is a Galois extension. Then for any two prime ideals $\mathfrak{P}_i, \mathfrak{P}_j \in \{\mathfrak{P}_1, \ldots, \mathfrak{P}_g\}$ there is $\sigma \in \mathrm{Gal}(L/K)$ such that $\mathfrak{P}_j = \sigma\mathfrak{P}_i$. Further, $e(\mathfrak{P}_1|\mathfrak{p}) = \cdots = e(\mathfrak{P}_g|\mathfrak{p})$ and $f(\mathfrak{P}_1|\mathfrak{p}) = \cdots = f(\mathfrak{P}_g|\mathfrak{p})$.

Proof. See Lang (1970), chapter 1, section 7, proposition 21, corollary 2.   □

Proposition 1.3.2 (transitivity in towers) *Let $M \supset L \supset K$ be a tower of finite field extensions. Further, let $\mathfrak{P}$ be a prime ideal of O_L in the prime ideal factorization of $\mathfrak{p}O_L$ and $\mathfrak{Q}$ a prime ideal in the prime ideal factorization of $\mathfrak{P}O_M$. Then*

$$e(\mathfrak{Q}|\mathfrak{p}) = e(\mathfrak{Q}|\mathfrak{P}) \cdot e(\mathfrak{P}|\mathfrak{p}), \quad f(\mathfrak{Q}|\mathfrak{p}) = f(\mathfrak{Q}|\mathfrak{P}) \cdot f(\mathfrak{P}|\mathfrak{p}).$$

Proof. See Lang (1970), chapter 1, section 7, proposition 20.   □

Let again K be an algebraic number field and L a finite extension of K. We define the *norm* over K of a prime ideal $\mathfrak{P}$ of O_L by $\mathfrak{N}_{L/K}(\mathfrak{P}) := \mathfrak{p}^{f(\mathfrak{P}|\mathfrak{p})}$, where $\mathfrak{p}$ is the prime ideal of O_K such that $\mathfrak{P}$ occurs in the prime ideal factorization of $\mathfrak{p}O_L$. Then the norm $\mathfrak{N}_{L/K}(\mathfrak{A})$ of an arbitrary fractional ideal $\mathfrak{A}$ of O_L is defined by multiplicativity, i.e.,

$$\mathfrak{N}_{L/K}(\mathfrak{A}) := \prod_{\mathfrak{p} \in \mathcal{P}(O_K)} \mathfrak{p}^{\sum_{\mathfrak{P}|\mathfrak{p}} f(\mathfrak{P}|\mathfrak{p}) \cdot \mathrm{ord}_{\mathfrak{P}}(\mathfrak{A})}, \tag{1.3.1}$$

where the sum is over all prime ideals of O_L lying above $\mathfrak{p}$. Thus, $\mathfrak{N}_{L/K}$ defines a homomorphism from the group of fractional ideals of O_L to the group of fractional ideals of O_K.

Below, we give some properties of the norm.

Proposition 1.3.3 *Let L be a finite extension of K.*

(i) *Let $\mathfrak{A}$ be a fractional ideal of O_L. Then $\mathfrak{N}_{L/K}(\mathfrak{A})$ is equal to the fractional ideal generated by the numbers $N_{L/K}(\alpha)$, $\alpha \in \mathfrak{A}$.*

(ii) *For every $\alpha \in L^*$ we have $\mathfrak{N}_{L/K}(\alpha O_L) = N_{L/K}(\alpha)O_K$.*

(iii) *Let $\mathfrak{p}$ be a prime ideal of O_K, and $\mathfrak{P}_1, \ldots, \mathfrak{P}_g$ the prime ideals of O_L dividing $\mathfrak{p}$. Then for every $\alpha \in O_L$,*

$$\operatorname{ord}_\mathfrak{p}(N_{L/K}(\alpha)) = \sum_{i=1}^{g} f(\mathfrak{P}_i | \mathfrak{p})\operatorname{ord}_{\mathfrak{P}_i}(\alpha).$$

(iv) *For every fractional ideal $\mathfrak{a}$ of O_K we have $\mathfrak{N}_{L/K}(\mathfrak{a} O_L) = \mathfrak{a}^{[L:K]}$.*

(v) *Let M be a finite extension of L. Then for every fractional ideal $\mathfrak{C}$ of O_M,*

$$\mathfrak{N}_{M/K}(\mathfrak{C}) = \mathfrak{N}_{L/K}(\mathfrak{N}_{M/L}(\mathfrak{C})).$$

Proof. For (i), (iv), (v) see Neukirch (1992), Kapitel III, Satz 1.6. Part (ii) is a consequence of (i), and part (iii) a consequence of (ii) and (1.3.1). $\qquad\square$

Let K be an algebraic number field. The norm $\mathfrak{N}_{K/\mathbb{Q}}(\mathfrak{a})$ of a fractional ideal $\mathfrak{a}$ of O_K is a fractional ideal of $\mathbb{Z}$. Hence there is a positive rational number a such that $\mathfrak{N}_{K/\mathbb{Q}}(\mathfrak{a}) = (a)$. This number a is called the *absolute norm* of $\mathfrak{a}$, notation $N_K(\mathfrak{a})$ (often written as $N(\mathfrak{a})$ if it is clear from the context which is the underlying number field). It is obvious that the absolute norm is multiplicative. From parts (ii) and (iv) of Proposition 1.3.3, we obtain at once:

$$N_K((\alpha)) = |N_{K/\mathbb{Q}}(\alpha)| \text{ for } \alpha \in K^*, \quad N_K((a)) = |a|^{[K:\mathbb{Q}]} \text{ for } a \in \mathbb{Q}^*.$$

If $\mathfrak{p}$ is a prime ideal of O_K dividing a prime number p, we have $N_K(\mathfrak{p}) = p^{f(\mathfrak{p}|p)} = |O_K/\mathfrak{p}|$. More generally, for any non-zero ideal $\mathfrak{a}$ of O_K we have

$$N_K(\mathfrak{a}) = |O_K/\mathfrak{a}|.$$

1.4 Discriminant, class number, unit group and regulator

Let K be an algebraic number field of degree d over $\mathbb{Q}$. There are d distinct isomorphic embeddings of K in $\mathbb{C}$, which we denote by $\sigma_1, \ldots, \sigma_d$; further, we will write $\alpha^{(i)} := \sigma_i(\alpha)$ for $\alpha \in K$. We assume that among these embeddings there are precisely r_1 real embeddings, i.e., embeddings σ with $\sigma(K) \subset \mathbb{R}$, and r_2 pairs of complex conjugate embeddings, i.e., pairs $\{\sigma, \overline{\sigma}\}$ where $\overline{\sigma}(\alpha) = \overline{\sigma(\alpha)}$ for $\alpha \in K$. Thus, $d = r_1 + 2r_2$ and after reordering the embeddings we may assume that σ_i $(i = 1, \ldots, r_1)$ are the real embeddings and $\{\sigma_i, \sigma_{i+r_2}\}$ $(i = r_1 + 1, \ldots r_1 + r_2)$ the pairs of complex conjugate embeddings.

Viewed as a $\mathbb{Z}$-module, O_K is free of rank d. Taking any $\mathbb{Z}$-basis $\{\omega_1, \ldots, \omega_d\}$ of O_K, we define the *discriminant* of K by

$$D_K := D_{K/\mathbb{Q}}(\omega_1, \ldots, \omega_d) = \left(\det \left(\omega_j^{(i)} \right)_{i,j=1,\ldots,d} \right)^2.$$

This is a non-zero rational integer which, by (1.1.3), is independent of the choice of the basis.

Denote as before by $I(O_K)$ the group of fractional ideals of O_K. Further, denote by $P(O_K)$ the subgroup of principal fractional ideals of O_K. The quotient group $\mathrm{Cl}(O_K) = I(O_K)/P(O_K)$ is called the *class group* of K.

Theorem 1.4.1 *The class group* $\mathrm{Cl}(O_K)$ *of* O_K *is finite.*

Proof. See Neukirch (1992), Kapitel I, Satz 6.3. □

The cardinality of the class group is called the *class number* of K, and we denote this by h_K.

We denote by W_K the multiplicative group consisting of all roots of unity in K. This is a finite, cyclic subgroup of K^*. We denote the number of roots of unity of K by ω_K.

We recall the following fundamental theorem of Dirichlet concerning the unit group O_K^* of O_K. Elements of O_K^* will usually be referred to as units of K. Recall that if V is an n-dimensional vector space over $\mathbb{R}$, then a *full lattice* in V is an additive subgroup

$$\{z_1 \mathbf{a}_1 + \cdots + z_n \mathbf{a}_n : z_1, \ldots, z_n \in \mathbb{Z}\},$$

where $\{\mathbf{a}_1, \ldots, \mathbf{a}_n\}$ is a basis of V.

Theorem 1.4.2 *The map*

$$LOG_K : \varepsilon \mapsto \left(e_1 \log |\varepsilon^{(1)}|, \ldots, e_{r_1+r_2} \log |\varepsilon^{(r_1+r_2)}| \right)$$

(where $e_j = 1$ for $j = 1, \ldots r_1$ and $e_j = 2$ for $j = r_1 + 1, \ldots, r_1 + r_2$) defines a surjective homomorphism from O_K^* to a full lattice in the real vector space given by

$$\{\mathbf{x} = (x_1, \ldots, x_{r_1+r_2}) \in \mathbb{R}^{r_1+r_2} : x_1 + \cdots + x_{r_1+r_2} = 0\}$$

with kernel W_K.

Proof. See Neukirch (1992), Kapitel I, Satz 7.1. □

The following consequence is immediate.

Corollary 1.4.3 *Put* $r = r_K := r_1 + r_2 - 1$. *Then*

$$O_K^* \cong W_K \times \mathbb{Z}^r.$$

More explicitly, there are $\varepsilon_1, \ldots, \varepsilon_r \in O_K^*$ *such that every* $\varepsilon \in O_K^*$ *can be expressed uniquely as*

$$\varepsilon = \zeta \varepsilon_1^{b_1} \ldots \varepsilon_r^{b_r}$$

where ζ *is a root of unity in* K *and* $b_1, \ldots b_r$ *are rational integers.*

The number r_K (denoted by r if it is clear to which number field it refers) is called the *unit rank* of K. A set of units $\{\varepsilon_1, \ldots, \varepsilon_r\}$ as above is called a *fundamental system of units* for K. Given such a system, we define the *regulator* of K by

$$R_K := \left| \det\left(e_j \log \left| \varepsilon_i^{(j)} \right| \right) \right)_{i,j=1,\ldots,r} \right|.$$

This regulator is non-zero, and independent of the choice of $\varepsilon_1, \ldots, \varepsilon_r$.

1.5 Explicit estimates

We have collected from the literature some estimates for the field parameters defined above. As before, K is an algebraic number field of degree d.

For the number ω_K of roots of unity of K we have the estimate

$$\omega_K \leq 20d \log \log d \quad \text{if } d \geq 3. \tag{1.5.1}$$

This follows from the observation that the degree of the maximal cyclotomic subfield of K, which is $\varphi(\omega_K)$ where φ is Euler's totient function, divides d, and from Rosser and Schoenfeld (1962), Theorem 15, which gives an explicit lower bound for $\varphi(n)$ of the order $n/\log \log n$.

For the class number and regulator of K we have

$$h_K R_K \leq |D_K|^{1/2} (\log^* |D_K|)^{d-1}. \tag{1.5.2}$$

The first inequality of this type was proved by Landau (1918). The above version follows from Louboutin (2000) and (1.5.1); see (59) in Győry and Yu (2006). The following lower bound for the regulator is due to Friedman (1989):

$$R_K > 0.2052. \tag{1.5.3}$$

We recall an important lower estimate for discriminants. By an inequality due to Minkowski (see, e.g., Lang (1970), chapter V, section 4, proof of corollary of theorem 4) we have

$$|D_K| > \left(\frac{\pi}{4}\right)^d \left(\frac{d^d}{d!}\right)^2. \tag{1.5.4}$$

Further, we need the following lemma.

Lemma 1.5.1 *Let $g \in \mathbb{Z}[X]$ be a monic polynomial of degree m with non-zero discriminant. Assume that the coefficients of g have absolute values at most M. Let $K = \mathbb{Q}(\theta)$, where θ is a zero of g. Then*

$$|D_K| \le m^{2m-1} M^{2m-2}.$$

Proof. The monic minimal polynomial, say f, of θ is in $\mathbb{Z}[X]$ and it divides g in $\mathbb{Z}[X]$. Suppose K has degree d. Using the expression of the discriminant of a monic polynomial as the product of the squares of the differences of its zeros, one easily shows that the discriminant $D(f)$ of f divides $D(g)$ in the ring of algebraic integers and so also in $\mathbb{Z}$. Further, by (1.1.2), we have $D(f) = D_{K/\mathbb{Q}}(1, \theta, \ldots, \theta^{d-1})$. Writing $1, \theta, \ldots, \theta^{d-1}$ as $\mathbb{Z}$-linear combinations of a $\mathbb{Z}$-basis of O_K, and using (1.1.3), we infer that D_K divides $D(f)$. Therefore, D_K divides $D(g)$. Using for instance an estimate from Lewis and Mahler (1961) (bottom of p. 335), which uses a determinantal expression for $D(g)$, one obtains

$$|D(g)| \le m^{2m-1} M^{2m-2}.$$

This proves our lemma.   □

Remark There is an analogue for this lemma where for g one can take any non-zero polynomial in $\mathbb{Z}[X]$, not necessarily monic or of non-zero discriminant. We will not work this out.

1.6 Absolute values: generalities

We have collected some facts on absolute values. Our basic reference is Neukirch (1992), Kapitel II.

Let K be an infinite field. An *absolute value* on K is a function $|\cdot| : K \to \mathbb{R}_{\ge 0}$ satisfying the following conditions:

$|xy| = |x| \cdot |y|$ for all $x, y \in K$;

there is $C \ge 1$ such that $|x + y| \le C \max(|x|, |y|)$ for all $x, y \in K$;

$|x| = 0 \iff x = 0$.

These conditions imply that $|1| = 1$. An absolute value $|\cdot|$ on K is called *trivial* if $|x| = 1$ for $x \in K^*$. Henceforth, all absolute values we will consider are non-trivial. Two absolute values $|\cdot|_1, |\cdot|_2$ on K are called equivalent if there is $c > 0$ such that $|x|_2 = |x|_1^c$ for all $x \in K$.

An absolute value $|\cdot|$ on K is called *non-archimedean* if it satisfies the *ultrametric inequality*

$$|x + y| \leq \max(|x|, |y|) \quad \text{for } x, y \in K$$

and *archimedean* if it does not satisfy this inequality.

A *valuation* on K is a function $v : K \rightarrow \mathbb{R} \cup \{\infty\}$ such that C^{-v} defines a non-archimedean absolute value on K, where C is any constant > 1. Equivalently,

$$v(0) = \infty, \quad v(x) \in \mathbb{R} \quad \text{for } x \in K^*,$$

$$v(xy) = v(x) + v(y), \quad v(x + y) \geq \min(v(x), v(y)) \quad \text{for } x, y \in K.$$

Notice that if v is a valuation on K, then

$$v(K^*) = \{v(x) : x \in K^*\}$$

is an additive subgroup of $\mathbb{R}$, called the *value group* of v. In this book, we agree that a *discrete valuation* on K is a valuation v on K for which $v(K^*) = \mathbb{Z}$. (In much of the literature, a discrete valuation on K is a valuation v for which $v(K^*)$ is a non-trivial discrete subgroup of $\mathbb{R}$; then a valuation v for which $v(K^*) = \mathbb{Z}$ is called a *normalized discrete valuation*).

A *field with absolute value* is a pair $(K, |\cdot|)$, where K is an infinite field, and $|\cdot|$ a non-trivial absolute value on K. An injective homomorphism/isomorphism of fields with absolute value $\varphi : (K_1, |\cdot|_1) \rightarrow (K_2, |\cdot|_2)$ is an injective homomorphism/isomorphism $\varphi : K_1 \rightarrow K_2$ such that $|x|_1 = |\varphi(x)|_2$ for $x \in K$.

Let $(K, |\cdot|)$ be a field with absolute value. A sequence $\{a_n\} = \{a_n\}_{n=0}^{\infty}$ in K is called a *convergent sequence* of $(K, |\cdot|)$, if there is $\alpha \in K$ such that $\lim_{n \to \infty} |a_n - \alpha| = 0$. A *Cauchy sequence* of $(K, |\cdot|)$ is a sequence $\{a_n\}$ in K with $\lim_{m,n \to \infty} |a_m - a_n| = 0$. We call $(K, |\cdot|)$ *complete* if every Cauchy sequence of $(K, |\cdot|)$ converges.

Suppose that $(K, |\cdot|)$ is a non-complete field with absolute value. Then we can extend this to a complete field with absolute value $(\widetilde{K}, |\cdot|)$, the *completion* of $(K, |\cdot|)$, as follows. Let R be the ring of Cauchy sequences of $(K, |\cdot|)$ with componentwise addition and multiplication, and M the ideal of sequences of $(K, |\cdot|)$ converging to 0. Then M is a maximal ideal of R and thus, R/M is a field which will be our $\widetilde{K}$. We view K as a subfield of $\widetilde{K}$ by identifying $\alpha \in K$ with the element of $\widetilde{K}$ represented by the constant sequence $\{\alpha\}$. We extend $|\cdot|$ to $\widetilde{K}$ by setting $|\alpha| := \lim_{n \to \infty} |a_n|$ for $\alpha \in \widetilde{K}$, where $\{a_n\}$ is any Cauchy sequence of $(K, |\cdot|)$ representing α. The field $\widetilde{K}$ is the smallest complete field containing K, in the sense that if there exists an injective homomorphism of $(K, |\cdot|)$ into a complete field with absolute value $(L, |\cdot|')$, say, then this can be extended in precisely one way to an injective homomorphism from $(\widetilde{K}, |\cdot|)$ into $(L, |\cdot|')$.

It is easy to see that notions such as convergence, Cauchy sequence, completeness, completion, depend on the equivalence class of an absolute value rather than the absolute value itself.

If K is a field with valuation v, then notions such as convergence, Cauchy sequence, completeness with respect to v are meant to be the corresponding notions with respect to the absolute value C^{-v}, where $C > 1$ is any constant.

Example 1: Absolute values on $\mathbb{Q}$. Define $M_{\mathbb{Q}} := \{\infty\} \cup \{\text{prime numbers}\}$. This is called the *set of places* of $\mathbb{Q}$. We call ∞ the infinite place, and the prime numbers the finite places of $\mathbb{Q}$. We define absolute values $|\cdot|_p$ ($p \in M_{\mathbb{Q}}$) by

$$|a|_\infty := \max(a, -a) \quad \text{for } a \in \mathbb{Q},$$
$$|a|_p := p^{-\mathrm{ord}_p(a)} \quad \text{for } a \in \mathbb{Q}$$

for every prime number p, where $\mathrm{ord}_p(a)$ is the exponent of p in the unique prime factorization of a, i.e., if $a = p^m b/c$ with $m, b, c \in \mathbb{Z}$ and $p \nmid bc$, then $\mathrm{ord}_p(a) = m$. We agree that $\mathrm{ord}_p(0) = \infty$ and $|0|_p = 0$.

The absolute value $|\cdot|_\infty$ is archimedean, while the other ones are non-archimedean. The completion of $\mathbb{Q}$ with respect to $|\cdot|_\infty$ is $\mathbb{Q}_\infty := \mathbb{R}$. For a prime number p, the completion of $\mathbb{Q}$ with respect to $|\cdot|_p$ is the field of *p-adic numbers*, denoted by $\mathbb{Q}_p$. The above absolute values satisfy the *Product Formula*

$$\prod_{p \in M_{\mathbb{Q}}} |a|_p = 1 \quad \text{for } a \in \mathbb{Q}^*.$$

By a theorem of Ostrowski (see Neukirch (1992), Kapitel II, Satz 3.7), every non-trivial absolute value on $\mathbb{Q}$ is equivalent to one of $|\cdot|_p$ ($p \in M_{\mathbb{Q}}$).

Example 2. By another theorem of Ostrowski (see Neukirch (1992), Kapitel II, Satz 4.2), if $(K, |\cdot|)$ is a complete field with archimedean absolute value, then up to isomorphism, $K = \mathbb{R}$ or $\mathbb{C}$, and $|\cdot|$ is equivalent to the ordinary absolute value.

We finish with recalling some facts about extensions of absolute values. Let $(K, |\cdot|)$ be a field with absolute value, and L an extension field of K. By an *extension* or *continuation* of $|\cdot|$ to L we mean an absolute value on L whose restriction to K is $|\cdot|$.

Proposition 1.6.1 *Let* $(K, |\cdot|)$ *be a complete field with absolute value.*

(i) Let L be a finite extension of K. Then there is precisely one extension of $|\cdot|$ to L, which is given by $|N_{L/K}(\cdot)|^{1/[L:K]}$. The field L is complete with respect to this extension.

(ii) Let $\overline{K}$ be an algebraic closure of K. Then $|\cdot|$ has a unique extension to $\overline{K}$. If we denote this extension also by $|\cdot|$, we have $|\tau(x)| = |x|$ for $x \in \overline{K}$, $\tau \in \mathrm{Gal}(\overline{K}/K)$.

Proof. See for instance Neukirch (1992), Kapitel II, Theorem 4.8. □

1.7 Absolute values and places on number fields

Let K be an algebraic number field. We introduce a collection of normalized absolute values $\{|\cdot|_v\}_{v \in M_K}$ on K by taking suitable powers of the extensions to K of the absolute values $|\cdot|_p$ ($p \in M_{\mathbb{Q}}$) defined in the previous subsection.

A *real place* of K is a set $\{\sigma\}$ where $\sigma : K \hookrightarrow \mathbb{R}$ is a real embedding of K. A *complex place* of K is a pair $\{\sigma, \overline{\sigma}\}$ of conjugate complex embeddings $K \hookrightarrow \mathbb{C}$. An *infinite place* is a real or complex place. Clearly, if r_1, r_2 denote the number of real and complex places of K, we have $r_1 + 2r_2 = [K : \mathbb{Q}]$. A *finite place* of K is a non-zero prime ideal of O_K. We denote by M_K^∞, M_K^0 the sets of infinite places and finite places, respectively, of K, and by M_K the set of all places of K, i.e., $M_K := M_K^\infty \cup M_K^0$.

With every place $v \in M_K$ we associate an absolute value $|\cdot|_v$ on K, which is defined as follows for $\alpha \in K$:

$$|\alpha|_v := |\sigma(\alpha)| \quad \text{if } v = \{\sigma\} \text{ is real};$$

$$|\alpha|_v := |\sigma(\alpha)|^2 = |\overline{\sigma}(\alpha)|^2 \quad \text{if } v = \{\sigma, \overline{\sigma}\} \text{ is complex};$$

$$|a|_v := N_K(\mathfrak{p})^{-\mathrm{ord}_\mathfrak{p}(a)} \quad \text{if } v = \mathfrak{p} \text{ is a prime ideal of } O_K,$$

where $N_K(\mathfrak{p}) = |O_K/\mathfrak{p}|$ is the absolute norm of $\mathfrak{p}$, and $\mathrm{ord}_\mathfrak{p}(\alpha)$ is the exponent of $\mathfrak{p}$ in the prime ideal factorization of (α), where we agree that $\mathrm{ord}_\mathfrak{p}(0) = \infty$. We denote by K_v the completion of K with respect to $|\cdot|_v$. Notice that $K_v = \mathbb{R}$ if v is real, $K_v = \mathbb{C}$ if v is complex, while K_v is a finite extension of $\mathbb{Q}_p$ if $v = \mathfrak{p}$ is a prime ideal of O_K, and p is the prime number with $\mathfrak{p} \cap \mathbb{Z} = (p)$. Combining the Product Formula over $\mathbb{Q}$ with the identity $N_K((\alpha)) = |N_{K/\mathbb{Q}}(\alpha)|$ for $\alpha \in K$, where the left-hand side denotes the absolute norm of (α), one easily deduces the *Product Formula* over K,

$$\prod_{v \in M_K} |\alpha|_v = 1 \quad \text{for } \alpha \in K^*. \tag{1.7.1}$$

To deal with archimedean and non-archimedean absolute values simultaneously, it is convenient to use

$$|x_1 + \cdots + x_n|_v \leq n^{s(v)} \max(|x_1|_v, \ldots, |x_n|_v) \tag{1.7.2}$$

for $v \in M_K$, $x_1, \ldots, x_n \in K$, where

$$s(v) = 1 \text{ if } v \text{ is real}, s(v) = 2 \text{ if } v \text{ is complex}, s(v) = 0 \text{ if } v \text{ is finite}.$$

Note that $\sum_{v \in M_K^\infty} s(v) = [K : \mathbb{Q}]$.

Let $\rho : K_1 \to K_2$ be an isomorphism of algebraic number fields, and v a place of K_2. We define a place $v \circ \rho$ on K_1 by

$$v \circ \rho := \begin{cases} \{\sigma\rho\} & \text{if } v = \{\sigma\} \text{ is real}, \\ \{\tau\rho, \overline{\tau}\rho\} & \text{if } v = \{\tau, \overline{\tau}\} \text{ is complex}, \\ \rho^{-1}(\mathfrak{p}) & \text{if } v = \mathfrak{p} \text{ is a prime ideal of } O_K. \end{cases}$$

Then

$$|\alpha|_{v \circ \rho} = |\rho(\alpha)|_v \quad \text{for } \alpha \in K_1, v \in M_{K_2}. \tag{1.7.3}$$

Let L be a finite extension of K and v, V places of K, L, respectively. We say that V lies above v or v below V, notation $V|v$, if the restriction of $| \cdot |_V$ to K is a power of $| \cdot |_v$. This is the case precisely if either both v, V are infinite and the embeddings in v are the restrictions to K of the embeddings in V; or if $v = \mathfrak{p}$, $V = \mathfrak{P}$ are prime ideals of O_K, O_L, respectively with $\mathfrak{P} \supset \mathfrak{p}$. In that case, the completion L_V of L with respect to $| \cdot |_V$ is a finite extension of K_v. In fact, $[L_V : K_v]$ is 1 or 2 if v, V are infinite, while if $v = \mathfrak{p}$, $V = \mathfrak{P}$ are finite, we have $[L_V : K_v] = e(\mathfrak{P}|\mathfrak{p})f(\mathfrak{P}|\mathfrak{p})$, where $e(\mathfrak{P}|\mathfrak{p})$, $f(\mathfrak{P}|\mathfrak{p})$ denote the ramification index and residue class degree of $\mathfrak{P}$ over $\mathfrak{p}$.

We say that two places V_1, V_2 of L are conjugate over K if there is a K-automorphism σ of L such that $V_2 = V_1 \circ \sigma$.

Proposition 1.7.1 *Let K be a number field, and L a finite extension of K. Further, let v be a place of K and $V_1, \ldots, V_g$ the places of L above v. Then*

$$|\alpha|_{V_k} = |\alpha|_v^{[L_{V_k}:K_v]} \quad \text{for } \alpha \in K, k = 1, \ldots, g, \tag{1.7.4}$$

$$\prod_{k=1}^g |\alpha|_{V_k} = |N_{L/K}(\alpha)|_v \quad \text{for } \alpha \in L, \tag{1.7.5}$$

$$\sum_{k=1}^g [L_{V_k} : K_v] = [L : K]. \tag{1.7.6}$$

Further, if L/K is Galois, then $V_1, \ldots, V_g$ are conjugate to each other, and we have $[L_{V_k} : K_v] = [L : K]/g$ for $k = 1, \ldots, g$.

Proof. The verification is completely straightforward if v is an infinite place. So assume that $v = \mathfrak{p}$ is a finite place; then $V_k = \mathfrak{P}_k$ ($k = 1, \ldots, g$) are the prime ideals of O_L containing $\mathfrak{p}$. The first identity (1.7.4) follows from the observation that for $\alpha \in K, \mathfrak{P} \in \{\mathfrak{P}_1, \ldots, \mathfrak{P}_g\}$,

$$|\alpha|_{\mathfrak{P}} = N_L(\mathfrak{P})^{-\mathrm{ord}_{\mathfrak{P}}(\alpha)} = N_K(\mathfrak{p})^{-f(\mathfrak{P}|\mathfrak{p})e(\mathfrak{P}|\mathfrak{p})\mathrm{ord}_{\mathfrak{p}}(\alpha)} = |\alpha|_{\mathfrak{p}}^{[L_{\mathfrak{P}}:K_{\mathfrak{p}}]}.$$

Identity (1.7.5) follows by expressing both sides of the identity as powers of $N_K(\mathfrak{p})$ and showing by means of Proposition 1.3.3 (iii) that the exponents are equal. Identity (1.7.6) follows by combining (1.7.4) with (1.7.5) with $\alpha \in K^*$. The last assertion follows from Proposition 1.3.1 (ii). □

1.8 *S*-integers, *S*-units and *S*-norm

Let S denote a finite subset of M_K containing all infinite places. We say that $\alpha \in K$ is an *S-integer* if $|\alpha|_v \le 1$ for all $v \in M_K \setminus S$. The S-integers form a ring in K, denoted by O_S. Its unit group, denoted O_S^*, is called the group of *S-units*.

For $S = M_K^\infty$ the ring of S-integers is just O_K and the group of S-units just O_K^*. Otherwise, we have $S = M_K^\infty \cup \{\mathfrak{p}_1, \ldots, \mathfrak{p}_t\}$, where $\mathfrak{p}_1, \ldots, \mathfrak{p}_t$ are prime ideals of O_K. Then $O_S = O_K[(\mathfrak{p}_1 \cdots \mathfrak{p}_t)^{-1}]$, and O_S^* consists of those elements α of K such that (α) is composed of prime ideals from $\mathfrak{p}_1, \ldots, \mathfrak{p}_t$. In the case $K = \mathbb{Q}, S = \{\infty, p_1, \ldots, p_t\}$ where $p_1, \ldots, p_t$ are prime numbers, we write $\mathbb{Z}_S$ for the ring of S-integers. Thus, $\mathbb{Z}_S = \mathbb{Z}[(p_1 \cdots p_t)^{-1}]$.

We define the *S-norm* of $\alpha \in K$ by

$$N_S(\alpha) := \prod_{v \in S} |\alpha|_v.$$

Notice that the S-norm is multiplicative. Let again $S = M_K^\infty \cup \{\mathfrak{p}_1, \ldots, \mathfrak{p}_t\}$. Take $\alpha \in K^*$ and write

$$(\alpha) = \mathfrak{p}_1^{k_1} \cdots \mathfrak{p}_t^{k_t} \mathfrak{a},$$

where $\mathfrak{a}$ is a fractional ideal of O_K composed of prime ideals outside S. Then by the Product Formula,

$$N_S(\alpha) = \prod_{v \notin S} |\alpha|_v^{-1} = \prod_{\mathfrak{p} \in \mathcal{P}(O_K) \setminus \{\mathfrak{p}_1, \ldots, \mathfrak{p}_t\}} N_K(\mathfrak{p})^{\mathrm{ord}_{\mathfrak{p}}(\alpha)}$$
$$= N_K(\mathfrak{a}).$$

Let L be a finite extension of K and T the set of places of L lying above the places of S. Then O_T, the ring of T-integers of L is the integral closure of O_S

in L. Further, by Proposition 1.7.1,

$$N_T(\alpha) = N_S(\alpha)^{[L:K]} \quad \text{for } \alpha \in K^*.$$

We recall the extension of Dirichlet's Unit Theorem to S-units.

Theorem 1.8.1 (*S-unit Theorem*) *Let $S = \{v_1, \ldots, v_s\}$ be a finite set of places of K, containing all infinite places. Then the map*

$$LOG_S : \varepsilon \mapsto ((\log |\varepsilon|_{v_1}, \ldots, \log |\varepsilon|_{v_s})$$

defines a surjective homomorphism from O_S^* to a full lattice in the real vector space

$$\{\mathbf{x} = (x_1, \ldots, x_s) \in \mathbb{R}^s : x_1 + \cdots + x_s = 0\}$$

with kernel W_K.

Proof. See Lang (1970), chapter V, section 1, Unit Theorem.   □

This implies at once:

Corollary 1.8.2 *We have*

$$O_S^* \cong W_K \times \mathbb{Z}^{s-1}.$$

More explicitly, there are $\varepsilon_1, \ldots, \varepsilon_{s-1} \in O_S^*$ such that every $\varepsilon \in O_S^*$ can be expressed uniquely as

$$\varepsilon = \zeta \varepsilon_1^{b_1} \ldots \varepsilon_{s-1}^{b_{s-1}}, \tag{1.8.1}$$

where ζ is a root of unity in K and $b_1, \ldots b_{s-1}$ are rational integers.

A system $\{\varepsilon_1, \ldots, \varepsilon_{s-1}\}$ as above is called a *fundamental system of S-units*. Analogously as for units of O_K, we define the *S-regulator* by

$$R_S := \left| \det \left(\log |\varepsilon_i|_{v_j} \right) \right)_{i,j=1,\ldots,s-1} \right|. \tag{1.8.2}$$

This quantity is non-zero, and independent of the choice of $\varepsilon_1, \ldots, \varepsilon_{s-1}$ and of the choice $v_1, \ldots, v_{s-1}$ from S. In the case that $S = M_K^\infty$, the S-regulator R_S is equal to the regulator R_K. More generally, we have

$$R_S = R_K \cdot [I(S) : P(S)] \cdot \prod_{i=1}^{t} \log N(\mathfrak{p}_i), \tag{1.8.3}$$

where $\mathfrak{p}_1, \ldots, \mathfrak{p}_t$ are the prime ideals in S, $I(S)$ is the group of fractional ideals of O_K composed of prime ideals from $\mathfrak{p}_1, \ldots, \mathfrak{p}_t$ and $P(S)$ is the group of principal fractional ideals of O_K composed of prime ideals from $\mathfrak{p}_1, \ldots, \mathfrak{p}_t$.

We note that the index $[I(S) : P(S)]$ is a divisor of the class number h_K. By combining (1.8.3) with (1.5.2) we obtain

$$R_S \leq h_K R_K \cdot \prod_{i=1}^{t} \log N_K(\mathfrak{p}_i)$$

$$\leq |D_K|^{1/2}(\log^* |D_K|)^{d-1} \cdot \prod_{i=1}^{t} \log N_K(\mathfrak{p}_i). \qquad (1.8.4)$$

By combining (1.8.3) with (1.5.3), we obtain

$$R_S \geq \begin{cases} (\log 2)(\log 3)^{s-2} & \text{if } K = \mathbb{Q}, s := |S| \geq 3, \\ 0.2052(\log 2)^{s-2} & \text{if } K \neq \mathbb{Q}, s \geq 3. \end{cases} \qquad (1.8.5)$$

1.9 Heights

There are various ways to define the height of an algebraic number, a vector with algebraic coordinates or a polynomial with algebraic coefficients. Here we have made a small selection. The other notions of height needed in this book will be defined on the spot. We fix an algebraic closure $\overline{\mathbb{Q}}$ of $\mathbb{Q}$.

1.9.1 Heights of algebraic numbers

The *absolute multiplicative height* of $\alpha \in \overline{\mathbb{Q}}$ is defined by

$$H(\alpha) := \prod_{v \in M_K} \max(1, |\alpha|_v)^{1/[K:\mathbb{Q}]},$$

where $K \subset \overline{\mathbb{Q}}$ is any number field containing α. It follows from Proposition 1.7.1 that this is independent of the choice of K. The *absolute logarithmic height* of α is given by

$$h(\alpha) := \log H(\alpha).$$

Below, we have brought together some properties of the absolute logarithmic height. These can easily be reformulated into properties of the absolute multiplicative height.

We start with a trivial but useful observation: if K is an algebraic number field, S any non-empty subset of M_K, and $\alpha \neq 0$, then

$$-h(\alpha) \leq \frac{1}{[K : \mathbb{Q}]} \sum_{v \in S} \log |\alpha|_v \leq h(\alpha). \qquad (1.9.1)$$

Indeed, the upper bound is obvious from $h(\alpha) = \frac{1}{[K:\mathbb{Q}]} \sum_{v \in M_K} \log \max(1, |\alpha|_v)$, and the lower bound follows in the same manner, applying the Product Formula $\sum_{v \in S} \log |\alpha|_v = -\sum_{v \notin S} \log |\alpha|_v$. In the case that S is a finite subset of M_K, containing the infinite places, (1.9.1) translates into

$$-h(\alpha) \leq \frac{1}{[K:\mathbb{Q}]} \log N_S(\alpha) \leq h(\alpha). \qquad (1.9.2)$$

The next lemma gives an estimate for the denominator of an algebraic number.

Lemma 1.9.1 *Let K be a number field and $\alpha \in K^*$. Then there is a positive integer d such that $d \leq H(\alpha)^{[K:\mathbb{Q}]}$ and $d\alpha \in O_K$.*

Proof. We take $d := \prod_{v \in M_K^0} \max(1, |\alpha|_v)$. It is clear that $d \leq H(\alpha)^{[K:\mathbb{Q}]}$. We show that d is a positive integer and $d\alpha \in O_K$. First observe that

$$d = \prod_{\mathfrak{p} \in \mathcal{P}(O_K)} N_K(\mathfrak{p})^{\max(0, -\mathrm{ord}_\mathfrak{p}(\alpha))} = \prod_p p^{\sum_{\mathfrak{p}|p} f(\mathfrak{p}|p) \max(0, -ord_\mathfrak{p}(\alpha))} \in \mathbb{Z}_{>0},$$

where the product is over the rational primes. Further, if $\mathfrak{p}$ is a prime ideal of O_K lying above the prime p, say, then

$$\mathrm{ord}_\mathfrak{p}(d) \geq \mathrm{ord}_p(d) \geq -\mathrm{ord}_p(\alpha),$$

implying $\mathrm{ord}_\mathfrak{p}(d\alpha) \geq 0$. This holds for every prime ideal $\mathfrak{p}$ of O_K, hence $d\alpha \in O_K$. $\qquad\square$

In the following lemma we have listed some further properties.

Lemma 1.9.2 *Let $\alpha, \alpha_1, \ldots, \alpha_n \in \overline{\mathbb{Q}}$, $m \in \mathbb{Z}$ and let σ be an automorphism of $\overline{\mathbb{Q}}$. Then*

$$h(\sigma(\alpha)) = h(\alpha);$$

$$h(\alpha_1 \cdots \alpha_n) \leq \sum_{i=1}^{n} h(\alpha_i);$$

$$h(\alpha_1 + \cdots + \alpha_n) \leq \log n + \sum_{i=1}^{n} h(\alpha_i);$$

$$h(\alpha^m) = |m| h(\alpha) \quad \text{if } \alpha \neq 0.$$

Proof. The first property is a consequence of (1.7.3), the third of (1.7.2), while the other two are obvious. See also Waldschmidt (2000), chapter 3. $\qquad\square$

The minimal polynomial of $\alpha \in \overline{\mathbb{Q}}$ over $\mathbb{Z}$, denoted by P_α, is by definition the polynomial $P \in \mathbb{Z}[X]$ of minimal degree, having positive leading coefficient and coefficients with greatest common divisor 1, such that $P(\alpha) = 0$. Writing

$P_\alpha = a_0(X - \alpha^{(1)}) \cdots (X - \alpha^{(d)})$ where $d = \deg \alpha$ and $\alpha^{(1)}, \ldots, \alpha^{(d)}$ are the conjugates of α in $\mathbb{C}$, we have

$$H(\alpha) = \left(|a_0| \prod_{i=1}^{d} \max \left(1, |\alpha^{(i)}| \right) \right)^{1/d},$$

i.e., $H(\alpha)$ is the d-th root of the *Mahler measure* of α (see Waldschmidt (2000), Lemma 3.10). Further, writing $P_\alpha = a_0 X^d + \cdots + a_d$, we have

$$-\frac{1}{2d} \log(d+1) + h(\alpha) \leq h(P_\alpha) \leq \log 2 + h(\alpha), \qquad (1.9.3)$$

where $h(P_\alpha) := \log \max(|a_0|, \ldots, |a_d|)$ (see Waldschmidt (2000), Lemma 3.11).

From this we deduce at once:

Theorem 1.9.3 (Northcott's Theorem) *Let D, H be positive reals. Then there are only finitely many $\alpha \in \overline{\mathbb{Q}}$ such that $\deg \alpha \leq D$ and $h(\alpha) \leq H$.*

1.9.2 v-adic norms and heights of vectors and polynomials

Let K be an algebraic number field, $v \in M_K$, and denote the unique extension of $|\cdot|_v$ to $\overline{K_v}$ also by $|\cdot|_v$. We define the v-adic norm of a vector $\mathbf{x} = (x_1, \ldots, x_n) \in \overline{K_v}^n$ by

$$|\mathbf{x}|_v = |x_1, \ldots, x_n|_v := \max(|x_1|_v, \ldots, |x_n|_v).$$

Let $\mathbf{x} = (x_1, \ldots, x_n) \in \overline{\mathbb{Q}}^n$ and choose an algebraic number field K such that $\mathbf{x} \in K^n$. Then the *multiplicative height* and *homogeneous multiplicative height* of $\mathbf{x}$ are defined by

$$H(\mathbf{x}) = H(x_1, \ldots, x_n) := \left(\prod_{v \in M_K} \max(1, |\mathbf{x}|_v) \right)^{1/[K:\mathbb{Q}]},$$

$$H^{\text{hom}}(\mathbf{x}) = H^{\text{hom}}(x_1, \ldots, x_n) := \left(\prod_{v \in M_K} |\mathbf{x}|_v \right)^{1/[K:\mathbb{Q}]},$$

respectively. By Proposition 1.7.1, these definitions are independent of the choice of K. We define the corresponding logarithmic heights by

$$h(\mathbf{x}) := \log H(\mathbf{x}), \quad h^{\text{hom}}(\mathbf{x}) := \log H^{\text{hom}}(\mathbf{x}),$$

respectively. For instance, for $\mathbf{x} = (x_1, \ldots, x_n) \in \mathbb{Z}^n \setminus \{0\}$ we have

$$\left. \begin{aligned} h(\mathbf{x}) &= \log \max(|x_1|, \ldots, |x_n|), \\ h^{\text{hom}}(\mathbf{x}) &= \log \left(\frac{\max(|x_1|, \ldots, |x_n|)}{\gcd(x_1, \ldots, x_n)} \right). \end{aligned} \right\} \qquad (1.9.4)$$

It is easy to verify that for $\mathbf{x} = (x_1, \ldots, x_n) \in \overline{\mathbb{Q}}^n, \lambda \in \overline{\mathbb{Q}}^*$ and for $\mathbf{x}_1, \ldots, \mathbf{x}_m \in \overline{\mathbb{Q}}^n$,

$$h^{\mathrm{hom}}(\mathbf{x}) \leq h(\mathbf{x}), \tag{1.9.5}$$

$$\max_{1 \leq i \leq n} h(x_i) \leq h(\mathbf{x}) \leq \sum_{i=1}^{n} h(x_i), \tag{1.9.6}$$

$$h(\mathbf{x}) - h(\lambda) \leq h(\lambda \mathbf{x}) \leq h(\mathbf{x}) + h(\lambda), \tag{1.9.7}$$

$$h^{\mathrm{hom}}(\lambda \mathbf{x}) = h^{\mathrm{hom}}(\mathbf{x}), \tag{1.9.8}$$

$$h(\mathbf{x}_1 + \cdots + \mathbf{x}_m) \leq \sum_{i=1}^{m} h(\mathbf{x}_i) + \log m. \tag{1.9.9}$$

We recall a few facts on heights and norms of polynomials. Let K be an algebraic number field and $v \in M_K$. Denote the unique extension of $|\cdot|_v$ to $\overline{K}_v$ also by $|\cdot|_v$. For a polynomial $P \in \overline{K}_v[X_1, \ldots, X_g]$, we denote by $|P|_v$ the v-adic norm of a vector, consisting of all non-zero coefficients of P. We write as before $s(v) = 1$ if v is real, $s(v) = 2$ if v is complex, and $s(v) = 0$ if v is finite.

Proposition 1.9.4 *Let $P_1, \ldots, P_m$ be non-zero polynomials in $\overline{K}_v[X_1, \ldots, X_g]$ and let n be the sum of the partial degrees of $P := P_1 \cdots P_m$. Then*

$$2^{-ns(v)} \leq \frac{|P|_v}{|P_1|_v \cdots |P_m|_v} \leq 2^{ns(v)}.$$

Proof. In the case that v is finite then the term $2^{ns(v)}$ is 1, and so this is Gauss' Lemma. In the case that v is infinite this is a version of a lemma of Gelfond. Proofs of both can be found for instance in Bombieri and Gubler (2006), Lemmas 1.6.3, 1.6.11. □

For a polynomial $P \in \overline{\mathbb{Q}}[X_1, \ldots, X_g]$, we denote by $H(P), H^{\mathrm{hom}}(P), h(P), h^{\mathrm{hom}}(P)$, the respective heights of a vector whose coordinates are the non-zero coefficients of P. Obviously, for polynomials we have similar inequalities as in (1.9.5)–(1.9.9). From Proposition 1.9.4 we deduce at once:

Corollary 1.9.5 *Let $P_1, \ldots, P_m$ be non-zero polynomials in $\overline{\mathbb{Q}}[X_1, \ldots, X_g]$ and let n be the sum of the partial degrees of $P := P_1 \cdots P_m$. Then*

$$\left| h^{\mathrm{hom}}(P) - \sum_{i=1}^{m} h^{\mathrm{hom}}(P_i) \right| \leq n \log 2.$$

Proof. Choose a number field K containing the coefficients of $P_1, \ldots, P_m$, apply Proposition 1.9.4 and take the product over $v \in M_K$. □

Corollary 1.9.6 *Let* $P = (X - \alpha_1) \cdots (X - \alpha_n) \in \overline{\mathbb{Q}}[X]$. *Then*

$$\left| h(P) - \sum_{i=1}^{n} h(\alpha_i) \right| \le n \log 2,$$

$$2^{-n} H(P) \le \prod_{i=1}^{n} H(\alpha_i) \le 2^n H(P).$$

Proof. The second assertion is an immediate consequence of the first one. To prove the first, observe that $h(\alpha) = h^{\text{hom}}(X - \alpha)$ for $\alpha \in \overline{\mathbb{Q}}$ and that $h(P) = h^{\text{hom}}(P)$ since P is monic. Applying Corollary 1.9.5 to the identity $P(X) = \prod_{i=1}^{n}(X - \alpha_i)$, the first assertion follows. $\square$

For monic irreducible polynomials P with coefficients in $\mathbb{Z}$, Corollary 1.9.6 gives a slightly weaker version of (1.9.3).

1.10 Effective computations in number fields

We have listed the basic algorithmic results for algebraic number fields that will be needed later. We shall not present the algorithms themselves, but refer to the literature for their description. Our main references are Borevich and Shafarevich (1967), Pohst and Zassenhaus (1989) and Cohen (1993, 2000).

When we say that for any given input from a specified set we can determine/compute effectively an output, we mean that there exists an algorithm (that is, a deterministic Turing machine) which, for any choice of input from the given set, computes the output in finitely many steps. We say that an object is *given effectively* if it is given in such a way that it can serve as input for an algorithm.

An *algebraic number field* K is said to be *effectively given* if $K = \mathbb{Q}(\theta)$ and the monic minimal polynomial $P \in \mathbb{Q}[X]$ of θ are given. Then $K \simeq \mathbb{Q}[X]/(P)$. Here we may assume that $P \in \mathbb{Z}[X]$ and that θ is an algebraic integer. Throughout this section we assume that K is effectively given in the form $K = \mathbb{Q}(\theta)$ with the monic minimal polynomial P of θ in $\mathbb{Z}[X]$. We denote by d the degree of P, that is the degree of K over $\mathbb{Q}$.

We say that an element α of K is *effectively given/computable* if in the representation

$$\alpha = a_0 + a_1 \theta + \cdots + a_{d-1} \theta^{d-1} \tag{1.10.1}$$

of α the coefficients $a_0, \ldots, a_{d-1} \in \mathbb{Q}$ are effectively given/computable. (1.10.1) is regarded as the standard representation of $\alpha \in K$ with respect to θ.

We shall need the following algorithmic results.

(I) If α, $\beta \in K$ are effectively given/computable then $\alpha \pm \beta$, $\alpha\beta$ and α/β ($\beta \neq 0$) are also effectively computable; see e.g. Cohen (1993), section 4.2.

(II) One can determine effectively an integral basis of K, that is a $\mathbb{Z}$-module basis $\{1, \omega_2, \ldots, \omega_d\}$ of the ring of integers O_K of K, and from that the discriminant D_K of K; see e.g. Cohen (1993), section 6.1. It is easy to see that if $\alpha \in K$ is effectively given then one can determine $b_1, \ldots, b_d$ in $\mathbb{Q}$ such that

$$\alpha = b_1 + b_2\omega_2 + \cdots + b_d\omega_d. \tag{1.10.2}$$

In particular, one can decide whether α is in O_K.

(III) For any given $F \in K[X]$, one can factorize F into irreducible polynomials over K; see Pohst and Zassenhaus (1989) or Cohen (1993), section 3.6. As a consequence, for given $F \in K[X]$ one can determine all zeros of F in K.

(IV) If $\alpha \in K$ is effectively given, then its characteristic polynomial relative to $K/\mathbb{Q}$ and its monic minimal polynomial over $\mathbb{Q}$ can be effectively determined. Conversely, if a monic, irreducible polynomial $P(X)$ over $\mathbb{Q}$ is given, then one can decide whether any of its zeros belongs to K, and if it is so then all zeros of $P(X)$ in K can be effectively determined. Consequently, if $K/\mathbb{Q}$ is normal, then all conjugates of any given $\alpha \in K$ can be effectively determined; see e.g. Győry (1983), remark 1.

(V) For given $C > 0$ one can determine a finite and effectively determinable subset $\mathcal{A}$ of K such that if $\alpha \in K$ and $h(\alpha) \leq C$ then $\alpha \in \mathcal{A}$. Indeed, representing α in the form (1.10.2) with an effectively given integral basis $\{1, \omega_2, \ldots, \omega_d\}$, taking conjugates with respect to $K/\mathbb{Q}$ and using Cramer's Rule, one can get an effective upper bound for $\max_i h(b_i)$. But, for such b_i, the numbers $b_1 + b_2\omega_2 + \cdots + b_d\omega_d$ form a finite and effectively computable subset of K.

(VI) If $\alpha \in K$ is effectively given then one can effectively compute an upper bound for $h(\alpha)$. Indeed, by (IV) we can compute the minimal polynomial $P_\alpha(X) \in \mathbb{Z}[X]$ of α with relatively prime coefficients. Then (1.9.3) provides an upper bound for $h(\alpha)$.

We say that a fractional ideal $\mathfrak{a}$ of O_K is effectively given/computable if a finite set of generators of $\mathfrak{a}$ over O_K is effectively given/computable. For other representations of fractional ideals we refer to Pohst and Zassenhaus (1989), section 6.3 and Cohen (1993), section 4.7.

(VII) If a fractional ideal $\mathfrak{a}$ of O_K is effectively given then it can be decided whether $\mathfrak{a}$ is principal. Further, if it is, one can compute an $\alpha \in K$ such that $\mathfrak{a} = (\alpha)$; see Cohen (1993), section 6.5.

(VIII) For effectively given fractional ideals of O_K one can compute their sum, product and their absolute norms. Further, one can test equality, inclusion (i.e. divisibility) and whether a given element of K is in a given fractional ideal; see e.g. Cohen (1993), section 4.7.

(IX) If $\mathfrak{a}$ is an effectively given fractional ideal of O_K then its prime ideal factorization can be effectively determined; see e.g. Cohen (2000), section 2.3. In particular, one can decide whether $\mathfrak{a}$ is an ideal of O_K or a prime ideal.

Let S be a finite set of places of K containing all infinite places. We say that S is *effectively given* if the prime ideals in S are effectively given. In what follows, we assume that S is effectively given. We recall that O_S (resp. O_S^*) denotes the ring of S-integers (resp. the group of S-units) in K.

(X) For an effectively given place $v \in M_K$ and an effectively given $\alpha \in K$, one can effectively compute $|\alpha|_v$. For the definition of an infinite place v being effectively given, and the computation of $|\alpha|_v$ see Cohen (1993), section 4.2. For the computation of $|\alpha|_v$ for a finite place v combine (IX) and (VIII) above.

(XI) In view of (IX) one can decide for any given $\alpha \in K^*$ whether $\alpha \in O_S$ or $\alpha \in O_S^*$.

(XII) A fundamental system of S-units can be effectively determined; see Cohen (2000), section 7.4. In particular, a fundamental system of units in K can be effectively found; see Borevich and Shafarevich (1967), chapter 2, section 5 or Pohst and Zassenhaus (1989), section 5.7. Further, the roots of unity in K can be effectively found; see e.g. Pohst and Zassenhaus (1989), section 5.4 or Cohen (1993), section 4.9.

(XIII) If $\varepsilon \in O_S^*$ and a fundamental system of S-units $\{\varepsilon_1, \ldots, \varepsilon_{s-1}\}$ are effectively given, then one can determine effectively rational integers $b_1, \ldots, b_{s-1}$ and a root of unity ζ in K such that (1.8.1) holds.

Proof. By Corollary 1.8.2, ε can be written in the form (1.8.1). Let $S = \{v_1, \ldots, v_s\}$ be as in Theorem 1.8.1. Then (1.8.1) implies that

$$\log |\varepsilon|_{v_i} = \sum_{j=1}^{s-1} b_j \log |\varepsilon_j|_{v_i} \quad \text{for } i = 1, \ldots, s.$$

Considering this as a system of linear equations in $b_1, \ldots, b_{s-1}$ and using Cramer's Rule, (1.8.5) and the fact that $|\log |\varepsilon|_{v_i}|$ and $|\log |\varepsilon_j|_{v_i}|$ can be effectively bounded above for each i and j, one can derive an effectively computable upper bound for $\max_j |b_j|$. Testing all possible values of $b_1, \ldots, b_{s-1}$, one can determine $b_1, \ldots, b_{s-1}$ such that $\varepsilon/\varepsilon_1^{b_1} \cdots \varepsilon_{s-1}^{b_{s-1}}$ is a root of unity. $\qquad \square$

1.11 *p*-adic numbers

Let p be a prime number. Recall that we have defined the absolute value $|\cdot|_p$ on $\mathbb{Q}$ by $|\cdot|_p := p^{-\mathrm{ord}_p(\cdot)}$, where $\mathrm{ord}_p(a)$ denotes the exponent of p in the unique prime factorization of $a \in \mathbb{Q}^*$, and $\mathrm{ord}_p(0) = \infty$. We denote by $\mathbb{Q}_p$ the completion of $\mathbb{Q}$ with respect to $|\cdot|_p$ or equivalently, with respect to ord_p. Clearly, ord_p defines a discrete valuation on $\mathbb{Q}$, and hence on $\mathbb{Q}_p$. We define the *ring of p-adic integers* by

$$\mathbb{Z}_p := \{x \in \mathbb{Q}_p : \mathrm{ord}_p(x) \geq 0\}.$$

Let L be a finite extension of $\mathbb{Q}_p$. There is precisely one absolute value on L that extends $|\cdot|_p$, given by $|N_{L/\mathbb{Q}_p}(\cdot)|_p^{1/[L:\mathbb{Q}_p]}$, and L is complete with respect to this absolute value. We can extend ord_p to a valuation on L, by defining $\mathrm{ord}_p(\alpha) := -\log |N_{L/\mathbb{Q}_p}(\alpha)|_p/[L : \mathbb{Q}_p] \log p$ for $\alpha \in L$. Clearly, $\mathrm{ord}_p(\alpha)$ is a rational number with denominator dividing $[L : \mathbb{Q}_p]$. As a consequence, the value set of ord_p on L^* is a cyclic subgroup of $\mathbb{Q}$ containing $\mathbb{Z}$, say of the shape $e^{-1}\mathbb{Z}$, where e is a positive integer. This integer e is called the *ramification index* of L over $\mathbb{Q}_p$. Any positive integer e may occur as ramification index; for instance if $\alpha^e = p$, then $\mathbb{Q}_p(\alpha)$ has ramification index e over $\mathbb{Q}_p$.

Now, let $\overline{\mathbb{Q}_p}$ denote an algebraic closure of $\mathbb{Q}_p$. Then the above considerations imply that ord_p extends uniquely to a valuation on $\overline{\mathbb{Q}_p}$, denoted also by ord_p, with value group $\mathrm{ord}_p(\overline{\mathbb{Q}_p}^*) = \mathbb{Q}$. It can be shown that $\overline{\mathbb{Q}_p}$ is *not* complete with respect to ord_p but that the completion of $\overline{\mathbb{Q}_p}$ is algebraically closed (see Koblitz (1984), pp. 71–73). The ring of integers of $\overline{\mathbb{Q}_p}$ and its unit group are given by

$$\overline{\mathbb{Z}_p} := \{x \in \overline{\mathbb{Q}_p} : \mathrm{ord}_p(x) \geq 0\}, \quad \overline{\mathbb{Z}_p}^* = \{x \in \overline{\mathbb{Q}_p} : \mathrm{ord}_p(x) = 0\}.$$

Let K be an algebraic number field. Then any discrete valuation v on K lying above ord_p corresponds to a prime ideal of O_K above p, that is,

$$\mathfrak{p} := \{x \in O_K : v(x) > 0\}.$$

Further, for $x \in K^*$, $v(x)$ is precisely the exponent of $\mathfrak{p}$ in the unique prime ideal factorization of (x), i.e., $v = \mathrm{ord}_{\mathfrak{p}}$. The completion of K with respect to $\mathrm{ord}_{\mathfrak{p}}$, denoted by $K_{\mathfrak{p}}$, is a finite extension of $\mathbb{Q}_p$, and the ramification index of $K_{\mathfrak{p}}$ over $\mathbb{Q}_p$ is precisely the ramification index $e(\mathfrak{p}|p)$ of $\mathfrak{p}$ over p. Further, there is an embedding $\sigma : K \hookrightarrow \overline{\mathbb{Q}_p}$ such that

$$\operatorname{ord}_{\mathfrak{p}}(x) = e(\mathfrak{p}|p)\operatorname{ord}_p(\sigma(x)) \quad \text{for } x \in K .$$

We are going to define the p-adic logarithm. We start with some preliminaries.

Lemma 1.11.1 *Let $\alpha \in \overline{\mathbb{Z}_p}^*$ and $c \in \mathbb{R}_{>0}$. Then there is a positive integer m such that $\operatorname{ord}_p(\alpha^m - 1) > c$. This integer m depends only on p, c and the field $\mathbb{Q}_p(\alpha)$.*

Proof. Let $L := \mathbb{Q}_p(\alpha)$ and define

$$\mathfrak{o} := \{x \in L : \operatorname{ord}_p(x) \geq 0\}, \quad \mathfrak{m} := \{x \in L : \operatorname{ord}_p(x) > c\}.$$

Then $\mathfrak{o}$ is the integral closure of $\mathbb{Z}_p$ in L and $\mathfrak{m}$ is an ideal of $\mathfrak{o}$. Let l be the smallest integer $\geq c$. Then $\mathfrak{m} \supseteq p^l\mathfrak{o}$. Since the additive structure of $\mathfrak{o}$ is that of a free $\mathbb{Z}_p$-module of rank $[L : \mathbb{Q}_p]$, the residue class ring $\mathfrak{o}/p^l\mathfrak{o}$ has cardinality $p^{l \cdot [L:\mathbb{Q}_p]}$. This shows that the residue class ring $\mathfrak{o}/\mathfrak{m}$ is finite. But then its unit group $(\mathfrak{o}/\mathfrak{m})^*$ is also finite, say of order m, and $\alpha^m - 1 \in \mathfrak{m}$. Clearly, m depends only on p, c and L. $\qquad\square$

Lemma 1.11.2

(i) *Let $\alpha \in \overline{\mathbb{Z}_p}^*$ with $\operatorname{ord}_p(\alpha - 1) > 0$. Then the series*

$$\log_p(\alpha) := \sum_{n=1}^{\infty} \frac{(-1)^{n-1}}{n} \cdot (\alpha - 1)^n$$

converges to a limit in the field $\mathbb{Q}_p(\alpha)$.

(ii) *Let $\alpha \in \overline{\mathbb{Z}_p}^*$ with $\operatorname{ord}_p(\alpha - 1) > 1/(p - 1)$. Then*

$$\operatorname{ord}_p(\log_p(\alpha)) = \operatorname{ord}_p(\alpha - 1).$$

(iii) *Let $\alpha, \beta \in \overline{\mathbb{Z}_p}^*$ with $\operatorname{ord}_p(\alpha - 1) > 0$, $\operatorname{ord}_p(\beta - 1) > 0$. Then*

$$\log_p(\alpha\beta) = \log_p(\alpha) + \log_p(\beta).$$

Proof. The proofs of (i) and (iii) can be found in Koblitz (1984), section 4.1. To prove (ii), put $\kappa := \operatorname{ord}_p(\alpha - 1)$. Then, since $\kappa > 1/(p - 1)$,

$$\operatorname{ord}_p(\log_p(\alpha)) = \min(\kappa, p\kappa - 1, p^2\kappa - 2, \ldots) = \kappa. \qquad\square$$

We now define $\log_p$ on the whole group $\overline{\mathbb{Z}_p}^*$ as follows: take $\alpha \in \overline{\mathbb{Z}_p}^*$, choose a positive integer m such that $\operatorname{ord}_p(\alpha^m - 1) > 0$ (which exists by

Lemma 1.11.1) and put

$$\log_p(\alpha) := \frac{1}{m} \log_p(\alpha^m).$$

By part (iii) of Lemma 1.11.2 this does not depend on the choice of m. Moreover,

$$\log_p(\alpha\beta) = \log_p(\alpha) + \log_p(\beta) \quad \text{for } \alpha, \beta \in \overline{\mathbb{Z}_p}^*. \tag{1.11.1}$$

Proposition 1.11.3

(i) $\log_p$ defines a surjective group homomorphism from $\overline{\mathbb{Z}_p}^*$ to the additive group of $\overline{\mathbb{Q}_p}$ with kernel the roots of unity in $\overline{\mathbb{Q}_p}$.

(ii) Let L be a finite extension of $\mathbb{Q}_p$. Then $\log_p$ defines a non-surjective homomorphism from $\overline{\mathbb{Z}_p}^* \cap L$ to the additive group of L.

Proof. (i) By (1.11.1), $\log_p$ defines a homomorphism from $\overline{\mathbb{Z}_p}^*$ to the additive group of $\overline{\mathbb{Q}_p}$. We determine the kernel of $\log_p$. First, let α be a root of unity from $\overline{\mathbb{Q}_p}$. Then $\alpha \in \overline{\mathbb{Z}_p}^*$. Further, there is a positive integer m with $\alpha^m = 1$ and so, $\log_p \alpha = m^{-1} \log_p(\alpha^m) = 0$. Now let $\alpha \in \overline{\mathbb{Z}_p}^*$ which is not a root of unity. By Lemma 1.11.1, there exists a positive integer m such that $\text{ord}_p(\alpha^m - 1) > 1/(p - 1)$. Then by part (ii) of Lemma 1.11.2,

$$\text{ord}_p(\log_p(\alpha)) = \text{ord}_p(\log_p(\alpha^m)) - \text{ord}_p(m)$$
$$= \text{ord}_p(\alpha^m - 1) - \text{ord}_p(m) < \infty \tag{1.11.2}$$

and thus $\log_p(\alpha) \neq 0$. This proves that the kernel of $\log_p$ consists of the roots of unity of $\overline{\mathbb{Q}_p}$.

To prove the surjectivity of $\log_p$, we use the p-adic exponential. By, e.g., Koblitz (1984), section 4.1, for $\alpha \in \overline{\mathbb{Q}_p}^*$ with $\text{ord}_p(\alpha) > p/(p - 1)$, the series

$$\exp_p(\alpha) := \sum_{n=0}^{\infty} \frac{\alpha^n}{n!}$$

converges to a limit in the field $\mathbb{Q}_p(\alpha)$. Moreover, again by Koblitz (1984), section 4.1, for these α we have $\text{ord}_p(\exp_p(\alpha) - 1) > 0$ and $\log_p(\exp_p(\alpha)) = \alpha$.

Now let $\beta \in \overline{\mathbb{Q}_p}$ be arbitrary. Choose $k \in \mathbb{Z}_{>0}$ such that $k + \text{ord}_p(\beta) > p/(p - 1)$ and then $\alpha \in \overline{\mathbb{Q}_p}$ with $\alpha^{p^k} = \exp_p(p^k \beta)$. We have $\alpha \in \overline{\mathbb{Z}_p}^*$ since $\text{ord}_p(\exp_p(p^k \beta) - 1) > 0$. Now, clearly,

$$\log_p(\alpha) = p^{-k} \log_p(\exp_p(p^k \beta)) = \beta.$$

This proves the surjectivity of $\log_p$.

(ii) Let $\alpha \in \overline{\mathbb{Z}_p}^* \cap L$. By Lemma 1.11.1, there exists a positive integer m, depending only on p and L, such that $\text{ord}_p(\alpha^m - 1) > 1/(p - 1)$. Then $\log_p(\alpha) = m^{-1}\log_p(\alpha^m) \in L$ and also, similarly to (1.11.2),

$$\text{ord}_p(\log_p(\alpha)) = \text{ord}_p(\alpha^m - 1) - \text{ord}_p(m) > \frac{1}{p-1} - \text{ord}_p(m),$$

which is independent of α. So $\log_p : \overline{\mathbb{Z}_p}^* \cap L \to L$ is certainly not surjective. $\qquad\square$

For the computation of p-adic logarithms of algebraic numbers, see de Weger (1989) and Smart (1998).

2

Algebraic function fields

By an algebraic function field in one variable over a field **k**, or, in short, function field over **k**, we mean a finitely generated field extension of transcendence degree 1 over **k**. We shall restrict ourselves to the case that **k** is algebraically closed and of characteristic 0. Thus, if K is a function field over **k** and z is any element from $K \setminus \mathbf{k}$, then K is a finite extension of the field of rational functions $\mathbf{k}(z)$.

We have collected here the concepts and results that are used in our book. For further details and proofs, we refer to the books Eichler (1966) and Mason (1984) and to the paper Schmidt (1978).

2.1 Valuations

Let **k** be an algebraically closed field of characteristic 0 and K an algebraic function field over **k**. By a *valuation* on K over **k** we mean a discrete valuation with value group $\mathbb{Z}$ such that $v(x) = 0$ for $x \in \mathbf{k}^*$, i.e., a surjective map $v : K \to \mathbb{Z} \cup \{\infty\}$ such that

$$v(x) = \infty \iff x = 0;$$
$$v(xy) = v(x) + v(y), \; v(x + y) \geq \min(v(x), v(y)) \quad \text{for } x, y \in K;$$
$$v(x) = 0 \quad \text{for } x \in \mathbf{k}^*.$$

The corresponding local ring and maximal ideal of v are given by

$$\mathfrak{o}_v := \{x \in K : v(x) \geq 0\}, \quad \mathfrak{m}_v := \{x \in K : v(x) > 0\},$$

respectively. The quotient $\mathfrak{o}_v/\mathfrak{m}_v$ is a field, called the *residue class field* of v. Since **k** is algebraically closed, we have $\mathfrak{o}_v/\mathfrak{m}_v = \mathbf{k}$.

By a *local parameter of* v we mean an element z_v of K such that $v(z_v) = 1$. Then the completion of K at v is the field of formal Laurent series $\mathbf{k}((z_v))$.

Analogously to number fields, we denote the set of valuations on K by M_K.

Let K be a function field over $\mathbf{k}$ and L a finite extension of K. Let v be a valuation on K. We say that a valuation w of L lies above v, notation $w|v$, if the restriction of w to K is a multiple of v. In that case, we have $w(x) = e(w|v)v(x)$ for $x \in K$, where $e(w|v)$ is a positive integer, called the *ramification index* of w over v.

First we describe the valuations on $\mathbf{k}(z)$. For every element a of $\mathbf{k}$, each non-zero element x of $\mathbf{k}(z)$ may be expanded as a formal Laurent series

$$\sum_{m=n}^{\infty} a_m (z - a)^m,$$

where a_m is an element of $\mathbf{k}$ and $a_n \neq 0$. Then ord_a defined by $\mathrm{ord}_a(x) := n$ is a valuation on $\mathbf{k}(z)$, and the field of Laurent series $\mathbf{k}((z - a))$ is the completion of $\mathbf{k}(z)$ at ord_a. Similarly, we define a valuation ord_∞ on $\mathbf{k}(z)$ expanding $x \in \mathbf{k}(z)$ as a Laurent series in z^{-1}. In particular, $\mathrm{ord}_\infty(x) = -\deg(x)$ for $x \in \mathbf{k}[z]$. The completion of $\mathbf{k}(z)$ at ord_∞ is $\mathbf{k}((z^{-1}))$. The valuations ord_a ($a \in \mathbf{k} \cup \{\infty\}$) provide all valuations on $\mathbf{k}(z)$. These valuations satisfy the *Sum Formula*

$$\sum_{a \in \mathbf{k} \cup \{i\infty\}} \mathrm{ord}_a(x) = 0 \quad \text{for } x \in \mathbf{k}(z)^*.$$

Now, let K be an algebraic function field over $\mathbf{k}$. We give a concrete description of the valuations on K by means of Puiseux expansions. To this end, fix $z \in K \setminus \mathbf{k}$, so that K is a finite extension of $\mathbf{k}(z)$. Put $d := [K : \mathbf{k}(z)]$. The function field K has a primitive element y over $\mathbf{k}(z)$ which satisfies an irreducible equation

$$P(y, z) = y^d + p_1(z)y^{d-1} + \cdots + p_d(z) = 0 \qquad (2.1.1)$$

with coefficients $p_i(z)$ in $\mathbf{k}(z)$. If $Q(z)$ is the common denominator of the rational functions $p_i(z)$, then $y_1 := Q(z)y$ satisfies an equation of the form (2.1.1) with coefficients from $\mathbf{k}[z]$. Replacing y by y_1, we may assume that in (2.1.1) y is a primitive element of K with polynomials $p_i(z) \in \mathbf{k}[z]$.

The field K may be embedded both in the field of fractional power series in $z - a$, where a is an arbitrary element of $\mathbf{k}$, and in the field of fractional power series in z^{-1}. These fields are all algebraically closed. Every element x of K may be expressed in a unique way in the form

$$x = \sum_{i=1}^{d} q_i(z)y^{i-1}$$

with some $q_1, \ldots, q_d$ in $\mathbf{k}(z)$. Hence, to expand the functions of K in power series in $z - a$ or z^{-1}, it suffices to so expand the single function y.

We recall Puiseux's classical theorem.

Theorem 2.1.1 *For each element a of $\mathbf{k}$, there are positive integers $r_a \leq d$ and $e_1, \ldots, e_{r_a}$ with $e_1 + \cdots + e_{r_a} = d$, and formal Puiseux series*

$$y_\rho = \sum_{m=n_\rho}^{\infty} a_{\rho m}(z - a)^{m/e_\rho}, \quad a_{\rho n_\rho} \neq 0, \quad \rho = 1, \ldots, r_a$$

with coefficients $a_{\rho m}$ in $\mathbf{k}$, that satisfy (2.1.1). Further, if ζ is a primitive e_ρ-th root of unity and

$$y_{\rho j} = \sum_{m=n_\rho}^{\infty} a_{\rho m} \zeta^{jm}(z - a)^{m/e_\rho} \quad (j = 1, \ldots, e_\rho - 1),$$

then the left-hand side of (2.1.1) is identical with

$$P(y, z) = \prod_{\rho, j}(y - y_{\rho j}).$$

A similar assertion holds with z^{-1} instead of $z - a$.

Proof. See, e.g., Eichler (1966) chapter III, section 1. □

For each a in $\mathbf{k}$ and each ρ, j as above, the map $\varphi_{\rho j} : y \mapsto y_{\rho j}$ determines uniquely an embedding of K into the field of formal Laurent expansions in powers of $(z - a)^{1/e_\rho}$, i.e., for $x \in K$ we have

$$\varphi_{\rho j}(x) = \sum_{m=n}^{\infty} a_m(z - a)^{m/e_\rho} \text{ with } a_m \in \mathbf{k} \text{ for } m \geq n \text{ and } a_n \neq 0.$$

For every ρ with $1 \leq \rho \leq r_a$, we construct a valuation v on K as follows: choose any j with $1 \leq j \leq e_\rho$. Then, for any $x \in K^*$, we define $v(x) := n$ in the above Laurent series expression for $\varphi_{\rho j}(x)$. Notice that the valuation v on K lies above $u := \mathrm{ord}_a$, that $(z - a)^{1/e_\rho}$ is a local parameter for v, and that e_ρ is the ramification index $e(v|u)$ of v over u. The above construction gives all extensions of $u = \mathrm{ord}_a$ to K.

One can construct in a similar way the extensions v of ord_∞ to K. Each of these v is defined as the order of vanishing of the Laurent expansion in a local parameter z^{-1/e_ρ}.

In this way all valuations v of K are described. For convenience we say that v lies above a ($a \in \mathbf{k} \cup \{\infty\}$) if it lies above ord_a and write $e(v|a)$ for $e(v|\mathrm{ord}_a)$. Notice that for all $a \in \mathbf{k} \cup \{\infty\}$ we have $\sum_{v|a} e(v|a) = [K : \mathbf{k}(z)]$, where the sum is taken over all valuations v of K lying above a.

We say that $v \in M_K$ is called *infinite with respect to* z if it lies above ∞, i.e., if $v(z) < 0$. We denote this by $v \mid \infty$; otherwise, we say that v is *finite with respect to* z.

To get a uniform notation, if v lies above a and corresponds to the pair (ρ, j), we write z_v for $(z - a)^{1/e_\rho}$ if $a \neq \infty$ and for z^{-1/e_ρ} if $a = \infty$, and y_v for $y_{\rho j}$. Thus, for every valuation v of K, z_v is a local parameter for v, and $y \mapsto y_v$ defines an isomorphic embedding $\varphi_v : K \hookrightarrow \mathbf{k}((z_v))$. The valuations defined above have the following properties:

$$v(x) = 0 \text{ for all } v \in M_K \iff x \in \mathbf{k}^*, \tag{2.1.2}$$

$$\sum_{v \in M_K} v(x) = 0 \quad \text{for } x \text{ in } K^* \quad \text{(Sum Formula)}. \tag{2.1.3}$$

For each non-zero $x \in K$, only finitely many summands are non-zero.

Let L be a finite extension of K. On L we define valuations in the same way as for K. Then

$$\sum_{w \mid v} e(w \mid v) = [L : K] \quad \text{for } v \in M_K, \tag{2.1.4}$$

where the sum is taken over all valuations w of L lying above v. More generally, we have the *Extension Formula*

$$\sum_{w \mid v} w(x) = v(N_{L/K}(x)) \quad \text{for } x \in L. \tag{2.1.5}$$

2.2 Heights

Let again K be an algebraic function field in one variable over an algebraically closed field $\mathbf{k}$ of characteristic 0 and M_K its set of valuations over $\mathbf{k}$. For a vector $\mathbf{x} = (x_1, \ldots, x_n) \in K^n \setminus \{\mathbf{0}\}$ we define

$$v(\mathbf{x}) := -\min(v(x_1), \ldots, v(x_n)) \quad \text{for } v \in M_K,$$

and then

$$H_K^{\mathrm{hom}}(\mathbf{x}) = H_K^{\mathrm{hom}}(x_1, \ldots, x_n) := \sum_v v(\mathbf{x}),$$

where as usual $\sum_v$ indicates that the sum is taken over all valuations $v \in M_K$. This is called the homogeneous height of $\mathbf{x}$ with respect to K. The height H_K may be viewed as the function field analogue of the logarithmic height $h_L^{\mathrm{hom}}(\mathbf{x}) := \sum_v \log \max_i |x_i|_v$ for $\mathbf{x} = (x_1, \ldots, x_n) \in L^n \setminus \{0\}$ relative to a number field L; this is $[L : \mathbb{Q}]$ times the absolute logarithmic height h^{hom} defined in Section 1.9. It has become common practice to denote function field heights by capital H. By the Sum Formula we have

$$H_K^{\text{hom}}(\lambda \mathbf{x}) = H_K^{\text{hom}}(\mathbf{x}) \quad \text{for } \lambda \in K^*. \tag{2.2.1}$$

For instance, let $p_1, \ldots, p_n \in \mathbf{k}[z]$ with $\gcd(p_1, \ldots, p_n) = 1$. Then

$$H_{\mathbf{k}(z)}^{\text{hom}}(p_1, \ldots, p_n) = \max(\deg p_1, \ldots, \deg p_n). \tag{2.2.2}$$

If L is a finite extension of K, the valuations on L may be constructed as above, and the height in L may be defined accordingly. Furthermore, for $\mathbf{x} = (x_1, \ldots, x_n) \in K^n \setminus \{\mathbf{0}\}$ we have

$$H_L^{\text{hom}}(\mathbf{x}) = [L : K] H_K^{\text{hom}}(\mathbf{x}). \tag{2.2.3}$$

We define a *height* for elements of K by

$$H_K(x) := H_K^{\text{hom}}(1, x) = -\sum_v \min(0, v(x)) \quad \text{for } x \in K, \tag{2.2.4}$$

For instance, if $K = \mathbf{k}(z)$ and $x = p/q$ where p, q are coprime polynomials from $\mathbf{k}[z]$, then $H_{\mathbf{k}(z)}(x) = \max(\deg p, \deg q)$. From (2.1.4) and (2.1.5), one deduces that for any finite extension L of K,

$$H_L(x) = [L : K] \cdot H_K(x) \quad \text{for } x \in K, \tag{2.2.5}$$

where $H_L(x)$ denotes the height of x with respect to L.

We mention some properties of the height H_K. It is evident from (2.1.2) and (2.1.3) that

$$H_K(x) \geq 0 \quad \text{for } x \in K, \qquad H_K(x) = 0 \iff x \in \mathbf{k}.$$

Further, from simple manipulations with valuations and from the Sum Formula it follows that

$$H_K(x^m) = |m| H_K(x) \quad \text{for } x \in K^*, m \in \mathbb{Z}, \tag{2.2.6}$$

and

$$\left.\begin{array}{c} H_K(x + y) \\ H_K(xy) \end{array}\right\} \leq H_K(x) + H_K(y) \quad \text{for } x, y \in K. \tag{2.2.7}$$

Next, from (2.2.4) and (2.2.6) it follows that

$$H_K(x) = \tfrac{1}{2}\big(H_K(x) + H_K(x^{-1})\big) = \tfrac{1}{2} \sum_{v \in M_K} |v(x)| \geq \tfrac{1}{2}|S| \quad \text{for } x \in K^*, \tag{2.2.8}$$

where S is the set of valuations $v \in M_K$ for which $v(x) \neq 0$.

Let $P \in K[X_1, \ldots, X_r]$ be a non-zero polynomial, and $\{p_1, \ldots, p_n\}$ the set of non-zero coefficients of P. We define

$$v(P) := v(p_1, \ldots, p_n) = -\min(v(p_1), \ldots, v(p_n)) \quad \text{for } v \in M_K,$$
$$H_K^{\text{hom}}(P) := \sum_v v(P).$$

We have obvious analogues of (2.2.1), (2.2.2) and (2.2.5) for polynomials. By Gauss' Lemma for valuations (the method of proof is similar to that of Bombieri and Gubler (2006), lemma 1.6.3) we have for any two polynomials $P, Q \in K[X_1, \ldots, K_r]$,

$$v(PQ) = v(P) + v(Q) \quad \text{for } v \in M_K,$$

hence

$$H_K^{\text{hom}}(PQ) = H_K^{\text{hom}}(P) + H_K^{\text{hom}}(Q). \tag{2.2.9}$$

Suppose that $P = f_0(X - \alpha_1) \cdots (X - \alpha_g)$ with $f_0, \alpha_1, \ldots, \alpha_g \in K$. Then by (2.2.9) and the Sum Formula, applied to f_0, we obtain

$$H_K^{\text{hom}}(P) = \sum_{i=1}^{g} H_K(\alpha_i) \geq \max_{1 \leq i \leq g} H_K(\alpha_i). \tag{2.2.10}$$

2.3 Derivatives and genus

For the moment, let L be any field extension, not necessarily of finite type, which has transcendence degree 1 over an algebraically closed field $\mathbf{k}$ of characteristic 0. The L-vector space $\Omega(L/\mathbf{k})$ of *differentials* of L over $\mathbf{k}$ may be constructed as follows. We start with taking a variable δ_x for every $x \in L$. Then let V be the L-vector space consisting of all finite formal linear combinations $\sum_i y_i \delta_{x_i}$ with $y_i, x_i \in L$, and let V_0 be the L-linear subspace of V generated by $\delta_{x+y} - \delta_x - \delta_y$ and $\delta_{xy} - x\delta_y - y\delta_x$ for $x, y \in L$ and δ_x for $x \in \mathbf{k}$. Then define $\Omega(L/\mathbf{k}) := V/V_0$. For $x \in L$ denote by $\mathrm{d}x$ the residue class of δ_x modulo V_0. Thus, $\Omega(L/\mathbf{k})$ consists of all finite linear combinations $\omega = \sum_i y_i \mathrm{d}x_i$, where $x_i, y_i \in L$, and we have $\mathrm{d}x \neq 0$ for $x \in L \setminus \mathbf{k}$, $\mathrm{d}x = 0$ for $x \in \mathbf{k}$, and $\mathrm{d}(x + y) = \mathrm{d}x + \mathrm{d}y$, $\mathrm{d}(xy) = x\mathrm{d}y + y\mathrm{d}x$ for $x, y \in L$. Consequently, $\mathrm{d}(\lambda x) = \lambda \mathrm{d}x$ for $x \in L$, $\lambda \in \mathbf{k}$.

It is clear that if L' is a subfield of L then up to isomorphism, $\Omega(L'/\mathbf{k})$ is contained in $\Omega(L/\mathbf{k})$.

For any $x, y \in L$ with $y \notin \mathbf{k}$, there is an irreducible polynomial $Q \in \mathbf{k}[X, Y]$ such that $Q(x, y) = 0$. Then we have

$$\frac{\partial Q}{\partial X}(x, y)\mathrm{d}x + \frac{\partial Q}{\partial Y}(x, y)\mathrm{d}y = 0.$$

Hence there exists a function in L (in fact in $\mathbf{k}(x, y)$), which we denote by $\mathrm{d}x/\mathrm{d}y$, such that

$$\mathrm{d}x = \frac{\mathrm{d}x}{\mathrm{d}y} \cdot \mathrm{d}y.$$

We call $\mathrm{d}x/\mathrm{d}y$ the *derivative of x with respect to y*. Notice that we have the *chain rule*

$$\frac{\mathrm{d}x}{\mathrm{d}z} = \frac{\mathrm{d}x}{\mathrm{d}y} \cdot \frac{\mathrm{d}y}{\mathrm{d}z}$$

for any $x, y, z \in L$ with $y, z \notin \mathbf{k}$. As a consequence, if we fix $z \in L \setminus \mathbf{k}$, then every differential $\omega \in \Omega(L/\mathbf{k})$ can be expressed as $(\omega/\mathrm{d}z) \cdot \mathrm{d}z$ with $\omega/\mathrm{d}z \in L$.

Now let again K be a function field in one variable over $\mathbf{k}$, i.e., a finite type extension of transcendence degree 1 over $\mathbf{k}$. For every valuation $v \in M_K$ we choose a local parameter z_v. Let $x \in K^*$. Then for $v \in M_K$ we can express x as a formal Laurent series $\sum_{i=n_0}^{\infty} a_i z_v^i$ with $n_0 = v(x)$, $a_i \in \mathbf{k}$ for $i \geq n_0$ and $a_{n_0} \neq 0$, and then $\mathrm{d}x/\mathrm{d}z_v = \sum_{i=n_0}^{\infty} i a_i z_v^{i-1}$. As a consequence,

$$v\left(\frac{\mathrm{d}x}{\mathrm{d}z_v}\right) = v(x) - 1 \quad \text{for any } x \text{ in } K \text{ with } v(x) \neq 0, \qquad (2.3.1)$$

$$v\left(\frac{\mathrm{d}x}{\mathrm{d}z_v}\right) \geq 0 \quad \text{if } v(x) = 0. \qquad (2.3.2)$$

This shows that $v(\mathrm{d}x/\mathrm{d}z_v)$ is independent of the choice of z_v. Indeed, let z_v' be another local parameter for v. Then $v(z_v') = 1$, hence $v(\mathrm{d}z_v/\mathrm{d}z_v') = 0$, and so $v(\mathrm{d}x/\mathrm{d}z_v) = v(\mathrm{d}x/\mathrm{d}z_v')$.

A differential ω of K over $\mathbf{k}$ is called *holomorphic* if $v(\omega/\mathrm{d}z_v) \geq 0$ for all $v \in M_K$; this notion is independent of the choice of the z_v. It can be shown that the holomorphic differentials of K over $\mathbf{k}$ form a finite dimensional $\mathbf{k}$-vector space. The dimension of this space is called the *genus* of K over $\mathbf{k}$, denoted by $g_{K/\mathbf{k}}$.

Let $x \in K \setminus \mathbf{k}$ be arbitrary. It follows from the Sum Formula (2.1.3) and the chain rule $\mathrm{d}x/\mathrm{d}z_v = (\mathrm{d}x/\mathrm{d}z) \cdot (\mathrm{d}z/\mathrm{d}z_v)$ that

$$\sum_v v\left(\frac{\mathrm{d}x}{\mathrm{d}z_v}\right) = \sum_v v\left(\frac{\mathrm{d}z}{\mathrm{d}z_v}\right)$$

is independent of x provided the right-hand side of the equality is finite. We need only the following special case of the Riemann–Roch Theorem.

Theorem 2.3.1 *We have*

$$\sum_v v\left(\frac{\mathrm{d}x}{\mathrm{d}z_v}\right) = 2g_{K/\mathbf{k}} - 2 \text{ for every } x \in K \setminus \mathbf{k}.$$

It is not difficult to check that $\mathbf{k}(z)$ has genus 0.

The following genus estimate will be useful.

Proposition 2.3.2 *Let $z \in K \setminus \mathbf{k}$, let $F = X^g + f_1 X^{g-1} + \cdots + f_g$ with coefficients $f_1, \ldots, f_g \in \mathbf{k}[z]$, and suppose that K is the splitting field of F over $\mathbf{k}(z)$. Then*

$$g_{K/\mathbf{k}} \leq (d-1)g \max(\deg f_1, \ldots, \deg f_g),$$

where $d := [K : \mathbf{k}(z)]$.

Proof. This is Lemma H in Schmidt (1978). □

2.4 Effective computations

In order to perform effective computations in the function field K, it is necessary to assume that the ground field $\mathbf{k}$ is *presented explicitly* in the sense of Fröhlich and Shepherdson (1956). This means here that there is an algorithm to determine the zeros of any polynomial with coefficients in $\mathbf{k}$. In particular, in this case we can perform all the field operations with elements of $\mathbf{k}$.

Further, we assume that K is presented explicitly. This means that K is given in the form $\mathbf{k}(z)(y)$, with z a variable, and y a primitive element of K over $\mathbf{k}(z)$, with an explicitly given defining polynomial $y^d + p_1(z)y^{d-1} + \cdots + p_d(z)$ over K. We may assume that y is integral over $\mathbf{k}[z]$, that is that $p_1(z), \ldots, p_d(z)$ are polynomials with coefficients in $\mathbf{k}$.

Every element x of K can be expressed uniquely in the form

$$\sum_{i=1}^{d} \frac{q_i(z)}{q(z)} \cdot y^{i-1},$$

where $q_1, \ldots, q_d, q$ are polynomials of $\mathbf{k}[z]$ such that $\gcd(q, q_1, \ldots, q_d) = 1$ and q_d is monic. We call $(q_1, \ldots, q_d, q)$ a *representation* for x, and we say that x is given explicitly if a representation for x is given explicitly, and that x can be determined effectively from certain given input data if there is an algorithm to determine a representation for x from these data. From representations for elements $x_1, x_2 \in K$ one can determine representations for $x_1 \pm x_2$, $x_1 x_2$ and x_1/x_2, if $x_2 \neq 0$.

One can easily compute a minimal polynomial of an explicitly given x over $\mathbf{k}[z]$, i.e., a polynomial $F = f_0 X^r + f_1 X^{r-1} + \cdots + f_r \in \mathbf{k}[z][X]$ of minimal degree such that $F(x) = 0$ and $\gcd(f_0, \ldots, f_r) = 1$. Indeed, one starts by computing representations for $x^2, \ldots, x^d$. Then by straightforward linear algebra one can determine the smallest r, which is $\leq d$, for which there exist $g_0, \ldots, g_r \in \mathbf{k}(z)$, not all 0, such that $g_0 + g_1 x + \cdots + g_r x^r = 0$, and having found such, one obtains a minimal polynomial of x by clearing denominators.

It is important to note that if $\mathbf{k}$ and K are presented explicitly, then the valuations of K can be described explicitly. Specifically, in Section 2.1 we gave, for every valuation v of K, a local parameter z_v for v as well as a Laurent series y_v in z_v, such that $y \mapsto y_v$ gives rise to an isomorphic embedding of K into $\mathbf{k}((z_v))$. The pair (z_v, y_v) can be determined from the defining polynomial of y and the element of $\mathbf{k} \cup \{\infty\}$ above which v lies. By determining y_v we mean that by an inductive procedure we can determine the coefficients of y_v one by one. We say that the valuation v is given explicitly, if the pair (z_v, y_v) is given, i.e., the inductive procedure to compute the coefficients of y_v is given.

If $x \in K$ and the valuation v are given, then we can express x as a Laurent series in z_v by substituting y_v for y in the expression $\sum_{i=1}^{d} (q_i(z)/q(z)) \cdot y^{i-1}$ for x and by expressing z as a Laurent series in z_v. Then we can compute $v(x)$ by searching for the first non-zero coefficient in the Laurent series expansion for x.

Further details may be found in Eichler (1966), chapter III, section 1 and Mason (1984), chapter V.

We recall a result of Mason (1984), p. 11, lemma 1.

Proposition 2.4.1 *Suppose that* $\mathbf{k}$, *K are presented explicitly, and a finite set S of valuations of K and integers n_v $(v \in S)$ are explicitly given. Then we can determine effectively whether there exists an element x in K such that*

$$v(x) = n_v \text{ for } v \in S, \quad v(x) = 0 \text{ for } v \in M_K \setminus S. \tag{2.4.1}$$

Moreover, if such an x exists then it may be computed, and it is unique up to a non-zero factor in $\mathbf{k}$.

Proof. We do not lose any generality by augmenting S with a finite set of explicitly given valuations and setting $n_v := 0$ for the added valuations. So, without loss of generality, we may assume that S contains all valuations that lie above $\{\infty, a_1, \ldots, a_t\}$, where $a_1, \ldots, a_t$ are certain elements of $\mathbf{k}$.

Further, it is enough to prove the assertion for the case when $n_v \geq 0$ for all finite $v \in S$, i.e., not lying above ∞; then the elements x under consideration are integral over $\mathbf{k}[z]$. For assume that x satisfies (2.4.1). Denote by e_v the ramification index of v over the element of $\mathbf{k} \cup \{\infty\}$ over which it lies. Choose an integer m such that $m > 0$ and $m + n_v \geq 0$ for $v \in S$, and put $q(z) := \prod_{j=1}^{t}(z - a_j)^m$. Then if $v|\infty$ we have $v(q(z)x) = n_v - e_v mt =: n'_v$ while if $v|a_i$ for some i we have $v(q(z)x) = n_v + e_v m =: n'_v \geq 0$. Further, for v outside S we have $v(q(z)) = 0$ and so $v(q(z)x) = 0$. Clearly, $q(z)$ can be determined effectively. So it suffices to prove our assertion with n'_v ($v \in S$) instead of n_v.

Recall that K is explicitly given in the form $\mathbf{k}(z)(y)$, with an explicitly given minimal polynomial of y over $\mathbf{k}(z)$ which is monic and has its coefficients in $\mathbf{k}[z]$. Consider now the system of equations (2.4.1) in x, where it is assumed that $n_v \geq 0$ for $v \in S$ with v finite. Thus, the elements $x \in K$ under consideration are integral over $\mathbf{k}[z]$. Each such x can be expressed in a unique way in the form

$$x = \sum_{i=1}^{d} q_i(z)y^{i-1}$$

with $q_i(z) \in \mathbf{k}(z)$, $i = 1, \ldots, d$. Denote by $\sigma_1, \ldots, \sigma_d$ the distinct embeddings of K in a fixed algebraic closure of K. Then we infer that

$$\sigma_j(x) = \sum_{i=1}^{d} q_i(z)(\sigma_j(y))^{i-1} \quad \text{for } j = 1, \ldots, d. \tag{2.4.2}$$

Let $D = \det(\sigma_j(y)^{i-1})^2 = \prod_{1 \leq i < j \leq d}(\sigma_i(y) - \sigma_j(y))^2$, i.e., D is the discriminant of y. It has an explicit expression as a polynomial with integer coefficients in terms of the coefficients of the minimal polynomial of y. Hence it belongs to $\mathbf{k}[z]$ and is effectively computable. Since by assumption $\mathbf{k}$ is presented explicitly, we can determine the zeros of D in $\mathbf{k}$ together with their multiplicities. Hence we can give all valuations of K lying above the zeros of D explicitly, and for each such v we can determine $v(D)$. We may augment S with these valuations. Thus, without loss of generality, $v(D) = 0$ for v outside S.

It follows from (2.4.2) that, for each i, $Dq_i(z)$ may be expressed as a polynomial with integer coefficients in the $\sigma_j(x)$ and $\sigma_j(y)$. But $\sigma_j(x)$ and $\sigma_j(y)$ are integral over $\mathbf{k}[z]$ for each j, hence $Dq_i(z) \in \mathbf{k}[z]$ for all i. By selecting $\sigma_1, \ldots, \sigma_d$ to be the embeddings corresponding to the infinite valuations on K, we may determine an integer u depending only on the integers n_v ($v|\infty$) such that $\text{ord}_\infty(Dq_i(z)) \geq -u$ for $1 \leq i \leq d$, and hence each $Dq_i(z)$ is a polynomial of degree at most u. Consequently, we may write

$$Dx = \sum_{j=0}^{u} \sum_{i=1}^{d} a_{ij} z^j y^{i-1} \tag{2.4.3}$$

with some a_{ij} in $\mathbf{k}$ to be determined.

We now prove that, assuming that x satisfies (2.4.3) and $x \neq 0$, we can replace (2.4.1) by the finite list of conditions

$$v(x) \geq n_v \quad \text{for } v \in S. \tag{2.4.4}$$

Indeed, by the Sum Formula we must have $\sum_{v \in S} n_v = 0$, otherwise, (2.4.1) is not solvable. By (2.4.3) we have $v(Dx) \geq 0$ for every finite valuation v. So $v(x) \geq 0$ for every valuation v outside S. Suppose that $v(x) > n_v$ for some $v \in S$ or $v(x) > 0$ for some v outside S. Then

$$0 = \sum_{v \in M_K} v(x) > \sum_{v \in S} n_v = 0,$$

a contradiction.

So (2.4.1) is equivalent to the combination of (2.4.3) and (2.4.4). By replacing in (2.4.3) z and y by their Laurent expansions in terms of z_v, we obtain for Dx an expansion $\sum_{m=m_v}^{\infty} L_{vm}(\mathbf{a}) z_v^m$ where every term $L_{vm}(\mathbf{a})$ is a linear form with coefficients in $\mathbf{k}$ in the coefficients a_{ij} in (2.4.3). Thus, x satisfies (2.4.3) and (2.4.4) if and only if the a_{ij} satisfy the finite system of linear equations

$$L_{vm}(\mathbf{a}) = 0 \quad \text{for } v \in S, \quad m = m_v, m_v + 1, \ldots, n_v + v(D) - 1.$$

Now we can decide whether this system of linear equations has a non-zero solution in $\mathbf{k}$, and if so, compute one. Consequently, we may determine whether there exists an element x in K with (2.4.1) and if so, compute such an x. Finally, if there are two elements x_1 and x_2 in K which satisfy (2.4.1), then $v(x_1/x_2) = 0$ for all v, whence $x_1/x_2 \in \mathbf{k}$. Thus, x is unique apart from a non-zero factor from $\mathbf{k}$. This completes the proof. $\quad\square$

Proposition 2.4.2 *Let $a_1, \ldots, a_r, b$ be explicitly given elements of K. Then it can be determined effectively whether*

$$\xi_1 a_1 + \cdots + \xi_r a_r = b \tag{2.4.5}$$

is solvable in $(\xi_1, \ldots, \xi_r) \in \mathbf{k}^r$. If so, the set of solutions in $\mathbf{k}^r$ of (2.4.5) is a linear variety of dimension $r - \operatorname{rank}_{\mathbf{k}}(a_1, \ldots, a_r)$, a parameter representation of which can be determined effectively.

Proof. First assume that $a_1, \ldots, a_r$ are linearly independent over $\mathbf{k}$. Recall that from any given $x \in K$, we can effectively determine its derivatives $x^{(j)} := d^j x / dz^j$ for all $j \geq 0$. Now, clearly, if $\xi_1, \ldots, \xi_r \in \mathbf{k}$ satisfy (2.4.5), then

$$\xi_1 a_1^{(j)} + \cdots + \xi_r a_r^{(j)} = b^{(j)} \quad (j = 0, \ldots, r - 1).$$

Since $a_1, \ldots, a_r$ are linearly independent over $\mathbf{k}$, the Wronskian determinant $\det(a_i^{(j-1)})_{i,j=1,\ldots,r}$ is non-zero. Hence the latter system has a unique solution

$(\xi_1, \ldots, \xi_r) \in K^r$ which can be determined effectively. Then it can be checked whether this solution belongs to $\mathbf{k}^r$.

Now suppose that $\mathrm{rank}_{\mathbf{k}}\{a_1, \ldots, a_r\} = m < r$. By means of the above procedure, we can select a $\mathbf{k}$-linearly independent subset of m elements from $\{a_1, \ldots, a_r\}$ and express the other elements as $\mathbf{k}$-linear combinations of this subset. Assume that $\{a_1, \ldots, a_m\}$ is $\mathbf{k}$-linearly independent. Check with the above procedure whether b is a $\mathbf{k}$-linear combination of $a_1, \ldots, a_m$. If so, express b and $a_{m+1}, \ldots, a_r$ as $\mathbf{k}$-linear combinations of $a_1, \ldots, a_m$, substitute these into (2.4.5) and compare the coefficients of $a_1, \ldots, a_m$. Thus, one can rewrite (2.4.5) as a system of linear equations with coefficients in $\mathbf{k}$, whose solution set is a linear variety of dimension $r - m$, and it is straightforward to compute a parameter representation of the latter. $\qquad\square$

3

Tools from Diophantine approximation and transcendence theory

In this chapter, we have collected some fundamental results from Diophantine approximation and transcendence theory on which the main results of this book are based. Section 3.1 is about Schmidt's Subspace Theorem and its variations. These will be applied in Chapter 6. In Section 3.2 we recall the best known effective estimates for linear forms in logarithms, which are used in Chapters 4 and 5. For more details and background, we refer to Schmidt (1980), Evertse and Schlickewei (2002), Bombieri and Gubler (2006), chapter 6, 7 and Baker and Wüstholz (2007).

3.1 The Subspace Theorem and some variations

In this section we formulate some versions of the Subspace Theorem that are used in Chapter 6. In particular, we recall the p-adic Subspace Theorem, the Parametric Subspace Theorem, and a quantitative version of a special case of the latter. We start with a brief introduction, taking as starting point Roth's celebrated Theorem on the approximation of algebraic numbers by rationals.

Theorem 3.1.1 *Let $\alpha \in \mathbb{R} \setminus \mathbb{Q}$ be an algebraic number and $\epsilon > 0$. Then there are only finitely many pairs $(x, y) \in \mathbb{Z}^2$ with $y > 0$ such that*

$$\left| \alpha - \frac{x}{y} \right| \leq \max(|x|, |y|)^{-2-\epsilon}. \tag{3.1.1}$$

Proof. See Roth (1955). Roth's proof consists of two steps: first the deduction of a non-vanishing result for polynomials, now known as Roth's Lemma; second, under the assumption that Theorem 3.1.1 is false the construction of an auxiliary polynomial that violates Roth's Lemma. $\square$

Weaker versions of Roth's Theorem were proved earlier by Thue (1909) with exponent $\frac{1}{2}d + 1$ instead of 2 where $d = \deg \alpha$, Siegel (1921) with exponent $2\sqrt{d}$, and Dyson (1947) and Gelfond (1960) (result proved in the late 1940s), both with exponent $\sqrt{2d}$. The proofs of Thue–Roth are all ineffective, in that they do not provide a method to determine the solutions of the inequality under consideration.

Extending earlier work of Ridout (1958), Lang (1960) proved a generalization of Roth's Theorem, usually referred to as the p-adic Roth's Theorem, where the underlying inequality takes its solutions from an algebraic number field and where various archimedean and non-archimedean absolute values from this number field are involved.

Roth's Theorem was generalized in another direction by W. M. Schmidt to simultaneous approximation. His work culminated in his so-called *Subspace Theorem*. Below, we denote by $\| \cdot \|$ the maximum norm on $\mathbb{C}^n$, i.e.,

$$\|\mathbf{x}\| := \max(|x_1|, \ldots, |x_n|) \text{ for } \mathbf{x} = (x_1, \ldots, x_n) \in \mathbb{C}^n.$$

Theorem 3.1.2 (Subspace Theorem) *Let $n \geq 2$ and let $L_i = \sum_{j=1}^{n} \alpha_{ij} X_j$ $(i = 1, \ldots, n)$ be linearly independent linear forms with algebraic coefficients in $\mathbb{C}$ and let $\epsilon > 0$. Then the set of solutions of the inequality*

$$|L_1(\mathbf{x}) \cdots L_n(\mathbf{x})| \leq \|\mathbf{x}\|^{-\epsilon} \text{ in } \mathbf{x} \in \mathbb{Z}^n \setminus \{\mathbf{0}\} \tag{3.1.2}$$

is contained in a finite union of proper linear subspaces of $\mathbb{Q}^n$.

Proof. See Schmidt (1972) or Schmidt (1980). $\qquad\square$

In general, inequalities of the shape (3.1.2) need not have finitely many solutions.

Theorem 3.1.2 $\Longrightarrow$ Theorem 3.1.1. Notice that if $(x, y) \in \mathbb{Z}^2$ with $y > 0$ is a solution of (3.1.1), then

$$|y(x - \alpha y)| \leq \max(|x|, |y|)^{-\epsilon}.$$

Now Theorem 3.1.2 implies that the solutions of the latter, hence of (3.1.1), lie in finitely many one-dimensional subspaces of $\mathbb{Q}^2$. But the solutions of (3.1.1) in a given one-dimensional subspace of $\mathbb{Q}^2$ are all of the shape $m(x_0, y_0)$ with (x_0, y_0) a fixed pair of integers with $\gcd(x_0, y_0) = 1$ and $y_0 > 0$, and $m \in \mathbb{Z}_{>0}$. By substituting this into (3.1.1) we see that m is bounded. This shows that a given one-dimensional subspace of $\mathbb{Q}^2$ contains only finitely many solutions of (3.1.1). $\qquad\square$

Schmidt (1975) generalized his Subspace Theorem to inequalities of which the unknowns are taken from an algebraic number field, and Schlickewei

(1977b) extended this further to inequalities involving both archimedean and non-archimedean absolute values, thus generalizing both Lang's p-adic Roth's Theorem mentioned above and Schmidt's Subspace Theorem. We give a reformulation of his result that is better adapted to our purposes.

Let K be an algebraic number field. We use the absolute values $|\cdot|_v$ ($v \in M_K$) defined in Section 1.7. Let S be a finite set of places of K, containing all infinite places. Recall that the ring of S-integers of K is given by $O_S = \{x \in K : |x|_v \leq 1 \text{ for } v \in M_K \setminus S\}$. We define the S-*height* of $\mathbf{x} = (x_1, \ldots, x_n) \in O_S^n$ by

$$H_S(\mathbf{x}) = H_S(x_1, \ldots, x_n) := \prod_{v \in S} \max(|x_1|_v, \ldots, |x_n|_v).$$

It follows easily from the Product Formula that $H_S(\varepsilon \mathbf{x}) = H_S(\mathbf{x})$ for $\varepsilon \in O_S^*$. We shall show below that for any $C > 0$ there are, up to multiplication with a scalar from O_S^*, only finitely many vectors $\mathbf{x} \in O_S^n$ with $H_S(\mathbf{x}) \leq C$.

Theorem 3.1.3 (p-adic Subspace Theorem) *For $v \in S$, let $L_{1v}, \ldots, L_{nv}$ be linearly independent linear forms in $X_1, \ldots, X_n$ with coefficients in K. Further, let $\epsilon > 0$. Then the set of solutions of*

$$\prod_{v \in S} |L_{1v}(\mathbf{x}) \cdots L_{nv}(\mathbf{x})|_v \leq H_S(\mathbf{x})^{-\epsilon} \quad in \ \mathbf{x} \in O_S^n \setminus \{\mathbf{0}\} \tag{3.1.3}$$

is contained in a union of finitely many proper linear subspaces of K^n.

Proof. This is a reformulation of a result of Schlickewei (1977b). His proof is based on his earlier papers Schlickewei (1976a, 1976b, 1976c). A special case of Schlickewei's result was proved independently by Dubois and Rhin (1975). A complete proof of Schlickewei's theorem can also be found in Bombieri and Gubler (2006), chapter 7. $\square$

Theorem 3.1.3 $\Longrightarrow$ *Theorem 3.1.2.* Let $L_1, \ldots, L_n$ be the linear forms from Theorem 3.1.2. Let K be the algebraic number field generated by the coefficients of $L_1, \ldots, L_n$ and their conjugates, and suppose that K has degree d. Let S be the set of infinite places of K. Recall that if v is an infinite place of K, then either $|\cdot|_v = |\sigma(\cdot)|$ if $v = \sigma$ is a real embedding of K or $|\cdot|_v = |\sigma(\cdot)|^2$ if $v = \{\sigma, \overline{\sigma}\}$ is a pair of conjugate complex embeddings of K. For either $v = \sigma$ a real embedding or $v = \{\sigma, \overline{\sigma}\}$ a pair of conjugate complex embeddings, we put $L_{iv} := \sigma^{-1}(L_i)$, where $\sigma^{-1}(L_i)$ is the linear form obtained by applying σ^{-1} to the coefficients of L_i. For $\mathbf{x} \in \mathbb{Z}^n$, the left- and right-hand sides of (3.1.3) are precisely the d-th powers of the left- and right-hand sides of (3.1.2). Thus, for $\mathbf{x} \in \mathbb{Z}^n$, inequality (3.1.2) implies (3.1.3), and then an application of

Theorem 3.1.3 implies that the solutions of (3.1.2) lie in a union of finitely many proper linear subspaces of $\mathbb{Q}^n$. □

Schmidt's proof of Theorem 3.1.2 is basically an extension of Roth's method, i.e., the construction of an auxiliary polynomial and an application of Roth's Lemma, combined with techniques from the geometry of numbers. The arguments of Schlickewei and Dubois and Rhin are essentially a "*p*-adization" of Schmidt's method. In their groundbreaking paper Faltings and Wüstholz (1994) gave a totally different proof of Theorem 3.1.3, where they avoided the use of geometry of numbers by applying a very powerful generalization of Roth's Lemma due to Faltings, his *Product Theorem*, see Faltings (1991). We mention that both the method of Schmidt and that of Faltings and Wüstholz are ineffective, in that they do not provide a method to determine the subspaces containing the solutions of the inequality under consideration.

Theorems 3.1.2 and 3.1.3 are very powerful tools to obtain finiteness results for various types of Diophantine equations, such as unit equations, norm form equations, decomposable form equations and exponential-polynomial equations, see Chapters 6, 9 and Section 10.11 in the present book. The proofs of these finiteness results are all ineffective, in the sense that they do not provide a method to determine the solutions. On the other hand, there are now good *quantitative* versions of Theorems 3.1.2 and 3.1.3, giving explicit upper bounds for the number of subspaces, that led to explicit upper bounds for the numbers of solutions of the above mentioned equations. Schmidt (1989) obtained a quantitative version of Theorem 3.1.2, giving an explicit upper bound for the number of subspaces containing the "large" solutions. Schlickewei (1992) generalized this, and obtained a quantitative version of Theorem 3.1.3. This was substantially improved by Evertse (1996), by using a quantitative version of Faltings' Product Theorem. Schlickewei made the important observation that a sufficiently good quantitative version of the so-called *Parametric Subspace Theorem* (see below), which deals with a parametrized class of twisted heights, would lead to much better bounds for the number of solutions of certain classes of Diophantine equations, than quantitative versions of Theorem 3.1.3. In Schlickewei (1996a), he proved a special case of such a quantitative version, and applied this to obtain sharper estimates for the zero multiplicity of a linear recurrence sequence (see Section 10.11). Evertse and Schlickewei (2002) sharpened and extended Schlickewei's result, and obtained a completely general quantitative version of the Parametric Subspace Theorem. For more historical information we refer to Evertse and Schlickewei (1999). Evertse and Schlickewei essentially followed Schmidt's proof of his Theorem 3.1.2, with the necessary refinements. Evertse and Ferretti (2013) obtained a further improvement, following also Schmidt's proof scheme, but inserting ideas from Faltings and Wüstholz (1994).

We first state the Parametric Subspace Theorem in a qualitative form and then give, in a special case relevant for our purposes, a quantitative version of the latter. The Parametric Subspace Theorem is in fact a generalization of Theorem 3.1.3, although this is not obvious at a first glance.

Let again K be a number field, S a finite set of places of K containing all infinite places, $\epsilon > 0$, and for $v \in S$, let $\{L_{1v}, \ldots, L_{nv}\}$ be a system of linearly independent linear forms in $K[X_1, \ldots, X_n]$. Take a solution $\mathbf{x} \in O_S^n \setminus \{\mathbf{0}\}$ of (3.1.3). Assume that the left-hand side of (3.1.3) is non-zero. Write

$$|L_{iv}(\mathbf{x})|_v = H_S(\mathbf{x})^{d_{iv}} \quad (v \in S, \ i = 1, \ldots, n),$$
$$L_{iv} := X_i, \ d_{iv} := 0 \quad (v \in M_K \setminus S, \ i = 1, \ldots, n),$$
$$\mathbf{d} = (d_{iv} : \ v \in M_K, \ i = 1, \ldots, n),$$
$$Q := H_S(\mathbf{x}).$$

Define the so-called *twisted height*:

$$H_{Q,\mathbf{d}}(\mathbf{x}) := \prod_{v \in M_K} \left(\max_{1 \le i \le n} |L_{iv}(\mathbf{x})|_v \ Q^{-d_{iv}} \right). \tag{3.1.4}$$

Notice that by (3.1.3) we have

$$\sum_{v \in M_K} \sum_{i=1}^{n} d_{iv} \le -\epsilon,$$

and that

$$H_{Q,\mathbf{d}}(\mathbf{x}) \le 1.$$

In the above observations, both $\mathbf{d}$ and Q vary with $\mathbf{x}$. The Parametric Subspace Theorem deals with inequalities involving twisted heights, where Q varies but $\mathbf{d}$ is fixed.

Theorem 3.1.4 (Parametric Subspace Theorem)   *Let K be an algebraic number field and S a finite set of places of K containing all infinite places. Further, let $n \ge 2$, let $\{L_{1v}, \ldots, L_{nv}\}$ ($v \in S$) be systems of linearly independent linear forms from $K[X_1, \ldots, X_n]$, and let $\mathbf{d} = (d_{iv} : v \in M_K, i = 1, \ldots, n)$ be a tuple of reals such that*

$$d_{iv} = 0 \ \text{for} \ v \in M_K \setminus S, \ i = 1, \ldots, n.$$

Put

$$\mu := \frac{1}{n} \sum_{v \in M_K} \sum_{i=1}^{n} d_{iv}.$$

Then for every $\delta > 0$ there are Q_0 and a finite collection $\{\mathcal{T}_1, \ldots, \mathcal{T}_t\}$ of proper linear subspaces of K^n such that for every $Q \geq Q_0$ there is $\mathcal{T} \in \{\mathcal{T}_1, \ldots, \mathcal{T}_t\}$ with

$$\{\mathbf{x} \in K^n : H_{Q,\mathbf{d}}(\mathbf{x}) \leq Q^{-\mu-\delta}\} \subseteq \mathcal{T}.$$

Proof. This was first formulated by Evertse and Schlickewei (2002), in a quantitative form with explicit upper bounds for Q_0 and t. In fact, in their paper, Evertse and Schlickewei proved an "Absolute Parametric Subspace Theorem", with solutions $\mathbf{x}$ taken from $\overline{\mathbb{Q}}^n$ instead of K^n. □

Below we deduce Theorem 3.1.3 from Theorem 3.1.4. We first prove a lemma. We keep our convention that K is a number field and S a finite set of places of K, containing all infinite places. Further, we set $d := [K : \mathbb{Q}]$, $s := |S|$. For $\mathbf{x} = (x_1, \ldots, x_n) \in K^n$, $v \in M_K$, we put $\|\mathbf{x}\|_v := \max(|x_1|_v, \ldots, |x_n|_v)$.

Lemma 3.1.5

(i) *There is a constant C depending only on K and S, such that for every $\mathbf{x} \in O_S^n \setminus \{\mathbf{0}\}$, there is $\varepsilon \in O_S^*$ with*

$$\|\varepsilon \mathbf{x}\|_v \leq C H_S(\mathbf{x})^{1/s} \text{ for } v \in M_K.$$

(ii) *For every $A > 0$ there are, up to multiplication with a scalar from O_S^*, only finitely many vectors $\mathbf{x} \in O_S^n$ with $H_S(\mathbf{x}) \leq A$.*

Proof. (i) Let $S = \{v_1, \ldots, v_s\}$ and

$$H := \{\mathbf{x} = (x_1, \ldots, x_s) \in \mathbb{R}^s : x_1 + \cdots + x_s = 0\}.$$

Then by the S-unit Theorem (see Theorem 1.8.1) the map

$$LOG_S : \varepsilon \mapsto (\log|\varepsilon|_{v_1}, \ldots, \log|\varepsilon|_{v_s})$$

maps O_S^* to an $(s-1)$-dimensional lattice in H.

Let $\mathbf{x} \in O_S^n \setminus \{\mathbf{0}\}$. Then the point

$$\mathbf{a} := (s^{-1} \log H_S(\mathbf{x}) - \log\|\mathbf{x}\|_{v_1}, \ldots, s^{-1} \log H_S(\mathbf{x}) - \log\|\mathbf{x}\|_{v_s})$$

lies in H. Choose $\varepsilon \in O_S^*$ such that the lattice point $LOG_S(\varepsilon)$ is closest to $\mathbf{a}$. Then in fact $\|\mathbf{a} - \log_S(\varepsilon)\| \leq \gamma$, where $\|\cdot\|$ is the maximum norm on $\mathbb{R}^s$ and γ is a constant depending only on K, S. This ε satisfies (i) with $C = e^\gamma$.

(ii) Consider $\mathbf{x} \in O_S^n$ with $H_S(\mathbf{x}) \leq A$. After multiplying $\mathbf{x}$ with a suitable S-unit, we can arrange that $\|\mathbf{x}\|_v \leq C \cdot A^{1/s}$ for $v \in S$. Then the coordinates $x_1, \ldots, x_n$ of $\mathbf{x}$ have absolute heights

$$H(x_i) := \prod_{v \in M_K} \max(1, |x_i|_v) \leq C^{s/d} A^{1/d} \text{ for } i = 1, \ldots, n.$$

By Northcott's Theorem (see Theorem 1.9.3) this leaves only finitely many possibilities for $x_1, \ldots, x_n$, hence for $\mathbf{x}$. □

Theorem 3.1.4 ⟹ *Theorem 3.1.3.* By Lemma 3.1.5 (i), it suffices to show that the solutions $\mathbf{x} \in O_S^n \setminus \{\mathbf{0}\}$ of (3.1.3) with the additional property

$$\|\mathbf{x}\|_v \le C H_S(\mathbf{x})^{1/s} \text{ for } v \in S \qquad (3.1.5)$$

lie in finitely many proper linear subspaces of K^n. Lemma 3.1.5 (ii) implies that by assuming $H_S(\mathbf{x})$ to be sufficiently large, we exclude at most finitely many one-dimensional subspaces of solutions $\mathbf{x}$. Solutions with (3.1.5) and with $H_S(\mathbf{x})$ sufficiently large, in fact satisfy

$$|L_{iv}(\mathbf{x})|_v \le H_S(\mathbf{x})^{2/s} \text{ for } v \in S, i = 1, \ldots, n. \qquad (3.1.6)$$

Hence it suffices to prove that the solutions of (3.1.3) with (3.1.6) lie in finitely many proper linear subspaces of K^n.

Let $\mathbf{x}$ be a solution of (3.1.3) with (3.1.6), and define

$$d_{iv}(\mathbf{x}) := \max\left(-2n - \epsilon, \frac{\log |L_{iv}(\mathbf{x})|_v}{\log H_S(\mathbf{x})}\right) \text{ for } v \in S, i = 1, \ldots, n;$$

taking $\log 0 := -\infty$, this is well-defined also if $L_{iv}(\mathbf{x}) = 0$. Then

$$\left.\begin{array}{l} -2n - \epsilon \le d_{iv}(\mathbf{x}) \le 2/s \text{ for } v \in S, i = 1, \ldots, n, \\[2mm] \displaystyle\sum_{v \in S}\sum_{i=1}^{n} d_{iv}(\mathbf{x}) \le -\epsilon. \end{array}\right\} \qquad (3.1.7)$$

We define a tuple $\mathbf{d} := (d_{iv} : v \in M_K, i = 1, \ldots, n)$ by $d_{iv} := 0$ for $v \in M_K \setminus S, i = 1, \ldots, n$ and

$$d_{iv} \in \frac{\epsilon}{2ns}\mathbb{Z}, \quad d_{iv} - \frac{\epsilon}{2ns} < d_{iv}(\mathbf{x}) \le d_{iv} \text{ for } v \in S, i = 1, \ldots, n. \qquad (3.1.8)$$

Then by (3.1.7),

$$-2n - \epsilon \le d_{iv} \le \frac{2}{s} + \frac{\epsilon}{2ns} \text{ for } v \in S, i = 1, \ldots, n. \qquad (3.1.9)$$

Notice that by (3.1.7) we have also

$$\mu := \frac{1}{n}\sum_{v \in M_K}\sum_{i=1}^{n} d_{iv} \le -\frac{\epsilon}{2n}.$$

Further, we have $|L_{iv}(\mathbf{x})|_v \le H_S(\mathbf{x})^{d_{iv}(\mathbf{x})} \le H_S(\mathbf{x})^{d_{iv}}$ for $v \in S, i = 1, \ldots, n$, hence with $Q := H_S(\mathbf{x})$ we have

$$H_{Q,\mathbf{d}}(\mathbf{x}) \le 1.$$

By Theorem 3.1.4, the solutions of (3.1.3) satisfying (3.1.8) for some fixed tuple **d** lie in finitely many proper linear subspaces of K^n. Further, by (3.1.8) and (3.1.9), the tuples **d** belong to a finite set independent of **x**. This proves Theorem 3.1.3. □

The general statement of the quantitative version of Theorem 3.1.4, with explicit upper bounds for Q_0, t, is quite technical. We give here only a special case, which is sufficient for our purposes. We keep our assumptions that K is an algebraic number field of degree d and S is a finite set of places of K, containing the infinite places.

Theorem 3.1.6 *Let L_{iv} ($v \in M_K$, $i = 1, \ldots, n$) be linear forms such that for every $v \in M_K$, the set $\{L_{1v}, \ldots, L_{nv}\}$ is linearly independent and*

$$\{L_{1v}, \ldots, L_{nv}\} \subset \{X_1, \ldots, X_n, X_1 + \cdots + X_n\}.$$

Let $\mathbf{d} = (d_{iv} : v \in M_K, i = 1, \ldots, n)$ be any tuple of reals such that $d_{iv} = 0$ for $v \in M_K \setminus S$, $i = 1, \ldots, n$. Put

$$\mu := \frac{1}{n} \sum_{v \in M_K} \sum_{i=1}^{n} d_{iv}$$

and suppose that

$$\sum_{v \in M_K} \max(d_{1v}, \ldots, d_{nv}) \le \lambda \quad \text{with } \lambda > \mu.$$

Let $0 < \delta < \lambda - \mu$ and put

$$\Theta := \frac{\lambda - \mu}{\delta}.$$

Let $H_{Q,\mathbf{d}}$ be defined by (3.1.4).

Then there is a finite collection $\{\mathcal{T}_1, \ldots, \mathcal{T}_t\}$ of proper linear subspaces of K^n of cardinality

$$t \le C(n, \Theta)$$

with $C(n, \Theta)$ effectively computable and depending only on n and Θ, such that for every Q with

$$Q > n^{2d/\delta} \tag{3.1.10}$$

there is $\mathcal{T} \in \{\mathcal{T}_1, \ldots, \mathcal{T}_t\}$ such that

$$\{\mathbf{x} \in K^n : H_{Q,\mathbf{d}}(\mathbf{x}) \le Q^{-\mu-\delta}\} \subseteq \mathcal{T}. \tag{3.1.11}$$

This was proved by Evertse and Schlickewei (2002), Theorem 1.1 in the special case $\mu = 0, \lambda = 1$, with

$$C(n, \Theta) = 4^{(n+9)^2} \Theta^{n+4}.$$

For our purposes, the precise value of $C(n, \Theta)$ will not matter, but we should mention here that Evertse and Ferretti (2013), Theorem 1.1 proved the same result with the better bound

$$C(n, \Theta) = 10^6 2^{2n} n^{10} \Theta^3 (\log(6n\Theta))^2,$$

again in the special case $\mu = 0, \lambda = 1$.

It is not difficult to reduce Theorem 3.1.6 to the special case $\mu = 0, \lambda = 1$. Put

$$d'_{iv} := \frac{1}{\lambda - \mu} \left(d_{iv} - \frac{1}{n} \sum_{j=1}^n d_{jv} \right) \quad (v \in M_K, \ i = 1, \ldots, n),$$

$$\mathbf{d}' := \left(d'_{iv} : \ v \in M_K, \ i = 1, \ldots, n \right),$$

$$Q' := Q^{\lambda - \mu}, \qquad \delta' := \frac{\delta}{\lambda - \mu}.$$

Then $d'_{iv} = 0$ for $v \in M_K \setminus S, i = 1, \ldots, n$,

$$\sum_{v \in M_K} \sum_{i=1}^n d'_{iv} = 0, \qquad \sum_{v \in M_K} \max \left(d'_{1v}, \ldots, d'_{nv} \right) \leq 1,$$

and (3.1.10) changes into $Q' > n^{2d/\delta'}$. Further,

$$H_{Q,\mathbf{d}}(\mathbf{x}) = H_{Q',\mathbf{d}'}(\mathbf{x}) Q^\mu,$$

hence (3.1.11) changes into

$$\{ \mathbf{x} \in K^n : \ H_{Q',\mathbf{d}'}(\mathbf{x}) \leq Q'^{-\delta'} \} \subseteq \mathcal{T}.$$

Thus, Theorem 3.1.6 follows from the special case $\mu = 0, \lambda = 1$. □

The Subspace Theorem and its generalizations and quantitative refinements have many applications. In this book we have focused on applications to unit equations and subsequent applications thereof, see Chapters 6, 9, 10, but there is much more, see for instance the survey papers Bilu (2008), Corvaja and Zannier (2008), Bugeaud (2011), and the book Zannier (2003). Somewhat surprisingly, from Theorem 3.1.3 one can derive extensions where the linear polynomials are replaced by higher degree polynomials and where the solutions are taken from an arbitrary algebraic variety instead of K^n, see Corvaja and Zannier (2004a) and Evertse and Ferretti (2002, 2008).

3.2 Effective estimates for linear forms in logarithms

In this section we present some results from Baker's theory of logarithmic forms that are used in Chapters 4 and 5. We formulate, without proof, the best known effective estimates for linear forms in logarithms, due to Matveev (2000) in the complex case and Yu (2007) in the p-adic case, as well as a common, uniform formulation of them which will be more convenient to apply.

We first give a brief introduction, starting with the famous Gelfond–Schneider Theorem on transcendental numbers. For the moment, $\overline{\mathbb{Q}}$ denotes the algebraic closure of $\mathbb{Q}$ in $\mathbb{C}$, and algebraic numbers are supposed to belong to $\overline{\mathbb{Q}}$. Here and below log denotes, except otherwise stated, any fixed determination of the logarithm, and for $\alpha, \beta \in \mathbb{C}$ with $\alpha \neq 0$ we define $\alpha^{\beta} := e^{\beta \log \alpha}$.

Theorem 3.2.1 *Suppose that α and β are algebraic numbers such that $\alpha \neq 0$, 1 and that β is not rational. Then α^{β} is transcendental.*

Proof. See Gelfond (1934) and Schneider (1934). The theorem was proved independently by Gelfond and Schneider. Their proofs are different, but both depend on the construction of an auxiliary function. Assuming that in Theorem 3.2.1 α^{β} is algebraic and following the arguments of Gelfond, one can construct a function $F(z)$ of a complex variable z which is a polynomial in α^z and $\alpha^{\beta z}$ with integral coefficients, not all zero, such that $F^{(m)}(l) = 0$ for all integers l, m with $1 \leq l \leq h$ and $0 \leq m < k$, where h, k are appropriate parameters. Then combining some arithmetic and analytic considerations and using induction on k, one can prove that $F^{(m)}(l) = 0$ for all m, which leads to a contradiction. $\qquad\square$

Theorem 3.2.1 provided an answer to Hilbert's seventh problem. An equivalent formulation of the theorem is that if α_1, α_2 are non-zero algebraic numbers such that $\log \alpha_1$ and $\log \alpha_2$ are linearly independent over $\mathbb{Q}$, then they are linearly independent over $\overline{\mathbb{Q}}$.

By means of a refinement of his method of proof, Gelfond (1935) gave a non-trivial effective lower bound for the absolute value of $\beta_1 \log \alpha_1 + \beta_2 \log \alpha_2$, where β_1, β_2 denote algebraic numbers, not both 0, and α_1, α_2 denote algebraic numbers different from 0 and 1 such that $\log \alpha_1 / \log \alpha_2$ is not rational.

Mahler (1935b) proved a p-adic analogue of the Gelfond–Schneider Theorem. A generalization to the p-adic absolute value was given in Gelfond (1940) in a quantitative form. In his book Gelfond (1960), Gelfond remarked that a generalization of his above results from two logarithms to arbitrary many would be of great significance for the solutions of many difficult problems in number theory.

In his celebrated series of papers, Baker (1966, 1967a, 1967b, 1968a) made a major breakthrough in transcendental number theory by generalizing the Gelfond–Schneider Theorem to arbitrary many logarithms. In Baker (1966, 1967b), he proved the following.

Theorem 3.2.2 *Let $\alpha_1, \ldots, \alpha_n$ denote non-zero algebraic numbers. If $\log \alpha_1, \ldots, \log \alpha_n$ are linearly independent over $\mathbb{Q}$, then $1, \log \alpha_1, \ldots, \log \alpha_n$ are linearly independent over $\overline{\mathbb{Q}}$.*

Further, Baker (1967a, 1967b, 1968a) gave non-trivial lower bounds for the absolute value of linear forms in logarithms of the form

$$\beta_1 \log \alpha_1 + \cdots + \beta_n \log \alpha_n,$$

where $\alpha_1, \ldots, \alpha_n$ are non-zero algebraic numbers such that $\log \alpha_1, \ldots, \log \alpha_n$ are linearly independent over $\mathbb{Q}$ and $\beta_1, \ldots, \beta_n$ are algebraic numbers, not all 0.

Proof of Theorem 3.2.2 (sketch; see Baker (1967b) for full details). To illustrate most of the principal ideas of Baker, we sketch the main steps of the proof of a slightly weaker assertion, which states that if $\alpha_1, \ldots, \alpha_n, \beta_1, \ldots, \beta_{n-1}$ are non-zero algebraic numbers such that $\alpha_1, \ldots, \alpha_n$ are multiplicatively independent, then $\alpha_1^{\beta_1} \cdots \alpha_{n-1}^{\beta_{n-1}} = \alpha_n$ cannot hold. Supposing the opposite and following the arguments of Baker, one can construct an auxiliary function $F(z_1, \ldots, z_{n-1})$ in $n-1$ complex variables, which generalizes the function of a single variable employed by Gelfond. The function is a polynomial in $\alpha_1^{z_1}, \ldots, \alpha_{n-1}^{z_{n-1}}$ and $\alpha_1^{\beta_1 z_1} \cdots \alpha_{n-1}^{\beta_{n-1} z_{n-1}}$, such that

$$F(z, \ldots, z) = \sum_{\lambda_1=0}^{L} \cdots \sum_{\lambda_n=0}^{L} p(\lambda_1, \ldots, \lambda_n) \alpha_1^{\lambda_1 z} \cdots \alpha_n^{\lambda_n z},$$

where L is a large parameter and $p(\lambda_1, \ldots, \lambda_n)$ are rational integers, not all 0. Then for every positive integer l, the number $F(l, \ldots, l)$ lies in the algebraic number field $\mathbb{Q}(\alpha_1, \ldots, \alpha_n)$. It follows from a well-known lemma on linear equations (known as Siegel's Lemma) that the $p(\lambda_1, \ldots, \lambda_n)$ can be chosen such that their absolute values are not too large and such that

$$F_{m_1, \ldots, m_{n-1}}(l, \ldots, l) = 0 \tag{3.2.1}$$

for all integers $l, m_1, \ldots, m_{n-1}$ with $1 \leq l \leq h$ and $m_1 + \cdots + m_{n-1} \leq k$, where $F_{m_1, \ldots, m_{n-1}}$ denotes the corresponding derivative of $F(z_1, \ldots, z_{n-1})$ and h, k are appropriate parameters. In this situation, the basic interpolation techniques used earlier by Gelfond and others do not work in general. Using some analytic considerations Baker applied an ingenious extrapolation procedure to extend (3.2.1) to a larger range of values for l, at the price of slightly diminishing the range of values for $m_1 + \cdots + m_{n-1}$. Repeating this procedure, one can get $F(l, \ldots, l) = 0$ for $1 \leq l \leq (L + 1)^n$. This can be regarded as a system of linear equations in the coefficients $p(\lambda_1, \ldots, \lambda_n)$ of F, which, because of the multiplicative independence of $\alpha_1, \ldots, \alpha_n$, cannot have a non-zero solution. This proves the assertion. □

Let again $\alpha_1, \ldots, \alpha_n$ be $n \geq 2$ non-zero algebraic numbers, and let $\log \alpha_1, \ldots, \log \alpha_n$ denote now the principal values of the logarithm.

Theorem 3.2.3 *Let $b_1, \ldots, b_n$ be rational integers and $0 < \varepsilon \leq 1$. Assume that*

$$0 < |b_1 \log \alpha_1 + \cdots + b_n \log \alpha_n| < e^{-\varepsilon B},$$

where $B = \max\{|b_1|, \ldots, |b_n|\}$. Then $B \leq B_0$, where B_0 is effectively computable in terms of $\alpha_1, \ldots, \alpha_n$ and ε.

This was proved in Baker (1968a) with

$$B_0 = (4^{n^2} \varepsilon^{-1} d^{2n} A)^{(2n+1)^2},$$

where $d \geq 4$ and $A \geq 4$ are upper bounds for the degrees and heights, respectively, of $\alpha_1, \ldots, \alpha_n$. Here, by the height of an algebraic number we mean the maximum of the absolute values of the coefficients in its minimal defining polynomial, which is chosen such that its coefficients are relatively prime integers.

Baker's general effective estimates led to significant applications in number theory. For applications to Diophantine equations, the inequalities of Baker (1968a, 1968b) in which $\beta_1, \ldots, \beta_n$ are rational integers proved to be particularly useful. Using his effective estimates, Baker (1968b, 1968c, 1969) gave the first explicit upper bounds for the solutions of Thue equations, Mordell equations, and superelliptic and hyperelliptic equations; see also Sections 9.6 and 9.7.

Later, several improvements and generalizations were established by Baker and others, including Feldman, Baker and Stark, Tijdeman, van der Poorten, Sprindžuk, Shorey, Wüstholz, Philippon and Waldschmidt, Waldschmidt, Baker and Wüstholz, Laurent, Mignotte, Nesterenko and Matveev and, in the p-adic case, Coates, Sprindžuk, Brumer, Vinogradov and Sprindžuk, van der Poorten, Bugeaud, Laurent and Yu. They have introduced various new ideas to improve or refine the previous bounds. Their results made it possible to obtain enormously many applications. For further applications to Diophantine problems, we refer to Győry (1980b, 2002, 2010), Sprindžuk (1982, 1993),

Shorey and Tijdeman (1986), Serre (1989), de Weger (1989), Bilu (1995), Wildanger (1997, 2000), Smart (1998), Gaál (2002) and Tzanakis (2013), to Chapters 4 and 5 of the present book, to our next book on discriminant equations, and to the references given there.

Using an elementary geometric lemma due to Bombieri (1993) and Bombieri and Cohen (1997), Bilu and Bugeaud (2000) showed that one does not need the full strength of Baker's theory to get, for $b_n = \pm 1$, an effective version of Theorem 3.2.3: it can be deduced from an estimate for linear forms in just two logarithms. However, the results of the theory of linear forms in $n \geq 2$ logarithms provide much better bounds for B.

For comprehensive accounts of Baker's theory, analogues for elliptic logarithms and algebraic groups and extensive bibliographies the reader can consult Baker (1975, 1988), Baker and Masser (1977), Lang (1978), Feldman and Nesterenko (1998), Waldschmidt (2000), Wüstholz (2002) and, for the state of the art as well, Baker and Wüstholz (2007).

We now state the results of Matveev and Yu and give a common, uniform formulation of them.

Let again K be an algebraic number field of degree d, and assume that it is embedded in $\mathbb{C}$. We put $\chi = 1$ if K is real, and $\chi = 2$ otherwise. Let

$$\Sigma = b_1 \log \alpha_1 + \cdots + b_n \log \alpha_n,$$

where $\alpha_1, \ldots, \alpha_n$ are n (≥ 2) non-zero elements of K with some fixed non-zero values of $\log \alpha_1, \ldots, \log \alpha_n$, and $b_1, \ldots, b_n$ are rational integers, not all zero. Let $A_1, \ldots, A_n$ be reals with

$$A_i \geq \max \{dh(\alpha_i), |\log \alpha_i|, 0.16\} \quad (i = 1, \ldots, n)$$

and put

$$B := \max \{1, \max \{|b_i| A_i / A_n : 1 \leq i \leq n\}\}.$$

The following deep result was proved by Matveev (2000).

Theorem 3.2.4 *Let* $K, \alpha_1, \ldots, \alpha_n, b_1, \ldots, b_n$ *and* Σ *be as above, and suppose that* $\Sigma \neq 0$. *Then*

$$\log |\Sigma| > -C_1(n, d) A_1 \cdots A_n \log(eB),$$

where

$$C_1(n, d) := \min \left\{ \frac{1}{\chi} \left(\frac{1}{2} en \right)^\chi 30^{n+3} n^{3.5}, 2^{6n+20} \right\} d^2 \log(ed).$$

Further, B may be replaced by $\max (|b_1|, \ldots, |b_n|)$.

Proof. This is Corollary 2.3 of Matveev (2000). □

We shall use the following consequence of Theorem 3.2.4. Let

$$\Lambda = \alpha_1^{b_1} \cdots \alpha_n^{b_n} - 1 \tag{3.2.2}$$

and

$$A_i' \geq \max\{dh(\alpha_i), \pi\}, \quad i = 1, \ldots, n.$$

Theorem 3.2.5 *Suppose that* $\Lambda \neq 0$, $b_n = \pm 1$ *and that* B' *satisfies*

$$B' \geq \max\left\{|b_1|, \ldots, |b_{n-1}|, 2e \max\left(\frac{n\pi}{\sqrt{2}}, A_1', \ldots, A_{n-1}'\right) A_n'\right\}. \tag{3.2.3}$$

Then we have

$$\log|\Lambda| > -C_2(n, d)A_1' \cdots A_n' \log\left(B'/(\sqrt{2}A_n')\right), \tag{3.2.4}$$

where

$$C_2(n, d) := \min\left\{1.451(30\sqrt{2})^{n+4}(n+1)^{5.5}, \pi 2^{6.5n+27}\right\}d^2 \log(ed).$$

Proof. Let log denote the principal value of the logarithm. There exists an even rational integer b_0 such that $|b_0| \leq |b_1| + \cdots + |b_n|$ and that $|\text{Im}(\Sigma')| \leq \pi$, where

$$\Sigma' := b_0 \log \alpha_0 + b_1 \log \alpha_1 + \cdots + b_n \log \alpha_n$$

and $\alpha_0 = -1$. The assumption $\Lambda \neq 0$ implies that $\Sigma' \neq 0$. We may assume that $|e^{\Sigma'} - 1| = |\Lambda| \leq 1/3$. Then it follows that $|\Sigma'| \leq 0.6$, whence

$$|\Lambda| \geq \frac{1}{2}|\Sigma'|. \tag{3.2.5}$$

Using $|\log|\alpha_i|| \leq dh(\alpha_i)$, it is easy to show that

$$|\log \alpha_i| \leq \sqrt{2} \max\{dh(\alpha_i), \pi\}, i = 1, \ldots, n.$$

Thus, setting $A_0' = \pi/\sqrt{2}$, we have

$$\sqrt{2}A_i' \geq \max\{dh(\alpha_i), |\log \alpha_i|, 0.16\}, i = 0, 1, \ldots, n.$$

Further, (3.2.3) implies

$$\left(\frac{B'}{\sqrt{2}A_n'}\right)^2 \geq e \max\left\{1, \max_{0 \leq i \leq n}\left(\frac{|b_i|A_i'}{A_n'}\right)\right\}.$$

By applying now Theorem 3.2.4 to $|\Sigma'|$ and using (3.2.5), we obtain (3.2.4).

$\square$

Remark 3.2.6 Since for any complex number z, $|e^z - 1| \leq |z|e^{|z|}$ holds, for $|\Sigma| \leq 1$ it follows that

$$|\Lambda| \leq e|\Sigma|.$$

Together with (3.2.5) this implies that if we have an effective and quantitative result for $|\Lambda|$ then we also have a similar one for the corresponding $|\Sigma|$ or $|\Sigma'|$, and conversely.

Consider again Λ defined by (3.2.2). Let now B and B_n be real numbers satisfying

$$B \geq \max\{|b_1|, \ldots, |b_n|\}, \quad B \geq B_n \geq |b_n|.$$

Let $\mathfrak{p}$ be a prime ideal of O_K and denote by $e_\mathfrak{p}$ and $f_\mathfrak{p}$ the ramification index and the residue class degree of $\mathfrak{p}$, respectively. Suppose that $\mathfrak{p}$ lies above the rational prime number p. Then the norm of $\mathfrak{p}$ is $N(\mathfrak{p}) = p^{f_\mathfrak{p}}$.

The following profound result is due to Yu (2007).

Theorem 3.2.7 *Assume that* $\mathrm{ord}_p b_n \leq \mathrm{ord}_p b_i$ *for* $i = 1, \ldots, n$, *and set*

$$h_i' := \max\{h(\alpha_i), 1/16e^2d^2\}, \quad i = 1, \ldots, n.$$

If $\Lambda \neq 0$, *then for any real* δ *with* $0 < \delta \leq 1/2$ *we have*

$$\mathrm{ord}_\mathfrak{p} \Lambda < C_3(n, d) \frac{e_\mathfrak{p}^n N(\mathfrak{p})}{(\log N(\mathfrak{p}))^2} \max\left\{h_1' \cdots h_n' \log(M\delta^{-1}), \frac{\delta B}{B_n C_4(n, d)}\right\}, \tag{3.2.6}$$

where

$$C_3(n, d) := (16ed)^{2(n+1)} n^{3/2} \log(2nd) \log(2d),$$

$$C_4(n, d) := (2d)^{2n+1} \log(2d) \log^3(3d),$$

and

$$M := B_n C_5(n, d) N(\mathfrak{p})^{n+1} h_1' \cdots h_{n-1}'$$

with

$$C_5(n, d) := 2e^{(n+1)(6n+5)} d^{3n} \log(2d).$$

Proof. This is the second consequence of the Main Theorem in Yu (2007). As is remarked there, for $p > 2$, the expression $(16ed)^{2(n+1)}$ can be replaced by $(10ed)^{2(n+1)}$. $\qquad\square$

For the proof of Theorem 4.2.1, it will be more convenient to use a uniform lower bound for $\log |\Lambda|_v$ which is valid both for infinite and for finite places v.

For a place $v \in M_K$, we write as above

$$N(v) := \begin{cases} 2 & \text{if } v \text{ is infinite,} \\ N(\mathfrak{p}) & \text{if } v = \mathfrak{p} \text{ is finite.} \end{cases}$$

The following theorem is a consequence of Theorems 3.2.5 and 3.2.7.

Theorem 3.2.8 *Let* $v \in M_K$. *Suppose that in (3.2.2)* $\Lambda \neq 0$, $b_n = \pm 1$ *and that* $\alpha_1, \ldots, \alpha_{n-1}$ *are not roots of unity. Let*

$$\Theta := h(\alpha_1) \cdots h(\alpha_{n-1}), \quad H := \max\{h(\alpha_n), 1\}.$$

If B is a real number such that

$$B \geq \max\{|b_1|, \ldots, |b_{n-1}|, 2e(3d)^{2n}\Theta H\}, \tag{3.2.7}$$

then

$$\log|\Lambda|_v > -C_6(n, d)\frac{N(v)}{\log N(v)}\Theta H \log^*\left(\frac{BN(v)}{H}\right). \tag{3.2.8}$$

where $C_6(n, d) := \lambda(16ed)^{3n+2}(\log^* d)^2$, *and* $\lambda = 1$ *or* 12 *according as* $n \geq 3$ *or* $n = 2$.

In the proof, we shall also need the following.

Proposition 3.2.9 *Let* α *be a non-zero algebraic number of degree* d *which is not a root of unity. Then*

$$dh(\alpha) \geq \begin{cases} \log 2 \text{ if } d = 1, \\ 2/(\log 3d)^3 \text{ if } d \geq 2. \end{cases}$$

Proof. This result is due to Voutier (1996). □

Remark 3.2.10 For $d \geq 2$ this lower bound may be replaced by the quantity $(1/4)(\log\log d / \log d)^3$; see Voutier (1996). It is a conjecture, inspired by a question of D. H. Lehmer (1933), that even $dh(\alpha) \geq c > 0$ should hold for some absolute constant c.

Proof of Theorem 3.2.8. First assume that v is infinite. There is an embedding $\sigma : K \hookrightarrow \mathbb{C}$ such that $|\Lambda|_v = |\sigma(\Lambda)|$ or $|\sigma(\Lambda)|^2$ according as σ is real or not. Observe further that $h(\sigma(\alpha)) = h(\alpha)$ for each $\alpha \in \overline{\mathbb{Q}}$. Hence it suffices to prove (3.2.8) for $|\Lambda|$. Suppose that in Theorem 3.2.5 $A'_i = \max\{dh(\alpha_i), \pi\}$ for $i = 1, \ldots, n$. Then, using Proposition 3.2.9, it is easy to see that

$$A'_1 \cdots A'_n \leq (2.52d)^{2n}\Theta H.$$

Further, we have $\sqrt{2}A'_n > H/N(v)$ and

$$2e \max \left\{ \frac{n\pi}{\sqrt{2}}, A'_1, \ldots, A'_{n-1} \right\} A'_n \leq 2e\,(3d)^{2n}\,\Theta H.$$

Now (3.2.7) implies (3.2.3), and (3.2.8) follows from the inequality (3.2.4) of Theorem 3.2.5.

Next assume that v is finite. Keeping the notation of Theorem 3.2.7 and using again Proposition 3.2.9, we infer that

$$h'_i = h(\alpha_i) \text{ for } i = 1, \ldots, n-1 \text{ and } h'_n \leq \max\{h(\alpha_n), 1\} = H.$$

Choosing $\delta = h'_1 \cdots h'_{n-1} H/B$ and $B_n = 1$ in Theorem 3.2.7, (3.2.7) implies that $\delta \leq \frac{1}{2}$. Using the fact that $|\Lambda|_v = N(\mathfrak{p})^{-\mathrm{ord}_\mathfrak{p}\,\Lambda}$, after some computation (3.2.8) follows from (3.2.6) of Theorem 3.2.7. $\qquad\square$

PART II

Unit equations and applications

4

Effective results for unit equations in two unknowns over number fields

In this chapter we present effective finiteness results in quantitative form on equations of the shape

$$a_1 x_1 + a_2 x_2 = 1, \tag{4.1}$$

where a_1, a_2 are non-zero elements of an algebraic number field K, and the unknowns x_1, x_2 are units, S-units or, more generally, elements of a finitely generated multiplicative subgroup Γ of K^*. We usually refer to such equations as "unit equations", also if the unknowns are taken from a group Γ that is not the unit group of a ring. In the case that the unknowns are S-units, we speak about an S-unit equation. In certain applications, it is more convenient to consider equation (4.1) in *homogeneous* form

$$a_1 x_1 + a_2 x_2 + a_3 x_3 = 0, \tag{4.2}$$

where a_1, a_2, a_3 denote non-zero elements of K, and the unknowns x_1, x_2, x_3 are units, S-units or elements of Γ.

For a long time equations (4.1) and (4.2) were utilized merely in special cases and in an implicit way. It was proved by Siegel (1921) (in an implicit form) for units of a number field, and by Mahler (1933a) for S-units in $\mathbb{Q}$ that equation (4.1) has only finitely many solutions. For S-unit equations over number fields, the finiteness of the number of solutions follows from work of Parry (1950). Extending results of Siegel, Mahler and Parry, Lang (1960) proved that equation (4.1) has only finitely many solutions in $x_1, x_2 \in \Gamma$ even in the case when K is any field of characteristic 0 and Γ is any finitely generated multiplicative subgroup of K^*. This implies that, up to a common proportional factor, (4.2) has also finitely many solutions. These results are ineffective.

In this chapter we restrict ourselves to the case when K is a number field. The general case will be discussed in Chapters 6 and 8.

Using Baker's theory of logarithmic forms, Győry (1972, 1973, 1974, 1979, 1979/1980) gave the first effective upper bounds for the heights of the solutions

61

of unit equations and S-unit equations over number fields. He systematically applied his results among others to decomposable form equations, polynomials and algebraic numbers of given discriminant, and irreducible polynomials, see Győry (1972, 1973, 1974, 1976, 1978a,b, 1980a,b, 1981a,b,c, 1982c). Győry's bounds have been improved by several people, for references see the Notes in Section 4.7.

In the present chapter, we derive effective upper bounds for the heights of the solutions of S-unit equations by means of the best known variants of the classical Baker's method. There are now other methods giving effective bounds for the solutions, see Bombieri (1993), Bombieri and Cohen (1997, 2003), Bugeaud (1998), Murty and Pasten (2013) and von Känel (2014b). A brief discussion of these methods, together with a comparison of the bounds they yield, is given in Section 4.5.

In Section 4.1, we present the best upper bounds to date for the heights of the solutions of (4.1) and (4.2) in units, S-units and, more generally, in an arbitrary finitely generated subgroup Γ of a number field K. These results will be used to prove the main results in Chapter 8 on unit equations over finitely generated integral domains, and in Section 9.6 on decomposable form equations over K. Further, they will be applied to discriminant equations in our next book. For these and other possible applications, we give the upper bounds in completely explicit form.

In Section 4.2 we state new effective and quantitative results on approximation of numbers from K^* by elements of a finitely generated subgroup of K^*. These are the hard core of our proofs. In Section 4.6, an application is presented in the direction of the abc-conjecture over number fields. Many other applications are mentioned in the Notes, Section 4.7 of this chapter and in Chapter 10.

Sections 3.2 and 4.3 contain the main tools needed in the proofs. We recalled in Section 3.2 the best known effective estimates, due to Matveev (2000) and Yu (2007), for linear forms in logarithms. Further, in Section 4.3 we prove a new result from the geometry of numbers and give height estimates for units/ S-units in a fundamental/maximal independent system of units/S-units. Finally, in Section 4.4 we prove the results from Sections 4.1 and 4.2.

4.1 Effective bounds for the heights of the solutions

4.1.1 Equations in units of a number field

Let K be an algebraic number field of degree d, O_K the ring of integers of K and O_K^* the group of units of O_K. We denote by R the regulator of K, by r

the rank of O_K^*, by M_K the set of (infinite and finite) places, and by M_K^∞ the set of infinite places of K. For $v \in M_K$, $|\cdot|_v$ denotes the absolute value corresponding to v, defined in Section 1.7.

We recall that the *absolute (multiplicative) height* $H(\alpha)$ of $\alpha \in K$ is defined by

$$H(\alpha) := \left(\prod_{v \in M_K} \max(1, |\alpha|_v) \right)^{1/d}$$

and the *absolute logarithmic height* $h(\alpha)$ by

$$h(\alpha) := \log H(\alpha).$$

More generally, we define the height $h(\alpha)$ of $\alpha \in \overline{\mathbb{Q}}$ by taking a number field K containing α and using the above definition; one can show that this is independent of the choice of K. For more details and for the most important properties of the height, we refer to Section 1.9. We shall frequently use these properties without any further reference.

Let a_1, a_2, a_3 be non-zero elements of K and let H be a real with

$$H \geq \max\{h(a_1), h(a_2), h(a_3)\}, \quad H \geq \max\{1, \pi/d\}.$$

Consider the *homogeneous unit equation*

$$a_1x_1 + a_2x_2 + a_3x_3 = 0 \quad \text{in } x_1, x_2, x_3 \in O_K^*. \tag{4.1.1}$$

The following theorem is due to Győry and Yu (2006).

Theorem 4.1.1 *All solutions x_1, x_2, x_3 of (4.1.1) satisfy*

$$\max_{i,j} h(x_i/x_j) \leq c_1 R(\log^* R)H, \tag{4.1.2}$$

where

$$c_1 := 4(r + 1)^{2r+9} 2^{3.2(r+12)} \log(2r + 2)(d \log^*(2d))^3.$$

In some applications, for instance in our book on discriminant equations, at least two of the unknowns x_1, x_2, x_3 are conjugate to each other over $\mathbb{Q}$. In these situations the following theorem will lead to much sharper quantitative results.

Let K_1 be a subfield of K with degree d_1, unit rank r_1 and regulator R_{K_1}. Assume that for some $\mathbb{Q}$-isomorphism σ of K_1, $\sigma(K_1)$ is also a subfield of K.

Theorem 4.1.2 *All solutions x_1, x_2, x_3 of (4.1.1) with $x_2 \in K_1$, $x_3 = \sigma(x_2)$ satisfy*

$$\max_{1 \leq i,j \leq 3} h(x_i/x_j) \leq c_2 R_{K_1} H \log \left(\frac{h(x_2)}{H} \right), \tag{4.1.3}$$

provided that

$$h(x_2) > c_3 R_{K_1} H, \qquad (4.1.4)$$

where

$$c_2 := 2^{5.5r_1+45} r_1^{2r_1+2.5}, \qquad c_3 := 320 d^2 r_1^{2r_1}.$$

It should be observed that in (4.1.3) the upper bound depends on $h(x_2)$.

In terms of d and r_1, Theorem 4.1.2 is an improvement of a result of Győry (1998).

In the next subsection we give more general versions of Theorem 4.1.1. A similar generalization of Theorem 4.1.2 is given in Győry (1998). But Theorems 4.1.1 and 4.1.2 provide, in the special situation they deal with, much better bounds in terms of d and r.

4.1.2 Equations with unknowns from a finitely generated multiplicative group

Let again K be an algebraic number field of degree d. Let Γ be a finitely generated multiplicative subgroup of K^* of rank $q > 0$, and Γ_∞ the torsion subgroup of Γ consisting of all elements of finite order. We recall that q is the smallest positive integer such that $\Gamma / \Gamma_{\text{tors}}$ has a system of q generators. Let S denote the smallest set of places of K such that S contains all infinite places, and $\Gamma \subseteq O_S^*$ where O_S^* denotes the group of S-units in K. Further, let $a_1, a_2 \in K^*$. We consider the equation

$$a_1 x_1 + a_2 x_2 = 1 \quad \text{in } x_1 \in \Gamma, x_2 \in O_S^*. \qquad (4.1.5)$$

In our first theorem below the following notation is used:

$H := \max\{1, h(a_1), h(a_2)\};$

$\{\xi_1, \ldots, \xi_m\}$ is a system of generators for $\Gamma / \Gamma_{\text{tors}}$ (not necessarily a basis) and

$$\Theta := h(\xi_1) \cdots h(\xi_m);$$

$s := |S|, \mathfrak{p}_1, \ldots, \mathfrak{p}_t$ are the prime ideals in S, and

$$P := \max\{2, N(\mathfrak{p}_1), \ldots, N(\mathfrak{p}_t)\},$$

where $N(\mathfrak{p}_i) := |O_K / \mathfrak{p}_i|$ denotes the norm of $\mathfrak{p}_i$; in the case that S consists only of the infinite places we put $t := 0$, $P := 2$.

Theorem 4.1.3 *If x_1, x_2 is a solution of (4.1.5), then*

$$\max\{h(x_1), h(x_2)\} < 6.5 \, c_4 s \frac{P}{\log P} \Theta H \max\{\log(c_4 s P), \log^* \Theta\}, \qquad (4.1.6)$$

where

$$c_4 := 11\lambda \cdot (m+1)(\log^* m)(16ed)^{3m+5}$$

$$\text{with } \lambda = 12 \text{ if } m = 1, \lambda = 1 \text{ if } m \geq 2.$$

For some of our applications it is essential that we allow $\xi_1, \ldots, \xi_m$ to be any set of generators of $\Gamma / \Gamma_{\text{tors}}$ and not necessarily a basis; see for instance the proof of Theorem 9.6.2 and the proofs of certain results on discriminant equations, to be discussed in our next book. Almost the same bounds as in (4.1.6) were obtained in Bérczes, Evertse and Győry (2009), but with c_4 replaced by a constant which, for $m > q > 0$, contains also the factor q^q. This improvement here will be important in our book on discriminant equations.

Theorem 4.1.3 implies in an effective way the finiteness of the number of solutions $x_1, x_2 \in \Gamma$ of (4.1.5). To formulate this in a precise form we recall that as in Section 1.10, K is said to be *effectively given* if the minimal polynomial over $\mathbb{Z}$ of a primitive element θ of K over $\mathbb{Q}$ is given. We may assume that θ is an algebraic integer. Further, an element α of K is said to be *given/effectively determinable* if it is expressed in the form

$$\alpha = (p_0 + p_1\theta + \cdots + p_{d-1}\theta^{d-1})/q$$

with rational integers $p_0, \ldots, p_{d-1}, q$ with $\gcd(p_0, \ldots, p_{d-1}, q) = 1$ that are given/can be effectively computed (see Section 1.10).

Corollary 4.1.4 *For given $a_1, a_2 \in K^*$, equation (4.1.5) has only finitely many solutions in $x_1, x_2 \in \Gamma$. Further, there exists an algorithm which, from effectively given K, a_1, a_2, a system of generators for $\Gamma / \Gamma_{\text{tors}}$ and Γ_{tors}, computes all solutions x_1, x_2.*

In the special case $\Gamma = O_S^*$, we obtain from Theorem 4.1.3 the following. Let S be a finite subset of M_K containing all infinite places, with the above parameters s, P. Denote by R_S the S-regulator (see (1.8.2)). Define

$$c_5 := 11\lambda s^2 (\log^* s)(16ed)^{3s+2} \text{ with } \lambda = 12 \text{ if } s = 2, \lambda = 1 \text{ if } s \geq 3,$$

$$c_6 := ((s-1)!)^2/(2^{s-2}d^{s-1}).$$

Corollary 4.1.5 *Every solution x_1, x_2 of*

$$a_1 x_1 + a_2 x_2 = 1 \quad \text{in } x_1, x_2 \in O_S^* \tag{4.1.7}$$

satisfies

$$\max(h(x_1), h(x_2)) < 6.5 c_5 c_6 (P/\log P) H R_S \max\{\log(c_5 P), \log^*(c_6 R_S)\}. \tag{4.1.8}$$

This was proved by Győry and Yu (2006) in a slightly sharper form in terms of d and s. Their proof is a more general variant of that of Theorem 4.1.1. In the special case $S = M_K^\infty$, Corollary 4.1.5 gives Theorem 4.1.1 but only with a weaker bound in terms of d and r. From Theorem 4.1.3, a weaker version of Theorem 4.1.2 can also be deduced.

We say that S is *effectively given* if the prime ideals in S are effectively given in the sense defined in Section 1.10. The next corollary follows both from Corollary 4.1.5 and from Corollary 4.1.4.

Corollary 4.1.6 *For given $a_1, a_2 \in K^*$, equation (4.1.7) has only finitely many solutions. Further, there exists an algorithm which, from effectively given K, a_1, a_2 and S, computes all solutions.*

If the number t of finite places in S exceeds $\log P$, then, in terms of S, s^s is the dominating factor in the bound occurring in (4.1.8). This factor is a consequence of the use of Proposition 4.3.9 concerning S-units whose proof is based on Minkowski's Theorem on successive minima. In the following version of Corollary 4.1.5 there is no factor of the form s^s or t^t. This improvement plays an important role in some applications, see e.g. Győry, Pink and Pintér (2004), Győry and Yu (2006), Győry (2006) and it is also applied in our next book on discriminant equations.

Let

$$\mathcal{R} := \max\{h, R\},$$

where h and R denote the class number and regulator of K, respectively. Further, let r denote the unit rank of K. From Theorem 4.2.1 below we shall deduce the following.

Theorem 4.1.7 *Let $t > 0$. Then every solution x_1, x_2 of equation (4.1.7) satisfies*

$$\max\{h(x_1), h(x_2)\} < \left(c_7 d^{r+3}\mathcal{R}\right)^{t+4} P H R_S, \qquad (4.1.9)$$

where c_7 is an effectively computable positive absolute constant.

This was established in Győry and Yu (2006) in a somewhat different and completely explicit form; for a slight improvement see Győry (2008a).

We note that in view of (1.5.2) and (1.5.3), $\mathcal{R}$ can be estimated from above in terms of d and the discriminant of K. Further, in view of (1.8.3) we have

$$R \prod_{i=1}^{t} \log N(\mathfrak{p}_i) \le R_S \le h R \prod_{i=1}^{t} \log N(\mathfrak{p}_i).$$

The linear dependence on H of the bounds in (4.1.6), (4.1.8) and (4.1.9) cannot be improved. Indeed, let $a_1 = 1 - \varepsilon$ with $\varepsilon \in O_S^*$ and $a_2 = 1$. Then equations (4.1.5) and (4.1.7) have the solution $x_1 = 1, x_2 = \varepsilon$, and it is easy to see that

$$H - \log 2 \le \max\{h(x_1), h(x_2)\} \le H + \log 2.$$

4.2 Approximation by elements of a finitely generated multiplicative group

We deduce Theorem 4.1.3 from the following Diophantine approximation theorem. Keeping the above notation, we put $N(v) := 2$ if v is an infinite place, and $N(v) := N(\mathfrak{p})$ if $v = \mathfrak{p}$ is a finite place, i.e., prime ideal of O_K.

Theorem 4.2.1 *Let Γ be a finitely generated multiplicative subgroup of K^* with system of generators $\{\xi_1, \ldots, \xi_m\}$ for $\Gamma / \Gamma_{\mathrm{tors}}$. Let $\alpha \in K^*$, and put*

$$H := \max(h(\alpha), 1), \quad \Theta := h(\xi_1) \cdots h(\xi_m).$$

Further, let $v \in M_K$. Then for every $\xi \in \Gamma$ with $\alpha\xi \neq 1$, we have

$$\log |1 - \alpha\xi|_v > -c_8 \frac{N(v)}{\log N(v)} \Theta H \log^* \left(\frac{N(v)h(\xi)}{H} \right), \qquad (4.2.1)$$

where

$$c_8 := 2\lambda \cdot (m + 1) \log^*(dm)(\log^* d)^2(16ed)^{3m+5}$$
$$\text{with } \lambda = 12 \text{ if } m = 1, \lambda = 1 \text{ if } m \geq 2.$$

The following theorem is an immediate consequence of Theorem 4.2.1. The estimate (4.2.3) below is of a similar flavour to results in Bombieri (1993), Bombieri and Cohen (1997, 2003) and Bugeaud (1998) (see also Bilu (2002), Bombieri and Gubler (2006), section 5.4), but, as will be seen in Section 4.5, inequality (4.2.3) below gives in many cases a better upper bound for $h(\xi)$.

Theorem 4.2.2 *Let $\alpha \in K^*$, $v \in M_K$ and $0 < \kappa \leq 1$. If $\xi \in \Gamma$ is such that $\alpha\xi \neq 1$ and*

$$\log |1 - \alpha\xi|_v < -\kappa h(\xi) \qquad (4.2.2)$$

then

$$h(\xi) < 6.4(c_8/\kappa) \frac{N(v)}{\log N(v)} \Theta H \max\{\log((c_8/\kappa)N(v)), \log^* \Theta\}. \qquad (4.2.3)$$

Similar results were proved in Bérczes, Evertse and Győry (2009) but with c_8 replaced by a constant which, for $m > q > 0$, contains also the factor q^q. Here q denotes the rank of Γ. It is crucial for some applications of Theorems 4.2.1 and 4.2.2, for example in Theorem 4.1.3 and Theorem 4.1.7, that no factor q^q occurs in c_8.

The main tool in the proofs of Theorems 4.2.1 and 4.2.2 is the theory of logarithmic forms, more precisely Theorem 3.2.8. It will be combined with some new results from the geometry of numbers and some estimates for fundamental/independent units.

4.3 Tools

4.3.1 Some geometry of numbers

Let V be a real vector space of finite dimension n. We endow V with a topology by choosing a linear isomorphism $\varphi : V \to \mathbb{R}^n$ and taking the inverse images under φ of the open sets of $\mathbb{R}^n$. This does not depend on the choice of φ.

By a *lattice* in V we mean an additive group of the shape

$$\mathcal{L} = \left\{ \sum_{i=1}^{q} z_i \mathbf{a}_i : z_1, \ldots, z_q \in \mathbb{Z} \right\},$$

where $\mathbf{a}_1, \ldots, \mathbf{a}_q$ are linearly independent vectors of V. We call $\{\mathbf{a}_1, \ldots, \mathbf{a}_q\}$ a basis of $\mathcal{L}$ and q the dimension of $\mathcal{L}$. Clearly, $q \leq n$. By a *full lattice* in V we mean a lattice in V of maximal dimension n.

A *norm* or *convex distance function* on V is a function $\|.\| : V \to \mathbb{R}_{\geq 0}$ such that

$$\|\mathbf{x} + \mathbf{y}\| \leq \|\mathbf{x}\| + \|\mathbf{y}\| \quad \text{for } \mathbf{x}, \mathbf{y} \in V;$$

$$\|\lambda \mathbf{x}\| = |\lambda| \cdot \|\mathbf{x}\| \quad \text{for } \mathbf{x} \in V, \lambda \in \mathbb{R};$$

$$\|\mathbf{x}\| = 0 \quad \text{if and only if } \mathbf{x} = \mathbf{0}.$$

The *unit ball* of $\|.\|$ is defined by

$$B_{\|.\|} = \{\mathbf{x} \in V : \|\mathbf{x}\| \leq 1\}.$$

It is a convex, compact, symmetric body in V, i.e., it is convex, symmetric about $\mathbf{0}$, and it is compact and has interior points with respect to the topology on V defined above. Conversely, with any convex, compact, symmetric body C in V one can associate a norm $\|.\|_C$ on V such that C is the unit ball of $\|.\|_C$: take $\|\mathbf{x}\|_C := \lambda$, where λ is the minimum of all reals $\mu \geq 0$ such that $\mathbf{x} \in \mu C := \{\mu \mathbf{y} : \mathbf{y} \in C\}$.

Let $\|\cdot\|$ be a norm on V, and $\mathcal{L}$ a q-dimensional lattice in V. For $i = 1, \ldots, q$ we define the i-th *successive minimum* $\lambda_i = \lambda_i(\|.\|, \mathcal{L})$ of $\|.\|$ with respect to $\mathcal{L}$, to be the minimum of all numbers λ such that $\{\mathbf{x} \in V : \|\mathbf{x}\| \leq \lambda\}$ contains at least i linearly independent vectors from $\mathcal{L}$.

We recall Minkowski's Theorem on successive minima. For technical simplicity we restrict ourselves to the special case of full lattices in $\mathbb{R}^q$. But note that the general case can be reduced to this special case by means of a linear isomorphism. We denote by "vol" the Lebesgue measure on $\mathbb{R}^q$, normalized such that the unit cube $[0, 1]^q$ has measure 1. If $\mathcal{L}$ is a full lattice in $\mathbb{R}^q$ with basis $\{\mathbf{a}_1, \ldots, \mathbf{a}_q\}$, say, we define the determinant of L by

$$d(\mathcal{L}) = |\det(\mathbf{a}_1, \ldots, \mathbf{a}_q)|.$$

This is independent of the choice of the basis.

Theorem 4.3.1 *Let* $\lambda_1, \ldots, \lambda_q$ *be the successive minima of a norm* $\|.\|$ *on* $\mathbb{R}^q$ *with respect to a full lattice* $\mathcal{L}$ *in* $\mathbb{R}^q$. *Then*

$$\frac{2^q}{q!} \leq \lambda_1 \cdots \lambda_q \frac{\text{vol}(B_{\|.\|})}{d(\mathcal{L})} \leq 2^q.$$

Proof. For a proof, see Cassels (1959), chapter VIII or Minkowski (1910). We note that both the upper bound and the lower bound are best possible. $\square$

Corollary 4.3.2 *Let* $\| \cdot \|$ *be a norm on* $\mathbb{R}^q$, *and* $\mathcal{L}$ *a full lattice in* $\mathbb{R}^q$ *such that* $\text{vol}(B_{\|.\|}) \geq 2^q d(\mathcal{L})$. *Then there is a non-zero* $\mathbf{x} \in \mathcal{L}$ *with* $\|\mathbf{x}\| \leq 1$.

Proof. By Theorem 4.3.1 we have $\lambda_1 \leq (\lambda_1 \cdots \lambda_q)^{1/q} \leq 1$. $\square$

Theorem 4.3.3 *Let* $\|.\|$ *be a norm on* $\mathbb{R}^q$, $\mathcal{L}$ *a full lattice in* $\mathbb{R}^q$, *and* $\lambda_1, \ldots, \lambda_q$ *the successive minima of* $\| \cdot \|$ *with respect to* $\mathcal{L}$. *Then* $\mathcal{L}$ *has a basis* $\{\mathbf{a}_1, \ldots, \mathbf{a}_q\}$ *such that* $\|\mathbf{a}_i\| \leq \max(1, i/2)\lambda_i$ *for* $i = 1, \ldots, q$.

Proof. See Cassels (1959), chapter V. The idea of the proof originates from Mahler. $\square$

We now prove a technical result, which will be applied later in combination with logarithmic forms estimates. Proposition 4.4.1 from Section 4.4, which is a consequence of Proposition 4.3.4 below, will play an important role in the proof of Theorem 4.2.1.

Proposition 4.3.4 *Let* V *be a real vector space,* $\mathcal{L}$ *a lattice in* V *of dimension* $q \geq 1$, *and* $\|.\|$ *a norm on* V, *such that* $\|\mathbf{x}\| \geq \theta > 0$ *for all* $\mathbf{x} \in \mathcal{L} \setminus \{\mathbf{0}\}$. *Further, let* $m \geq q$ *be an integer, and let* $\mathbf{a}_1, \ldots, \mathbf{a}_m$ *be vectors in* $\mathcal{L} \setminus \{\mathbf{0}\}$ *for which*

$$\mathbf{a}_1, \ldots, \mathbf{a}_m \text{ generate } \mathcal{L} \text{ as a } \mathbb{Z}\text{-module,}$$

and among all systems of m *vectors that generate* $\mathcal{L}$,

$$\prod_{i=1}^{m} \|\mathbf{a}_i\| \quad \text{is minimal.} \tag{4.3.1}$$

Then for every $\mathbf{x} \in \mathcal{L}$ *there are* $b_1, \ldots, b_m \in \mathbb{Z}$ *such that*

$$\mathbf{x} = b_1\mathbf{a}_1 + \cdots + b_m\mathbf{a}_m \quad \text{with } |b_i| \leq q^{2q}\frac{\|\mathbf{x}\|}{\theta} \quad \text{for } i = 1, \ldots, m.$$

It is crucial for applications that in Proposition 4.3.4 $\mathbf{a}_1, \ldots, \mathbf{a}_m$ do not have to form a basis of $\mathcal{L}$.

We assume that $V = \mathbb{R}^q$, $\mathcal{L} = \mathbb{Z}^q$ which is no loss of generality. Indeed, we may assume without loss of generality that V is the real vector space spanned by $\mathcal{L}$. Let $\varphi : \mathbb{R}^q \to V$ be a linear isomorphism such that $\varphi(\mathbb{Z}^q) = \mathcal{L}$. Define a norm $\|.\|_\varphi$ on $\mathbb{R}^q$ by $\|\mathbf{x}\|_\varphi := \|\varphi(\mathbf{x})\|$. Then clearly, it suffices to prove Proposition 4.3.4 with $\mathbb{R}^q$, $\mathbb{Z}^q$, $\varphi^{-1}(\mathbf{a}_1), \ldots, \varphi^{-1}(\mathbf{a}_m)$, $\|.\|_\varphi$ instead of $\mathcal{L}$, $\mathbf{a}_1, \ldots, \mathbf{a}_m, \|.\|$.

For the proof of Proposition 4.3.4 (with $V = \mathbb{R}^q$, $\mathcal{L} = \mathbb{Z}^q$) we make some preparations. Since we assume $\mathcal{L} = \mathbb{Z}^q$, the factor $d(\mathcal{L})$ in these results disappears. Denote by $\lambda_1, \ldots, \lambda_q$ the successive minima of $\|.\|$ with respect to $\mathbb{Z}^q$. By assumption, we have $\lambda_1 \geq \theta$. We define

$$V := \text{vol}(B_{\|.\|}) = \text{vol}(\{\mathbf{x} \in \mathbb{R}^q : \|\mathbf{x}\| \leq 1\}).$$

We need a number of lemmas.

Lemma 4.3.5 *Let $\mathbf{f}_0, \mathbf{f}_1, \ldots, \mathbf{f}_m$ be vectors in $\mathbb{Z}^q$ such that $\mathbf{f}_1, \ldots, \mathbf{f}_m$ generate $\mathbb{Z}^q$. Then there are integers $b_1, \ldots, b_m$ such that*

$$\mathbf{f}_0 = \sum_{i=1}^{m} b_i \mathbf{f}_i, \quad |b_i| \leq M(\mathbf{f}_0, \ldots, \mathbf{f}_m) \, for \, i = 1, \ldots, m,$$

where

$$M(\mathbf{f}_0, \ldots, \mathbf{f}_m) = \max_{0 \leq i_1 < \cdots < i_q \leq m} |\det(\mathbf{f}_{i_1}, \ldots, \mathbf{f}_{i_q})|.$$

Proof. This is a result of Borosh, Flahive, Rubin and Treybig (1989). □

Lemma 4.3.6 *Let $\mathbf{f}_1, \ldots, \mathbf{f}_q \in \mathbb{R}^q$. Then*

$$|\det(\mathbf{f}_1, \ldots, \mathbf{f}_q)| \leq \frac{q!}{2^q} V \|\mathbf{f}_1\| \cdots \|\mathbf{f}_q\|.$$

Proof. We assume without loss of generality that $\mathbf{f}_1, \ldots, \mathbf{f}_q$ are linearly independent. Put $\mathbf{g}_i := \|\mathbf{f}_i\|^{-1} \mathbf{f}_i$ for $i = 1, \ldots, q$, and denote by D the convex hull of the points $\pm \mathbf{g}_i$ $(i = 1, \ldots, q)$. Then our lemma follows at once from the observations $D \subset B_{\|.\|}$ and

$$\text{vol}(D) = \frac{2^q}{q!} \cdot |\det(\mathbf{g}_1, \ldots, \mathbf{g}_q)| = \frac{2^q}{q!} \cdot \frac{|\det(\mathbf{f}_1, \ldots, \mathbf{f}_q)|}{\|\mathbf{f}_1\| \cdots \|\mathbf{f}_q\|}. \qquad \square$$

In what follows, let $\mathbf{a}_1, \ldots, \mathbf{a}_m$ be as in Proposition 4.3.4, and assume again that $\mathcal{L} = \mathbb{Z}^q$.

Lemma 4.3.7 *Let $i_1, \ldots, i_q$ be any distinct indices from $\{1, \ldots, m\}$. Then*

$$\prod_{j=1}^{q} \|\mathbf{a}_{i_j}\| \leq \frac{q!}{2^{q-1}} \lambda_1 \cdots \lambda_q.$$

Proof. For convenience, we put

$$\mu_1 = \cdots = \mu_{m-q+1} := \lambda_1, \quad \mu_{m-q+2} := \lambda_2, \ldots, \mu_m := \lambda_q.$$

By Theorem 4.3.3, the lattice $\mathbb{Z}^q$ has a basis $\{\mathbf{y}_1, \ldots, \mathbf{y}_q\}$ such that $\|\mathbf{y}_i\| \leq \max(1, i/2)\lambda_i$ for $i = 1, \ldots, q$. This implies

$$\|\mathbf{a}_1\| \cdots \|\mathbf{a}_m\| \leq \|\mathbf{y}_1\|^{m-q+1} \|\mathbf{y}_2\| \cdots \|\mathbf{y}_q\|$$
$$\leq \frac{q!}{2^{q-1}} \lambda_1^{m-q+1} \lambda_2 \cdots \lambda_q = \frac{q!}{2^{q-1}} \mu_1 \cdots \mu_m, \quad (4.3.2)$$

where we have used (4.3.1).

Without loss of generality we may assume that $\|\mathbf{a}_1\| \leq \cdots \leq \|\mathbf{a}_m\|$. Let $i_0 := 0$ and for $j = 1, \ldots, q$ define i_j to be the largest index i such that $\mathrm{rank}\{\mathbf{a}_1, \ldots, \mathbf{a}_i\} = j$. Then

$$\|\mathbf{a}_i\| \geq \lambda_j \quad \text{for } i_{j-1} + 1 \leq i \leq i_j, j = 1, \ldots, q,$$

and so

$$\|\mathbf{a}_i\| \geq \mu_i \text{ for } i = 1, \ldots, m.$$

Together with (4.3.2) this implies that for any subset I of $\{1, \ldots, m\}$,

$$\prod_{i \in I} \frac{\|\mathbf{a}_i\|}{\mu_i} \leq \frac{q!}{2^{q-1}}.$$

Hence for any q distinct indices $i_1, \ldots, i_q$ from $\{1, \ldots, m\}$,

$$\prod_{j=1}^{q} \|\mathbf{a}_{i_j}\| \leq \|\mathbf{a}_{m-q+1}\| \cdots \|\mathbf{a}_m\| \leq \frac{q!}{2^{q-1}} \mu_{m-q+1} \cdots \mu_m \leq \frac{q!}{2^{q-1}} \lambda_1 \cdots \lambda_q,$$

which is our lemma. $\qquad\square$

Proof of Proposition 4.3.4. Without loss of generality, we assume that $\mathbf{x} \neq \mathbf{0}$. In view of Lemma 4.3.5, it suffices to show that

$$M(\mathbf{x}, \mathbf{a}_1, \ldots, \mathbf{a}_m) \leq q^{2q} \cdot \frac{\|\mathbf{x}\|}{\theta}. \quad (4.3.3)$$

First, let $i_1, \ldots, i_q$ be any q distinct indices from $\{1, \ldots, m\}$. Then by Lemmas 4.3.6 and 4.3.7, Theorem 4.3.1 (Minkowski's Theorem on successive minima) and our assumption $\|\mathbf{x}\| \geq \theta$, we have

$$|\det(\mathbf{a}_{i_1}, \ldots, \mathbf{a}_{i_q})| \leq \frac{q!}{2^q} \cdot V \cdot \|\mathbf{a}_{i_1}\| \cdots \|\mathbf{a}_{i_q}\|$$
$$\leq \frac{(q!)^2}{2^{2q-1}} \cdot V\lambda_1 \cdots \lambda_q$$
$$\leq \frac{(q!)^2}{2^{q-1}} \leq q^{2q} \cdot \frac{\|\mathbf{x}\|}{\theta}.$$

Next, let $i_1, \ldots, i_{q-1}$ be any $q - 1$ distinct indices from $\{1, \ldots, m\}$. Using Lemma 4.3.6, our assumption $\|\mathbf{a}_i\| \geq \theta$ for $i = 1, \ldots, m$, and Lemma 4.3.7 and Theorem 4.3.1 (Minkowski's Theorem), we get

$$
\begin{aligned}
|\det(\mathbf{x}, \mathbf{a}_{i_1}, \ldots, \mathbf{a}_{i_{q-1}})| &\leq \frac{q!}{2^q} \cdot V \|\mathbf{x}\| \cdot \|\mathbf{a}_{i_1}\| \cdots \|\mathbf{a}_{i_{q-1}}\| \\
&\leq \frac{q!}{2^q} \cdot V \|\mathbf{a}_{i_1}\| \cdots \|\mathbf{a}_{i_q}\| \cdot \frac{\|\mathbf{x}\|}{\theta} \\
&\qquad (\text{with } i_q \in \{1, \ldots, m\} \setminus \{i_1, \ldots, i_{q-1}\}) \\
&\leq \frac{(q!)^2}{2^{2q-1}} \cdot V \lambda_1 \cdots \lambda_q \cdot \frac{\|\mathbf{x}\|}{\theta} \\
&\leq \frac{(q!)^2}{2^{q-1}} \cdot \frac{\|\mathbf{x}\|}{\theta} \leq q^{2q} \cdot \frac{\|\mathbf{x}\|}{\theta}.
\end{aligned}
$$

This clearly proves (4.3.3) and Proposition 4.3.4. $\qquad\square$

4.3.2 Estimates for units and S-units

Let K be an algebraic number field of degree d with ring of integers O_K, unit rank r and regulator R. Denote by ω_K the number of roots of unity in K. We determine upper bounds for the heights of units and S-units in a fundamental/maximal independent system. We start with some auxiliary results. The first is due to Loher and Masser.

Proposition 4.3.8 *For $n \geq 1$, let $\alpha_1, \ldots, \alpha_n$ be multiplicatively independent non-zero elements of K. Then we have*

$$
58(n! e^n / n^n) d^{n+1} (\log^* d) h(\alpha_1) \cdots h(\alpha_n) \geq \omega_K.
$$

Proof. This is a consequence of Loher and Masser (2004), Theorem 3. $\qquad\square$

As is known, $n! e^n / n^n$ is asymptotic to $\sqrt{2\pi n}$ and $n! e^n / n^n \leq e\sqrt{n}$. Hence Proposition 4.3.8 gives

$$
58 e\sqrt{n}\, d^{n+1} (\log^* d) h(\alpha_1) \cdots h(\alpha_n) \geq \omega_K. \tag{4.3.4}
$$

For simplicity, we shall apply the consequence (4.3.4) of Proposition 4.3.8.

Let $S = \{v_1, \ldots, v_s\}$ be a finite set of places on K which contains the set M_K^∞ of the infinite places. Denote by O_S, O_S^* and R_S the ring of S-integers, the group of S-units and the S-regulator of K, respectively. If in particular $S = M_K^\infty$, then $s = r + 1$, $O_S = O_K$, O_S^* is just the unit group O_K^* of K, and $R_S = R$.

We define the constants

$$c_9 := ((s-1)!)^2/(2^{s-2}d^{s-1}),$$
$$c_9' := (s-1)!/d^{s-1},$$
$$c_{10} := 29e\sqrt{s-2}\,d^{s-1}(\log^* d)\,c_9 \quad (s \geq 3),$$
$$c_{10}' := 29e\sqrt{s-2}\,d^{s-1}(\log^* d)c_9' \quad (s \geq 3),$$
$$c_{11} := (((s-1)!)^2/2^{s-1})(\log(3d))^3.$$

Proposition 4.3.9 *Let* $s \geq 2$. *There exists in* K *a fundamental (respectively independent) system* $\{\varepsilon_1, \ldots, \varepsilon_{s-1}\}$ *of* S-units *with the following properties:*

(i) $\displaystyle\prod_{i=1}^{s-1} h(\varepsilon_i) \leq c_9 R_S$ (resp. $c_9' R_S$);

(ii) $\displaystyle\max_{1 \leq i \leq s-1} h(\varepsilon_i) \leq c_{10} R_S$ (resp. $c_{10}' R_S$) *if* $s \geq 3$;

(iii) *for such a fundamental system* $\{\varepsilon_1, \ldots, \varepsilon_{s-1}\}$, *the absolute values of the entries of the inverse matrix of* $(\log |\varepsilon_i|_{v_j})_{i,j=1,\ldots,s-1}$ *do not exceed* c_{11}.

Remark A similar result was proved earlier by Siegel (1969) for ordinary units, i.e., in the case $S = M_K^\infty$. The proof given below, which is a straightforward extension of Siegel's argument, is due to Győry and Yu (2006) and, in slightly weaker forms Hajdu (1993) and Bugeaud and Győry (1996a).

Recently, for multiplicatively independent S-units, Vaaler (2014), theorems 1, 2 obtained the slightly better upper bound $s!/(2d)^{s-1}$ instead of c_9'.

Proof. For $\alpha \in K \setminus \{0\}$, put

$$\mathbf{v}(\alpha) := \left(\log |\alpha|_{v_1}, \ldots, \log |\alpha|_{v_{s-1}}\right).$$

The full lattice $\mathcal{L}$ in $\mathbb{R}^{s-1}$ spanned by the vectors $\mathbf{v}(\eta)$ with $\eta \in O_S^*$ has determinant R_S; see Section 1.8.

The function $\| \cdot \| : \mathbb{R}^{s-1} \to \mathbb{R}$ defined by

$$\|\mathbf{x}\| := |x_1| + \cdots + |x_{s-1}|$$

for $\mathbf{x} = (x_1, \ldots, x_{s-1}) \in \mathbb{R}^{s-1}$ is a norm; see Section 4.3.1. Denote by V the volume of the unit ball $\{\mathbf{x} \in \mathbb{R}^{s-1} : \|\mathbf{x}\| \leq 1\}$. It is easy to check that

$$V = 2^{s-1}/(s-1)!.$$

By Theorem 4.3.1 (Minkowski's Theorem on successive minima) the successive minima $\lambda_1, \ldots, \lambda_{s-1}$ of $\mathcal{L}$ with respect to $\| \cdot \|$ have the property

$$\lambda_1 \cdots \lambda_{s-1} \leq 2^{s-1} R_S/V = (s-1)! R_S. \tag{4.3.5}$$

Further, there are multiplicatively independent S-units $\eta_1, \ldots, \eta_{s-1}$ in O_S for which

$$\|\mathbf{v}(\eta_i)\| = \lambda_i, \quad i = 1, \ldots, s - 1. \tag{4.3.6}$$

However, for every $\eta \in O_S^*$ we have $\prod_{j=1}^s |\eta|_{v_j} = 1$, hence

$$
\begin{aligned}
h(\eta) &= \frac{1}{d} \sum_{j=1}^s \max\left\{0, \log|\eta|_{v_j}\right\} = \frac{1}{2d} \sum_{j=1}^s \left|\log|\eta|_{v_j}\right| \\
&= \frac{1}{2d} \left(\sum_{j=1}^{s-1} \left|\log|\eta|_{v_j}\right| + \left|\sum_{i=1}^{s-1} \log|\eta|_{v_i}\right| \right),
\end{aligned}
$$

which implies that

$$\frac{1}{2d} \|\mathbf{v}(\eta)\| \le h(\eta) \le \frac{1}{d} \|\mathbf{v}(\eta)\|. \tag{4.3.7}$$

We infer from (4.3.5), (4.3.6) and (4.3.7) that

$$\prod_{i=1}^{s-1} h(\eta_i) \le \frac{(s-1)!}{d^{s-1}} \cdot R_S,$$

i.e. (i) holds for $\eta_1, \ldots, \eta_{s-1}$.

It follows from Theorem 4.3.3 that there exists a fundamental system of S-units $\{\varepsilon_1, \ldots, \varepsilon_{s-1}\}$ in O_S such that

$$\|\mathbf{v}(\varepsilon_i)\| \le \max\{1, i/2\} \|\mathbf{v}(\eta_i)\|, \quad i = 1, \ldots, s - 1. \tag{4.3.8}$$

Further, by (4.3.7), (4.3.8), (4.3.6) and (4.3.5) we have

$$
\begin{aligned}
\prod_{i=1}^{s-1} h(\varepsilon_i) &\le \frac{1}{d^{s-1}} \prod_{i=1}^{s-1} \|\mathbf{v}(\varepsilon_i)\| \le \frac{(s-1)!}{2^{s-2} d^{s-1}} \prod_{i=1}^{s-1} \|\mathbf{v}(\eta_i)\| \\
&\le \frac{((s-1)!)^2}{2^{s-2} d^{s-1}} \cdot R_S,
\end{aligned} \tag{4.3.9}
$$

which proves (i).

(ii) is an immediate consequence of (i) and (4.3.4).

It remains to prove (iii). Putting $E := (\log|\varepsilon_i|_{v_j})_{i,j=1,\ldots,s-1}$ we have $|\det(E)| = R_S$. If $s = 2$, then (1.5.3) and (1.8.5) prove (iii). Now let $s > 2$ and $e_{ij} := \det(E_{ij})/\det(E)$, where E_{ij} denotes the matrix obtained from E by omitting the i-th row and j-th column. It follows from (4.3.9) and Hadamard's inequality that

$$|\det(E_{ij})| \le \prod_{\substack{p=1 \\ p \neq i}}^{s-1} \sqrt{\sum_{\substack{q=1 \\ q \neq j}}^{s-1} \left(\log |\varepsilon_p|_{v_q}\right)^2} \le \prod_{\substack{p=1 \\ p \neq i}}^{s-1} \|\mathbf{v}(\varepsilon_p)\|$$

$$\le \frac{((s-1)!)^2}{2^{s-2}} \cdot \frac{R_S}{\|\mathbf{v}(\varepsilon_i)\|}.$$

Together with (4.3.7), $|\det(E)| = R_S$ and Proposition 3.2.9, this proves (iii). $\qquad \square$

For $s \ge 3$, let

$$c_{12} := 29e\sqrt{s-2} \cdot \frac{((s-1)!)^2}{2^{s-2}} \cdot \pi^{s-2} d \log^* d.$$

When we apply Theorem 3.2.5 to unit equations, we shall get better bounds by using the following version of (i), Proposition 4.3.9.

Lemma 4.3.10 *Let* $\{\varepsilon_1, \ldots, \varepsilon_{s-1}\}$ *be a fundamental system of S-units in* K *with the properties specified in Proposition 4.3.9. Then*

$$\prod_{i=1}^{s-1} \max(dh(\varepsilon_i), \pi) \le \begin{cases} \max(R_S, \pi), & \text{if } s = 2, \\ c_{12}R_S, & \text{if } s \ge 3. \end{cases} \qquad (4.3.10)$$

Proof. The case $s = 2$ is trivially true by Proposition 4.3.9. Suppose $s \ge 3$. Let k denote the number of indices i with $1 \le i \le s-1$ such that $dh(\varepsilon_i) < \pi$.

Suppose first $1 \le k \le s - 2$. Without loss of generality, we may assume $dh(\varepsilon_i) < \pi$ for $i = 1, \ldots, k$ and $dh(\varepsilon_j) \ge \pi$ for $j = k+1, \ldots, s-1$. Thus, using (4.3.4) and Proposition 4.3.9, we infer that

$$\prod_{i=1}^{s-1} \max(dh(\varepsilon_i), \pi) = \left(\pi^k / d^k h(\varepsilon_1) \cdots h(\varepsilon_k)\right) d^{s-1} h(\varepsilon_1) \cdots h(\varepsilon_{s-1})$$

$$\le 29e\sqrt{k} \cdot \frac{((s-1)!)^2}{2^{s-2}} \cdot \pi^k d(\log^* d) R_S \le c_{12}R_S,$$

which proves (4.3.10).

If $k = 0$, then (4.3.10) immediately follows from (i) of Proposition 4.3.9. For $k = s - 1$, we have $dh(\varepsilon_i) < \pi$ for each i. Further, if $s \ge 3$ then (1.8.5) gives a lower bound for R_S and (4.3.10) follows. $\qquad \square$

Let $\mathfrak{p}_1, \ldots, \mathfrak{p}_t$ be the prime ideals in S, and put

$$Q := N(\mathfrak{p}_1 \cdots \mathfrak{p}_t) \quad \text{if } t > 0, \qquad Q := 1 \quad \text{if } t = 0.$$

Let h_K denote the class number of K, and let

$$c_{13} := \begin{cases} 0, & \text{if } r = 0, \\ 1/d, & \text{if } r = 1, \\ 29er!r\sqrt{r-1}\log d, & \text{if } r \geq 2. \end{cases}$$

Proposition 4.3.11 *Let θ_v ($v \in S$) be reals with $\sum_{v \in S} \theta_v = 0$. Then there exists $\varepsilon \in O_S^*$ such that*

$$\sum_{v \in S} |\log |\varepsilon|_v - \theta_v| \leq c_{13}dR + h_K \log Q. \tag{4.3.11}$$

Remark As will follow from the proof, in the special case $t = 0$ the unit $\varepsilon \in O_K^*$ occurring in Proposition 4.3.11 can be chosen from the group generated by independent units having the properties specified in (i) and (ii) of Proposition 4.3.9.

Proof. We start with the case $t = 0$. Then $S = \{v_1, \ldots, v_{r+1}\}$, where $v_1, \ldots, v_{r+1}$ are the infinite places of K. Write θ_i for θ_{v_i}. If $r = 0$, then $S = \{v_1\}$, hence $\theta_1 = 0$, and thus the assertion holds with $\varepsilon = 1$. Assume that $r > 0$. Choose a system of independent units $\varepsilon_1, \ldots, \varepsilon_r$ in K with the properties specified in Proposition 4.3.9. Consider the system of linear equations

$$\sum_{j=1}^{r} \left(\log |\varepsilon_j|_{v_i} \right) x_j = \theta_i \quad i = 1, \ldots, r+1$$

in $x_1, \ldots, x_r$. The equations with $i = 1, \ldots, r$ have a unique solution $(x_1, \ldots, x_r) \in \mathbb{R}^r$, since $\det(\log |\varepsilon_j|_{v_i})_{i,j=1,\ldots,r} = R \neq 0$. This solution satisfies also the equation with $i = r+1$, since $\sum_{i=1}^{r+1} \log |\varepsilon_j|_{v_i} = 0$ for $j = 1, \ldots, r$, and $\sum_{i=1}^{r+1} \theta_i = 0$. Let $b_1, \ldots, b_r$ be the rational integers with

$$-\tfrac{1}{2} < b_j - x_j \leq \tfrac{1}{2} \quad \text{for } j = 1, \ldots, r$$

and take $\varepsilon = \varepsilon_1^{b_1} \cdots \varepsilon_r^{b_r}$. Then

$$\sum_{i=1}^{r+1} \left| \log |\varepsilon|_{v_i} - \theta_i \right| = \sum_{i=1}^{r+1} \left| \sum_{j=1}^{r} (b_j - x_j) \log |\varepsilon_j|_{v_i} \right|$$

$$\leq \tfrac{1}{2} \sum_{i=1}^{r+1} \sum_{j=1}^{r} \left| \log |\varepsilon_j|_{v_i} \right| \leq \sum_{j=1}^{r} \sum_{i=1}^{r} \left| \log |\varepsilon_j|_{v_i} \right|.$$

We assert that if $r > 1$, then the inner sum over i in the last expression is at most $(d/r)c_{13}R$. This can be seen by using (4.3.5), (4.3.6), the second inequality of (4.3.7) and by applying Proposition 4.3.8 to any $r - 1$ of the

$\varepsilon_i (1 \le i \le r)$. Thus Proposition 4.3.11 is proved for $r > 1$. If $r = 1$, we can use (i) of Proposition 4.3.9 to prove the assertion.

Now let $t > 0$. Recall that $S = M_K^\infty \cup S_0$, where $S_0 = \{\mathfrak{p}_1, \ldots, \mathfrak{p}_t\}$ is the set of finite places, i.e., prime ideals in S. For $\mathfrak{p} \in S_0$, let $k_\mathfrak{p}$ be the integer such that

$$-\tfrac{1}{2} < k_\mathfrak{p} + \frac{\theta_\mathfrak{p}}{h_K \log N\mathfrak{p}} \le \tfrac{1}{2}.$$

There is $\alpha \in K^*$ such that $(\alpha) = (\prod_{\mathfrak{p} \in S_0} \mathfrak{p}^{k_\mathfrak{p}})^{h_K}$. We have $\alpha \in O_S^*$ and

$$\left| \log |\alpha|_\mathfrak{p} - \theta_\mathfrak{p} \right| = \left| k_\mathfrak{p} h_K \log N\mathfrak{p} - \theta_\mathfrak{p} \right| \le \tfrac{1}{2} h_K \log N\mathfrak{p} \quad (\mathfrak{p} \in S_0). \quad (4.3.12)$$

By the Product Formula, $\sum_{v \in S} \log |\alpha|_v = 0$. Together with what we just proved, this implies that there is $\eta \in O_K^*$ such that

$$A := \sum_{v \in M_K^\infty} \left| \log |\eta|_v + \log |\alpha|_v - \theta_v + \frac{B}{r+1} \right| \le c_{13} dR, \quad (4.3.13)$$

where

$$B := \sum_{\mathfrak{p} \in S_0} (\log |\alpha|_\mathfrak{p} - \theta_\mathfrak{p}).$$

Now take $\varepsilon := \eta\alpha$. Clearly, (4.3.12) holds with ε instead of α. Hence

$$\sum_{\mathfrak{p} \in S_0} |\log |\varepsilon|_\mathfrak{p} - \theta_\mathfrak{p}| \le \tfrac{1}{2} h_K \log Q.$$

Further, (4.3.13) and (4.3.12) imply

$$\sum_{v \in M_K^\infty} |\log |\varepsilon|_v - \theta_v| \le A + |B| \le c_{13} dR + \tfrac{1}{2} h_K \log Q.$$

Now (4.3.11) follows by a simple addition. $\qquad\square$

Recall that we have defined the S-norm $N_S(\alpha) := \prod_{v \in S} |\alpha|_v$ for $\alpha \in K$; see Section 1.8. In the case of $S = M_K^\infty$, this is just $|N_{K/\mathbb{Q}}(\alpha)|$. In addition, we define

$$M_S(\alpha) := \max \left(\prod_{v \in M_K \setminus S} \max(1, |\alpha|_v), \prod_{v \in M_K \setminus S} \max(1, |\alpha|_v^{-1}) \right)$$

for $\alpha \in K^*$. By the Product Formula we have

$$M_S(\alpha) = \prod_{v \in M_K \setminus S} |\alpha|_v^{-1} = N_S(\alpha) \quad \text{for } \alpha \in O_S \setminus \{0\}.$$

Proposition 4.3.12 *Let $\alpha \in K^*$ and let n be a positive integer. Then there exists $\varepsilon \in O_S^*$ such that*

$$h(\varepsilon^n \alpha) \leq \frac{1}{d} \log M_S(\alpha) + n \left(c_{13} R + \frac{h_K}{d} \log Q \right). \qquad (4.3.14)$$

In particular, if $\alpha \in O_S \setminus \{0\}$ then there exists $\varepsilon \in O_S^*$ such that

$$h(\varepsilon^n \alpha) \leq \frac{1}{d} \log N_S(\alpha) + n \left(c_{13} R + \frac{h_K}{d} \log Q \right). \qquad (4.3.15)$$

Proof. Inequality (4.3.15) is an immediate consequence of (4.3.14). We prove (4.3.14). We assume that $N_S(\alpha) \geq 1$. This is no loss of generality since both the height and M_S are invariant under $x \mapsto x^{-1}$, and $N_S(\alpha) \geq 1$ can be achieved by replacing α by α^{-1} if necessary.

By Proposition 4.3.11, there is $\varepsilon \in O_S^*$ such that

$$B := \sum_{v \in S} \left| \log |\varepsilon|_v + \frac{1}{n} \log |\alpha|_v - \frac{1}{sn} \log N_S(\alpha) \right| \leq c_{13} d R + h_K \log Q,$$

where $s = |S|$. Hence

$$
\begin{aligned}
\frac{1}{d} \log \prod_{v \in S} \max(1, |\varepsilon^n \alpha|_v) &= \frac{1}{d} \sum_{v \in S} \max(0, n \log |\varepsilon|_v + \log |\alpha|_v) \\
&\leq \frac{n}{d} \cdot B + \frac{1}{d} \log N_S(\alpha) \\
&\leq n \left(c_{13} R + \frac{h_K}{d} \log Q \right) + \frac{1}{d} \log N_S(\alpha)
\end{aligned}
$$

since by assumption $N_S(\alpha) \geq 1$. By adding $\frac{1}{d} \log \prod_{v \in M_K \setminus S} \max(1, |\varepsilon^n \alpha|_v)$ on both sides and observing that by the Product Formula

$$
\begin{aligned}
N_S(\alpha) \prod_{v \in M_K \setminus S} \max \left(1, |\varepsilon^n \alpha|_v\right) &= N_S(\alpha) \prod_{v \in M_K \setminus S} \max(1, |\alpha|_v) \\
&= \prod_{v \in M_K \setminus S} \left(|\alpha|_v^{-1} \cdot \max(1, |\alpha|_v) \right) \leq M_S(\alpha),
\end{aligned}
$$

our Proposition follows. $\square$

4.4 Proofs

4.4.1 Proofs of Theorems 4.1.1 and 4.1.2

We keep the notation from Section 4.1. Thus, K is an algebraic number field of degree d, R is the regulator of K and r the rank of O_K^*. Further, H is a real with $H \geq \max(h(a_1), h(a_2), h(a_3))$ and $H \geq \max(1, \pi/d)$.

Proof of Theorem 4.1.1. For $r = 0$ the assertion is trivial, hence we assume that $r \geq 1$. Let x_1, x_2, x_3 be a solution of (4.1.1). Assume without loss of generality that $h(x_1/x_3) \geq h(x_i/x_j)$ for $1 \leq i < j \leq 3$. Put

$$\alpha := -a_1/a_3, \quad \beta := -a_2/a_3, \quad x := x_1/x_3, \quad y := x_2/x_3. \quad (4.4.1)$$

Then

$$\alpha x + \beta y = 1, \quad x, y \in O_K^*, \quad (4.4.2)$$

$$\max\{h(\alpha), h(\beta)\} \leq 2H. \quad (4.4.3)$$

Clearly, $h(x) \geq h(y)$. We give an upper bound for the height of x. Let $\varepsilon_1, \ldots, \varepsilon_r$ be a fundamental system of units in K with the properties specified in Proposition 4.3.9. Then y can be written in the form

$$y = \zeta \varepsilon_1^{b_1} \cdots \varepsilon_r^{b_r}, \quad (4.4.4)$$

where ζ is a root of unity in K and $b_1, \ldots, b_r$ are rational integers. Denote by $v_1, \ldots, v_{r+1}$ the infinite places of K. We infer from (4.4.4) that

$$\log |y|_{v_j} = \sum_{i=1}^{r} b_i \log |\varepsilon_i|_{v_j}, \quad j = 1, \ldots, r,$$

whence, using (iii) of Proposition 4.3.9 and the fact that y is a unit, we get

$$\max\{|b_1|, \ldots, |b_r|\} \leq c_{11}' \sum_{j=1}^{r} \left|\log |y|_{v_j}\right| = 2c_{11}'dh(y) \leq 2c_{11}'dh(x), \quad (4.4.5)$$

where c_{11}' denotes the constant c_{11} with $s - 1$ replaced by r. Set $\alpha_{r+1} := \zeta\beta$ and $b_{r+1} := 1$. Let v be an infinite place for which $|x|_v$ is minimal. Then, from (4.4.2) we deduce that

$$\log \left|\varepsilon_1^{b_1} \cdots \varepsilon_r^{b_r} \alpha_{r+1}^{b_{r+1}} - 1\right|_v = \log |\alpha x|_v \leq -\frac{d}{r+1}h(x) + 2dH. \quad (4.4.6)$$

We shall prove that

$$h(x) < c_{14}R(\log^* R)H, \quad (4.4.7)$$

where

$$c_{14} := \min\{(r+1)^{2r+9}2^{3.2r+38.4}, (r+1)^{2r+3.5}2^{4.3r+44.3}\}$$
$$\times 2\log(2r+2)(d\log^*(2d))^3,$$

which is somewhat stronger than (4.1.2). Set

$$\left.\begin{array}{l} A'_i = \max(dh(\varepsilon_i), \pi), \ i = 1, \ldots, r, \\ A'_{r+1} = 2dH \geq \max(dh(\alpha_{r+1}), \pi) \end{array}\right\}. \qquad (4.4.8)$$

We may assume that $h(x) > 4(r+1)H$ and

$$2c'_{11}dh(x) > 2e\max\left(\frac{(r+1)\pi}{\sqrt{2}}, A'_1, \ldots, A'_r\right)A'_{r+1},$$

since otherwise, using (1.5.3), Proposition 4.3.9 and (4.3.4), the upper bound (4.4.7) easily follows. In view of (4.4.5), we can apply Theorem 3.2.5 with $B' = 2c'_{11}dh(x)$. Combining this with Lemma 4.3.10, and using (4.4.6) and (4.4.8), we infer that

$$\log|\alpha x|_v > \begin{cases} -2d_vC_2(2,d)dH\max\{R,\pi\}\log\left(\frac{2c'_{11}h(x)}{2\sqrt{2}H}\right) & \text{if } r = 1, \\ -2d_vC_2(r+1,d)c'_{12}dHR\log\left(\frac{2c'_{11}h(x)}{2\sqrt{2}H}\right) & \text{if } r \geq 2, \end{cases}$$

where $C_2(r+1,d)$ is the constant occurring in Theorem 3.2.5 and c'_{12} denotes the constant c_{12} with $s-1$ replaced by r. Together with (4.4.6) this implies (4.4.7), hence (4.1.2). $\qquad\Box$

Proof of Theorem 4.1.2. We follow the arguments of the proof of Theorem 4.1.1. Let x_1, x_2, x_3 be an arbitrary but fixed solution of (4.1.1) with $x_2 \in K_1$ and $x_3 = \sigma(x_2)$. The cases $x_3 = x_2$ and $r_1 = 0$ being trivial, we assume that $x_3 \neq x_2$ and $r_1 \geq 1$. Then $d \geq d_1 \geq 2$. We define again $\alpha := -a_1/a_3$, $\beta := -a_2/a_3$, $x := x_1/x_3$, $y := x_2/x_3$ so that we have again (4.4.2), (4.4.3).

Let $\{\varepsilon_1, \ldots, \varepsilon_{r_1}\}$ be a fundamental system of units in K_1 with the properties specified in Proposition 4.3.9. Then

$$x_2 = \zeta\varepsilon_1^{b_1}\cdots\varepsilon_{r_1}^{b_{r_1}}$$

with a root of unity ζ in K_1 and with rational integers $b_1, \ldots, b_{r_1}$. We obtain as in the proof of Theorem 4.1.1 that

$$\max\{|b_1|, \ldots, |b_{r_1}|\} \leq 2c_{15}d_1h(x_2), \qquad (4.4.9)$$

where

$$c_{15} := 2\left((r_1!)^2/2^{r_1}\right)(\log(3d_1))^3.$$

Consider the infinite place v on K for which $|x|_v$ is minimal. Setting $\alpha_{r_1+1} = -\alpha_2\zeta/(\alpha_3\sigma(\zeta))$ and $\eta_i = \varepsilon_i/\sigma(\varepsilon_i)$ for $i = 1, \ldots, r_1$, we deduce from (4.4.2) that

$$\log\left|\eta_1^{b_1}\cdots\eta_{r_1}^{b_{r_1}}\alpha_{r_1+1} - 1\right|_v = \log\left|\frac{\alpha_1 x_1}{\alpha_3 x_3}\right|_v$$

$$\leq -\frac{d}{r+1}h(x_1/x_3) + 2dH. \quad (4.4.10)$$

We have $h(\alpha_{r_1+1}) \leq 2H$ and $h(\eta_i) \leq 2h(\varepsilon_i)$ for $i = 1, \ldots, r_1$.

To apply Theorem 3.2.5 to the left-hand side of (4.4.10), set

$$A_i' := \max\{dh(\varepsilon_i), \pi\}, \quad i = 1, \ldots, r_1, \quad A_{r_1+1}' := 2dH. \quad (4.4.11)$$

These imply

$$\max\{dh(\eta_i), \pi\} \leq 2A_i', \quad i = 1, \ldots, r_1, \quad (4.4.12)$$

$$\max\left\{dh(\alpha_{r_1+1}), \pi\right\} \leq A_{r_1+1}'. \quad (4.4.13)$$

We may assume that $h(x_1/x_3) > 4(r+1)H$ and

$$2c_{15}d_1h(x_2) > 2e\max\left\{\frac{(r_1+1)\pi}{\sqrt{2}}, 2A_1', \ldots, 2A_{r_1}'\right\}A_{r_1+1}'$$

since otherwise, using (1.5.3), Proposition 4.3.9 and (4.4.11), (4.4.13), we obtain $h(x_2) < 320d^2r_1R_{K_1}H$ which contradicts our assumption (4.1.4). Applying now Theorem 3.2.5 and using (4.4.9), (4.4.10), (4.4.12) and (4.4.13), we obtain

$$\log\left|\frac{a_1x_1}{a_3x_3}\right|_v > -C_2(r_1+1, d)2^{r_1}A_1'\cdots A_{r_1+1}'\log\left(\frac{c_{15}d_1h(x_2)}{d\sqrt{2}H}\right), \quad (4.4.14)$$

where $C_2(r_1+1, d)$, coming from Theorem 3.2.5, is

$$C_2(r_1+1, d) = \min\left\{1.451(30\sqrt{2})^{r_1+5}(r_1+2)^{5.5}, \pi 2^{6.5r_1+33.5}\right\}d^2\log(ed).$$

Comparing (4.4.14) with (4.4.10) and using (4.4.11), (4.4.13), Lemma 4.3.10, (4.1.4) and (1.5.3) we deduce first for $h(x_1/x_3)$ and then, by (4.4.1)–(4.4.3), for each $h(x_i/x_j)$ the estimate (4.1.3). $\square$

4.4.2 Proofs of Theorems 4.2.1 and 4.2.2

The proof of Theorem 4.2.1 will be based on Theorem 3.2.8. Theorem 4.2.2 is a simple corollary of Theorem 4.2.1. We need also the following.

Proposition 4.4.1 *Let Γ be a finitely generated multiplicative subgroup of K^* with rank $\Gamma = q > 0$. Let $m \geq q$ be a given integer, and let $\{\xi_1, \ldots, \xi_m\}$ be* *a system of generators for* $\Gamma / \Gamma_{\text{tors}}$ *such that the product*

$$h(\xi_1) \cdots h(\xi_m) \text{ is minimal} \tag{4.4.15}$$

among all systems of m elements that generate $\Gamma / \Gamma_{\text{tors}}$. *Then for every* $\xi \in \Gamma$ *there are rational integers* $b_1, \ldots, b_m$ *and a root of unity* ζ *such that* $\xi = \zeta \xi_1^{b_1} \cdots \xi_m^{b_m}$ *and*

$$\max(|b_1|, \ldots, |b_m|) \le c_{16}(d/2)(\log 3d)^3 h(\xi), \tag{4.4.16}$$

where $c_{16} := q^{2q}$.

Proof. Let $S = \{v_1, \ldots, v_s\}$ be a finite subset of M_K such that $S \supseteq M_K^\infty$ and Γ is a subgroup of O_S^*. Then for $\xi \in \Gamma$ we have

$$h(\xi) = \frac{1}{2d} \sum_{i=1}^{s} |\log |\xi|_{v_i}|.$$

Let $\{\eta_1, \ldots, \eta_q\}$ be a basis for $\Gamma / \Gamma_{\text{tors}}$. Then every $\xi \in \Gamma$ can be expressed uniquely as

$$\xi = \zeta \eta_1^{x_1} \cdots \eta_q^{x_q} \text{ with } x_1, \ldots, x_q \in \mathbb{Z} \text{ and a root of unity } \zeta. \tag{4.4.17}$$

We define a norm on $\mathbb{R}^q$ by

$$\|\mathbf{x}\| := \frac{1}{2d} \sum_{i=1}^{s} \left| \sum_{j=1}^{q} x_j \log |\eta_j|_{v_i} \right| \text{ for } \mathbf{x} = (x_1, \ldots, x_q) \in \mathbb{R}^q.$$

Then if ξ and $\mathbf{x} = (x_1, \ldots, x_q) \in \mathbb{Z}^q$ are related by (4.4.17), we have

$$h(\xi) = \|\mathbf{x}\|. \tag{4.4.18}$$

Further, by Proposition 3.2.9,

$$\|\mathbf{x}\| \ge \theta := \frac{2}{d(\log 3d)^3} \quad \text{for } \mathbf{x} \in \mathbb{Z} \setminus \{\mathbf{0}\}.$$

Define vectors $\mathbf{a}_i = (a_{i1}, \ldots, a_{iq}) \in \mathbb{Z}^q$ by

$$\xi_i = \zeta_i \eta_1^{a_{i1}} \cdots \eta_q^{a_{iq}}, \quad i = 1, \ldots, m,$$

where ζ_i is a root of unity. Then $\mathbf{a}_1, \ldots, \mathbf{a}_m$ generate $\mathbb{Z}^q$ and by (4.4.15) and (4.4.18), the product $\|\mathbf{a}_1\| \cdots \|\mathbf{a}_m\|$ is minimal. Further, if ξ and the vector $\mathbf{x} = (x_1, \ldots, x_q) \in \mathbb{Z}^q$ are related by (4.4.17), it follows that

$$\xi = \zeta \xi_1^{b_1} \cdots \xi_m^{b_m} \quad \text{with } \zeta \in \Gamma_{\text{tors}}, b_1, \ldots, b_m \in \mathbb{Z} \tag{4.4.19}$$

holds for some root of unity ζ if and only if

$$\mathbf{x} = \sum_{i=1}^{m} b_i \mathbf{a}_i.$$

Now Proposition 4.4.1 follows immediately by applying Proposition 4.3.4 to the norm $\|.\|$ defined above. □

Proof of Theorem 4.2.1. We may assume without loss of generality that $\xi_1, \ldots,$ ξ_m have been chosen so that $h(\xi_1) \cdots h(\xi_m)$ is minimal among all systems of m elements that generate Γ. By Proposition 4.4.1, there are rational integers $b_1, \ldots, b_m$ and a root of unity ζ in K for which (4.4.19) and (4.4.16) hold. Then we have

$$1 - \alpha\xi = 1 - \alpha'\xi_1^{b_1} \cdots \xi_m^{b_m} \text{ with } \alpha' = \zeta\alpha.$$

Let

$$c_{17} := m^{2m}(d/2)(\log 3d)^3.$$

We distinguish two cases.

First assume that $c_{17}h(\xi) \geq 2e(3d)^{2(m+1)}\Theta H$. Then we can apply Theorem 3.2.8 with $B = c_{17}h(\xi)$ and (4.2.1) follows.

Next consider the case when $c_{17}h(\xi) < 2e(3d)^{2(m+1)}\Theta H$. Then, by the Product Formula we have the following Liouville type inequality

$$|1 - \alpha\xi|_v = \prod_{\substack{w \in M_K \\ w \neq v}} |1 - \alpha\xi|_w^{-1} \geq 2^{-d} \prod_{\substack{w \in M_K \\ w \neq v}} \max(1, |\alpha\xi|_w)^{-1}$$

$$\geq 2^{-d} \exp(-dh(\alpha\xi)) \geq 2^{-d} \exp\left(-d\left(H + \frac{2e(3d)^{2(m+1)}\Theta H}{c_{17}}\right)\right).$$

In view of Proposition 3.2.9 we have

$$\Theta \geq \left(\frac{2}{d(\log 3d)^3}\right)^m \text{ if } d \geq 2 \text{ and } \Theta \geq (\log 2)^m \text{ if } d = 1, \qquad (4.4.20)$$

hence we obtain again (4.2.1). □

Proof of Theorem 4.2.2. Let $\xi \in \Gamma$ be such that $\alpha\xi \neq 1$ and (4.2.2) holds. By Theorem 4.2.1 we have (4.2.1). Then, with the notation $X := N(v)h(\xi)/H$ and $b := c_8\kappa^{-1}\Theta N(v)^2/(\log N(v))$, it follows that $X \leq b \log X$. In view of (4.4.20) we have $b \geq e^2$. We use now that if a, b, X are real numbers with $a \geq 0, b \geq 1$, $X \geq 1$ and $X \leq b \log X + a$ then

$$X \leq \begin{cases} 2(b \log b + a) & \text{if } b > e^2 \\ 2(2e^2 + a) & \text{if } b \leq e^2 \end{cases} \qquad (4.4.21)$$

(see Pethő and de Weger (1986)). This gives (4.2.3). □

4.4.3 Proofs of Theorem 4.1.3 and its corollaries

We keep the notation from Section 4.1. In particular, K is an algebraic number field of degree d, Γ is a finitely generated subgroup of K^* of rank $q > 0$, and S is the smallest set of places of K containing all infinite places, such that $\Gamma \subseteq O_S^*$.

Proof of Theorem 4.1.3. Let x_1, x_2 be a solution of (4.1.5). It follows from (4.1.5) that

$$h(x_1) \leq 3H + h(x_2) + \log 2. \tag{4.4.22}$$

First assume that $h(x_2) < 400s H$. Then (4.4.22) gives

$$h(x_1) < 404s H,$$

whence $P \cdot h(x_1)/H \leq 404s P$. Using now Proposition 3.2.9 and the fact that the function $X/\log X$ is monotone increasing for $X > e$, (4.1.6) easily follows.
  Now assume that

$$h(x_2) \geq 400s H. \tag{4.4.23}$$

Pick $v \in S$ for which $|x_2|_v$ is minimal. Then we deduce from (4.1.5) that

$$\log|1 - a_1 x_1|_v = \log|a_2 x_2|_v \leq -\frac{d}{s}h(x_2) + dH. \tag{4.4.24}$$

Further, (4.4.22) and (4.4.23) imply that

$$h(x_1) \leq 1.01h(x_2).$$

Hence we infer from (4.4.23) and (4.4.24) that

$$\log|1 - a_1 x_1|_v < -\kappa h(x_1)$$

with the choice $\kappa = d/(2.02s)$. By applying Theorem 4.2.2, for $h(x_1)$ we get the upper bound occurring in (4.1.6) with 6.5 replaced by 6.4 in the bound. Finally, (4.1.5) implies $h(x_2) \leq 3H + h(x_1) + \log 2$. But, in view of Proposition 3.2.9 the bound obtained for $h(x_1)$ is much larger than $3H + \log 2$, hence we get (4.1.6) for $h(x_2)$ as well. $\qquad\square$

Proof of Corollary 4.1.4. For given $a_1, a_2 \in K^*$, the finiteness of the number of solutions of equation (4.1.5) in $x_1, x_2 \in \Gamma$ immediately follows from Theorem 4.1.3 and Theorem 1.9.3.
  The group of roots of unity in K being cyclic, Γ_{tors} is also cyclic. Suppose that K, a_1, a_2, a system of generators $\{\xi_1, \ldots, \xi_m\}$ of $\Gamma/\Gamma_{\text{tors}}$ and a generator

ζ of Γ_{tors} are effectively given. We shall utilize some algorithmic results from algebraic number theory, references to the literature of which are listed in Section 1.10. The factorizations of $\xi_1, \ldots, \xi_m$ into prime ideals can be effectively determined. Then the set S consisting of all infinite places of K and of all prime ideals occurring in these factorizations can be effectively determined. If $x_1, x_2 \in \Gamma$ is a solution of (4.1.5), then x_1, x_2 belong to O_S^*, the group of S-units and Theorem 4.1.3 provides an effectively computable upper bound for $h(x_1)$ and $h(x_2)$. Therefore x_1, x_2 are contained in a finite and effectively computable subset, say $\mathcal{H}$, of K^*.

We can select those pairs (x_1, x_2) from $\mathcal{H} \times \mathcal{H}$ that satisfy (4.1.5). From the remaining x_1, x_1 one can select those x_1, x_2 that are S-units. We have still to decide whether such x_1, x_2 are contained in Γ or not, that is that

$$x_1 = \zeta^{z_0} \xi_1^{z_1} \cdots \xi_m^{z_m} \quad \text{with some } z_0, \ldots, z_m \in \mathbb{Z}, \qquad (4.4.25)$$

and similarly for x_2.

One can determine a fundamental system $\{\varepsilon_1, \ldots, \varepsilon_{s-1}\}$ of S-units in K where $s = |S|$, and a generator ρ of the group of roots of unity in K. Further, for any effectively given $\varepsilon \in O_S^*$, one can determine effectively rational integers $b_1, \ldots, b_{s-1}$ and b with $0 \le b < w_K$ such that

$$\varepsilon = \rho^b \varepsilon_1^{b_1} \cdots \varepsilon_{s-1}^{b_{s-1}}, \qquad (4.4.26)$$

where w_K denotes the number of roots of unity in K. In (4.4.25) we represent now $x, \zeta, \xi_1, \ldots, \xi_m$ in the form (4.4.26) and compare the representations of the left- and right-hand sides of (4.4.25). Then we arrive at a system of linear equations in $z_0, z_1, \ldots, z_m$. But one can decide whether this system of equations is solvable in $\mathbb{Z}$ or not, that is, whether $x_1 \in \Gamma$ or not. In case of x_2 one can proceed in the same way. □

Proof of Corollary 4.1.5. Let $\{\varepsilon_1, \ldots, \varepsilon_{s-1}\}$ be a fundamental system of S-units in K with the properties described in Proposition 4.3.9. Then, putting

$$\Theta := h(\varepsilon_1) \cdots h(\varepsilon_{s-1}),$$

we have $\Theta \le c_9 R_S$ with $c_9 = ((s-1)!)^2/(2^{s-2} d^{s-1})$. Now the assertion follows from Theorem 4.1.3. □

Proof of Corollary 4.1.6. Corollary 4.1.6 can be deduced both from Corollary 4.1.5 and from Corollary 4.1.4. The finiteness of the number of solutions of (4.1.7) follows immediately from Corollary 4.1.4 with the choice $\Gamma = O_S^*$. If S is effectively given, a fundamental system of S-units and the roots of unity in K are effectively determinable. Hence the effective part of Corollary 4.1.6 is also an immediate consequence of Corollary 4.1.4.   □

Theorem 4.1.7 will be deduced from Theorem 4.2.1.

Proof of Theorem 4.1.7. Let x_1, x_2 be a solution of (4.1.7). We infer as in the proof of Theorem 4.1.3 that, for some $v \in S$,

$$0 < \log|1 - a_1 x_1|_v < -(d/2.02s)h(x_1), \qquad (4.4.27)$$

whence $|1 - a_1 x_1|_v \leq 1$. This implies that $|a_1 x_1|_v \leq 1$ or 4 according as v is finite or not. Consequently, we have

$$
\begin{aligned}
|1 - (a_1 x_1)^{h_K}|_v &= |1 - a_1 x_1|_v \cdot |1 + a_1 x_1 + \cdots + (a_1 x_1)^{h_K - 1}|_v \\
&\leq c_{18}|1 - a_1 x_1|_v, \qquad\qquad\qquad (4.4.28)
\end{aligned}
$$

where $c_{18} := 1$ or 4^{h_K}, according as v is finite or not. Here h_K denotes the class number of K.

We shall give a lower bound for $|1 - (a_1 x_1)^{h_K}|_v$ by means of Theorem 4.2.1. We first construct a subgroup Γ of O_S^* such that $x_1^{h_K} \in \Gamma$. Denote by $\mathfrak{p}_1, \ldots, \mathfrak{p}_t$ the prime ideals in S. There are $\pi_1, \ldots, \pi_t$ in O_K such that $(\pi_i) = \mathfrak{p}_i^{h_K}$ and by Proposition 4.3.12, they can be chosen so that

$$h(\pi_i) \leq c_{19} d^r \mathcal{R} \log N(\mathfrak{p}_i), \quad i = 1, \ldots, t.$$

Here c_{19} and $c_{20}, \ldots, c_{25}$ below denote effectively computable absolute constants. By Proposition 4.3.9, there exists a fundamental system of units $\{\varepsilon_1, \ldots, \varepsilon_r\}$ such that $h(\varepsilon_1) \cdots h(\varepsilon_r) \leq c_{20} d^r R$. Then

$$
\begin{aligned}
\Theta &:= h(\varepsilon_1) \cdots h(\varepsilon_r) h(\pi_1) \cdots h(\pi_t) \\
&\leq \left(c_{21} d^r \right)^{t+1} \mathcal{R}^{t+1} \prod_{i=1}^{t} \log N(\mathfrak{p}_i). \qquad (4.4.29)
\end{aligned}
$$

Since x_1 is an S-unit in K, we can write $(x_1) = \mathfrak{p}_1^{u_1} \cdots \mathfrak{p}_t^{u_t}$ with appropriate integers $u_1, \ldots, u_t$. Consequently, we can write

$$x_1^{h_K} = \zeta \varepsilon_1^{b_1} \cdots \varepsilon_r^{b_r} \pi_1^{u_1} \cdots \pi_t^{u_t},$$

where ζ is a root of unity and $b_1, \ldots, b_r$ are integers. That is, $x_1^{h_K}$ belongs to the multiplicative subgroup Γ of O_S^* generated by $\varepsilon_1, \ldots, \varepsilon_r, \pi_1, \ldots, \pi_t$ and the roots of unity of K. Further, putting $H' := \max(h(a_1), 1)$, we have

$$h_K H' \geq \max\left(h(a_1^{h_K}), 1 \right) \geq H'.$$

If $a_1^{h_K} x_1^{h_K} = 1$, together with (4.1.7) and (1.8.5) this implies immediately (4.1.9). If $a_1^{h_K} x_1^{h_K} \neq 1$, then Theorem 4.2.1 gives

$$\log\left|1 - a_1^{h_K} x_1^{h_K}\right|_v > -c'(d, h_K, s) \frac{P}{\log P} \Theta H' \log^* \left(\frac{P h(x_1)}{H'}\right), \qquad (4.4.30)$$

where $c'(d, h_K, s) = (c_{22}d)^{3s+2}(\log^* d)^3 \log^* h_K$. We may assume that $h(x_1) \geq 12 \log(4 s h_K / d)$, since otherwise by (4.1.7) we are done. Then (4.4.27) and (4.4.28) imply that $\log |1 - a_1^{h_K} x_1^{h_K}|_v$ can be estimated from above by $-(d/4s)h(x)$. Comparing this with (4.4.30), we infer that

$$h(x_1) \leq c''(d, h_K, s) \frac{P}{\log P} \Theta H' \log^*(P\Theta),$$

where $c''(d, h_K, s) = (c_{23}d)^{3s+2}(\log^* h_K)^2$. In view of (4.4.29) we get

$$\log^*(P\Theta) \leq c_{24} t d (\log^* d)(\log^* \mathcal{R}) \log P.$$

Finally, using again (4.4.29), (1.5.3) and (1.8.3), we obtain

$$h(x_1) \leq \left(c_{25}d^{r+3}\mathcal{R}\right)^{t+4} P H \prod_{i=1}^{t} \log N(\mathfrak{p}_i) \leq \left(c_{25}d^{r+3}\mathcal{R}\right)^{t+4} P H R_S,$$

which gives (4.1.9) for $h(x_1)$. An upper bound of the same form follows for $h(x_2)$ from the equation (4.1.7). $\qquad \square$

4.5 Alternative methods, comparison of the bounds

Baker's theory of logarithmic forms made it possible to derive effective bounds for the solutions of S-unit equations. Later, some alternative methods were developed to obtain effective results for such equations. We briefly discuss these methods.

4.5.1 The results of Bombieri, Bombieri and Cohen, and Bugeaud

Bombieri (1993) and Bombieri and Cohen (1997, 2003) developed an effective method in Diophantine approximation, based on an extended version of the Thue–Siegel principle, the Dyson Lemma and some geometry of numbers, to prove an earlier, weaker version of Theorem 4.2.2. Bugeaud (1998), following their approach and combining it with estimates for linear forms in logarithms, obtained results which are in certain parameters sharper than those of Bombieri and Cohen. This improvement is partly due to the use of linear forms in at most three logarithms. It follows from Bugeaud's results that if (4.2.2) holds with $0 < \kappa \leq 1$, then

$$h(\xi) \leq \begin{cases} 10T \max\{H, T\} & \text{if } v \text{ is infinite,} \\ 8c_{26}T \max\{H, 40T\} & \text{if } v \text{ is finite ,} \end{cases} \quad (4.5.1)$$

where

$$T := (2mc_{26})^m N(v)(\log N(v))\Theta,$$

with

$$c_{26} := \begin{cases} 8 \times 10^{19}(d^4(\log 3d)^7/\kappa)\log^*(2d/\kappa) & \text{if } v \text{ is infinite,} \\ 8 \times 10^6(d^5/\kappa)(\log^*(2d/\kappa))^2 & \text{if } v \text{ is finite.} \end{cases}$$

It is easily seen that the bound in (4.2.3) has a better dependence on each parameter than the bound in (4.5.1), except possibly Θ and H. In fact, the bound in (4.5.1) is smaller than that in (4.2.3) precisely when both Θ and $H \log \Theta/\Theta$ are large relative to d, κ and $N(v)$, and in that case, the bound (4.5.1) is at most a factor $\log \Theta$ better than (4.2.3). It is important to observe that in contrast with (4.5.1), the bound (4.2.3) does not contain the factor m^m.

Bugeaud (1998) used his result (4.5.1) to derive the bound

$$\max\{h(x_1), h(x_2)\} < c_{27}P(\log^* P)R_S \max\{c_{27}P(\log^* P)R_S, H\} \quad (4.5.2)$$

for the solutions x_1, x_2 of the S-unit equation (4.1.7), where

$$c_{27} := (10^{23}s^4(\log^* s)^2 d^3)^s.$$

Observe that (4.1.8) is better than (4.5.2) in terms of each parameter, except possibly R_S and H. The bound in (4.5.2) is smaller than that in (4.1.8) precisely if both R_S and $H \log R_S/R_S$ are large relative to P, d and s, and in that case the bound in (4.5.2) is at most a factor $\log^* R_S$ better than that in (4.1.8). If $H > c_{27}P(\log^* P)R_S$, then there is no $\log^* R_S$ factor in (4.5.2), however this bound contains s^s. Our bound in (4.1.9) contains neither $\log^* R_S$ nor s^s, but it depends on $\mathcal{R}^t$.

4.5.2 The results of Murty, Pasten and von Känel

Let $S \subset M_{\mathbb{Q}}$ consist of the infinite place and the prime numbers $p_1, \ldots, p_t$. Then the corresponding ring of S-integers is $\mathbb{Z}_S = \mathbb{Z}[(p_1 \cdots p_t)^{-1}]$. Murty and Pasten (2013) (see also Pasten (2014)) developed a new effective method to bound the heights of the solutions of special S-unit equations of the form

$$x_1 + x_2 = 1 \quad \text{in } x_1, x_2 \in \mathbb{Z}_S^*. \quad (4.5.3)$$

A similar method was obtained later and independently by von Känel (2014b). The basic idea behind the approach of Murty and Pasten and that of von Känel is an observation by Frey, that the now proved Shimura–Taniyama Conjecture, which states that the L-function of an elliptic curve is equal to that of an associated modular form, implies that (4.5.3) has only finitely many solutions. Murty and Pasten and von Känel observed that this can be made effective, and used this to obtain an explicit upper bound for the heights of the solutions of (4.5.3). We formulate the result of Murty and Pasten, which is slightly sharper. To be precise, let S be as above, and put $Q := p_1 \cdots p_t$. Further, assume that $t \geq 2$ and $2 \in S$. Murty and Pasten proved that any solution x_1, x_2 of the S-unit equation (4.5.3) satisfies

$$h(x_1), h(x_2) \leq 4.8Q \log Q + 13Q + 25. \tag{4.5.4}$$

In the special case of equation (4.1), where the number field is $\mathbb{Q}$ and the coefficients a_1, a_2 are equal to 1, (4.5.4) can be compared with the estimates obtained in Theorem 4.1.3 and Corollary 4.1.5, and even with the slightly sharper Theorem 2 of Győry and Yu (2006). In this case this latter result gives the estimate

$$h(x_1), h(x_2) \leq 2^{10t+2} t^4 (P/\log P) \prod_{i=1}^{t} \log p_i \tag{4.5.5}$$

for the solutions x_1, x_2 of (4.5.3), where $P := \max_i p_i$. It is easily seen that (4.5.4) improves (4.5.5) if Q is small, in particular if $Q \leq 2^{30}$. However, if t is small and Q is large then (4.5.5) gives a better bound for the solutions.

It should be remarked that for most applications of S-unit equations, more general results concerning equations of the form (4.1) $a_1 x_1 + a_2 x_2 = 1$ over number fields are needed.

4.6 The abc-conjecture

An extremely important S-unit equation is the abc-equation

$$a + b = c,$$

where S is a finite set of primes, and a, b, c are coprime positive integers not divisible by primes outside S. Then Corollary 4.1.5 and Theorem 4.1.7 provide explicit upper bounds for c in terms of S. However, these bounds are far from being best possible.

The *radical* of (a, b, c) is defined as

$$Q(a, b, c) := \prod_{p | abc} p.$$

Oesterlé and, in a refined form, Masser (1985) proposed the following.

abc-conjecture *For every $\varepsilon > 0$, there is a positive number $C(\varepsilon)$ such that if a, b, c are coprime positive integers with*

$$a + b = c \quad \text{and radical } Q = Q(a, b, c), \tag{4.6.1}$$

then

$$c < C(\varepsilon)Q^{1+\varepsilon}. \tag{4.6.2}$$

This is already sharp in the sense that (4.6.2) does not remain valid for $\varepsilon = 0$.

On August 30, 2012, Shinichi Mochizuki (Kyoto University), posted on the internet a sequence of four papers on Inter-universal Teichmüller theory in which he claims to prove the abc-conjecture. For recent updates of these papers, see Mochizuki's home page

www.kurims.kyoto-u.ac.jp/~motizuki/top-english.html.

At the moment of completion of this book, his proof had not yet been checked.

There are several refinements or modifications of the abc-conjecture; for references see e.g. Robert, Stewart and Tenenbaum (2014) where the authors propose and motivate the following **conjecture**. We denote by $\log_n$ the n times iterated natural logarithm.

There exists a real number C_1 such that if a, b, c are coprime positive integers as in (4.6.1) *then*

$$c < Q \exp(4\sqrt{3 \log Q / \log_2 Q}(1 + (\log_3 Q + C_1)/2\log_2 Q)).$$

Furthermore, there exists a real number C_2 and infinitely many pairs of coprime positive integers a, b, c with (4.6.1) *such that*

$$c > Q \exp(4\sqrt{3 \log Q / \log_2 Q}(1 + (\log_3 Q + C_2)/2\log_2 Q)).$$

For any positive integer m we denote by $\omega(m)$ the number of distinct prime factors of m. Baker (1998, 2004) and Granville (1998) formulated such refinements of the abc-conjecture which involve also $\omega(abc)$. The following **completely explicit refined version** is due to Baker (2004).

If a, b, c are coprime positive integers with (4.6.1) *then*

$$c < \frac{6}{5t!}Q(\log Q)^t,$$

where $t = \omega(abc)$.

The abc-conjecture has a very extensive literature. It unifies and motivates a number of results and problems in number theory. Further, it has several striking consequences. We mention here only some of them.

- It is easy to show that the abc-conjecture implies Fermat's Last Theorem for every sufficiently large exponent. Indeed, assume that x, y, z are relatively prime positive integers, $n > 3$ and $x^n + y^n = z^n$. Then, for $\varepsilon = 1$, the abc-conjecture gives $z^n < C(1)Q^2$, where

$$Q := \prod_{p \mid x^n y^n z^n} p = \prod_{p \mid xyz} p \leq xyz < z^3,$$

whence $z^n < C(1)z^6$. This proves that there exists n_0 such that $n \leq n_0$ or, in other words, Fermat's Last Theorem is asymptotically true. The weaker version of the abc-conjecture when $\varepsilon = 1$, $C(1) = 1$ implies in the same way that $n \leq 5$. As is known, Fermat's Last Theorem is now proved by Wiles (1995), Taylor and Wiles (1995).
- It follows in a similar manner from the abc-conjecture that the generalized Fermat equation $Ax^k + By^m + Cz^n = 0$, where A, B, C are given non-zero integers, has finitely many solutions in relatively prime integers x, y, z greater that 1, and positive integers k, m, n which satisfy $1/k + 1/m + 1/n < 1$.
- Elkies (1991) proved that the abc-conjecture implies Roth's Approximation Theorem, that is Theorem 3.1.1, and that an effective abc-theorem would make Roth's Theorem effective. See also Langevin (1999).
- Confirming a conjecture of Mordell (1922a), Faltings (1983) proved that a geometrically irreducible smooth projective curve of genus $g \geq 2$, defined over $\mathbb{Q}$, has only finitely many rational points. Falting's Theorem is ineffective. Elkies (1991) showed that the abc-conjecture implies this theorem of Faltings, and in fact even an effective version of this if an effective version of the abc-conjecture is available.
- By a result of Lagarias and Soundararajan (2011), the abc-conjecture implies that for any fixed $\kappa < 1$, there are only finitely many coprime positive integers a, b, c such that

$$a + b = c \text{ and } P(abc) \leq (\log c)^\kappa.$$

On the other hand, under the Generalized Riemann Hypothesis the authors proved that for $\kappa \geq 8$ there are infinitely many triples a, b, c satisfying these properties.

Some weaker versions of the abc-conjecture have been proved in an effective way. By means of the theory of logarithmic forms, Stewart and Tijdeman (1986), Stewart and Yu (1991, 2001), Győry and Yu (2006), Surroca (2007) and Győry (2008a) obtained upper bounds for c as a function of $Q(a, b, c)$. Stewart and Yu (2001) proved that

$$c < \exp\left(c_{28}Q^{1/3}(\log Q)^3\right), \tag{4.6.3}$$

where $Q = Q(a, b, c)$ and c_{28} is an effectively computable positive absolute constant. Further, Győry (2008a) deduced from a slightly improved and completely explicit version of Theorem 4.1.7 that

$$c < \exp(2^{10t+22}/t^{t-4})Q(\log Q)^t). \tag{4.6.4}$$

For a deeper connection between the *abc*-conjecture and the theory of logarithmic forms, we refer to Baker (2004).

We present now a **number field version** of the abc-conjecture. Let K be an algebraic number field of degree d, and M_K the set of places on K; see Section 1.7. The *height* of $(a, b, c) \in (K^*)^3$ is defined as

$$H_K(a, b, c) = \prod_{v \in M_K} \max(|a|_v, |b|_v, |c|_v)$$

and the *radical* of (a, b, c) as

$$Q_K(a, b, c) := \prod_{\mathfrak{p}} (N_K(\mathfrak{p}))^{e(\mathfrak{p})}.$$

Here the product is taken over all prime ideals $\mathfrak{p}$ for which $|a|_{\mathfrak{p}}, |b|_{\mathfrak{p}}, |c|_{\mathfrak{p}}$ are not all equal and $e(\mathfrak{p})$ is the ramification index of $\mathfrak{p}$ over the rational prime below it. Denote by D_K the absolute value of the discriminant of K.

Vojta (1987) proposed a very general conjecture, and, as a consequence, suggested the first number field version of the abc-conjecture. Later, several refinements of Vojta's version were suggested, see Elkies (1991), Broberg (1999), Vojta (2000), Granville and Stark (2000), Browkin (2000) and Masser (2002). The following uniform version is due to Masser.

ABC-conjecture for the number field K *For every $\epsilon > 0$ there exists $C(\epsilon) > 0$, such that if*

$$a + b + c = 0 \text{ with } a, b, c \in K^*, Q_K = Q_K(a, b, c), \tag{4.6.5}$$

then

$$H_K(a, b, c) < C(\epsilon)^d(|D_K| \cdot Q_K)^{1+\epsilon}.$$

For $K = \mathbb{Q}$, this reduces to the Oesterlé–Masser Conjecture. The upper bound is again best possible in term of ϵ. This general conjecture has also a very rich literature, and has many profound implications; see the abc-conjecture home page mentioned below.

The bounds obtained for the solutions of S-unit equations can be used to derive weaker but unconditional upper bounds for $H_K(a, b, c)$. Let a, b, c be non-zero elements of K with $a + b + c = 0$, and let

$$S = M_K^\infty \cup \{\text{finite } v \in M_K \text{ such that } |a|_v, |b|_v, |c|_v \text{ are not all equal}\}.$$

Then $x = -a/c$, $y = -b/c$ is a solution of the S-unit equation

$$x + y = 1 \text{ in } x, y \in O_S^*.$$

Every bound for $h(x)$, $h(y)$ gives a bound for $H_K(a, b, c)$. Using a result of Bugeaud and Győry (1996a) concerning S-unit equations, Surroca (2007) derived a bound for $H_K(a, b, c)$. By means of a slightly improved and explicit version of Theorem 4.1.7, Győry (2008a) considerably improved Surroca's bound by showing that if $\epsilon > 0$ and (4.6.5) holds then $H_K(a, b, c)$ can be estimated from above by

$$\exp\left(c_{29}(d, D_K, \epsilon)Q_K^{1+\epsilon}\right) \tag{4.6.6}$$

and, if

$$Q_K = Q_K(a, b, c) > \max\left\{|D_K|^{2/\epsilon}, \exp\exp(\max(|D_K|, e))\right\},$$

by

$$\exp\left(c_{30}(d, \epsilon)(|D_K| \cdot Q_K)^{1+\epsilon}\right), \tag{4.6.7}$$

where c_{29}, c_{30} are effectively computable constants depending only on the parameters occurring in the parentheses. Clearly, the bounds in (4.6.3), (4.6.4), (4.6.6) and (4.6.7) are still far from the conjectured best bounds.

For other details, including generalizations and applications of the abc-conjecture, we refer the reader to Bombieri and Gubler (2006), Baker and Wüstholz (2007), the *abc-conjecture home page* created and maintained by Nitaj, www.math.unicaen.fr/~nitaj/abc.html, and the references given there.

4.7 Notes

4.7.1 Historical remarks and some related results

- In the special case of S-unit equations over $\mathbb{Q}$, effective finiteness results can be deduced for the solutions from a theorem of Coates (1969) on the greatest prime factor of binary forms and also from a result of Sprindžuk (1969) on ternary exponential equations. In the general case, for unit and S-unit equations over number fields, various effective bounds for the solutions were established in several papers and books, including Győry (1972, 1973, 1974, 1979, 1979/1980, 1980b, 2008a), Sprindžuk (1973, 1976, 1982, 1993), Lang (1978), Kotov and Trelina (1979), Schmidt (1992), Bugeaud and Győry (1996a), and Haristoy (2003). The best known bounds can be found in Győry and Yu (2006) and, in a more general form, for the solutions from a finitely generated multiplicative subgroup of $\overline{\mathbb{Q}}^*$, in Section 4.1 of the present book. Later, Bérczes, Evertse and Győry (2009) gave effective bounds for the heights and degrees of the solutions from the division group of a finitely generated multiplicative subgroup of $\overline{\mathbb{Q}}^*$. We note that in certain applications of Baker's theory, Bilu systematically used so-called functional units instead of applying unit equations; see Bilu (2002) and the references given there.

- Corollary 4.1.6 states that over a number field K, the S-unit equation $x_1 + x_2 = 1$ in $x_1, x_2 \in O_S^*$ has only finitely many solutions, and all of them can be, at least in principle, effectively determined. An equivalent statement is that the set of S-integral points of $\mathbb{P}^1(K) \setminus \{0, 1, \infty\}$ is finite, and these points can be, at least in principle, effectively computed. Here $\mathbb{P}^1(K)$ denotes the projective line over K. For this and other equivalent statements, see e.g. Section 9.2 and LeVesque and Waldschmidt (2011).

- Let $p_1, \ldots, p_t$ be distinct rational primes, and $S = \{\infty, p_1, \ldots, p_t\}$ and denote by $\mathbb{Z}_S^*$ the group of S-unit equations in $\mathbb{Q}$. As a common generalization of S-unit equations and binomial Thue equations Győry and Pintér (2008) considered over $\mathbb{Q}$ the equation

$$u_1^n x_1 + u_2^n x_2 = 1 \quad \text{in } u_1, u_2 \in \mathbb{Z} \setminus \{0\}, n \geq 3, x_1, x_2 \in \mathbb{Z}_S^*$$
$$\text{with } \gcd(u_1 u_2, p_1 \cdots p_t) = 1. \tag{4.7.1}$$

They proved that the heights of u_1^n, u_2^n, x_1 and x_2 can be effectively bounded above in terms of S. This implies that there are only finitely many u_1^n, u_2^n with the given properties for which equation (4.7.1) can have a solution x_1, x_2, and these u_1^n, u_2^n, together with the possible solutions x_1, x_2, can be, at least in principle, effectively determined. All the results mentioned above were proved by means of the theory of logarithmic forms.

4.7.2 Some notes on applications

- The effective results concerning equations (4.1) and (4.2) led to a great number of applications, among others to
  - *Thue equations, Thue–Mahler equations* and *decomposable form equations*, see Section 9.6 and the Notes in Chapter 9,
  - *discriminant form* and *index form equations*, see Section 9.6, the Notes in Chapter 9 and our book on discriminant equations,
  - *discriminant equations* and *power integral bases*, see Section 10.6 and our book on discriminant equations,
  - *binary forms* and *decomposable forms of given discriminant*, see Section 10.7 and our book on discriminant equations,
  - *irreducible polynomials* and *arithmetic graphs*, see Section 10.5,

– *unit equations* in two unknowns *over finitely generated domains*, see Chapter 8,
– *bounding the number of solutions of S-unit equations*, see the Notes in Chapter 6.

For applications of the so-obtained results and for references, the reader should consult Chapters 9 and 10, the books and survey papers Győry (1980b, 1992a, 2002, 2010), Sprindžuk (1982, 1993), Shorey and Tijdeman (1986), Evertse, Győry, Stewart and Tijdeman (1988b), Bombieri and Gubler (2006), Baker and Wüstholz (2007), and our book on discriminant equations.

• In many cases, the applicability of Baker's theory can be considerably extended by reducing the Diophantine problem under consideration to the study of such systems of unit equations in which the equations possess certain graph-theoretic connectedness properties. Then a combination of the effective results concerning equation (4.1) with some combinatorial arguments enables one to derive a bound for the solutions of the initial Diophantine problem; see Sections 9.6, 10.5, 10.6, Győry (1980c, 1981a, 1981c, 1982c) and Evertse, Győry, Stewart and Tijdeman (1988b).

• We now mention some recent applications of the results presented in this chapter. There are many important applications to polynomials and binary forms of given discriminant; these will be discussed in full detail in our book on discriminant equations. Further, Theorem 1 of Győry and Yu (2006), that is Corollary 4.1.5 of the present chapter, has been recently used to obtain among others the following effective results.

– In von Känel (2011, 2014a), an effective version of Shafarevich's conjecture/Faltings' Theorem is proved for hyperelliptic curves. This has been worked out in our book on discriminant equations.

– In de Jong and Rémond (2011), the authors give an effective version of Shafarevich's conjecture/Faltings' Theorem for cyclic covers of the projective line of prime degree.

– Finally, in von Känel (2013) a generalization of Szpiro's Discriminant conjecture concerning elliptic curves is formulated for hyperelliptic curves, and a completely explicit exponential version of Szpiro's Generalized conjecture is established.

5

Algorithmic resolution of unit equations in two unknowns

Let K be an algebraic number field, Γ_1, Γ_2 two finitely generated multiplicative subgroups of K^*, and a_1, a_2 two non-zero elements of K. It follows from the results of the preceding chapter that the equation

$$a_1 x_1 + a_2 x_2 = 1 \quad \text{in } (x_1, x_2) \in \Gamma_1 \times \Gamma_2$$

has only finitely many solutions, and effective upper bounds can be given for the heights of the solutions. These bounds are, however, too large for practical use, for finding all solutions of concrete equations of the above form. In this chapter a practical method will be provided to locate all the solutions to such equations, subject to the conditions that a_1, a_2 and the generators of Γ_1 and Γ_2 are effectively given and that the ranks of Γ_1 and Γ_2 are not too large, presently the bound is about 12. In particular, we present an efficient algorithm for solving completely S-unit equations in two unknowns.

The unknowns x_1 and x_2 can be represented as a power product of the generators of Γ_1 and Γ_2, respectively. Assuming that the generators of infinite order are multiplicatively independent, these representations are unique up to powers of roots of unity. Thus, we arrive at an exponential Diophantine equation of the form (5.1.3) below which has to be solved. As in Chapter 4, we first derive an explicit upper bound for the absolute values of the unknown exponents, using the best known Baker's type inequalities concerning linear forms in logarithms. In this way the existence of "large" solutions will be excluded. This part is an adaptation of Győry's method (Győry (1979)) who was the first to give explicit bounds for the solutions in case of S-unit equations over number fields. Then, in concrete cases, we can considerably reduce the obtained bound by means of de Weger's reduction techniques (de Weger (1987, 1989)) based on the LLL lattice basis reduction algorithm. This means that even "medium" sized solutions do not exist. Finally, some enumeration procedures due to Wildanger (1997, 2000) and Smart (1999) can be utilized to determine the "small" solutions under the reduced bound. We shall briefly illustrate the resolution process on two concrete equations. Of course, during the process some standard algebraic number-theoretical concepts and algorithms will also be needed, references to which, for convenience, are collected in Section 1.10.

For further details, related results, methods, applications and examples we refer to de Weger (1987, 1989), Wildanger (1997, 2000), Smart (1998, 1999), Győry (2002), Gaál (2002) and Baker and Wüstholz (2007).

5.1 Application of Baker's type estimates

Let K be an algebraic number field of degree d, given by the minimal polynomial of a primitive integral element θ of K over $\mathbb{Q}$. Assume that two non-zero algebraic numbers a_1, a_2 are explicitly given, as defined in Section 1.10, that is,

$$a_i = (p_{i,0} + p_{i,1}\theta + \cdots + p_{i,d-1}\theta^{d-1})/q_i$$

with given rational integers $p_{i,0}, \ldots, p_{i,d-1}, q_i$ with $\gcd(p_{i,0}, \ldots, p_{i,d-1}, q_i) = 1$ for $i = 1, 2$. Further, for $i = 1, 2$, let Γ_i be a multiplicative subgroup of rank r_i in K^*, and $\Gamma_{i,\infty}$ the torsion subgroup of Γ_i. We assume that for $i = 1, 2$, a system of generators $\xi_{i,1}, \ldots, \xi_{i,r_i}$, that is a basis of $\Gamma_i / \Gamma_{i,\infty}$ is explicitly given. We consider the equation

$$a_1 x_1 + a_2 x_2 = 1 \quad \text{in } (x_1, x_2) \in \Gamma_1 \times \Gamma_2. \tag{5.1.1}$$

To avoid trivialities, we deal only with the case $r_1, r_2 \geq 1$. Then each solution x_1, x_2 of (5.1.1) can be written uniquely in the form

$$x_i = \zeta_i \prod_{j=1}^{r_i} \xi_{i,j}^{b_{i,j}}, \quad i = 1, 2, \tag{5.1.2}$$

where ζ_i is a root of unity in K and the $b_{i,j}$ are rational integers. Hence (5.1.1) takes the form

$$(a_1 \zeta_1) \prod_{j=1}^{r_1} \xi_{1,j}^{b_{1,j}} + (a_2 \zeta_2) \prod_{j=1}^{r_2} \xi_{2,j}^{b_{2,j}} = 1 \tag{5.1.3}$$

with unknown integer exponents $b_{i,j}$.

Let $B := \max_{i,j} |b_{i,j}|$. We are going to derive an upper bound for B. Such a bound could be deduced from the general effective results of Chapter 4. However, as will be seen, in concrete cases it is more profitable to reduce (5.1.3) to Baker's type inequalities; see (5.1.10), (5.1.13) and (5.1.15) below. Then we can apply Baker's method to the left-hand sides of these inequalities to get a bound for B, and then we can use the LLL-algorithm to reduce the bound so obtained.

Let M_K denote the set of places on K. Further, for $i = 1, 2$ let S_i be the support of the group Γ_i, that is the subset of M_K which consists of the infinite places and of those finite places v for which

$$|\alpha|_v \neq 1 \quad \text{for some } \alpha \in \Gamma_i.$$

In view of the assumptions made on Γ_1 and Γ_2, the sets S_1 and S_2 can be effectively determined in the sense defined in Section 1.10. In what follows, we assume some implicit fixed order for the real and complex infinite places and for the finite places in M_K. This gives an order on S_1 and S_2.

We first consider the case when $B = \max_j |b_{1,j}|$. We infer from (5.1.2) that

$$\log |x_1|_v = \sum_{j=1}^{r_1} b_{1,j} \log |\xi_{1,j}|_v$$

for all $v \in M_K$. Let $S_1 = \{v_1, \ldots, v_{s_1}\}$, and choose $k, l \in \{1, \ldots, s_1\}$ such that

$$|\log |x_1|_{v_k}| = \max_{v \in S_1} |\log |x_1|_v|$$

$$|x_1|_{v_l} = \min_{v \in S_1} |x_1|_v.$$

We need to perform our calculations for each possible l. Using the algorithms (XII) and (XIII) mentioned in Section 1.10, one can determine a fundamental system $\{\varepsilon_1, \ldots, \varepsilon_{s_1-1}\}$ of S_1-units, and write $\xi_{1,j} = \zeta_{1,j}\varepsilon_1^{a_{1,j}} \cdots \varepsilon_{s_1-1}^{a_{s_1-1,j}}$ with a root of unity $\zeta_{1,j}$ and with rational integers $a_{i,j} \in \mathbb{Z}$. In view of the multiplicative independence of $\xi_{1,1}, \ldots, \xi_{1,r_1}$ it follows that the rank of the matrix $(a_{i,j})_{\substack{1 \leq i \leq s_1-1 \\ 1 \leq j \leq r_1}}$ is r_1. But the matrix $(\log |\varepsilon_i|_{v_j})_{\substack{1 \leq i \leq s_1-1 \\ 1 \leq j \leq s_1-1}}$ is invertible, hence it is easy to show that the matrix $(\log |\xi_{1,i}|_{v_j})_{\substack{1 \leq i \leq r_1 \\ 1 \leq j \leq s_1-1}}$ is also of rank r_1. Consequently, there is a subset $\{u_1, \ldots, u_{r_1}\}$ of S_1 such that the matrix

$$M = \begin{pmatrix} \log |\xi_{1,1}|_{u_1} & \cdots & \log |\xi_{1,r_1}|_{u_1} \\ \vdots & \vdots & \vdots \\ \log |\xi_{1,1}|_{u_{r_1}} & \cdots & \log |\xi_{1,r_1}|_{u_{r_1}} \end{pmatrix}$$

is invertible. Thus we have

$$\begin{pmatrix} b_{1,1} \\ \vdots \\ b_{1,r_1} \end{pmatrix} = M^{-1} \begin{pmatrix} \log |x_1|_{u_1} \\ \vdots \\ \log |x_1|_{u_{r_1}} \end{pmatrix}.$$

This gives

$$B \leq c_1 |\log |x_1|_{v_k}|, \tag{5.1.4}$$

where c_1 is the row norm of M^{-1}, that is the maximum, taken over all rows, of the sum of the absolute values of the elements of a row of M^{-1}.

Remark 5.1.1 The value of c_1 depends not only on Γ_1 but also on the choice of the generators $\xi_{1,1}, \ldots, \xi_{1,r_1}$ and the matrix M. As will be seen later, the bounds that will be derived for B depend heavily on c_1.

When $B = \max_j |b_{2,j}|$, we obtain an inequality similar to (5.1.4). Thus we can compute a constant $c_1^* \geq c_1$ such that

$$\max_{i,j} |b_{i,j}| \leq c_1^* \max_{v \in S_1 \cup S_2} \{|\log |x_1|_v|, |\log |x_2|_v|\}. \tag{5.1.5}$$

This constant c_1^* will be needed in Section 5.3. For later purpose it is worth keeping c_1 and c_1^* as small as possible.

Remark 5.1.2 In the case of S-unit equations, Hajdu (2009) proved that there is a system of fundamental S-units which is optimal with respect to c_1 and such a system can be constructed.

Choose now c_2 such that $0 < c_2 < 1/c_1(s_1 - 1)$. We shall see that an appropriate choice for c_2 is $0.999/c_1(s_1 - 1)$, provided that s_1 is not too large. We show that

$$|x_1|_{v_l} \leq \exp\{-c_2 B\}. \tag{5.1.6}$$

Assuming the contrary, in view of (5.1.4) there are two possibilities. If $|x_1|_{v_k} \geq \exp\{\frac{1}{c_1} B\}$, then using the Product Formula (1.7.1) we get

$$\exp\left\{\frac{1}{c_1} B\right\} \leq \prod_{\substack{j=1 \\ j \neq k}}^{s_1} |x_1|_{v_j}^{-1} < \exp\{c_2(s_1 - 1)B\},$$

which is impossible because of $1/c_1 > c_2(s_1 - 1)$. On the other hand, if $|x_1|_{v_k} \leq \exp\{-\frac{1}{c_1} B\}$ then

$$\exp\{-c_2 B\} < |x_1|_{v_l} \leq |x_1|_{v_k} \leq \exp\left\{-\frac{1}{c_1} B\right\}$$

which is again a contradiction. This proves (5.1.6).

Set

$$\Lambda := 1 - a_2 x_2 = 1 - a_2' \prod_{j=1}^{r_2} \xi_{2,j}^{b_{2,j}}, \tag{5.1.7}$$

where $a_2' := a_2\zeta_2$ and, in view of (5.1.1), $\Lambda \neq 0$. The following computations must be performed for all roots of unity ζ_2 in K. Let $c_3 := \max_{v \in S_1} |a_1|_v$. Then it follows from (5.1.1) and (5.1.6) that

$$|\Lambda|_{v_l} \leq c_3 \exp\{-c_2 B\}. \tag{5.1.8}$$

We shall give a lower bound for $|\Lambda|_{v_l}$ which, together with (5.1.8), will yield an upper bound for B. We could apply here Theorem 3.2.8 which is valid for each v. We shall, however, get a slightly better bound for B if we use Theorem 3.2.4 or Theorem 3.2.7 according as v_l is infinite or finite.

5.1.1 Infinite places

First consider the case when v_l is infinite. There is an embedding σ of K in $\mathbb{C}$ such that $|\Lambda|_{v_l} = |\sigma(\Lambda)|$ if v_l is real and $|\sigma(\Lambda)|^2$ otherwise. Since $h(\sigma(\alpha)) = h(\alpha)$ for each $\alpha \in K$, in applying Theorem 3.2.4 we omit v_l and σ and we write simply

$$|\Lambda| \leq c_3' \exp\{-c_2' B\}, \tag{5.1.9}$$

in place of (5.1.8), where $c_3' := c_3$, $c_2' := c_2$ if v_l is real and $c_3' := \sqrt{c_3}$, $c_2' := c_2/2$ otherwise.

v_l **is real.** Using the inequality $|\log z| \leq 2|z - 1|$ which holds for $|z - 1| < 0.795$, we deduce from (5.1.3), (5.1.7) and (5.1.9) that putting

$$\Sigma := \log |a_2'| + b_{2,1} \log |\xi_{2,1}| + \cdots + b_{2,r_2} \log |\xi_{2,r_2}|,$$

we have

$$|\Sigma| = |\log |a_2' x_2|| \leq 2|1 - |a_2' x_2|| \leq 2|1 - a_2 x_2| = 2|\Lambda|$$
$$\leq 2c_3 \exp\{-c_2 B\}. \tag{5.1.10}$$

Further, $\Lambda \neq 0$ implies that $\Sigma \neq 0$. Let

$$H := \max\{dh(\alpha_2'), |\log \alpha_2'|, 0.16\}$$

and $c_4 := \max_{1 \leq j \leq r_2}\{dh(|\xi_{2,j}|), |\log |\xi_{2,j}||, 0.16\}$.

We recall that $B = \max_j |b_{1,j}|$. Applying Theorem 3.2.4 to $|\Sigma|$ with B replaced by $c_4 B/H$, we can compute explicit constants c_5 and c_6 such that either

$$B \leq \frac{1}{c_4} H$$

or

$$|\Sigma| > \exp\left\{-c_5 H \log\left(\frac{c_6 B}{H}\right)\right\}. \tag{5.1.11}$$

In the second case (5.1.10) and (5.1.11) imply that

$$\frac{c_6 B}{H} < \frac{c_5 c_6}{c_2} \log\left(\frac{c_6 B}{H}\right) + \frac{c_6 \log(2c_3)}{c_2 H}. \tag{5.1.12}$$

Hence (5.1.12) and (4.4.21) give

$$B \le 2H \max\left\{\frac{c_5}{c_2} \log\left(\frac{c_5 c_6}{c_2}\right), \frac{2e^2}{c_6}\right\} + 2\frac{\log(2c_3)}{c_2} =: c_7.$$

Thus we get

$$B_0(v_l) := \max\left\{\frac{1}{c_4} H, c_7\right\}$$

as an upper bound for B.

v_l **is complex.** Let log denote the principal value of the logarithm. There exists an even rational integer $b_{2,0}$ such that

$$|b_{2,0}| \le 1 + |b_{2,1}| + \cdots + |b_{2,r_2}| \le (r_2 + 1) \max_j |b_{2,j}|$$

and that $|\text{Im}(\Sigma)| \le \pi$, where now

$$\Sigma = \log a_2' + b_{2,0} \log \xi_{2,0} + \cdots + b_{2,r_2} \log \xi_{2,r_2}$$

with $\xi_{2,0} = -1$. It follows from $\Lambda \ne 0$ that $\Sigma \ne 0$. We infer from (5.1.9) that either

$$B < \frac{\log(3c_3')}{c_2'} =: c_8$$

or $|e^\Sigma - 1| = |\Lambda| \le 1/3$. In the latter case $|\Sigma| \le 0.6$, whence $|\Sigma| \le 2|\Lambda|$ and so, by (5.1.9),

$$|\Sigma| \le 2c_3' \exp\{-c_2' B\}. \tag{5.1.13}$$

We apply now Theorem 3.2.4 to $|\Sigma|$. We can compute explicit constants c_9, c_{10} and c_{11} such that either

$$B \le c_9 H$$

or

$$|\Sigma| > \exp\left\{-c_{10} H \log\left(\frac{c_{11} B}{H}\right)\right\}. \tag{5.1.14}$$

Comparing now (5.1.13) and (5.1.14), we deduce that

$$c_{11} \frac{B}{H} < \frac{c_{10}c_{11}}{c_2'} \log\left(\frac{c_{11}B}{H}\right) + \frac{c_{11} \log(2c_3')}{c_2' H},$$

whence, using (4.4.21), we infer that

$$B < 2H \max\left\{\frac{c_{10}}{c_2'} \log\left(\frac{c_{10}c_{11}}{c_2'}\right), \frac{2e^2}{c_{11}}\right\} + \frac{2 \log(2c_3')}{c_2'} =: c_{12}.$$

Thus

$$B_0(v_l) := \max\{c_8, c_9 H, c_{12}\}$$

is an upper bound for B.

5.1.2 Finite places

Suppose now that in (5.1.8) v_l is finite. Let $\mathfrak{p}$ denote the prime ideal of K which corresponds to v_l. Using

$$\log |\Lambda|_{v_l} = -(\mathrm{ord}_{\mathfrak{p}} \Lambda) \log N(\mathfrak{p}),$$

we infer from (5.1.8) that

$$\mathrm{ord}_{\mathfrak{p}} \Lambda \geq (c_2 B - \log c_3)/\log N(\mathfrak{p}). \tag{5.1.15}$$

Recall that the generators $\xi_{2,1}, \ldots, \xi_{2,r_2}$ are not roots of unity. Taking into consideration Proposition 3.2.9, we can apply Theorem 3.2.7 to $\mathrm{ord}_{\mathfrak{p}} \Lambda$ with the choice $h_j' = h(\xi_{2,j})$, $j = 1, \ldots, r_2$, $H = \max(h(a_2'), 1)$, $B_n = 1$ and $\delta = h(\xi_{2,1}) \cdots h(\xi_{2,r_2}) H/B$. We can compute explicit constants c_{13}, c_{14}, c_{15} which depend among others on v_l such that $2h(\xi_{2,1}) \cdots h(\xi_{2,r_2}) \leq c_{13}$ and either

$$B \leq c_{13} H$$

or $B > c_{13} H$, which guarantees that in Theorem 3.2.7, $\delta \leq 1/2$. In the second case Theorem 3.2.7 gives

$$\mathrm{ord}_{\mathfrak{p}} \Lambda < c_{14} H \log\left(c_{15} \frac{B}{H}\right). \tag{5.1.16}$$

Now (5.1.15) and (5.1.16) imply that

$$c_{15} \frac{B}{H} < \frac{c_{15}c_{16}}{c_2} \log\left(c_{15} \frac{B}{H}\right) + c_{15} \frac{\log c_3}{c_2 H},$$

where $c_{16} := c_{14} \log N(\mathfrak{p})$. In view of (4.4.21) this gives

$$B < 2H \max\left\{\frac{c_{16}}{c_2} \log\left(\frac{c_{15}c_{16}}{c_2}\right), 2e^2\right\} + \frac{2 \log c_3}{c_2} =: c_{17},$$

whence we obtain

$$B_0(v_l) := \max\{c_{13}H, c_{17}\}$$

as an upper bound for B.

When $B = \max_j |b_{2,j}|$, we can get in the same way an upper bound for B for each $v \in S_2$. So in all cases we have a bound on $B = \max_{i,j} |b_{i,j}|$.

5.2 Reduction of the bounds

The bounds obtained above for B are too large for practical use, to find all solutions of (5.1.3) in $b_{i,j}$. We now show how to reduce these bounds by means of the LLL-algorithm. For the LLL-algorithm, we refer to Section 5.6. Further details on the applications of the LLL-algorithm to reduce Baker's type bounds can be found in de Weger (1989), Smart (1998) and, in case of infinite places, in Gaál (2002).

We first consider the case when $B = \max_j |b_{1,j}|$. We shall distinguish again two cases according as v_l is infinite or finite. We illustrate the reduction procedure on the inequalities (5.1.10) (v_l infinite and real), (5.1.13) (v_l infinite and complex) and (5.1.15) (v_l finite), reducing the corresponding bounds $B_0(v_l)$ to much smaller ones.

5.2.1 Infinite places

The inequalities (5.1.10) and (5.1.13) are of the form

$$|b_1\vartheta_1 + \cdots + b_t\vartheta_t| < c_{18}\exp\{-c_{19}B\}, \tag{5.2.1}$$

where $\vartheta_1, \ldots, \vartheta_t$ are logarithms of some non-zero algebraic numbers, c_{18}, c_{19} are given explicit positive constants, and $b_1, \ldots, b_t$ are unknown rational integers such that

$$0 < \max(|b_1|, \ldots, |b_t|) \le B \quad \text{and} \quad B \le B_0$$

with some explicit constant B_0.

Remark We could also work with an inhomogeneous version of (5.1.1), when $b_1 = 1$.

We want to considerably reduce the upper bound B_0 in the following way. Consider the inequality (5.2.1), where $\vartheta_1, \ldots, \vartheta_t$ are real or complex numbers. Denote by $\mathcal{L}$ the t-dimensional lattice spanned by the columns of the

$(t + 2) \times t$ matrix

$$\begin{pmatrix} 1 & 0 & \cdots & 0 \\ 0 & 1 & \cdots & 0 \\ \vdots & \vdots & & \vdots \\ 0 & 0 & \cdots & 1 \\ C\mathrm{Re}(\vartheta_1) & C\mathrm{Re}(\vartheta_2) & \cdots & C\mathrm{Re}(\vartheta_t) \\ C\mathrm{Im}(\vartheta_1) & C\mathrm{Im}(\vartheta_2) & \cdots & C\mathrm{Im}(\vartheta_t) \end{pmatrix}$$

where C is a large constant to be specified in numerical cases. The last row can be omitted if $\vartheta_1, \ldots, \vartheta_t$ are all reals. Let $\mathbf{a}_1$ denote the first vector of an LLL-reduced basis of $\mathcal{L}$.

Lemma 5.2.1 *If in* (5.2.1) $\max_i |b_i| \leq B \leq B_0$ *and*

$$\|\mathbf{a}_1\| \geq \sqrt{(t + 1)2^{t-1}} B_0, \tag{5.2.2}$$

then

$$B \leq \frac{\log C + \log c_{18} - \log B_0}{c_{19}}. \tag{5.2.3}$$

This is a slight extension of a result of Gaál and Pohst (2002) where it is assumed that $B = \max(|b_1|, \ldots, |b_t|)$. Our version will be important below, applying Lemma 5.2.1 to (5.1.10) and (5.1.13).

Proof. Following the proof of Lemma 1 in Gaál and Pohst (2002), we denote by $\mathbf{a}_0$ the shortest non-zero vector in $\mathcal{L}$. Then it follows from the inequality (iv) of Proposition 5.6.1 that $\|\mathbf{a}_1\|^2 \leq 2^{t-1}\|\mathbf{a}_0\|^2$. Using (5.2.1) and the assumptions of our lemma, we infer that

$$2^{1-t}\left((t+1)2^{t-1}B_0^2\right) \leq 2^{1-t}\|\mathbf{a}_1\|^2 \leq \|\mathbf{a}_0\|^2 \leq tB_0^2 + C^2 c_{18}^2 \exp\{-2c_{19}B\}.$$

This gives

$$B_0 \leq C \cdot c_{18} \exp\{-c_{19}B\},$$

whence (5.2.3) follows. □

We note that if in (5.2.1) the numbers $\vartheta_1, \ldots, \vartheta_t$ are linearly dependent over $\mathbb{Q}$, then the number of unknowns can be reduced and we can apply Lemma 5.2.1 to a lower dimensional lattice.

We expect our Lemma 5.2.1 to reduce our upper bound B_0 for B, because it is believed that the logarithms of algebraic numbers behave as random complex numbers. To ensure (5.2.2) we have to choose C sufficiently large. A suitable value of C is usually of magnitude B_0^t. Then the bound B_0 is reduced almost to its logarithm. If Lemma 5.2.1 does not reduce our upper bound, a larger C can be chosen and we repeat the procedure.

Keeping the notation of Section 5.1, we apply Lemma 5.2.1, for each infinite v_l, to the corresponding inequality

$$\left|\log|a_2'| + b_{2,1}\log|\xi_{2,1}| + \cdots + b_{2,r_2}\log|\xi_{2,r_2}|\right| \le 2c_3\exp\{-c_2 B\} \quad (5.1.10)$$

or

$$\left|\log a_2' + b_{2,0}\log(-1) + b_{2,1}\log\xi_{2,1} + \cdots + b_{2,r_2}\log\xi_{2,r_2}\right|$$
$$\le 2\sqrt{c_3}\exp\left\{-\frac{c_2}{2(r_2+1)}B'\right\}, \quad (5.1.13)$$

according as v_l is real or not. We recall that in the first case $\max_{1\le j\le r_2}|b_{2,j}| \le B$, while in the second case $\max_{0\le j\le r_2}|b_{2,j}| \le B'$, where $B' = (r_2+1)B$. In the previous section we derived in each case an explicit upper bound $B_0(v_l)$ for B. Lemma 5.2.1 can be applied to (5.1.10) and (5.1.13) repeatedly. In every step we take as B_0 the previous bound, initially the bound $B_0(v_l)$, to get smaller and smaller bounds for B. The reduction is very efficient in the first and second steps. After about $4-5$ steps the procedure stabilizes, that is does not yield an improvement any more. The final reduced bound is usually between 100 and 1000.

5.2.2 Finite places

Now let v_l be finite, and $\mathfrak{p}$ the prime ideal of O_K corresponding to v_l. We recall that

$$a_2' = \zeta_2 a_2 \quad \text{and} \quad x_2 = \zeta_2 \prod_{j=1}^{r_2} \xi_{2,j}^{b_{2,j}}.$$

Consider now (5.1.15) in the form

$$\mathrm{ord}_{\mathfrak{p}}\left(a_2' \prod_{j=1}^{r_2} \xi_{2,j}^{b_{2,j}} - 1\right) \ge c_{20}B - c_{21}, \quad (5.2.4)$$

where $b_{2,1}, \ldots, b_{2,r_2}$ are rational integers with $\max_{1\le j\le r_2}|b_{2,j}| \le B$ and $c_{20} := c_2/\log N(\mathfrak{p})$, $c_{21} := \log c_3/\log N(\mathfrak{p})$. In Section 5.1.2 we derived an upper bound $B_0(v_l)$ for B. We are now going to reduce this bound to a much smaller one by means of the LLL-reduction algorithm. To apply the reduction procedure we have to convert (5.2.4) into a linear form estimate. This will be done by using p-adic logarithms.

We shall proceed in several steps. We follow the arguments of de Weger (1989) and Smart (1998).

Step 1. Firstly we show that a_2x_2 can be written in the form

$$a_2x_2 = \eta_0 \prod_{j=1}^{q_2} \eta_j^{d_j}, \tag{5.2.5}$$

where d_j are rational integers with $|d_j| \leq |b_{2,j}| \leq B$ and η_j, $j \geq 1$, are multiplicatively independent elements of K^* with $\mathrm{ord}_p(\eta_j) = 0$ for $j = 0, \ldots, q_2$, $q_2 = r_2$ if $\mathrm{ord}_p(a_2) = 0$ and $v_l \notin S_2$, and $q_2 = r_2 - 1$ otherwise.

It follows from (5.2.4) and (5.1.1) that $\mathrm{ord}_p(a_1x_1) > 0$ if $B > c_{21}/c_{20}$. Hence (5.1.1) implies that $\mathrm{ord}_p(a_2x_2) = 0$. Put

$$m_0 := \mathrm{ord}_p(a_2') \quad \text{and } m_j := \mathrm{ord}_p(\xi_{2,j}) \quad \text{for } j = 1, \ldots, r_2.$$

Then we infer that

$$m_0 + \sum_{j=1}^{r_2} m_j b_{2,j} = 0. \tag{5.2.6}$$

If $m_j = 0$ for each j with $0 \leq j \leq r_2$, then we may take $\eta_0 = a_2'$, $\eta_j = \xi_{2,j}$ for $j = 1, \ldots, r_2$ and we are done. Next assume that not all m_j are zero. Choose $k > 0$ such that $|m_k|$ is minimal among the non-zero numbers $|m_j|$, $j = 1, \ldots, r_2$. Let

$$\eta_j := \xi_{2,j}^{m_k} \xi_{2,k}^{-m_j} \quad \text{for } j = 1, \ldots, r_2.$$

Then $\eta_k = 1$, the other η_j are multiplicatively independent and $\mathrm{ord}_p(\eta_j) = 0$ for $j \geq 1$, $j \neq k$. Let d_j, t_j be rational integers such that

$$b_{2,j} = m_k d_j + t_j \quad \text{with } 0 \leq t_j < |m_k|, \quad j = 1, \ldots, r_2 \tag{5.2.7}$$

and let

$$m = -\left(m_0 + \sum_{\substack{j=1 \\ j \neq k}}^{r_2} m_j t_j\right), \quad \eta_0 = a_2'\left(\prod_{\substack{j=1 \\ j \neq k}}^{r_2} \xi_{2,j}^{t_j}\right) \xi_{2,k}^{m/m_k}. \tag{5.2.8}$$

It follows from (5.2.6), (5.2.7) and (5.2.8) that

$$m = m_k \sum_{j=1}^{r_2} m_j d_j + m_k t_k, \qquad (5.2.9)$$

whence $m \equiv 0 \pmod{m_k}$ which implies that $\eta_0 \in K^*$. Further, we obtain that

$$\eta_0 \prod_{\substack{j=1 \\ j \neq k}}^{r_2} \eta_j^{d_j} = a_2' \left(\prod_{\substack{j=1 \\ j \neq k}}^{r_2} \xi_{2,j}^{t_j} \right) \xi_{2,k}^{m/m_k} \prod_{\substack{j=1 \\ j \neq k}}^{r_2} \left(\xi_{2,j}^{m_k} \xi_{2,k}^{-m_j} \right)^{d_j} = a_2' \prod_{j=1}^{r_2} \xi_{2,j}^{b_{2,j}} = a_2 x_2,$$

whence (5.2.5) follows. Finally, in view of (5.2.7), (5.2.8) and (5.2.9) we have $\mathrm{ord}_{\mathfrak{p}}(\eta_0) = 0$ which proves our claim.

Remark 5.2.2 It is important to note that $m_0, m_1, \ldots, m_{r_2}$ and hence the numbers $\eta_0, \eta_1, \ldots, \eta_{r_2}$ can be explicitly determined. However, we get different η_0 for each possible choice of $t_1, \ldots, t_{r_2}$ with $0 \leq t_j < |m_k|$, $j = 1, \ldots, r_2$ and $m_0 + \sum_{j=1}^{r_2} m_j t_j \equiv 0 \pmod{m_k}$, and we have to perform our computations for each η_0.

Step 2. We reduce (5.2.4) to an inequality concerning linear form in p-adic logarithms. In view of (5.2.4) and (5.2.5) we have

$$\mathrm{ord}_p(\Lambda) \geq c_{20} B - c_{21}, \qquad (5.2.10)$$

where

$$\Lambda = 1 - \eta_0 \prod_{j=1}^{q_2} \eta_j^{d_j}.$$

Then (5.2.10) implies that $\mathrm{ord}_p(\Lambda) > \frac{1}{p-1}$ whenever

$$B > \frac{1}{c_{20}} \left(c_{21} + \frac{1}{p-1} \right) =: c_{22}.$$

Using p-adic logarithms in $\overline{\mathbb{Q}}_p$, the algebraic closure of $\mathbb{Q}_p$, and applying Lemma 1.11.2 and (1.11.1), we infer that

$$\mathrm{ord}_p \Sigma = \mathrm{ord}_p(\Lambda) \geq c_{20} B - c_{21}, \qquad (5.2.11)$$

where

$$\Sigma = \log_p \eta_0 + d_1 \log_p \eta_1 + \cdots + d_{q_2} \log_p \eta_{q_2}.$$

We note that here $\mathrm{ord}_p(\eta_j) = 0$ for $j = 0, \ldots, q_2$, hence the p-adic logarithms of $\eta_0, \ldots, \eta_{q_2}$ are well-defined (see Section 1.11). Further, $\log_p \eta_0, \ldots, \log_p \eta_{q_2}$ are elements of $K_{\mathfrak{p}}$, the $\mathfrak{p}$-adic completion of K (see Proposition 1.11.3), and they can be approximated with any desired accuracy; see de Weger (1989) and Smart (1998), chapter 5.

Step 3. We consider now (5.2.11) in the following more general form:

$$\infty > \text{ord}_p(b_1 \vartheta_1 + \cdots + b_t \vartheta_t) \geq c_{23} B - c_{24}, \qquad (5.2.12)$$

where $\vartheta_1, \ldots, \vartheta_t$ are given elements of K_p, $c_{23} > 0$, c_{24} are given explicit constants, $b_1, \ldots, b_t \in \mathbb{Z}$ with $|b_j| \leq B$ for $j = 1, \ldots, t$ and $B \leq B_0$ for some explicit constant B_0. This is the p-adic analogue of the inequality (5.2.1).

We first show that (5.2.12) can be reduced to the case when, in (5.2.12), all ϑ_i are integers in $\mathbb{Q}_p$. The field $\mathbb{Q}_p(\vartheta_1, \ldots, \vartheta_t)$ is a finite extension of degree m, say, of $\mathbb{Q}_p$. Using standard arguments, we can determine an element δ which is integral over $\mathbb{Q}_p$, and p-adic numbers ϑ_{ij} ($i = 1, \ldots, t$, $j = 0, \ldots, m-1$), such that $\mathbb{Q}_p(\vartheta_1, \ldots, \vartheta_t) = \mathbb{Q}_p(\delta)$, and

$$\vartheta_i = \sum_{j=0}^{m-1} \vartheta_{ij} \delta^j, \quad i = 1, \ldots, t.$$

Putting

$$\Phi(\mathbf{b}) := \sum_{i=1}^{t} b_i \vartheta_i \text{ and } \Phi_j(\mathbf{b}) = \sum_{i=1}^{t} b_i \vartheta_{ij} \quad (j = 0, \ldots, m-1),$$

we have

$$\Phi(\mathbf{b}) = \sum_{j=0}^{m-1} \Phi_j(\mathbf{b}) \delta^j. \qquad (5.2.13)$$

We claim that

$$\text{ord}_p(\Phi_j(\mathbf{b})) \geq c_{23} B - c_{24} - \tfrac{1}{2} \text{ord}_p(D(\delta)) \qquad (5.2.14)$$

for $j = 0, \ldots, m-1$, where $D(\delta)$ denotes the discriminant of δ over $\mathbb{Q}_p$.

To prove (5.2.14), consider the conjugates $\delta^{(1)} = \delta, \ldots, \delta^{(m)}$ of δ over $\mathbb{Q}_p$. Taking the corresponding conjugates in (5.2.13) we get

$$\Phi^{(i)}(\mathbf{b}) = \sum_{j=0}^{m-1} \Phi_j(\mathbf{b})(\delta^{(i)})^j, \quad i = 1, \ldots, m. \qquad (5.2.15)$$

Put $\Delta(\delta) := \prod_{1 \leq i < j \leq m}(\delta^{(i)} - \delta^{(j)})$. It follows from (5.2.15) that there are p-adic algebraic numbers κ_{ij} such that $\text{ord}_p(\kappa_{ij}) \geq 0$ and

$$\Delta(\delta)\Phi_j(\mathbf{b}) = \sum_{i=1}^{m} \kappa_{ij} \Phi^{(i)}(\mathbf{b}) \quad \text{for } j = 0, \ldots, m-1. \qquad (5.2.16)$$

Since $\mathrm{ord}_p(\Phi^{(i)}(\mathbf{b}))$ does not depend on i, we infer from (5.2.12), (5.2.16) and $2\mathrm{ord}_p\Delta(\delta) = \mathrm{ord}_p(D(\delta))$ that (5.2.14) holds.

We could consider here the linear form estimates (5.2.14) simultaneously, as is done in Smart (1995, 1998). This would give a better reduced bound than using only one linear form. Nevertheless, we work only with one form, say $\Phi_{j_0}(\mathbf{b}) = \sum_{i=1}^{t} b_i \vartheta_{ij_0}$ such that $\Phi_{j_0}(\mathbf{b}) \neq 0$. On one hand, this case is simpler to apply. On the other hand it will enable us to apply the LLL-reduction algorithm similarly to that used in the complex case.

For simplicity we omit the index j_0. Then, in view of (5.2.14), we arrive at (5.2.12) under the assumption that now $\vartheta_1, \ldots, \vartheta_t$ are elements of $\mathbb{Q}_p$ and c_{24} is replaced by $c_{25} := c_{24} + \frac{1}{2}\mathrm{ord}_p(D(\delta))$. We may assume without loss of generality that $\min_i \mathrm{ord}_p(\vartheta_i) = \mathrm{ord}_p(\vartheta_t) =: c_{26}$. Then

$$\vartheta_i' := -\vartheta_i/\vartheta_t \in \mathbb{Z}_p \quad \text{for } i = 1, \ldots, t-1$$

and (5.2.12) implies that

$$\infty > \mathrm{ord}_p(-b_1\vartheta_1' - \cdots - b_{t-1}\vartheta_{t-1}' + b_t) \geq c_{23}B - c_{27}, \qquad (5.2.17)$$

where $c_{27} := c_{25} + c_{26}$.

Step 4. We apply now the LLL-reduction algorithm to (5.2.17) to reduce the bound B_0.

For any $\vartheta \in \mathbb{Z}_p$ and for any positive integer u denote by $\vartheta^{\{u\}}$ the unique rational integer such that

$$\vartheta \equiv \vartheta^{\{u\}} \pmod{p^u} \quad \text{and} \quad 0 \leq \vartheta^{\{u\}} < p^u.$$

Let $\mathcal{L}_u$ denote the t-dimensional lattice generated by the columns of the matrix

$$A = \begin{pmatrix} 1 & 0 & \cdots & 0 & 0 \\ 0 & 1 & \cdots & 0 & 0 \\ \vdots & \vdots & \ddots & & \vdots \\ 0 & 0 & \cdots & 1 & 0 \\ \vartheta_1'^{\{u\}} & \vartheta_2'^{\{u\}} & \cdots & \vartheta_{t-1}'^{\{u\}} & p^u \end{pmatrix}.$$

For any $\mathbf{b} = (b_1, \ldots, b_t) \in \mathbb{Z}^t$, write

$$\Sigma(\mathbf{b}) = -b_1\vartheta_1' - \cdots - b_{t-1}\vartheta_{t-1}' + b_t.$$

We claim that

$$\mathcal{L}_u = \left\{ \mathbf{b}^{\mathrm{T}} : \mathbf{b} = (b_1, \ldots, b_t) \in \mathbb{Z}^t \text{ and } \mathrm{ord}_p\Sigma(\mathbf{b}) \geq u \right\}.$$

Indeed, if $\mathbf{b}^{\mathrm{T}} \in \mathcal{L}_u$ then $\mathbf{b}^{\mathrm{T}} = A\mathbf{x}^{\mathrm{T}}$ for some $\mathbf{x} = (x_1, \ldots, x_t) \in \mathbb{Z}^t$. This implies that $b_i = x_i$ for $i = 1, \ldots, t - 1$ and

$$b_t = \sum_{i=1}^{t-1} x_i \vartheta_i'^{\{u\}} + x_t p^u \equiv \sum_{i=1}^{t-1} b_i \vartheta_i' \pmod{p^u},$$

whence $\operatorname{ord}_p \Sigma(\mathbf{b}) \geq u$ follows. Conversely, if $\operatorname{ord}_p(\Sigma(\mathbf{b})) \geq u$ for some $\mathbf{b} \in \mathbb{Z}^t$ then there exists $\mathbf{x} \in \mathbb{Z}^t$ such that $\mathbf{b}^{\mathrm{T}} = A\mathbf{x}^{\mathrm{T}}$, that is $\mathbf{b}^{\mathrm{T}} \in \mathcal{L}_u$ which proves our claim.

We recall that in (5.2.17) we have a bound B_0 for B. Choose an integer constant u such that $p^u \geq B_0^{t+1}$. We may expect that u is large enough to bound B using the following lemma. If it is not sufficiently large then we make u a little larger and apply the lemma again.

Let $\mathbf{a}_1$ denote the first vector of an LLL-reduced basis of $\mathcal{L}_u$. We prove the following analogue of Lemma 5.2.1.

Lemma 5.2.3 *If in* (5.2.17) $\max_i |b_i| \leq B \leq B_0$ *and*

$$\|\mathbf{a}_1\| > \sqrt{t2^{t-1}} B_0, \tag{5.2.18}$$

then

$$B \leq \frac{1}{c_{23}}(u - 1 + c_{27}). \tag{5.2.19}$$

Proof. Using (iv) of Proposition 5.6.1 and (5.2.18) we infer that

$$\|\mathbf{a}_0\|^2 \geq 2^{1-t}\|\mathbf{a}_1\|^2 > tB_0^2,$$

where $\mathbf{a}_0$ denotes the shortest non-zero vector in the lattice $\mathcal{L}_u$. Hence we infer that

$$\|\mathbf{a}_0\| > \sqrt{t}B_0.$$

This means that for $\mathbf{b} = (b_1 \ldots, b_t) \in \mathbb{Z}^t$ with $\max_i |b_i| \leq B \leq B_0$ which satisfies (5.2.17), $(b_1, \ldots, b_t)^{\mathrm{T}}$ cannot be a lattice point in $\mathcal{L}_u$. Hence, for such a $\mathbf{b}$, $\operatorname{ord}_p \Sigma(\mathbf{b}) \leq u - 1$, which, together with (5.2.17), gives (5.2.19). $\qquad\square$

Similarly to Lemma 5.2.1, Lemma 5.2.3 also reduces the bound B_0 to almost its logarithm. Of course, Lemma 5.2.3 can also be applied repeatedly until we get a better bound than the previous one.

Finally we note that performing the above procedure in (5.2.11) for all $v \in S_1$ (when $B = \max_j |b_{1j}|$), then proceeding similarly for all $v \in S_2$ (when $B = \max_j |b_{2j}|$), and denoting by B_R the maximum of the reduced bounds obtained, we get

$$B \leq B_R$$

in our equation (5.1.3).

5.3 Enumeration of the "small" solutions

In the first section we gave an upper bound for the solutions $b_{i,j}$ of the equation (5.1.3). Further, in the previous section we considerably reduced this bound to a new bound B_R. A crucial problem in the resolution of equation (5.1.3) is now to check the remaining $(2B_R + 1)^{r_1+r_2}$ cases for the exponents, where r_1, r_2 denote the ranks of Γ_1 and Γ_2, respectively. Even if the bound B_R is moderate (say < 100) the direct enumeration is almost hopeless whenever the number $r_1 + r_2$ of exponents is greater than eight.

We now present an efficient algorithm for finding all solutions of (5.1.3) under the reduced bound B_R. This algorithm has been established by Wildanger (1997, 2000) for the case when both Γ_1 and Γ_2 are the unit group of O_K, and by Smart (1999) in the general case. We follow the presentation of Smart (1999) with certain simplifications.

For any real number $H > 1$ and for any finite set S of places of K containing the infinite places, we define the set

$$\langle\!\langle H, S \rangle\!\rangle := \left\{ \alpha \in K : \frac{1}{H} \leq |\alpha|_v \leq H \text{ for all } v \in S \right\}.$$

Denote by $\mathscr{S}$ the set of solutions $(x_1, x_2) \in \Gamma_1 \times \Gamma_2$ of (5.1.1). Writing x_1, x_2 in the form (5.1.2), we consider (5.1.1) as the exponential equation (5.1.3) in integers $b_{i,j}$ with $B = \max_{i,j} |b_{i,j}|$. For a positive integer B_k, denote by $\mathscr{S}_{B_k}$ the set of solutions of (5.1.1) such that the absolute values of the corresponding exponents $b_{i,j}$ is at most B_k. Then $\mathscr{S} = \mathscr{S}_{B_R}$. We define

$$\mathscr{S}_{B_k}(H) := \{(x_1, x_2) \in \mathscr{S}_{B_k} : x_1 \in \langle\!\langle H, S_1 \rangle\!\rangle\}.$$

We first show that for

$$H_0 := \max_{v \in S_1} \exp\left(B_R \sum_{j=1}^{r_1} \left|\log |\xi_{1,j}|_v\right| \right),$$

we have

$$\mathscr{S} = \mathscr{S}_{B_R}(H_0). \tag{5.3.1}$$

Indeed, using (5.1.2), for every solution (x_1, x_2) of (5.1.1) and for each $v \in S_1$ we infer that

$$|\log |x_1|_v| = \left| \sum_{j=1}^{r_1} b_{1,j} \log |\xi_{1,j}|_v \right| \le B \sum_{j=1}^{r_1} |\log |\xi_{1,j}|_v|$$

$$\le \max_{v \in S_1} \left(B_R \sum_{j=1}^{r_1} |\log |\xi_{1,j}|_v| \right) = \log H_0.$$

This means that

$$\frac{1}{H_0} \le |x_1|_v \le H_0,$$

whence (5.3.1) follows.

In what follows, we shall proceed in several steps.

Step 1. We first *decompose* the solution set $\mathscr{S}$ into appropriate subsets.
   Set

$$t_i := \max_{v \in S_1 \cup S_2} \max \left(|a_i|_v, |a_i^{-1}|_v \right) \quad \text{for } i = 1, 2,$$

and

$$t_3 := \max_{v \in S_1 \cup S_2} \min \left(|a_2|_v, |a_2^{-1}|_v \right).$$

For $k \ge 0$, let B_k be a positive number with the choice $B_0 = B_R$, and let H_k, H_{k+1} be real numbers such that

$$\max \left(t_1, t_2, t_3, \frac{t_3 - 1}{t_1} \right) < H_{k+1} < H_k.$$

Note that $H_{k+1} > 1$. We intend to find a positive number $B_{k+1} < B_k$ and then decompose the set $\mathscr{S}_{B_k}(H_k)$ into the union of $\mathscr{S}_{B_{k+1}}(H_{k+1})$ and a union of some subsets, each containing a few elements which can be easily determined. If, starting with $k = 0$, that is with (5.3.1), this process can be repeated, finally it will remain to enumerate a set of the form $\mathscr{S}_{B_{k_0}}(H_{k_0})$ for some small values of B_{k_0} and H_{k_0}.

We define the sets $\mathcal{T}_{j,v} = \mathcal{T}_{j,v}(B_k, H_k, H_{k+1})$, $j = 1, 2, 3, 4$, in the following way:

$$\mathcal{T}_{1,v} := \left\{ (x_1, x_2) \in \mathscr{S}_{B_k}(H_k) : |a_1 x_1 - 1|_v < \frac{1}{1 + t_1 H_{k+1}} \right\} (v \in S_2),$$

$$\mathcal{T}_{2,v} := \left\{ (x_1, x_2) \in \mathscr{S}_{B_k}(H_k) : \left| \frac{1}{a_1 x_1} - 1 \right|_v < \frac{1}{1 + t_1 H_{k+1}} \right\} (v \in S_1 \cup S_2),$$

$$\mathcal{T}_{3,v} := \left\{ (x_1, x_2) \in \mathscr{S}_{B_k}(H_k) : |a_2 x_2 - 1|_v < \frac{t_1}{H_{k+1}}, \right.$$
$$\left. a_2 x_2 \in \langle\!\langle 1 + t_1 H_k, S_2 \rangle\!\rangle \right\} (v \in S_1),$$

$$\mathcal{T}_{4,v} := \left\{ (x_1, x_2) \in \mathscr{S}_{B_k}(H_k) : \left| -\frac{a_2 x_2}{a_1 x_1} - 1 \right|_v < \frac{t_1}{H_{k+1}}, \right.$$
$$\left. \frac{a_2 x_2}{a_1 x_1} \in \langle\!\langle 1 + t_1 H_k, S_1 \cup S_2 \rangle\!\rangle \right\} (v \in S_1).$$

Further, let

$$\mathcal{T}_1(B_k, H_k, H_{k+1}) := \bigcup_{v \in S_2} \mathcal{T}_{1,v}(B_k, H_k, H_{k+1}),$$

$$\mathcal{T}_2(B_k, H_k, H_{k+1}) := \bigcup_{v \in S_1 \cup S_2} \mathcal{T}_{2,v}(B_k, H_k, H_{k+1}),$$

$$\mathcal{T}_3(B_k, H_k, H_{k+1}) := \bigcup_{v \in S_1} \mathcal{T}_{3,v}(B_k, H_k, H_{k+1}),$$

$$\mathcal{T}_4(B_k, H_k, H_{k+1}) := \bigcup_{v \in S_1} \mathcal{T}_{4,v}(B_k, H_k, H_{k+1}).$$

We recall that c_1^* denotes the constant occurring in (5.1.5).

Lemma 5.3.1 *Let*

$$c_{28} := \max \left(\log \left(\frac{t_1 H_{k+1} + 1}{t_3} \right), \log(H_{k+1}) \right)$$

and $B_{k+1} := c_1^* c_{28}$. *Then*

$$\mathscr{S}_{B_k}(H_k) = \mathscr{S}_{B_{k+1}}(H_{k+1}) \bigcup_{j=1}^{4} \mathcal{T}_j(B_k, H_k, H_{k+1}). \tag{5.3.2}$$

Proof. Assume that $(x_1, x_2) \in \mathscr{S}_{B_k}(H_k)$ and that $(x_1, x_2) \notin \mathscr{S}_{B_k}(H_{k+1})$. Then there is a $v \in S_1$ such that either $|x_1|_v < 1/H_{k+1}$ or $|x_1|_v > H_{k+1}$. In the first case we infer that

$$|a_2 x_2 - 1|_v = |a_1 x_1|_v < \frac{t_1}{H_{k+1}}.$$

If $(x_1, x_2) \notin \mathcal{T}_1(B_k, H_k, H_{k+1})$ then, for each $u \in S_2$,

$$|a_2 x_2|_u = |a_1 x_1 - 1|_u \geq \frac{1}{1 + t_1 H_{k+1}}.$$

Further, we have

$$|a_2 x_2|_u = |a_1 x_1 - 1|_u \leq 1 + |a_1 x_1|_u \leq \begin{cases} 1 + t_1 H_k \text{ if } u \in S_1, \\ 1 + t_1 \text{ if } u \in S_2 \setminus S_1. \end{cases}$$

Consequently, for each $u \in S_2$ we have

$$\begin{aligned} |\log |a_2 x_2|_u| &\leq \max\{\log(1 + t_1), \log(1 + t_1 H_{k+1}), \log(1 + t_1 H_k)\} \\ &= \log(1 + t_1 H_k). \end{aligned}$$

This implies that if $|x_1|_v < 1/H_{k+1}$ for some $v \in S_1$ and $(x_1, x_2) \notin \mathcal{T}_1(B_k, H_k, H_{k+1})$, then $(x_1, x_2) \in \mathcal{T}_3(B_k, H_k, H_{k+1})$.

Next consider the case when $|x_1|_v > H_{k+1}$. Then it follows that

$$\left| -\frac{a_2 x_2}{a_1 x_1} - 1 \right|_v = \left| \frac{1}{a_1 x_1} \right|_v < \frac{t_1}{H_{k+1}}.$$

Assume that $(x_1, x_2) \notin \mathcal{T}_2(B_k, H_k, H_{k+1})$. Then for each $u \in S_1 \cup S_2$ we have

$$\left| \frac{a_2 x_2}{a_1 x_1} \right|_u = \left| \frac{1}{a_1 x_1} - 1 \right|_u \geq \frac{1}{1 + t_1 H_{k+1}}.$$

Further, it follows that

$$\left| \frac{a_2 x_2}{a_1 x_1} \right|_u = \left| \frac{1}{a_1 x_1} - 1 \right|_u \leq \left| \frac{1}{a_1 x_1} \right|_u + 1 \leq \begin{cases} 1 + t_1 H_k \text{ if } u \in S_1, \\ 1 + t_1 \text{ if } u \in S_2 \setminus S_1. \end{cases}$$

This implies that $(x_1, x_2) \in \mathcal{T}_4(B_k, H_k, H_{k+1})$. Hence

$$\mathcal{S}_{B_k}(H_k) = \mathcal{S}_{B_k}(H_{k+1}) \bigcup \left(\bigcup_{j=1}^{4} \mathcal{T}_j(B_k, H_k, H_{k+1}) \right). \tag{5.3.3}$$

Now consider the case when $(x_1, x_2) \in \mathcal{S}_{B_k}(H_{k+1})$. Then for each $v \in S_1 \cup S_2$ we have $|\log |x_1|_v| \leq \log H_{k+1}$ and

$$|x_2|_v = \left| \frac{a_1 x_1 - 1}{a_2} \right|_v \leq \frac{|a_1 x_1|_v + 1}{|a_2|_v} \leq \frac{t_1 H_{k+1} + 1}{t_3}.$$

If $(x_1, x_2) \notin \mathcal{T}_1(B_k, H_k, H_{k+1})$, then for each $v \in S_2$

$$|x_2|_v = \left| \frac{a_1 x_1 - 1}{a_2} \right|_v \geq \frac{t_3}{t_1 H_{k+1} + 1}.$$

Clearly, this inequality holds for $v \in S_1 \setminus S_2$ as well. Thus, we deduce that if $(x_1, x_2) \in \mathscr{S}_{B_k}(H_{k+1}) \setminus \mathcal{T}_1(B_k, H_k, H_{k+1})$, then for each $v \in S_1 \cup S_2$

$$|\log |x_1|_v| \le c_{28}, \quad |\log |x_2|_v| \le c_{28}.$$

However, in view of (5.1.5) we must have

$$B \le c_1^* c_{28} = B_{k+1}.$$

Thus

$$\mathscr{S}_{B_k}(H_{k+1}) = \mathscr{S}_{B_{k+1}}(H_{k+1}) \bigcup \mathcal{T}_1(B_k, H_k, H_{k+1})$$

which, together with (5.3.3), completes the proof. $\qquad\square$

Remark 5.3.2 Applying Lemma 5.3.1 we need to choose a value H_{k+1} such that the algorithm prescribed below allows us to deduce that the sets $\mathcal{T}_{j,v}(B_k, H_k, H_{k+1})$ are easy to enumerate for each j and v under consideration. Wildanger (2000) provides a heuristic method to find the best value for H_{k+1} in the case when $\Gamma_1 = \Gamma_2$ is the unit group of O_K. In the general case the analysis appears similar, and the choice of Wildanger for H_{k+1} seems to be sufficient.

Step 2. We are going to show how to enumerate, for each j and v in question, all the possible elements in $\mathcal{T}_{j,v}(B_k, H_k, H_{k+1})$.

In all cases our problem can be reformulated as trying to enumerate all non-trivial solutions of the following problem. Let $\alpha, \xi_1, \ldots, \xi_r$ be explicitly given elements of K^* such that $\xi_1, \ldots, \xi_r$ are multiplicatively independent. Further, let $S = \{v_1, \ldots, v_s\}$ be the support of the multiplicative group generated by $\xi_1, \ldots, \xi_r$. Set

$$x = \zeta^{b_0} \prod_{j=1}^{r} \xi_j^{b_j}, \tag{5.3.4}$$

where ζ is a primitive root of unity in K, and $b_0, \ldots, b_r$ are rational integers with $0 \le b_0 < w$, where w denotes the number of roots of unity in K. We have, for some $H > 1$ and for all $v \in S$,

$$\frac{1}{H} \le |\alpha x|_v \le H. \tag{5.3.5}$$

We wish to determine all x for which (5.3.4), (5.3.5) and, for some $v \in S$ and some given $\varepsilon \in (0, 1)$

$$|\alpha x - 1|_v < \epsilon \tag{5.3.6}$$

hold.

We shall distinguish two cases according as v is infinite or finite in (5.3.6).

v **is infinite.** Let $v = v_i$. For any $z \in \mathbb{C}$, $|z - 1| < \varepsilon$ implies that $|\log |z|| < \log(1/(1 - \epsilon))$. Hence we deduce from (5.3.6) that

$$
|\log |\alpha x|_v| \le \epsilon' := \begin{cases} \log \frac{1}{1-\epsilon}, & v \text{ real,} \\ \frac{1}{2}\log \frac{1}{1-\sqrt{\epsilon}}, & v \text{ complex.} \end{cases}
$$

Further, we have, with obvious notation

$$
\left| \mathrm{Arg}\left((\alpha x)^{(i)} \right) \right| \le \arccos\sqrt{1 - \epsilon} =: \varepsilon''.
$$

For simplicity, for any $\beta \in K^*$ we denote by $\beta^{(i)}$ the conjugate of β over $\mathbb{Q}$ corresponding to v_i.

Consider the sublattice of $\mathbb{R}^{s+1}$ which is generated by the columns of the matrix M obtained from the $(s + 1) \times (r + 1)$ matrix

$$
\frac{1}{\log H} \begin{pmatrix} \log |\xi_1|_{v_1} & \cdots & \log |\xi_r|_{v_1} & 0 \\ \vdots & & \vdots & \vdots \\ \log |\xi_1|_{v_s} & \cdots & \log |\xi_r|_{v_s} & 0 \\ 0 & \cdots & 0 & 0 \end{pmatrix}
$$

by replacing the i-th row by

$$
\frac{1}{\epsilon'}\left(\log |\xi_1|_{v_i}, \ldots, \log |\xi_r|_{v_i}, 0 \right)
$$

and the last row by

$$
\frac{1}{\epsilon''}\left(\mathrm{Arg}\left(\xi_1^{(i)} \right), \ldots, \mathrm{Arg}\left(\xi_r^{(i)} \right), \mathrm{Arg}\left(\zeta^{(i)} \right) \right).
$$

We expect the i-th and last row of M to have much larger entries than the other rows.

Let $\mathbf{x} = M\mathbf{b}$, where $\mathbf{b} = (b_1, \ldots, b_r, b_0)^{\mathrm{T}}$, and consider the vector $\mathbf{y}$ obtained from the vector

$$
\frac{1}{\log H}\left(-\log |\alpha|_{v_1}, \ldots, -\log |\alpha|_{v_r}, 0 \right)^{\mathrm{T}} \in \mathbb{R}^{r+1}
$$

by replacing the i-th coordinate by $-\log |\alpha|_{v_i}/\epsilon'$ for $i = 1, \ldots, s$, and the last coordinate by $\mathrm{Arg}((1/\alpha)^{(i)})/\epsilon''$. Then we have

$$
\|\mathbf{x} - \mathbf{y}\|^2 = \frac{\log^2 |\alpha x|_{v_i}}{\epsilon'^2} + \frac{\mathrm{Arg}^2\left((\alpha x)^{(i)} \right)}{\epsilon''^2} + \sum_{\substack{v \in S \\ v \ne v_i}} \frac{\log^2 |\alpha x|_v}{\log^2 H} \le s + 1. \quad (5.3.7)
$$

Hence we have proved that for any $(b_0, b_1, \ldots, b_r) \in \mathbb{Z}^{r+1}$ which corresponds by (5.3.4) to a solution x of (5.3.5) and (5.3.6), inequality (5.3.7) holds. The inequality (5.3.7) defines an *ellipsoid* with center **y**. The lattice points contained in this ellipsoid can be enumerated by means of the algorithm of Fincke and Pohst (1985). The enumeration is usually very fast. However, it is essential that the improved version (see Fincke and Pohst (1985)) of the algorithm must be used, involving LLL reduction.

v **is finite.** Let again $\mathfrak{p}$ be the prime ideal of O_K that corresponds to v. We proceed as above with the following modifications. Then (5.3.6) implies that $\mathrm{ord}_{\mathfrak{p}}(\alpha x) = 0$. As in Step 1 of Section 5.2.2, we can reduce (5.3.4) to a similar problem where $\alpha x = \eta_0 \prod_{j=1}^{q} \eta_j^{d_j}$ with multiplicatively independent $\eta_1, \ldots, \eta_q$ such that $\mathrm{ord}_{\mathfrak{p}}(\eta_j) = 0$ for $j = 0, \ldots, q$. We recall that η_0 may assume finitely many values, each of which can be determined. We have to perform our computations for every possible value of η_0. Suppose that $\mathfrak{p}$ has residue degree f and that it lies above the rational prime p. Choose a positive integer n such that

$$\epsilon \le p^{-fn}.$$

Then in view of (5.3.6) we have

$$\eta_0 \prod_{j=1}^{q} \eta_j^{d_j} \equiv 1 \pmod{\mathfrak{p}^n}. \tag{5.3.8}$$

First assume that $\eta_0, \ldots, \eta_q$ are multiplicatively independent. Let G denote the subgroup of K^* generated by $\eta_0, \ldots, \eta_q$. Since $\mathrm{ord}_{\mathfrak{p}}(\eta_j) = 0$ for all j, we can consider the image of G in $(O_K/\mathfrak{p}^n)^*$ under reduction mod $\mathfrak{p}^n$. The order of $\eta_j \pmod{\mathfrak{p}^n}$ can be computed very quickly, as it is a divisor of the order of $(O_K/\mathfrak{p}^n)^*$ which is $p^{(n-1)f}(p^f - 1)$. All $\mathbf{d} = (d_0, \ldots, d_q) \in \mathbb{Z}^{q+1}$ for which $\eta_0^{d_0} \cdots \eta_q^{d_q} \equiv 1 \pmod{\mathfrak{p}^n}$ form a full lattice in $\mathbb{Z}^{q+1}$, a basis of which can be computed by using the algorithm MINIMIZE; see Teske (1998). This algorithm computes such a basis in the form

$$\mathbf{d}_j = (d_{0j}, \ldots, d_{jj}, 0, \ldots, 0) \quad \text{for } j = 0, \ldots, q,$$

where $d_{ij} \in \mathbb{Z}$ and d_{jj} is the smallest positive integer for which $\eta_j^{d_{jj}}$ belongs to the subgroup of $G/\mathfrak{p}^n$ generated by $\{\eta_0, \ldots, \eta_{j-1}\}$. Then putting

$$\eta'_j := \eta_0^{d_{0j}} \cdots \eta_j^{d_{jj}},$$

we can write

$$\alpha x = \prod_{j=0}^{q} {\eta'_j}^{n_j}$$

with suitable integers $n_0, \ldots, n_q$. Let $S' = \{u_1, \ldots, u_{s'}\}$ denote the support of the group generated by $\eta'_0, \ldots, \eta'_q$. Obviously $S' \subset S$. We can now proceed in a similar way as in the case of infinite places. Consider the sublattice of $\mathbb{R}^{s'}$ generated by the vectors

$$\eta'_i = \frac{1}{\log H} \left(\log |\eta'_i|_{u_1}, \ldots, \log |\eta'_i|_{u_{s'}} \right)$$

for $i = 0, \ldots, q$. We have

$$\| n_0 \eta'_0 + \cdots + n_q \eta'_q \|^2 = \sum_{u \in S'} \frac{\log^2 |\alpha x|_u}{\log^2 H} \leq s' + 1.$$

Then, as above, we can determine all $(n_0, \ldots, n_q)$ and hence all x under consideration using the Fincke–Pohst algorithm.

Next consider the case when $\eta_0, \ldots, \eta_q$ are multiplicatively dependent, and let h be a positive integer for which η_0^h is contained in the multiplicative group generated by $\{\eta_1, \ldots, \eta_q\}$. Then it follows from (5.3.8) that

$$(\alpha x)^h = \prod_{j=1}^{q} \eta_j^{d'_j} \equiv 1 \pmod{\mathfrak{p}^n}$$

with some rational integers $d'_1, \ldots, d'_q$. Then we infer as above that

$$(\alpha x)^h = \prod_{j=1}^{q} \eta''^{n'_j}_j$$

with some rational integers $n'_1, \ldots, n'_q$. Thus, following the above procedure, we can determine all $(n'_1, \ldots, n'_q)$ and hence, up to a factor of a root of unity in K, all x can also be found. The factor in question can be easily determined from equation (5.1.1).

Step 3. By means of the above process we can determine all elements of the set $\mathcal{T}_{j,v}(B_k, H_k, H_{k+1})$ for $j = 1, 2, 3, 4$, and for all v in question.

Finally, at the end of the repeated procedure described in Steps 1 and 2 of this section we arrive at a set of the form $\mathscr{S}_{B_{k_0}}(H_{k_0})$ for some small values of B_{k_0} and H_{k_0}. Consider the lattice in $\mathbb{R}^{s_1}$ generated by the vectors

$$\xi_i := \frac{1}{\log H_{k_0}} \left(\log |\xi_{1,i}|_{v_1}, \ldots, \log |\xi_{1,i}|_{v_{s_1}} \right)^{\mathrm{T}}$$

for $i = 1, \ldots, r_1$, where $\{v_1, \ldots, v_{s_1}\} = S_1$. Then the set $\mathscr{S}_{B_{k_0}}(H_{k_0})$ is contained in the ellipsoid

$$\| b_{1,1} \xi_1 + \cdots + b_{1,r_1} \xi_{r_1} \|^2 \leq s_1$$

whose points can be found by using again the Fincke–Pohst algorithm.

5.4 Examples

We now briefly illustrate the use of the above presented algorithm on two concrete S-unit equations. In these examples $\Gamma_1 = \Gamma_2$ holds, and this group is the unit group of the ring of integers, respectively an S-unit group of the underlying number field. In our book on discriminant equations, we will meet examples where Γ_1, Γ_2 are distinct.

We note that in the examples below the fundamental units were computed by the KANT package; see Daberkow et al. (1997). For an alternative package, we mention MAGMA; see Bosma et al. (1997).

Example 5.4.1 (Smart (1997, 1999)). Let K_{16} denote the 16th cyclotomic field generated by ζ, where ζ is a 16th primitive root of unity. There is a prime ideal $\mathfrak{p}$ in K_{16} such that $\mathfrak{p}^8 = (2)$. This ideal is principal and we can take $\pi = 1 - \zeta$ as a generator for $\mathfrak{p}$. Let S denote the set of places of K_{16} which consists of all infinite places and the single finite place $v = \mathfrak{p}$. Consider the S-unit equation

$$x_1 + x_2 = 1 \quad \text{in } S\text{-units } x_1, x_2 \text{ of } K_{16}. \qquad (5.4.1)$$

The degree and unit rank of K_{16} are 8 and 3, respectively. One can take

$$\varepsilon_1 = \zeta^2 + \zeta^4 + \zeta^6, \quad \varepsilon_2 = -\left(\zeta^2 + \zeta^3 + \zeta^4\right), \quad \varepsilon_3 = 1 + \zeta^3 - \zeta^5$$

as generators for the unit group of K_{16}. Then the solutions of (5.4.1) can be uniquely represented in the form

$$x_1 = \zeta^{b_{1,0}} \varepsilon_1^{b_{1,1}} \varepsilon_2^{b_{1,2}} \varepsilon_3^{b_{1,3}} \pi^{b_{1,4}}, \quad x_2 = \zeta^{b_{2,0}} \varepsilon_1^{b_{2,1}} \varepsilon_2^{b_{2,2}} \varepsilon_3^{b_{2,3}} \pi^{b_{2,4}}.$$

with rational integer exponents $b_{i,j}$. Obviously one can assume that

$$0 \leq b_{1,0}, \quad b_{2,0} \leq 15. \qquad (5.4.2)$$

Using Baker's method and reduction techniques, it was shown in Smart (1997, 1999) that

$$\max_{i,j} |b_{i,j}| \leq 1066 = B_R,$$

where B_R denotes the reduced bound as defined in Section 5.2.

We note that in Smart (1997, 1999) some earlier versions of Theorems 3.2.4, 3.2.7 and of the reduction algorithm were utilized. Using our versions described in Sections 5.1 and 5.2 we could get a slightly better value for B_R, but this is in fact irrelevant for the last part of the computations.

The enumeration process was applied repeatedly with the initial values $B_0 = B_R$, $H_0 = 10^{3598}$ and with $c_1^* = 1.63189$. Then $\mathscr{S}_{H_0}(B_0)$ is just the set of solutions of (5.4.1). Smart (1999) then chose

$$H_1 = 10^{90}, \quad H_2 = 10^{30}, \quad H_3 = 10^{15}, \quad H_4 = 10^6, \quad H_5 = 10^3.$$

After the necessary computations it turned out that the sets $\mathcal{T}_{j,v}(B_k, H_k, H_{k+1})$ are empty both for $0 \le k \le 4$, $1 \le j \le 4$ and all $v \in S$ infinite, and for $0 \le k \le 2$, $1 \le j \le 4$ and v finite. For the finite v and for $k = 3, 4$, $1 \le j \le 4$, the solutions in $\mathcal{T}_{j,v}(B_k, H_k, H_{k+1})$ were determined by the Fincke–Pohst algorithm.

Finally, it remained to enumerate the set $\mathscr{S}_{B_5}(H_5)$ for $B_5 = 11$ which was accomplished again by means of the Fincke–Pohst method.

The equation (5.4.1) has exactly 795 solutions, each of which satisfies (5.4.2) and

$$\max_{1 \le j \le 4} \left(|b_{1,j}|, |b_{2,j}| \right) \le 11.$$

This result was needed in Smart (1997) for the calculation of curves of genus 2 with good reduction away from 2.

Example 5.4.2 (Wildanger (2000)). Let K_{19} be the 19th cyclotomic field generated by $\zeta = \exp(2\pi i/19)$, and denote by K_{19}^+ the maximal real subfield of K_{19}. Then $K_{19}^+ = \mathbb{Q}(\theta)$, where $\theta = \zeta + \zeta^{-1}$. Consider the unit equation

$$x_1 + x_2 = 1 \quad \text{in units } x_1, x_2 \text{ of the ring of integers of } K_{19}^+. \tag{5.4.3}$$

The number field K_{19}^+ is totally real, its degree is 9 and its unit rank is 8. Further,

$$\varepsilon_1 = 1 - 4\theta - 10\theta^2 + 10\theta^3 + 15\theta^4 - 6\theta^5 - 7\theta^6 + \theta^7 + \theta^8,$$

$$\varepsilon_2 = 3\theta - \theta^3, \quad \varepsilon_3 = 1 - 2\theta - 3\theta^2 + \theta^3 + \theta^4, \quad \varepsilon_4 = 2 - 9\theta^2 + 6\theta^4 - \theta^6,$$

$$\varepsilon_5 = \theta, \quad \varepsilon_6 = 2 - \theta^2, \quad \varepsilon_7 = 2 - 4\theta^2 + \theta^4,$$

$$\varepsilon_8 = -5\theta + 5\theta^2 + 10\theta^3 - 5\theta^4 - 6\theta^5 + \theta^6 + \theta^7$$

is a system of fundamental units in K_{19}^+. The solutions of (5.4.3) can be written uniquely in the form

$$x_1 = \pm \prod_{j=1}^{8} \varepsilon_j^{b_{1,j}}, \quad x_2 = \pm \prod_{j=1}^{8} \varepsilon_j^{b_{2,j}},$$

where $b_{i,j}$ are rational integers.

By means of Baker's method Wildanger proved that

$$\max_{i,j} |b_{i,j}| \le 10^{38}.$$

Further, using reduction techniques he showed that

$$\max_{i,j} |b_{i,j}| \le 2076 = B_{\mathrm{R}},$$

B_{R} being the reduced bound.

Finally, Wildanger's variant of the enumeration algorithm was used repeatedly with the initial values $B_0 = B_{\mathrm{R}}$, $H_0 = 6.9 \times 10^{4843}$ and then with the values

$$H_1 = 1.49 \times 10^{30}, \quad H_2 = 3.89 \times 10^{11}, \quad H_3 = 5.52 \times 10^7, \quad H_4 = 982\,337.37,$$

$$H_5 = 73\,360.74, \quad H_6 = 9896.88, \quad H_7 = 1780.14, \quad H_8 = 365.36,$$

$$H_9 = 74.25, \quad H_{10} = 11.47.$$

At the final enumeration all the 28 398 solutions of (5.4.3) were found.

5.5 Exceptional units

The units ε of the ring of integers of a number field K such that $1 - \varepsilon$ is also a unit of this ring of integers are called *exceptional units of K*, see Nagell (1970). Nagell (1964, 1968b, 1970) determined all exceptional units in number fields of unit rank 1 and in certain number fields of unit rank 2.

As was mentioned above, the number field K_{19}^+ has exactly 28 398 exceptional units. For a positive integer m for which $m \not\equiv 2 \pmod 4$, denote by K_m the m-th cyclotomic field and by K_m^+ its maximal real subfield. Using the above method, Wildanger (2000) determined all exceptional units in the number fields K_m^+ for $m \le 23$. Further, by means of the next lemma he extended his result to the number fields K_m as well.

A number field is called a *CM-field* if it is a totally imaginary quadratic extension of a totally real number field. For example, the imaginary quadratic number fields and the cyclotomic fields are all CM-fields.

Lemma 5.5.1 *Let K be a CM-field. Then all non-real exceptional units in K are of the form*

$$\frac{1 - \zeta_2}{\zeta_1 - \zeta_2},$$

where ζ_1, ζ_2 are roots of unity in K.

Proof. See Győry (1971). $\qquad\square$

Denote by $\mathscr{S}_m$ and $\mathscr{S}_m^+$ the set of exceptional units in K_m and K_m^+, respectively, and let $|\mathscr{S}_m|$ and $|\mathscr{S}_m^+|$ be their cardinalities. The following table given by Wildanger contains the values of $|\mathscr{S}_m|$ and $|\mathscr{S}_m^+|$ for those m for which the unit rank of K_m^+ is at most 10.

m	$[K_m^+ : \mathbb{Q}]$	$|\mathscr{S}_m^+|$	$[K_m : \mathbb{Q}]$	$|\mathscr{S}_m|$
1	1	0	1	0
3	1	0	2	2
4	1	0	2	0
5	2	6	4	18
7	3	42	6	72
8	2	0	4	0
9	3	18	6	38
11	5	570	10	660
12	2	0	4	14
13	6	1830	12	1962
15	4	90	8	440
16	4	0	8	0
17	8	11 700	16	11 940
19	9	28 398	18	28 704
20	4	54	8	138
21	6	1416	12	2192
23	11	130 812	22	131 274
24	4	0	8	86
25	10	47 766	20	48 078
27	9	8676	18	8858
28	6	678	12	888
32	8	0	16	0
33	10	73 110	20	75 242
36	6	354	12	710
40	8	4398	16	4914
44	10	30 030	20	30 660
48	8	0	16	422
60	8	14 274	16	16 340

We remark that for $m = 8, 16, 24, 32, 48$ there is a prime ideal in $O_{K_m^+}$ of norm 2. Hence these number fields K_m^+ cannot have exceptional units. This implies that in the solutions of the equation (5.4.1) in Example 5.4.1 one of the exponents $b_{1,4}$ and $b_{2,4}$ must be different from zero.

For each $d \in \{2, \ldots, 8\}$, $r \in \{2, \ldots, d-1\}$, Wildanger (2000) considered the number fields of degree d and unit rank r having one of the five discriminants with smallest absolute values, and computed for each of them all exceptional units.

Finally we note that Wildanger's method was implemented in KANT, see Daberkow et al. (1997).

5.6 Supplement: LLL lattice basis reduction

Let n be an integer ≥ 2. The standard inner product on $\mathbb{R}^n$ is defined by

$$\langle \mathbf{a}, \mathbf{b} \rangle = \sum_{i=1}^{n} a_i b_i \text{ for } \mathbf{a} = (a_1, \ldots, a_n), \ \mathbf{b} = (b_1, \ldots, b_n) \in \mathbb{R}^n.$$

We use $\| \cdot \|$ to denote the Euclidean norm on $\mathbb{R}^n$. Thus for $\mathbf{a} = (a_1, \ldots, a_n) \in \mathbb{R}^n$ we have

$$\|\mathbf{a}\| = \langle \mathbf{a}, \mathbf{a} \rangle^{1/2} = \sqrt{a_1^2 + \cdots + a_n^2}.$$

Let $\mathcal{L}$ be a t-dimensional lattice in $\mathbb{R}^n$, i.e.,

$$\mathcal{L} := \{z_1 \mathbf{a}_1 + \cdots z_t \mathbf{a}_t : z_1, \ldots, z_t \in \mathbb{Z}\},$$

where $\mathbf{a}_1, \ldots, \mathbf{a}_t$ are linearly independent vectors in $\mathbb{R}^n$. Then the *determinant* $d(\mathcal{L})$ of $\mathcal{L}$ is given by

$$d(\mathcal{L}) = \left| \det \left(\langle \mathbf{a}_i, \mathbf{a}_j \rangle \right)_{1 \leq i, j \leq t} \right|^{1/2}.$$

If in particular $\mathcal{L}$ is a full lattice in $\mathbb{R}^n$, i.e., with $t = n$, then

$$d(\mathcal{L}) = |\det(\mathbf{a}_1, \ldots, \mathbf{a}_n)|.$$

The determinant of $\mathcal{L}$ is independent of the choice of $\mathbf{a}_1, \ldots, \mathbf{a}_t$.

In this section, by a basis of a lattice or vector space we mean an *ordered* tuple of vectors $\mathbf{a}_1, \ldots, \mathbf{a}_t$ and not just a set $\{\mathbf{a}_1, \ldots, \mathbf{a}_t\}$, since the outcome of the LLL-algorithm depends on the order in which the vectors of the initial basis are inserted.

Let $\mathbf{a}_1, \ldots, \mathbf{a}_t$ be a basis of a t-dimensional lattice $\mathcal{L}$ in $\mathbb{R}^n$, where $1 \leq t \leq n$. To define an LLL-reduced basis of $\mathcal{L}$ we need an appropriate orthogonal basis in the subspace of $\mathbb{R}^n$ spanned by $\mathcal{L}$. By means of the Gram–Schmidt orthogonalization process such an orthogonal basis $\mathbf{a}_1^*, \ldots, \mathbf{a}_t^*$ can be defined inductively by

$$\mathbf{a}_i^* = \mathbf{a}_i - \sum_{j=1}^{i-1} \mu_{ij} \mathbf{a}_j^*, \quad 1 \leq i \leq t, \tag{5.6.1}$$

where

$$\mu_{ij} = \langle \mathbf{a}_i, \mathbf{a}_j^* \rangle / \|\mathbf{a}_j^*\|^2, \quad 1 \leq j < i \leq t. \tag{5.6.2}$$

A. K. Lenstra, H. W. Lenstra and Lovász (1982) introduced the notion of what is nowadays called an **LLL-reduced basis** of a lattice. A basis $\mathbf{a}_1, \ldots, \mathbf{a}_t$ of a lattice $\mathcal{L}$ in $\mathbb{R}^n$ is called **LLL-reduced** if $\mathbf{a}_1, \ldots, \mathbf{a}_t$ and the vectors $\mathbf{a}_1^*, \ldots, \mathbf{a}_t^*$ of the corresponding orthogonal basis satisfy

$$|\mu_{ij}| \leq \tfrac{1}{2}, \quad 1 \leq j < i \leq t$$

and

$$\|\mathbf{a}_i^* + \mu_{i,i-1}\mathbf{a}_{i-1}^*\|^2 \geq \frac{3}{4}\|\mathbf{a}_{i-1}^*\|^2, \quad 1 < i \leq t. \tag{5.6.3}$$

Clearly, (5.6.3) can be rewritten as

$$\|\mathbf{a}_i^*\|^2 \geq \left(\frac{3}{4} - \mu_{i,i-1}^2\right) \|\mathbf{a}_{i-1}^*\|^2.$$

Lenstra, Lenstra and Lovász proved that every lattice in $\mathbb{R}^n$ has such a basis. Further, they developed a very practical algorithm, which from any given lattice and any basis of this lattice computes an LLL-reduced basis of this lattice. (In fact, Lenstra, Lenstra and Lovász formally stated their results only for full lattices, but the generalization to arbitrary lattices is implicit in their proof; see also Pohst (1993)).

LLL-reduced bases have several useful properties. In our book the inequality (iv) below plays an important role in solving concrete Diophantine equations.

Proposition 5.6.1 *Let $\mathbf{a}_1, \ldots, \mathbf{a}_t$ be an LLL-reduced basis of a lattice $\mathcal{L}$ in $\mathbb{R}^n$ with associated orthogonal basis $\mathbf{a}_1^*, \ldots, \mathbf{a}_t^*$ defined in (5.6.1). Then we have*

(i) $\|\mathbf{a}_j\|^2 \leq 2^{i-1}\|\mathbf{a}_i^*\|$ *for* $1 \leq j \leq i \leq t$;

(ii) $d(\mathcal{L}) \leq \prod_{i=1}^{t} \|\mathbf{a}_i\| \leq 2^{t(t-1)/4}d(\mathcal{L})$;

(iii) $\|\mathbf{a}_1\| \leq 2^{(t-1)/4}d(\mathcal{L})^{1/t}$;

(iv) $\|\mathbf{a}_1\|^2 \leq 2^{t-1}\|\mathbf{x}\|^2$ *for every* $\mathbf{x} \in \mathcal{L}$, $\mathbf{x} \neq 0$;

(v) $\|\mathbf{a}_j\|^2 \leq 2^{t-1} \max\left(\|\mathbf{x}_1\|^2, \ldots, \|\mathbf{x}_s\|^2\right)$ *for* $1 \leq j \leq s$, *where* $1 \leq s \leq t$, *and* $\mathbf{x}_1, \ldots, \mathbf{x}_s$ *are linearly independent vectors of* $\mathcal{L}$.

Proof. See Lenstra, Lenstra and Lovász (1982) for $t = n$, and Pohst (1993) in the case $t \leq n$. □

Following Lenstra, Lenstra and Lovász (1982) and Pohst (1993), we now briefly present the LLL-lattice *basis reduction algorithm*, that transforms a given basis $\mathbf{a}_1, \ldots, \mathbf{a}_t$ of a given lattice $\mathcal{L}$ in $\mathbb{R}^n$ into an LLL-reduced one. First the constants μ_{ij} and the orthogonal basis vectors $\mathbf{a}_i^*$ are calculated using

(5.6.1) and (5.6.2). Then an LLL-reduced basis can be constructed by induction on the number of reduced basis vectors. The vectors $\mathbf{a}_1, \ldots, \mathbf{a}_t$ will be changed several times. However, the $\mathbf{a}_i^*$ and μ_{ij} will be updated at each step so that (5.6.1) and (5.6.2) remain valid.

Assume that for some m with $2 \leq m \leq t + 1$, the vectors $\mathbf{a}_1, \ldots, \mathbf{a}_{m-1}$ are already LLL-reduced, that is form an LLL-reduced basis of the lattice generated by them. In other words, we assume that

$$|\mu_{ij}| \leq \frac{1}{2} \quad \text{for } 1 \leq j < i < m$$

and

$$\|\mathbf{a}_i^* + \mu_{i,i-1}\mathbf{a}_{i-1}^*\|^2 \geq \frac{3}{4}\|\mathbf{a}_{i-1}^*\|^2 \quad \text{for } 1 < i < m.$$

These inequalities trivially hold if $m = 2$. For $m = t + 1$ the algorithm terminates because then the full basis $\mathbf{a}_1, \ldots, \mathbf{a}_t$ is reduced. Next consider the case $m \leq t$.

The major steps are as follows.

(a) Reduce $\mu_{m,m-1}$ to $|\mu_{m,m-1}| \leq 1/2$, subtracting an appropriate multiple of $\mathbf{a}_{m-1}$ from $\mathbf{a}_m$. After these changes all the vectors $\mathbf{a}_i^*$ remain unchanged.
(b) If (5.6.3) holds for $i = m$, one can proceed to (c). Otherwise interchange $\mathbf{a}_{m-1}$ and $\mathbf{a}_m$ and, if $m > 2$, replace m by $m - 1$. Then one can go on with (a).
(c) Reduce μ_{mj} as in (a) to $|\mu_{mj}| \leq 1/2$ for $j = m - 2, m - 3, \ldots, 1$. Then take $m + 1$ in place of m. If $m > t$, the algorithm terminates, otherwise we can go on with (a).

The vectors $\mathbf{a}_i^*$ are not used explicitly in the algorithm, only the squares of their norms

$$A_i := \|\mathbf{a}_i^*\|^2.$$

LLL-reduction algorithm (Pohst (1993)).
Input: *A basis* $\mathbf{a}_1, \ldots, \mathbf{a}_t$ *of a t-dimensional lattice* $\mathcal{L} \subseteq \mathbb{R}^n$.
Output: *A basis* $\mathbf{a}_1, \ldots, \mathbf{a}_t$ *of* $\mathcal{L}$ *which is LLL-reduced.*

(a) *(Initialization)*
For $i = 1, \ldots, t$ *set:*

$$\mu_{ij} \leftarrow \langle \mathbf{a}_i, \mathbf{a}_j^* \rangle / A_j \quad (1 \leq j \leq i - 1),$$

$$\mathbf{a}_i^* \leftarrow \mathbf{a}_i - \sum_{j=1}^{i-1} \mu_{ij}\mathbf{a}_j^*, \quad A_i \leftarrow \langle \mathbf{a}_i, \mathbf{a}_i^* \rangle.$$

Then set $m \leftarrow 2$.

(b) *(Set l). Set* $l \leftarrow m - 1$.

(c) *(Change* μ_{ml} *in the case* $|\mu_{ml}| > \frac{1}{2}$). *If* $|\mu_{ml}| > \frac{1}{2}$ *set* r *to the closest rational integer to* μ_{ml} *and*

$$\mathbf{a}_m \leftarrow \mathbf{a}_m - r\mathbf{a}_l,$$

$$\mu_{mj} \leftarrow \mu_{mj} - r\mu_{lj} \quad (1 \le j \le l - 1), \quad \mu_{ml} \leftarrow \mu_{ml} - r.$$

For $l = m - 1$ *go to (d), else to (e)*.

(d) *For* $A_m < \left(\frac{3}{4} - \mu_{m,m-1}^2\right) A_{m-1}$ *go to (f)*.

(e) *(Decrease l). Set* $l \leftarrow l - 1$. *For* $l > 0$ *go to (c). For* $m = t$ *terminate; else set* $m \leftarrow m + 1$ *and go to (b)*.

(f) *(Interchange* $\mathbf{a}_{m-1}, \mathbf{a}_m$). *Set* $\mu \leftarrow \mu_{m,m-1}$, $A \leftarrow A_m + \mu^2 A_{m-1}$, $\mu_{m,m-1} \leftarrow \mu A_{m-1}/A$, $A_m \leftarrow A_{m-1} A_m/A$, $A_{m-1} \leftarrow A$; *then set for* $1 \le j \le m - 2$ *and* $m + 1 \le i \le t$

$$\begin{pmatrix} \mathbf{a}_{m-1} \\ \mathbf{a}_m \end{pmatrix} \leftarrow \begin{pmatrix} \mathbf{a}_m \\ \mathbf{a}_{m-1} \end{pmatrix}, \quad \begin{pmatrix} \mu_{m-1,j} \\ \mu_{mj} \end{pmatrix} \leftarrow \begin{pmatrix} \mu_{mj} \\ \mu_{m-1,j} \end{pmatrix},$$

$$\begin{pmatrix} \mu_{i,m-1} \\ \mu_{im} \end{pmatrix} \leftarrow \begin{pmatrix} 1 & \mu_{m,m-1} \\ 0 & 1 \end{pmatrix} \begin{pmatrix} 0 & 1 \\ 1 & -\mu \end{pmatrix} \begin{pmatrix} \mu_{i,m-1} \\ \mu_{im} \end{pmatrix}.$$

For $m > 2$ *decrease* m *by 1. Then go to (b)*.

As is proved in Lenstra, Lenstra and Lovász (1982), see also Pohst (1993), the above algorithm always terminates. Further, it is shown that if $\mathcal{L}$ is a sublattice of $\mathbb{Z}^n$ of rank n with basis $\mathbf{a}_1, \ldots, \mathbf{a}_n$ with $\|\mathbf{a}_i\|^2 \le A$ for $i = 1, \ldots, n$, where $A \ge 2$, then the algorithm uses $O(n^4 \log A)$ arithmetic operations, while the integers occurring in the algorithm have binary lengths $O(n \log A)$.

For more detailed treatments of the LLL-algorithm as well as for some refinements, we refer to the books de Weger (1989), Pohst (1993), Cohen (1993), Smart (1998) and Gaál (2002).

5.7 Notes

- In the inhomogeneous version of (5.2.1), the first reduction algorithm was established in Baker and Davenport (1969). Generalizations of this algorithm to the case of several variables were given in Pethő and Schulenberg (1987) and de Weger (1989).
- We note that the enumeration algorithm presented in Section 5.3 can be made even more efficient by combining it with some sieving procedure with appropriate prime ideals; see e.g. Smart (1998).

- Exceptional units have several applications. An important application was given by Lenstra (1977) who showed that if a number field K contains a "large" subset $\{\varepsilon_1, \ldots, \varepsilon_m\}$ of integers of K such that $\varepsilon_i - \varepsilon_j$ is a unit for each $i \neq j$ then (the ring of integers in) K is Euclidean (with respect to the norm). This was used by Lenstra and others, see, e.g., Lenstra (1977), Mestre (1981), Leutbecher and Martinet (1982), Leutbecher (1985), Leutbecher and Niklasch (1989), Houriet (2007) to obtain several hundreds of new examples of Euclidean number fields.
- There is also a link between exceptional units, Lenstra's result and the dynamics of iterated polynomial mappings; see Zieve (1996).

6

Unit equations in several unknowns

In the previous chapters we considered equations

$$a_1 x_1 + a_2 x_2 = 1, \qquad (6.1)$$

where the unknowns x_1, x_2 are taken from the group of S-units, or more generally from a finitely generated multiplicative group in a number field. We proved effective finiteness results, which enable one to determine all solutions at least in principle. In fact, in several cases there are even practical algorithms to solve such equations. Our proofs are based on Baker-type inequalities for linear forms in ordinary or p-adic logarithms of algebraic numbers.

In this chapter, we consider equations

$$a_1 x_1 + \cdots + a_n x_n = 1 \qquad (6.2)$$

in an arbitrary number of unknowns $x_1, \ldots, x_n$, which again may be S-units of a number field, or elements from a finitely generated multiplicative group. It should be noticed that equations of the type (6.2) in $n > 2$ unknowns may have infinitely many solutions. For instance, consider (6.2) with solutions taken from an infinite multiplicative group Γ, and let $(x_1, \ldots, x_n)$ be a solution of this equation, with $a_1 x_1 + \cdots + a_m x_m = 0$, say, where $2 \leq m < n$. Then one obtains an infinite family of solutions by taking $(u x_1, \ldots, u x_m, x_{m+1}, \ldots, x_n)$ with u an arbitrary element of Γ. To obstruct such obvious constructions of infinite families, we usually consider only *non-degenerate* solutions of (6.2), i.e., with

$$\sum_{i \in I} a_i x_i \neq 0 \quad \text{for each non-empty subset } I \text{ of } \{1, \ldots, n\}.$$

We mention that for equations of type (6.2) in more than two unknowns, we can prove only ineffective finiteness results, as the only available methods to deal with such equations are ineffective. On the other hand, these methods make

128

it possible to give an explicit upper bound for the number of non-degenerate solutions of equation (6.2). The first method, which is the one followed in this book, is based on the p-adic Subspace Theorem of Schmidt and Schlickewei. The second method, originating from ideas in Faltings (1991) and further developed in Rémond (2002), is independent of the Subspace Theorem but uses instead Faltings' Product Theorem. We should mention here that the second method has a wider applicability but that the first method based on the Subspace Theorem leads to smaller upper bounds for the number of non-degenerate solutions of (6.2).

We give a quick overview of the results proved in this chapter. Our first theorem is a so-called "semi-effective" result, which is a reformulation of a result from Evertse (1984b). Let K be an algebraic number field, and S a finite set of places of K, containing the infinite places. For a vector $\mathbf{x} = (x_0, \ldots, x_n) \in O_S^{n+1}$, define

$$H_S(x_0, \ldots, x_n) := \prod_{v \in S} \max_i |x_i|_v, \quad N_S(x_0 \cdots x_n) := \prod_{v \in S} |x_0 \cdots x_n|_v.$$

Then for every $\epsilon > 0$, and every $\mathbf{x} \in O_S^{n+1}$ with

$$x_0 + \cdots + x_n = 0$$

and $\sum_{i \in I} x_i \neq 0$ for each proper, non-empty subset I of $\{0, \ldots, n\}$, we have

$$H_S(x_0, \ldots, x_n) \ll_{K,S,n,\epsilon} N_S(x_0 \cdots x_n)^{1+\epsilon},$$

where the implied constant depends only on K, S, n, ϵ. This constant is not effectively computable from our method of proof. We deduce this result from Theorem 3.1.3 (the p-adic Subspace Theorem).

A consequence of this result is that equation (6.2) has only finitely many non-degenerate solutions in S-units $x_1, \ldots, x_n$.

More generally, we consider equation (6.2) as an equation with unknowns from a multiplicative group of finite rank Γ, contained in any field K of characteristic 0. Taking as starting point Theorem 3.1.6 (a quantitative version of the Parametric Subspace Theorem), we prove a result from Evertse, Schlickewei and Schmidt (2002), stating that equation (6.2) has only finitely many non-degenerate solutions in $x_1, \ldots, x_n \in \Gamma$, whose number is bounded above by $C(n)^{r+1}$, where $r = \text{rank} \, \Gamma$, and $C(n)$ is an effectively computable number depending only on n.

Next, we consider again equation (6.1), in unknowns $x_1, x_2 \in \Gamma$. We have included a proof of the result of Beukers and Schlickewei (1996), implying that for every pair of non-zero coefficients a_1, a_2, equation (6.1) has at most C^{r+1} solutions, where $r = \text{rank} \, \Gamma$, and C is an effectively computable constant. We should mention here that the approach of Evertse, Schlickewei and Schmidt (2002) gives a similar result, but with a much larger constant C. Further, we prove a result from Evertse, Győry, Stewart and Tijdeman (1988a), which states in a precise way that for most pairs (a_1, a_2), equation (6.1) has at most two solutions.

We finish with some results concerning lower bounds for the number of solutions of (6.1) and (6.2). In particular, we have included a result by Konyagin and Soundararajan (2007) which implies that for every $\beta < 2 - \sqrt{2}$ there are groups Γ of arbitrarily large rank r and a_1, a_2 such that (6.1) has at least $\exp(r^\beta)$ solutions.

In Section 6.7, the Notes of this chapter, we give an overview of some historical developments, and some related results.

The results presented in this chapter have applications in Chapter 9 to decomposable form equations. Further, they will be applied in our book on discriminant equations.

6.1 Results

6.1.1 A semi-effective result

Let K be an algebraic number field, and S a finite set of places of K containing all infinite places. We define the S-height of $\mathbf{x} = (x_0, \ldots, x_n) \in O_S^{n+1}$ by

$$H_S(\mathbf{x}) = H_S(x_0, \ldots, x_n) := \prod_{v \in S} \max(|x_0|_v, \ldots, |x_n|_v),$$

where the absolute value $|\cdot|_v$ is normalized as in Section 1.7. Recall that the S-norm of $a \in O_S$ is defined by

$$N_S(a) := \prod_{v \in S} |a|_v.$$

Our first result is as follows.

Theorem 6.1.1 *Let* $\epsilon > 0, n \geq 1$. *There is a constant* $C^{\mathrm{ineff}}(K, S, n, \epsilon)$ *depending only on* K, S, n, ϵ *for which the following holds. For any non-zero* $x_0, x_1, \ldots, x_n \in O_S$ *with*

$$x_0 + x_1 + \cdots + x_n = 0,$$

$$\sum_{i \in I} x_i \neq 0 \quad \text{for each proper, non-empty subset } I \text{ of } \{0, \ldots, n\}$$

we have

$$H_S(x_0, x_1, \ldots, x_n) \leq C^{\text{ineff}}(K, S, n, \epsilon) N_S(x_0 \cdots x_n)^{1+\epsilon}. \tag{6.1.1}$$

This is in fact an equivalent formulation to Evertse (1984b), theorem 1. We have indicated by means of the superscript "ineff" that the constant C^{ineff} is not effectively computable by means of our method of proof. We view Theorem 6.1.1 as a "semi-effective result", since it is effective in terms of $N_S(x_0 \cdots x_n)$, but ineffective in terms of n, K, S, ϵ.

From Theorem 6.1.1 we deduce a finiteness result on the equation

$$a_1 x_1 + \cdots + a_n x_n = 1 \quad \text{in } x_1, \ldots, x_n \in O_S^*, \tag{6.1.2}$$

where $n \geq 2$ and $a_1, \ldots, a_n$ are non-zero elements of K. Recall that a solution $(x_1, \ldots, x_n)$ of (6.1.2) is called *non-degenerate* if

$$\sum_{i \in I} a_i x_i \neq 0 \quad \text{for each non-empty subset } I \text{ of } \{1, \ldots, n\}$$

and *degenerate* otherwise. Theorem 6.1.1 implies the following finiteness result.

Corollary 6.1.2 *Equation* (6.1.2) *has only finitely many non-degenerate solutions in* $x_1, \ldots, x_n \in O_S^*$.

This result was proved independently in Evertse (1984b) and van der Poorten and Schlickewei (1982). It was announced in van der Poorten and Schlickewei (1982) and then proved in Evertse and Győry (1988b) and later in van der Poorten and Schlickewei (1991) that Corollary 6.1.2 is valid in the more general situation as well when, in (6.1.2), K is any field of characteristic 0, and O_S^* is replaced by any finitely generated multiplicative subgroup Γ of K^*. Further, in Evertse and Győry (1988b) it was shown that the number of non-degenerate solutions can be estimated from above by a number depending only on n and Γ, but with the method of proof in that paper it is not possible to effectively compute this number.

In Section 6.2 we deduce Theorem 6.1.1 from the p-adic Subspace Theorem and then deduce from this Corollary 6.1.2. Here we follow Evertse (1984b).

6.1.2 Upper bounds for the number of solutions

In this subsection we consider a generalization of (6.1.2) and give an upper bound for the number of its solutions. We say that a multiplicatively written abelian group Γ is of finite rank r if Γ has a free subgroup Γ_0 of rank r such that for every $x \in \Gamma$ there is a positive integer m such that $x^m \in \Gamma_0$. We say that Γ is of rank 0 if every element of Γ has finite order.

Let now K be any field of characteristic 0, let $n \geq 2$, and denote by $(K^*)^n$ the n-fold direct product of the multiplicative group K^* of K, endowed with coordinatewise multiplication $(x_1, \ldots, x_n)(y_1, \ldots, y_n) = (x_1 y_1, \ldots, x_n y_n)$ and exponentiation $(x_1, \ldots, x_n)^m = (x_1^m, \ldots, x_n^m)$. The following result was established in Evertse, Schlickewei and Schmidt (2002).

Theorem 6.1.3 *Let K be a field of characteristic 0, let $n \geq 2$, let $a_1, \ldots, a_n \in K^*$ and let Γ be a subgroup of $(K^*)^n$ of finite rank r. Then the number of non-degenerate solutions of*

$$a_1 x_1 + \cdots + a_n x_n = 1 \quad in \ (x_1, \ldots, x_n) \in \Gamma \tag{6.1.3}$$

can be estimated from above by a quantity $A(n, r)$ depending on n and r only. For $A(n, r)$ one may take $exp((6n)^{3n}(r + 1))$.

The main ingredients of the proof of this result are a specialization argument, to make a reduction to the case that K is a number field and Γ is finitely generated, a version of the Quantitative Subspace Theorem (Evertse and Schlickewei (2002)) and an estimate of Schmidt (1996) for the number of points of very small height on an algebraic subvariety of a linear torus. This estimate of Schmidt was recently improved substantially by Amoroso and Viada (2009). By going through the proof of Evertse, Schlickewei and Schmidt, but replacing Schmidt's estimate by theirs, they obtained a stronger version of the above Theorem 6.1.3 with

$$A(n, r) = (8n)^{4n^4(n+r+1)}. \tag{6.1.4}$$

We note that by a different approach, based on Faltings' Product Theorem instead of the Subspace Theorem, Rémond (2002) proved a general quantitative result for subvarieties of tori (see Section 10.10), which gives as a special case that equation (6.1.3) has at most $exp(n^{4n^2}(r + 1))$ non-degenerate solutions.

If $n = 2$, then every solution is non-degenerate. In that case we have the following sharper result, which was proved by Beukers and Schlickewei (1996).

Theorem 6.1.4 *Let K be a field of characteristic 0 and Γ a subgroup of $K^* \times K^*$ of finite rank r. Then the equation*

$$x_1 + x_2 = 1 \quad in \ (x_1, x_2) \in \Gamma \tag{6.1.5}$$

has at most $2^{8(r+1)}$ solutions.

We immediately obtain the following corollary.

Corollary 6.1.5 *Let K, Γ be as in Theorem 6.1.4 and $a_1, a_2 \in K^*$. Then the equation*

$$a_1 x_1 + a_2 x_2 = 1 \quad in \ (x_1, x_2) \in \Gamma \qquad (6.1.6)$$

has at most $2^{8(r+2)}$ solutions.

Proof. Apply Theorem 6.1.4 with instead of Γ the group Γ' generated by Γ and (a_1, a_2). $\quad\square$

In most cases, the bound $2^{8(r+2)}$ in Corollary 6.1.5 can be improved. Let K, Γ be as in this corollary. Two pairs (a_1, a_2), $(b_1, b_2) \in K^* \times K^*$ are called Γ-equivalent if there is $(u_1, u_2) \in \Gamma$ such that $(b_1, b_2) = (a_1, a_2)(u_1, u_2)$. Obviously, the number of solutions of (6.1.5) does not change if (a_1, a_2) is replaced by a Γ-equivalent pair. Then we have the following result, which was proved by Evertse, Győry, Stewart and Tijdeman (1988a).

Theorem 6.1.6 *Let K, Γ be as in Theorem 6.1.4. There is a collection of at most finitely many Γ-equivalence classes of pairs in $K^* \times K^*$, such that for every pair $(a_1, a_2) \in K^* \times K^*$ outside the union of these classes, equation (6.1.6) has at most two solutions. The number of these Γ-equivalence classes is bounded above by a function $B(r)$ depending on the rank r of Γ only.*

In fact, the method of proof gives

$$B(r) = 12A(5, 2r) + 24A(3, 2r) + 60A(2, 2r)^2, \qquad (6.1.7)$$

where $A(n, r)$ is any upper bound depending only on n and r for the number of non-degenerate solutions of (6.1.3). By using (6.1.4) we obtain $B(r) = e^{20000(r+3)}$. For earlier bounds for $B(r)$, see Győry (1992b) and Bérczes (2000).

The bound 2 in Theorem 6.1.6 is optimal. For suppose that the set $\widehat{\Gamma} := \{(u_1, u_2) \in \Gamma : u_1 \neq u_2\}$ is infinite. Then for any $(u_1, u_2) \in \widehat{\Gamma}$, the equation

$$\frac{1 - u_2}{u_1 - u_2} x_1 + \frac{u_1 - 1}{u_1 - u_2} x_2 = 1$$

has two solutions in Γ, namely $(1, 1)$ and (u_1, u_2). But, by Corollary 6.1.5, the Γ-equivalence class of such an equation can have only finitely many equations with solutions $(1, 1)$. Hence if (u_1, u_2) runs through $\widehat{\Gamma}$, then $(\frac{1-u_2}{u_1-u_2}, \frac{u_1-1}{u_1-u_2})$ runs through infinitely many Γ-equivalence classes.

In Section 6.3 we sketch a proof of the following: equation (6.1.3) has at most $c(n)^{r+1}$ non-degenerate solutions, where $c(n)$ is an effectively computable constant depending only on n. For a detailed proof of Theorem 6.1.3 we refer to Evertse, Schlickewei and Schmidt (2002) (see also Rémond (2002)).

Theorem 6.1.4 is proved in Section 6.4. In Section 6.5 we deduce Theorem 6.1.6 from Theorem 6.1.3. Here we follow the proof of Evertse, Győry, Stewart and Tijdeman (1988a) and Bérczes (2000). For further historical comments related to the theorems in this subsection, we refer to Section 6.7.

6.1.3 Lower bounds

Erdős, Stewart and Tijdeman were the first to consider lower bounds for the number of solutions of S-unit equations. For a set of distinct primes $S = \{p_1, \ldots, p_t\}$, denote by $N(S)$ the number of solutions of

$$x_1 + x_2 = 1 \quad \text{in } x_1, x_2 \in \left\{ \pm p_1^{z_1} \cdots p_t^{z_t} : z_1, \ldots, z_t \in \mathbb{Z} \right\}. \tag{6.1.8}$$

Then, in Erdős, Stewart and Tijdeman (1988), it was shown that for every $\epsilon > 0$ and every sufficiently large t, there is a set of primes S of cardinality t such that

$$N(S) \geq \exp((4 - \epsilon)(t/\log t)^{1/2}).$$

Recall that Theorem 6.1.6 implies $N(S) \leq C^{t+1}$ with C a constant > 1.

Stewart conjectured that there are absolute constants $c_1, c_2 > 1$, such that for every $t > 0$ and every set of primes S of cardinality t we have $N(S) \leq c_1^{t^{2/3}}$, while conversely, for arbitrarily large t there is a set of primes S of cardinality t such that $N(S) \geq c_2^{t^{2/3}}$.

Konyagin and Soundararajan (2007) obtained the following result, which is a small further step towards Stewart's conjecture.

Theorem 6.1.7 *For every $\beta < 2 - \sqrt{2} = 0.586\ldots$, there are sets of primes S of arbitrarily large cardinality t, such that $N(S) \geq \exp(t^\beta)$.*

In Section 6.6 we have included the ingenious proof of Konyagin and Soundararajan.

In their paper mentioned above, Konyagin and Soundararajan proved also that for every sufficiently large t there are distinct primes $p_1, \ldots, p_t$ such that the equation

$$x - y = 1$$

has at least $\exp(t^{1/16})$ solutions in positive integers x, y composed of $p_1, \ldots, p_t$. We omit the proof of this result, which is based on much deeper analytic number theory.

There are also results on lower bounds for the number of solutions of S-unit equations in an arbitrary number of unknowns. Let again $S = \{p_1, \ldots, p_t\}$

be a finite set of primes, let $n \geq 2$, and denote by $N(n, S)$ the number of non-degenerate solutions to

$$x_1 + \cdots + x_n = 1 \quad \text{in } x_1, \ldots, x_n \in \{\pm p_1^{z_1} \cdots p_t^{z_t} : z_1, \ldots, z_t \in \mathbb{Z}\}. \quad (6.1.9)$$

In Evertse, Moree, Stewart and Tijdeman (2003), the authors proved that for every $n \geq 2$, $\epsilon > 0$, and every sufficiently large t there is a set of primes S of cardinality t such that

$$N(n, S) \geq \exp\left((1 - \epsilon)\frac{n^2}{n-1} t^{1-1/n}(\log t)^{-1/n}\right).$$

This is a slight improvement of an unpublished result by Granville. It would be of interest to improve this further, by extending the approach of Konyagin and Soundararajan.

We introduce another quantity, which more or less measures how much algebraic structure there is in the set of solutions of (6.1.9). Denote by $g(n, S)$ the smallest integer g such that there is a non-zero polynomial $P \in \mathbb{C}[X_1, \ldots, X_n]$ of total degree g, not divisible by $X_1 + \cdots + X_n - 1$, such that

$$P(x_1, \ldots, x_n) = 0 \quad \text{for every solution } (x_1, \ldots, x_n) \text{ of (6.1.9)}.$$

In other words, the set of solutions of (6.1.9) cannot be contained in a hypersurface of $\mathbb{C}^n$ of degree $< g(n, S)$.

It is not hard to show that

$$g(n, S) \leq 2^{n-1} - n + (n - 1)N(n, S)^{1/(n-1)}. \quad (6.1.10)$$

Indeed, let $N := N(n, S)$ and let g be the smallest integer such that $\binom{n+g-1}{n-1} > N$. Then there is a non-zero polynomial $P_1 \in \mathbb{C}[X_1, \ldots, X_{n-1}]$ of total degree $\leq g$ such that $P_1(x_1, \ldots, x_{n-1}) = 0$ for each non-degenerate solution $(x_1, \ldots, x_n)$ of (6.1.9). This can be seen by viewing the relations $P_1(x_1, \ldots, x_{n-1}) = 0$ as linear equations in the coefficients of P_1. Thus, we obtain a system of N linear equations in $\binom{n+g-1}{n-1}$ unknowns and by our choice of g it has a non-trivial solution. Our choice of g implies

$$\left(\frac{g}{n-1}\right)^{n-1} \leq \binom{n+g-2}{n-1} \leq N,$$

hence $g \leq (n - 1)N^{1/(n-1)}$. Now let P be the product of P_1 and of all polynomials $\sum_{i \in I} X_i$ with I a subset of $\{1, \ldots, n - 1\}$ of cardinality at least 2. Then P has total degree $g + 2^{n-1} - n \leq 2^{n-1} - n + (n - 1)N^{1/(n-1)}$, P is not divisible by $X_1 + \cdots + X_n - 1$ since it depends only on $X_1, \ldots, X_{n-1}$, and every solution of (6.1.9), degenerate or non-degenerate, is a zero of P. This implies (6.1.10).

Again in Evertse, Moree, Stewart and Tijdeman (2003), it was shown that for every $n \geq 2$, $\epsilon > 0$ and every sufficiently large t there is a set of primes S of cardinality t such that

$$g(n, S) \geq \exp((4 - \epsilon)(t/\log t)^{1/2}).$$

Using the above theorem of Konyagin and Soundararajan we improve this as follows.

Theorem 6.1.8 *For every $n \geq 2$, $\beta < 2 - \sqrt{2}$, there are sets of primes S of arbitrarily large cardinality t, such that $g(n, S) \geq \exp(t^\beta)$.*

The proof of this result is given in Section 6.6.

6.2 Proofs of Theorem 6.1.1 and Corollary 6.1.2

We take as starting point Theorem 3.1.3. As before, K is an algebraic number field, S a finite set of places of K containing all infinite places and n an integer with $n \geq 2$.

We prove the following result, which is in fact equivalent to Theorem 6.1.1.

Proposition 6.2.1 *Let T be a subset of S and $\epsilon > 0$. There is a constant $C^{\mathrm{ineff}}(K, S, n, \epsilon) > 0$ such that for all vectors $\mathbf{x} = (x_1, \ldots, x_n) \in O_S^n$ satisfying*

$$\sum_{i \in I} x_i \neq 0 \quad \text{for each non-empty subset } I \text{ of } \{1, \ldots, n\} \tag{6.2.1}$$

we have

$$\left(\prod_{v \in S} \prod_{i=1}^n |x_i|_v \right) \prod_{v \in T} |x_1 + \cdots + x_n|_v$$

$$\geq C^{\mathrm{ineff}}(K, S, n, \epsilon) \left\{ \prod_{v \in T} \max(|x_1|_v, \ldots, |x_n|_v) \right\} H_S(\mathbf{x})^{-\epsilon}. \tag{6.2.2}$$

Since $|x_1 + \cdots + x_n|_v \ll \max(|x_1|_v, \ldots, |x_n|_v)$ for $v \in M_K$, the special case of Proposition 6.2.1 with $T = S$ implies the general case of arbitrary subsets T of S. On the other hand, Proposition 6.2.1 with $T = S$ is a reformulation of Theorem 6.1.1. Indeed, writing $x_0 := -(x_1 + \cdots + x_n)$, we see that (6.2.2) with $T = S$ can be rewritten as

$$N_S(x_0 \cdots x_n) \gg H_S(x_0, \ldots, x_n)^{1-\epsilon}$$

(with implied constant depending on K, S, n, ϵ), which in turn is equivalent to inequality (6.1.1) in Theorem 6.1.1.

A weaker version of Proposition 6.2.1 was proved earlier by van der Poorten and Schlickewei (unpublished).

The proof of Proposition 6.2.1 is by induction on n. For $n = 1$ the assertion is trivially true. Assume Proposition 6.2.1 is true for vectors with fewer than n coordinates, where $n \geq 2$. We proceed to prove (6.2.2) under this induction hypothesis. Henceforth we restrict ourselves to vectors $\mathbf{x} = (x_1, \ldots, x_n) \in O_S^n$ satisfying (6.2.1) and

$$\left(\prod_{v \in S} \prod_{i=1}^{n} |x_i|_v \right) \prod_{v \in T} |x_1 + \cdots + x_n|_v \leq \left\{ \prod_{v \in T} \max \left(|x_1|_v, \ldots, |x_n|_v \right) \right\} H_S(\mathbf{x})^{-\epsilon}.$$

$$\tag{6.2.3}$$

This is obviously no loss of generality.

We start with a lemma.

Lemma 6.2.2 *The set of solutions* $\mathbf{x} \in O_S^n \setminus \{0\}$ *of (6.2.3) is contained in a union of finitely many proper linear subspaces of* K^n.

Proof. Let $\mathbf{x} = (x_1, \ldots, x_n)$ be a solution of (6.2.3). Define the linear form $X_0 := -(X_1 + \cdots + X_n)$ and put $x_0 := -(x_1 + \cdots + x_n)$. For $v \in S \setminus T$, let

$$L_{1v} = X_1, \ldots, L_{nv} = X_n.$$

For $v \in T$, let $i_v \in \{0, \ldots, n\}$ with $|x_{i_v}|_v = \max(|x_0|_v, \ldots, |x_n|_v)$, and let $L_{1v}, \ldots, L_{nv}$ be the linear forms from $\{X_0, \ldots, X_n\} \setminus \{X_{i_v}\}$ in some order. Then

$$|x_{i_v}|_v \ll \max(|x_1|_v, \ldots, |x_n|_v) \quad \text{for } v \in T,$$

so (6.2.3) implies

$$\prod_{v \in S} |L_{1v}(\mathbf{x}) \cdots L_{nv}(\mathbf{x})|_v \ll H_S(\mathbf{x})^{-\epsilon}.$$

By Theorem 3.1.3, the solutions $\mathbf{x} \in O_S^n \setminus \{0\}$ of the latter inequality, and hence the solutions of (6.2.3) with $|x_{i_v}|_v = \max(|x_0|_v, \ldots, |x_n|_v)$ for $v \in T$, lie in a union of finitely many proper linear subspaces of K^n. By applying this to all tuples $(i_v : v \in T)$, Lemma 6.2.2 follows. $\square$

Let $\mathcal{T}_1, \ldots, \mathcal{T}_t$ be the subspaces from Lemma 6.2.2 and let $\mathcal{T} \in \{\mathcal{T}_1, \ldots, \mathcal{T}_t\}$. Then $\mathcal{T}$ can be given by an equation

$$x_1 + \cdots + x_n = \beta_{i_1} x_{i_1} + \cdots + \beta_{i_m} x_{i_m}, \tag{6.2.4}$$

where $\beta_{i_1}, \ldots, \beta_{i_m} \in K$ and $m < n$. Let $\mathbf{x} \in O_S^n$ be a vector with (6.2.1), (6.2.3) and with $\mathbf{x} \in \mathcal{T}$. So $\mathbf{x}$ satisfies (6.2.4). Let J be a minimal subset of $\{i_1, \ldots, i_m\}$

such that $\sum_{j\in J}\beta_j x_j \neq 0$. By re-indexing, we may assume that

$$\left.\begin{aligned}
& x_1 + \cdots + x_n = \sum_{i=1}^{u}\beta_i x_i, \quad u < n, \\
& \sum_{i\in I}\beta_i x_i \neq 0 \text{ for each non-empty subset } I \text{ of } \{1,\ldots,u\}.
\end{aligned}\right\} \tag{6.2.5}$$

We now consider vectors $\mathbf{x} \in O_S^n$ with (6.2.1), (6.2.3), (6.2.5) and show that these satisfy (6.2.2) with an appropriate constant C^{ineff}. Below, constants implied by the Vinogradov symbols $\ll, \gg$ will be ineffective, and will depend only on $K, S, n, \epsilon, \beta_1, \ldots, \beta_u$. But notice that $\beta_1, \ldots, \beta_u$ in turn depend only on K, S, n, ϵ, as they are coming from Lemma 6.2.2. So the constants implied by $\ll, \gg$ ultimately depend only on K, S, n, ϵ. Choose $\delta \in O_S \setminus \{0\}$ such that $\delta\beta_i \in O_S$ for $i = 1, \ldots, u$, and for a solution $\mathbf{x} \in O_S^n$ of (6.2.1), (6.2.3) and (6.2.5), write $z_i := \delta\beta_i x_i$ $(i = 1, \ldots, u)$, $\mathbf{z} = (z_1, \ldots, z_u)$.

Let $\mathbf{x} = (x_1, \ldots, x_n) \in O_S^n$ be a vector with (6.2.1), (6.2.3) and (6.2.5). Then $|x_i|_v \gg |z_i|_v$ for $v \in S$, $i = 1, \ldots, u$, and $|x_1 + \cdots + x_n|_v \gg |z_1 + \cdots + z_u|_v$ for $v \in T$. Hence

$$\left(\prod_{v\in S}\prod_{i=1}^{n}|x_i|_v\right)\prod_{v\in T}|x_1 + \cdots + x_n|_v$$

$$\gg \left(\prod_{v\in S}\prod_{i=u+1}^{n}|x_i|_v\right)\left(\prod_{v\in S}\prod_{i=1}^{u}|z_i|_v\right)\prod_{v\in T}|z_1 + \cdots + z_u|_v.$$

Now a first application of the induction hypothesis gives

$$\left(\prod_{v\in S}\prod_{i=1}^{n}|x_i|_v\right)\prod_{v\in T}|x_1 + \cdots + x_n|_v$$

$$\gg \left(\prod_{v\in S}\prod_{i=u+1}^{n}|x_i|_v\right)\left(\prod_{v\in T}\max\left(|z_1|_v, \ldots, |z_u|_v\right)\right)H_S(\mathbf{z})^{-\epsilon/2}$$

$$\gg \left(\prod_{v\in S}\prod_{i=u+1}^{n}|x_i|_v\right)\left(\prod_{v\in T}\max\left(|x_1|_v, \ldots, |x_u|_v\right)\right)H_S(\mathbf{x})^{-\epsilon/2}.$$

Let

$$T_1 = \{v \in T : \max(|x_1|_v, \ldots, |x_n|_v) = \max(|x_1|_v, \ldots, |x_u|_v)\},$$
$$T_2 = T \setminus T_1.$$

If $T_1 = T$, we obtain at once (6.2.2), since $\prod_{v \in S} \prod_{i=u+1}^{n} |x_i|_v \geq 1$. Suppose $T_1 \subsetneq T$. Then

$$\max(|x_1|_v, \dots, |x_u|_v) = \max(|x_1|_v, \dots, |x_n|_v) \quad \text{for } v \in T_1,$$

$$\max(|x_1|_v, \dots, |x_u|_v) \gg |(\beta_1 - 1)x_1 + \cdots + (\beta_u - 1)x_u|_v$$

$$= |x_{u+1} + \cdots + x_n|_v \quad \text{for } v \in T_2.$$

Now a second application of the induction hypothesis yields

$$\left(\prod_{v \in S} \prod_{i=1}^{n} |x_i|_v \right) \prod_{v \in T} |x_1 + \cdots + x_n|_v$$

$$\gg \left(\prod_{v \in S} \prod_{i=u+1}^{n} |x_i|_v \right) \left(\prod_{v \in T_2} |x_{u+1} + \cdots + x_n|_v \right)$$

$$\times \left(\prod_{v \in T_1} \max(|x_1|_v, \dots, |x_n|_v) \right) H_S(\mathbf{x})^{-\epsilon/2}$$

$$\gg \left(\prod_{v \in T_2} \max(|x_{u+1}|_v, \dots, |x_n|_v) \right) H_S(x_{u+1}, \dots, x_n)^{-\epsilon/2}$$

$$\times \left(\prod_{v \in T_1} \max(|x_1|_v, \dots, |x_n|_v) \right) H_S(\mathbf{x})^{-\epsilon/2}$$

$$\gg \left(\prod_{v \in T} \max(|x_1|_v, \dots, |x_n|_v) \right) H_S(\mathbf{x})^{-\epsilon},$$

as required. This proves Proposition 6.2.1, hence Theorem 6.1.1.

Proof of Corollary 6.1.2. Let T be the smallest set of places such that $S \subseteq T$ and $a_1, \dots, a_n$ are T-units. Then for every non-degenerate solution $\mathbf{x} \in (O_S^*)^n$ of (6.1.2), we have

$$H(a_i x_i) \leq \left(\prod_{v \in T} \max(1, |a_1 x_1|_v, \dots, |a_n x_n|_v) \right)^{1/[K:\mathbb{Q}]}$$

$$= H_T(1, a_1 x_1, \dots, a_n x_n)^{1/[K:\mathbb{Q}]} \leq C^{\text{ineff}}$$

for some ineffective constant C^{ineff}. This leaves only finitely many possibilities for $a_i x_i$ for $i = 1, \dots, n$. This implies Corollary 6.1.2 at once. $\qquad\square$

6.3 A sketch of the proof of Theorem 6.1.3

We outline a proof of the following result.

Theorem 6.3.1 *Let K be a field of characteristic 0, $n \geq 2$, Γ a subgroup of $(K^*)^n$ of finite rank r, and $a_1, \ldots, a_n \in K^*$. Then the equation*

$$a_1 x_1 + \cdots + a_n x_n = 1 \quad in \ (x_1, \ldots, x_n) \in \Gamma \tag{6.1.3}$$

has at most $c_1(n)^{r+1}$ non-degenerate solutions, where $c_1(n)$ is an effectively computable number depending only on n.

Constants $c_2(n), c_3(n), \ldots$ introduced below will also be effectively computable and depending only on n.

6.3.1 A reduction

We reduce Theorem 6.3.1 to the following apparently weaker result.

Theorem 6.3.2 *Let $n \geq 2$ and let Γ be a finitely generated subgroup of $(\overline{\mathbb{Q}}^*)^n$ of rank r. Then the set of solutions of*

$$x_1 + \cdots + x_n = 1 \quad in \ (x_1, \ldots, x_n) \in \Gamma \tag{6.3.1}$$

is contained in a union of at most $c_2(n)^{r+1}$ proper linear subspaces of $\overline{\mathbb{Q}}^n$.

In the proof of Theorem 6.3.1, we have to make a reduction from the case that Γ is contained in an arbitrary field K of characteristic 0 to the case $\Gamma \subset (\overline{\mathbb{Q}}^*)^n$, and for this, we need the following specialization result from algebraic geometry.

Lemma 6.3.3 *Let K be a field of characteristic 0 with $K \supset \overline{\mathbb{Q}}$, and $u_1, \ldots, u_m$ ($m \geq 1$) non-zero elements of K. Then there exists a ring homomorphism $\varphi : \overline{\mathbb{Q}}[u_1, \ldots, u_m] \to \overline{\mathbb{Q}}$, leaving $\overline{\mathbb{Q}}$ invariant.*

Proof. Define the ideal

$$I := \{f \in \overline{\mathbb{Q}}[X_1, \ldots, X_m] : f(u_1, \ldots, u_m) = 0\}$$

and let

$$Z(I) := \left\{ \mathbf{x} = (x_1, \ldots, x_m) \in \overline{\mathbb{Q}}^m : f(\mathbf{x}) = 0 \text{ for all } f \in I \right\}.$$

Obviously, $I \neq (1)$, so by the Weak Nullstellensatz (see, e.g., Harris (1992), Theorem 5.17), the set $Z(I)$ is not empty. Choose $\mathbf{c} = (c_1, \ldots, c_m) \in Z(I)$. Then there is a well-defined ring homomorphism $\varphi : \overline{\mathbb{Q}}[u_1, \ldots, u_m] \to \overline{\mathbb{Q}}$, mapping u_i to c_i for $i = 1, \ldots, m$, and mapping the elements of $\overline{\mathbb{Q}}$ to itself. $\square$

Proof of Theorem 6.3.1. First suppose that Γ is a finitely generated subgroup of $(\overline{\mathbb{Q}}^*)^n$ of rank r, and that $a_1, \ldots, a_n \in \overline{\mathbb{Q}}^*$. By applying Theorem 6.3.2 to the group Γ' generated by Γ and $(a_1, \ldots, a_n)$, we infer that the set of solutions of (6.1.3) is contained in a union of at most $c_2(n)^{r+2}$ proper linear subspaces of $\overline{\mathbb{Q}}^n$.

By induction on n, it now follows that (6.1.3) has at most $c_1(n)^{r+1}$ non-degenerate solutions. We give the argument. For $n = 2$ Theorem 6.3.1 is obviously true. Let $n \geq 3$ and assume Theorem 6.3.1 is true for equations in fewer than n unknowns. Consider the solutions of (6.1.3) lying in a fixed proper linear subspace of $(\overline{\mathbb{Q}}^*)^n$, given by a non-trivial linear equation $\beta_1 x_1 + \cdots + \beta_n x_n = 0$, say. By combining this with (6.1.3) we can eliminate one of the unknowns, and obtain an equation $\sum_{i \in I} \gamma_i x_i = 1$, where I is a proper subset of $\{1, \ldots, n\}$. For each subset J of I, consider the solutions of the latter equation such that $\sum_{i \in J} \gamma_i x_i = 1$ but $\sum_{i \in J'} \gamma_i x_i \neq 0$ for each non-empty subset J' of J. Assuming J has cardinality m, by the induction hypothesis the latter equation has at most $c_1(m)^{r+1}$ solutions $(x_i : i \in J)$. Substituting a tuple $(x_i : i \in J)$ in (6.1.3) we obtain $\sum_{i \in J^c} a_i x_i = b$, where $J^c := \{1, \ldots, n\} \setminus J$ and $b := 1 - \sum_{i \in J} a_i x_i$, and b, as well as each proper subsum of the left-hand side, are non-zero. By applying again the induction hypothesis, we see that there are at most $c_1(n - m)^{r+1}$ possibilities for the remaining tuple $(x_i : i \in J^c)$. So for given $\beta_1, \ldots, \beta_n$ and J, we have at most $(c_1(m)c_1(n - m))^{r+1}$ solutions $(x_1, \ldots, x_n)$. By summing over $(\beta_1, \ldots, \beta_n)$ (the number of which is bounded by $c_2(n)^{r+2}$) and over all J, we obtain an upper bound $c_1(n)^{r+1}$ for the total number of non-degenerate solutions of (6.1.3). This completes the induction step.

We now consider the general case that Γ is a subgroup of $(K^*)^n$ of finite rank r, and that $a_1, \ldots, a_n \in K^*$, where K is any field of characteristic 0. We assume without loss of generality that $\overline{\mathbb{Q}}$ is contained in K.

Assume that (6.1.3) has $M > c_1(n)^{r+1}$ non-degenerate solutions, $\mathbf{x}_1, \ldots, \mathbf{x}_M$, say, where $\mathbf{x}_i = (x_{i1}, \ldots, x_{in})$ for $i = 1, \ldots, M$. We apply Lemma 6.3.3 with the set $\{u_1, \ldots, u_m\}$ consisting of $a_1, \ldots, a_n, x_{11}, \ldots, x_{Mn}$, the subsums $\sum_{j \in I} a_j x_{ij}$ ($i = 1, \ldots, M, I \subset \{1, \ldots, n\}$), the non-zero numbers among $x_{i_1 j} - x_{i_2 j}$ ($1 \leq i_1 < i_2 \leq M, j = 1, \ldots, n$), and also the multiplicative inverses of all of these numbers. Thus, the images under φ of $u_1, \ldots, u_m$ are all non-zero. Put $a'_j := \varphi(a_j)$, $x'_{ij} := \varphi(x_{ij})$ ($i = 1, \ldots, M, j = 1, \ldots, n$). Then $a'_j \neq 0$ for $j = 1, \ldots, n$ and $\mathbf{x}'_i = (x_{i1}, \ldots, x_{in})$ ($i = 1, \ldots, M$) are distinct, non-degenerate solutions of

$$a'_1 x'_1 + \cdots + a'_n x'_n = 1. \tag{6.3.2}$$

Let Γ' be the group generated by $\mathbf{x}'_1, \ldots, \mathbf{x}'_M$. Then Γ' is a finitely generated subgroup of $(\overline{\mathbb{Q}}^*)^n$, and Γ' has rank at most r since it is a homomorphic image of a subgroup of Γ. So (6.3.2) has at least $M > c_1(n)^{r+1}$ non-degenerate solutions in Γ', contrary to what has been established above. This shows that in the general case, (6.1.3) cannot have more than $c_1(n)^{r+1}$ non-degenerate solutions. $\quad\square$

The remainder of this section is devoted to the proof of Theorem 6.3.2.

6.3.2 Notation

Assume henceforth that Γ is a finitely generated subgroup of $(\overline{\mathbb{Q}}^*)^n$ of rank r. Let K be an algebraic number field such that $\Gamma \subset (K^*)^n$, and let S be a finite set of places of K, containing all infinite places, such that $\Gamma \subseteq (O_S^*)^n$. Put

$$d := [K : \mathbb{Q}], \quad s := |S|.$$

For $\mathbf{x} = (x_1, \ldots, x_n) \in K^n$ define the heights

$$h(\mathbf{x}) := \frac{1}{d} \sum_{v \in M_K} \max(0, \log |x_1|_v, \ldots, \log |x_n|_v),$$

$$\widehat{h}(\mathbf{x}) := \sum_{i=1}^{n} h(x_i) = \frac{1}{d} \sum_{i=1}^{n} \sum_{v \in M_K} \max(0, \log |x_i|_v),$$

where $h(x)$ denotes the absolute logarithmic height of an algebraic number x. These heights can be extended in the usual manner to $(\overline{\mathbb{Q}}^*)^n$ by picking any number field K containing $x_1, \ldots, x_n$ and applying the above definitions. It is straightforward to show that for $\mathbf{x} = (x_1, \ldots, x_n) \in (\overline{\mathbb{Q}}^*)^n$ we have

$$\frac{1}{n}\widehat{h}(\mathbf{x}) \leq h(\mathbf{x}) \leq \widehat{h}(\mathbf{x}). \tag{6.3.3}$$

Further, for $\mathbf{x} = (x_1, \ldots, x_n) \in \Gamma$ we have

$$\left. \begin{aligned} h(\mathbf{x}) &= \frac{1}{d} \sum_{v \in S} \max(0, \log |x_1|_v, \ldots, \log |x_n|_v), \\ \widehat{h}(\mathbf{x}) &= \frac{1}{d} \sum_{v \in S} \sum_{i=1}^{n} \max(0, \log |x_i|_v) = \frac{1}{2d} \sum_{v \in S} \sum_{i=1}^{n} |\log |x_i|_v|, \end{aligned} \right\} \tag{6.3.4}$$

where the last equality is a consequence of the Product Formula.

6.3.3 Covering results

We treat this in more detail since we can simplify the argument in Evertse, Schlickewei and Schmidt (2002).

Lemma 6.3.4 *Let V be an r-dimensional real vector space and $\| \ \|$ a norm on V. Let C, δ be positive reals, and let $\mathscr{S}$ be a subset of*

$$\{ \mathbf{x} \in V : \| \mathbf{x} \| \leq C \}.$$

Then $\mathscr{S}$ has a subset $\mathscr{S}_0$ such that

$$|\mathscr{S}_0| \leq \left(1 + \frac{2C}{\delta} \right)^r, \tag{6.3.5}$$

for every $\mathbf{x} \in \mathscr{S}$ there is $\mathbf{x}_0 \in \mathscr{S}_0$ with $\| \mathbf{x} - \mathbf{x}_0 \| \leq \delta$. (6.3.6)

Proof. Let $\mathscr{S}_0$ be any subset of $\mathscr{S}$ with the property that

$$\| \mathbf{x}' - \mathbf{x}'' \| > \delta \quad \text{for any two distinct } \mathbf{x}', \mathbf{x}'' \in \mathscr{S}_0.$$

We show that $\mathscr{S}_0$ satisfies (6.3.5). Knowing this, we can choose $\mathscr{S}_0$ of maximal cardinality; then it satisfies (6.3.6) as well.

For $\mathbf{u} \in V$, define $B_{\mathbf{u}} := \{ \mathbf{x} \in V : \| \mathbf{x} - \mathbf{u} \| \leq \delta/2 \}$. Then by the triangle inequality, the balls $B_{\mathbf{u}}$ ($\mathbf{u} \in \mathscr{S}_0$) are pairwise disjoint, and are all contained in $B := \{ \mathbf{x} \in V : \| \mathbf{x} \| \leq C + \delta/2 \}$. Let μ be the Lebesgue measure on V normalized such that the unit ball $\{ \mathbf{x} \in V : \| \mathbf{x} \| \leq 1 \}$ has measure 1. Then, by comparing measures,

$$|\mathscr{S}_0| \left(\frac{\delta}{2} \right)^r = \sum_{\mathbf{u} \in \mathscr{S}_0} \mu(B_{\mathbf{u}}) \leq \mu(B) = \left(C + \frac{\delta}{2} \right)^r,$$

which implies (6.3.5). $\qquad\qquad\square$

Write vectors in $\mathbb{R}^{ns}$ as $\mathbf{u} = (u_{iv} : v \in S, i = 1, \dots, n)$ and define the following homomorphism from Γ to the additive group of $\mathbb{R}^{ns}$:

$$\varphi : (x_1, \dots, x_n) \mapsto \left(\frac{\log |x_i|_v}{d} : v \in S, i = 1, \dots, n \right).$$

Then the kernel of φ is the torsion subgroup Γ_{tors} of Γ. Let V be the real vector space generated by $\varphi(\Gamma)$. Then V has dimension r. Define a norm on $\mathbb{R}^{ns}$ by

$$\| \mathbf{u} \| := \frac{1}{2} \sum_{v \in S} \sum_{i=1}^{n} |u_{iv}|. \tag{6.3.7}$$

Then by (6.3.4) we have

$$\widehat{h}(\mathbf{x}) = \| \varphi(\mathbf{x}) \| \quad \text{for } \mathbf{x} \in \Gamma. \tag{6.3.8}$$

By combining this with Lemma 6.3.4 we obtain the following.

Lemma 6.3.5 *Let C, δ be positive reals, and let $\mathcal{S}$ be a non-empty subset of $\{\mathbf{x} \in \Gamma : \widehat{h}(\mathbf{x}) \leq C\}$. Then $\mathcal{S}$ has a subset $\mathcal{S}_0$ such that*

$$|\mathcal{S}_0| \leq \left(1 + \frac{2C}{\delta}\right)^r, \tag{6.3.9}$$

for every $\mathbf{x} \in \mathcal{S}$ there is $\mathbf{x}_0 \in \mathcal{S}_0$ with $\widehat{h}(\mathbf{x} \cdot \mathbf{x}_0^{-1}) \leq \delta$. (6.3.10)

Proof. Let $\mathcal{S}^* := \varphi(\mathcal{S})$. Choose $\mathcal{S}_0^* \subset \mathcal{S}^*$ as in Lemma 6.3.4. Then choose for each $\mathbf{u}_0 \in \mathcal{S}_0^*$ precisely one element $\mathbf{x}_0 \in \mathcal{S}$ with $\varphi(\mathbf{x}_0) = \mathbf{u}_0$, and let $\mathcal{S}_0$ be the set of all elements thus chosen. Then clearly, $\mathcal{S}_0$ satisfies (6.3.9). To show that it also satisfies (6.3.10), let $\mathbf{x} \in \mathcal{S}$, choose $\mathbf{u}_0 \in \mathcal{S}_0^*$ with $\|\varphi(\mathbf{x}) - \mathbf{u}_0\| \leq \delta$, and then $\mathbf{x}_0 \in \mathcal{S}_0$ with $\varphi(\mathbf{x}_0) = \mathbf{u}_0$. Then by (6.3.8), $\widehat{h}(\mathbf{x} \cdot \mathbf{x}_0^{-1}) \leq \delta$. $\square$

We give another application.

Lemma 6.3.6 *Let $\theta > 0$. There is a subset $\mathcal{S}_1$ of V of cardinality*

$$|\mathcal{S}_1| \leq \left(1 + \frac{4n}{\theta}\right)^r$$

such that for every $\mathbf{x} = (x_1, \dots, x_n) \in \Gamma$ there is $\mathbf{c} = (c_{iv} : v \in S, i = 1, \dots, n) \in \mathcal{S}_1$ with

$$\sum_{v \in S} \sum_{i=1}^{n} \left| \frac{\log |x_i|_v}{dh(\mathbf{x})} - c_{iv} \right| \leq \theta. \tag{6.3.11}$$

Proof. We apply Lemma 6.3.4 to the set $\mathcal{S}$ of vectors

$$\mathbf{u}(\mathbf{x}) = \left(\frac{\log |x_i|_v}{dh(\mathbf{x})} : v \in S, i = 1, \dots, n \right) \quad (\mathbf{x} \in \Gamma)$$

which is contained in V, and with the norm given by (6.3.7). By (6.3.8) we have for $\mathbf{x} \in \Gamma$,

$$\|\mathbf{u}(\mathbf{x})\| = \frac{1}{h(\mathbf{x})} \|\varphi(\mathbf{x})\| = \frac{\widehat{h}(\mathbf{x})}{h(\mathbf{x})} \leq n.$$

So $\mathcal{S} \subseteq \{\mathbf{u} \in V : \|\mathbf{u}\| \leq n\}$. By Lemma 6.3.4 with $\delta = \theta/2$ there is a subset $\mathcal{S}_1$ of $\mathcal{S}$ of cardinality at most $(1 + \frac{2n}{\theta/2})^r = (1 + \frac{4n}{\theta})^r$ such that for every $\mathbf{x} \in \Gamma$ there is $\mathbf{c} \in \mathcal{S}_1$ with $\|\mathbf{u}(\mathbf{x}) - \mathbf{c}\| \leq \theta/2$. This implies (6.3.11). $\square$

6.3.4 The large solutions

We give an upper bound for the number of subspaces containing the solutions $\mathbf{x}$ of (6.3.1) for which $h(\mathbf{x})$ is large. Our main tool is Theorem 3.1.6, i.e., the quantitative version of the Parametric Subspace Theorem stated in Section 3.1.

We apply Lemma 6.3.6 with $\theta = c_3(n)^{-1}$, where $c_3(n)$ is a sufficiently large function of n. Let $\mathcal{S}_1$ be the set from Lemma 6.3.6. With our choice of θ, we have

$$|\mathcal{S}_1| \leq c_4(n)^r. \tag{6.3.12}$$

Pick a tuple $\mathbf{c} = (c_{iv} : v \in S, i = 1, \ldots, n)$ from $\mathcal{S}_1$, and consider the solutions $\mathbf{x} = (x_1, \ldots, x_n) \in \Gamma$ of (6.3.1) with

$$\sum_{v \in M_K} \sum_{i=1}^n \left| \frac{\log |x_i|_v}{dh(\mathbf{x})} - c_{iv} \right| \leq \theta = c_3(n)^{-1}. \tag{6.3.13}$$

Put $x_0 := 1$, $X_0 := X_1 + \cdots + X_n$, $c_{0v} := 0$ for $v \in S$, and for $v \in S$, choose $i_v \in \{0, \ldots, n\}$ such that

$$c_{i_v, v} = \max(c_{0v}, \ldots, c_{nv}).$$

Let $L_{1v}, \ldots, L_{nv}$ be the linear forms X_i ($i \in \{0, \ldots, n\} \setminus \{i_v\}$) in some order, and put $d_{jv} = c_{iv}$ if $L_{jv} = X_i$. Further, for $v \in M_K \setminus S, i = 1, \ldots, n$ put $L_{iv} = X_i, d_{iv} = 0$. Finally, put

$$Q := \exp(dh(\mathbf{x})),$$
$$\mathbf{d} := (d_{iv} : v \in M_K, i = 1, \ldots, n)$$

and define the twisted height $H_{Q,\mathbf{d}}$ by (3.1.4). By (6.3.13) we have

$$\sum_{v \in M_K} \sum_{j=1}^n \left| \frac{\log |L_{jv}(\mathbf{x})|_v}{\log Q} - d_{jv} \right| \leq c_3(n)^{-1},$$

and this implies

$$H_{Q,\mathbf{d}}(\mathbf{x}) \leq Q^{c_3(n)^{-1}}. \tag{6.3.14}$$

Let

$$\xi_{iv} := \frac{\log |x_i|_v}{dh(\mathbf{x})} \quad (v \in S, \quad i = 1, \ldots, n), \quad \xi_{0v} := 0 \quad (v \in S).$$

Then

$$\sum_{v \in S} \left(\sum_{i=0}^n \xi_{iv} - \max(\xi_{0v}, \ldots, \xi_{nv}) \right)$$
$$= \frac{1}{dh(\mathbf{x})} \left(\log \left(\prod_{v \in S} |x_1 x_2 \cdots x_n|_v \right) - \log \left(\prod_{v \in S} \max(1, |x_1|_v, \ldots, |x_n|_v) \right) \right)$$
$$= \frac{-dh(\mathbf{x})}{dh(\mathbf{x})} = -1$$

and

$$\sum_{v \in M_K} \max(\xi_{0v}, \ldots, \xi_{nv}) = 1.$$

In combination with (6.3.13) this implies, assuming that $c_3(n)$ is sufficiently large, that

$$\mu := \frac{1}{n} \sum_{v \in M_K} \sum_{j=1}^{n} d_{jv} = \frac{1}{n} \sum_{v \in S} \left(\sum_{i=0}^{n} c_{iv} - \max(c_{0v}, \ldots, c_{nv}) \right)$$

is approximately equal to

$$\frac{1}{n} \sum_{v \in S} \left(\sum_{i=0}^{n} \xi_{iv} - \max(\xi_{0v}, \ldots, \xi_{nv}) \right) = -\frac{1}{n};$$

more precisely,

$$-\frac{1}{n} - c_5(n)^{-1} \leq \mu \leq -\frac{1}{n} + c_5(n)^{-1},$$

where $c_5(n) = 2c_3(n)$. Likewise,

$$\lambda := \sum_{v \in M_K} \max(d_{1v}, \ldots, d_{nv}) \leq 1 + c_5(n)^{-1}.$$

Together with (6.3.14) this implies

$$H_{Q,\mathbf{d}}(\mathbf{x}) \leq Q^{-\mu-\delta} \tag{6.3.15}$$

where, provided that $c_3(n)$ is sufficiently large,

$$\delta = c_6(n)^{-1} := \frac{1}{n} - c_5(n)^{-1} + c_3(n)^{-1} > 0.$$

Recall that every solution $\mathbf{x}$ of (6.3.1) with (6.3.13) implies (6.3.15) with $Q = \exp(dh(\mathbf{x}))$. Now Theorem 3.1.6 (the Quantitative Parametric Subspace Theorem) implies that the set of solutions of (6.3.1) with (6.3.13) and with

$$h(\mathbf{x}) = \frac{1}{d} \log Q \geq 2c_6(n) \log n =: c_7(n)$$

is contained in a union of at most $c_8(n)$ proper linear subspaces of K^n. Taking into account the upper bound (6.3.12) for the cardinality of $\mathscr{S}_1$, which is an upper bound for the number of different inequalities (6.3.13), we arrive at the following.

Lemma 6.3.7 *The set of solutions* $\mathbf{x} = (x_1, \ldots, x_n) \in \Gamma$ *of* (6.3.1) *with*

$$h(\mathbf{x}) \geq c_7(n)$$

is contained in a union of at most $c_8(n)c_4(n)^r$ *proper linear subspaces of* K^n.

6.3.5 The small solutions, and conclusion of the proof

It remains to consider the solutions $\mathbf{x}$ of (6.3.1) with $h(\mathbf{x}) < c_7(n)$. The crucial tool is the following, which we state without proof.

Proposition 6.3.8 *Let* $b_1, \ldots, b_n \in \overline{\mathbb{Q}}^*$. *Then the equation*

$$b_1 y_1 + \cdots + b_n y_n = 1 \ in \ \mathbf{y} = (y_1, \ldots, y_n) \in (\overline{\mathbb{Q}}^*)^n \qquad (6.3.16)$$

has at most $c_9(n)$ *non-degenerate solutions with* $\widehat{h}(\mathbf{y}) \leq c_{10}(n)^{-1}$.

This result is a special case of more general estimates for the number of points of small height lying on an arbitrary subvariety of a linear torus. From a result of Schmidt (1996), theorem 4, which was obtained by an elementary method, one can deduce the above Proposition with

$$c_9(n) = c_{10}(n) = \exp((4n)^{2n+2}).$$

From David and Philippon (1999), Theorem 1.3 and errata, which is much deeper and uses difficult commutative algebra, it follows that Proposition 6.3.8 holds with

$$c_9(n) = 2^{(n+26)7^{n-1}}, \quad c_{10}(n) = c_9(n)^{3/4}.$$

Finally, a result of Amoroso and Viada (2009) implies that Proposition 6.3.8 holds with

$$c_9(n) = (400n^5 \log n)^{n^2(n-1)^2}, \quad c_{10}(n) = 2(400n^5 \log n)^{n(n-1)^2}.$$

The proof of Amoroso and Viada also uses commutative algebra but it is not as difficult as that of David and Philippon.

Consider the set $\mathscr{S}$ of solutions $\mathbf{x}$ of (6.3.1) with $h(\mathbf{x}) < c_7(n)$. By (6.3.3), these solutions satisfy $\widehat{h}(\mathbf{x}) < nc_7(n)$. Apply Lemma 6.3.5 with $C = nc_7(n)$, $\delta = c_{10}(n)^{-1}$. According to that lemma, there is a subset $\mathscr{S}_0 \subseteq \mathscr{S}$ of cardinality

$$|\mathscr{S}_0| \leq (1 + 2nc_7(n)c_{10}(n))^r \leq c_{11}(n)^r$$

such that for every $\mathbf{x} \in \mathscr{S}$ there is $\mathbf{x}_0 \in \mathscr{S}_0$ with

$$\widehat{h}\left(\mathbf{x} \cdot \mathbf{x}_0^{-1}\right) \leq c_{10}(n)^{-1}. \qquad (6.3.17)$$

Write $\mathbf{x}_0 = (b_1, \ldots, b_n), \mathbf{y} = (y_1, \ldots, y_n) := \mathbf{x} \cdot \mathbf{x}_0^{-1}$. Then clearly, the number of non-degenerate solutions $\mathbf{x}$ of (6.3.1) with (6.3.17) is at most the number of non-degenerate solutions $\mathbf{y}$ of (6.3.16), hence at most $c_{10}(n)$. Taking into account the cardinality of $\mathscr{S}_0$, it follows that (6.3.1) has at most

$$c_9(n)c_{11}(n)^r$$

non-degenerate solutions with $\widehat{h}(\mathbf{x}) < c_7(n)$.

The degenerate solutions of (6.3.1) lie in at most 2^n proper linear subspaces of K^n, each given by a vanishing subsum. We infer that the solutions $\mathbf{x}$ of (6.3.1) (non-degenerate or not) with $h(\mathbf{x}) < c_7(n)$ lie in a union of at most

$$2^n + c_9(n)c_{11}(n)^r$$

proper linear subspaces.

Adding to this the quantity from Lemma 6.3.7, it follows that the complete set of solutions of (6.3.1) is contained in a union of at most

$$c_8(n)c_4(n)^r + 2^n + c_9(n)c_{11}(n)^r \le c_2(n)^{r+1}$$

proper linear subspaces of K^n. This completes the proof of Theorem 6.3.2. □

6.4 Proof of Theorem 6.1.4

We follow Beukers and Schlickewei (1996). Let K be a field of characteristic 0, and Γ a subgroup of $K^* \times K^*$ of rank r. We first show that Theorem 6.1.4 can be reduced to the following special case.

Theorem 6.4.1 *Let Γ be a finitely generated subgroup of $\overline{\mathbb{Q}}^* \times \overline{\mathbb{Q}}^*$ of rank r. Then the equation*

$$x_1 + x_2 = 1 \quad in\ (x_1, x_2) \in \Gamma \tag{6.4.1}$$

has at most $2^{8(r+1)}$ solutions.

Proof of Theorem 6.1.4. We use again specializations. Let K be any field of characteristic 0, and let Γ be a subgroup of $K^* \times K^*$ of rank r. We have to prove that any finite subset of the set of solutions of (6.4.1) has cardinality at most $2^{8(r+1)}$. Let $\{(x_{i1}, x_{i2}) : i = 1, \dots, N\}$ be such a finite subset. We apply Lemma 6.3.3 with $\{u_1, \dots, u_m\}$ consisting of the numbers x_{ik} ($i = 1, \dots, N$, $k = 1, 2$), the non-zero numbers among $x_{ik} - x_{jk}$ ($1 \le i < j \le N, k = 1, 2$), and the multiplicative inverses of all these numbers. Let φ be the ring homomorphism from Lemma 6.3.3, and put $y_{ik} := \varphi(x_{ik})$ for $i = 1, \dots, N, k = 1, 2$. Since the images under φ of the numbers listed above are all non-zero, (y_{i1}, y_{i2}) ($i = 1, \dots, N$) are distinct pairs from $\overline{\mathbb{Q}}^* \times \overline{\mathbb{Q}}^*$. In fact, they yield N distinct solutions of

$$y_1 + y_2 = 1 \quad in\ (y_1, y_2) \in \Gamma',$$

where Γ' is the group generated by (y_{i1}, y_{i2}) ($i = 1, \dots, N$). But Γ' is a subgroup of $\varphi(\Gamma)$, hence it has rank $r' \le r$. Now it follows from Theorem 6.4.1 that $N \le 2^{8(r'+1)} \le 2^{8(r+1)}$. □

In the remainder of this section, we prove Theorem 6.4.1. Instead of the Quantitative Parametric Subspace Theorem, we can now use a much simpler method from Diophantine approximation, based on certain polynomial identities, going back to Thue and Siegel. For $N \in \mathbb{Z}_{>0}$ define the binary form

$$W_N(X, Y) := \sum_{m=0}^{N} \binom{2N - m}{N - m}\binom{N + m}{m} X^{N-m}(-Y)^m,$$

and set $Z := -X - Y$, so that $X + Y + Z = 0$.

Lemma 6.4.2 *We have the following polynomial identities, valid for every positive integer N:*

$$W_N(Y, X) = (-1)^N W_N(X, Y); \tag{6.4.2}$$

$$X^{2N+1} W_N(Y, Z) + Y^{2N+1} W_N(Z, X) + Z^{2N+1} W_N(X, Y) = 0; \tag{6.4.3}$$

$$\begin{vmatrix} Z^{2N+1} W_N(X, Y) & Y^{2N+1} W_N(Z, X) \\ Z^{2N+3} W_{N+1}(X, Y) & Y^{2N+3} W_{N+1}(Z, X) \end{vmatrix}$$

$$= c_N (XYZ)^{2N+1}(X^2 + XY + Y^2) \quad \text{with } c_N \neq 0. \tag{6.4.4}$$

Proof. Identity (6.4.2) is obvious.

Identity (6.4.3) can be deduced from classical relations between hypergeometric functions (see, for instance, Bombieri and Gubler (2006), chapter 5), but we give a direct proof. Fix a positive integer N, let x, y, z be non-zero complex numbers with $x + y + z = 0$ and consider the rational function

$$f(t) := \frac{1}{(t(1 - xt)(1 + yt))^{N+1}}.$$

This function has poles of order $N + 1$ at $t = 0$, $t = 1/x$, $t = -1/y$ and no other poles on $\mathbb{C} \cup \{\infty\}$. The sum of the residues at these poles is 0. We compute the residues. The residue of f at $t = 0$ is the coefficient of t^N in the power series expansion of $(1 - xt)^{-N-1}(1 + yt)^{-N-1}$, which is

$$\sum_{m=0}^{N} \binom{-N-1}{N - m}(-x)^{N-m}\binom{-N-1}{m} y^m$$

$$= \sum_{m=0}^{N} \binom{2N - m}{N - m}\binom{N + m}{m} x^{N-m}(-y)^m = W_N(x, y).$$

The residue of f at $t = 1/x$ is equal to the residue at $t = 0$ of

$$f(t + 1/x) = \frac{1}{((t + 1/x)(-xt)(-z/x + yt))^{N+1}}$$

$$= \frac{(x/z)^{N+1}}{(t(1 - xyt/z)(1 + xt))^{N+1}}.$$

This residue is equal to $(x/z)^{N+1} W_N(xy/z, x) = (x/z)^{2N+1} W_N(y, z)$. A similar computation gives that the residue of f at $t = -1/y$ equals $(y/z)^{2N+1} W_N(z, x)$. Summing the residues and multiplying with z^{2N+1} shows that (6.4.3) holds for all non-zero $x, y, z \in \mathbb{C}$ with $x + y + z = 0$.

Denote the left-hand side of (6.4.4) by Δ_N. We first show that $\Delta_N \not\equiv 0$. Indeed, by (6.4.2), the value of Δ_N at $X = 2$, $Y = Z = -1$ is

$$\begin{vmatrix} W_N(2, -1) & W_N(-1, 2) \\ W_{N+1}(2, -1) & W_{N+1}(-1, 2) \end{vmatrix} = \pm 2 W_N(2, -1) W_{N+1}(2, -1),$$

which is easily seen to be non-zero. By (6.4.3), up to sign, Δ_N is invariant under the substitutions $(X, Y) \mapsto (Y, Z)$, $(X, Y) \mapsto (Z, X)$. Hence Δ_N is divisible by $(XYZ)^{2N+1}$. Since Δ_N is homogeneous of degree $6N + 5$, the quotient $\Delta_N/(XYZ)^{2N+1}$ is a quadratic form, which is up to sign invariant under the above mentioned substitutions. So this quadratic form must be a scalar multiple of $X^2 + XY + Y^2$. $\qquad\square$

Recall that the homogeneous logarithmic height of $\mathbf{x} = (x_1, \ldots, x_n) \in \overline{\mathbb{Q}}^n$ is given by

$$h^{\text{hom}}(\mathbf{x}) = h^{\text{hom}}(x_1, \ldots, x_n) := \frac{1}{[K:\mathbb{Q}]} \log \prod_{v \in M_K} \max_{1 \le i \le n} |x_i|_v,$$

where K is a number field with $\mathbf{x} \in K^n$ (see Section 1.9). The other heights used in this chapter are related to this by

$$h(\mathbf{x}) = h^{\text{hom}}(1, x_1, \ldots, x_n), \quad \widehat{h}(\mathbf{x}) = \sum_{i=1}^{n} h^{\text{hom}}(1, x_i).$$

Lemma 6.4.3 *Let a, b, c be non-zero elements of $\overline{\mathbb{Q}}$, and let (x_i, y_i, z_i) $(i = 1, 2)$ be two linearly independent vectors from $\overline{\mathbb{Q}}^3$ such that $ax_i + by_i + cz_i = 0$ for $i = 1, 2$. Then*

$$h^{\text{hom}}(a, b, c) \le h^{\text{hom}}(x_1, y_1, z_1) + h^{\text{hom}}(x_2, y_2, z_2) + \log 2.$$

Proof. The vector (a, b, c) is proportional to the exterior product of (x_1, y_1, z_1), (x_2, y_2, z_2), which is $(y_1 z_2 - y_2 z_1, z_1 x_2 - x_1 z_2, x_1 y_2 - x_2 y_1)$. So

$$h^{\text{hom}}(a, b, c) = h^{\text{hom}}(y_1 z_2 - y_2 z_1, z_1 x_2 - x_1 z_2, x_1 y_2 - x_2 y_1).$$

Choose a number field K containing x_i, y_i, z_i for $i = 1, 2$. Let $s(v) := 1$ if v is a real place, $s(v) := 2$ if v is a complex place, and $s(v) := 0$ if v is a finite place of K. Recall that $\sum_{v \in M_K} s(v) = [K : \mathbb{Q}]$. Now the lemma follows easily by observing that

$$\max(|y_1 z_2 - y_2 z_1|_v, |z_1 x_2 - z_2 x_1|_v, |x_1 y_2 - x_2 y_1|_v)$$
$$\leq 2^{s(v)} \max(|x_1|_v, |y_1|_v, |z_1|_v) \max(|x_2|_v, |y_2|_v, |z_2|_v)$$

for $v \in M_K$, and then taking the product over $v \in M_K$, taking logarithms, and dividing by $[K : \mathbb{Q}]$. $\qquad\square$

Lemma 6.4.4 *Let* $\mathbf{x}_i = (x_{i1}, x_{i2}) \in \overline{\mathbb{Q}}^* \times \overline{\mathbb{Q}}^*$ *with* $x_{i1} + x_{i2} = 1$ *for* $i = 1, 2$ *and with* $\mathbf{x}_1 \neq \mathbf{x}_2$. *Then*

$$h(\mathbf{x}_1) \leq h\left(\mathbf{x}_2 \mathbf{x}_1^{-1}\right) + \log 2.$$

Proof. Apply Lemma 6.4.3 with $(a, b, c) = (x_{11}, y_{11}, -1)$, $(x_1, y_1, z_1) = (1, 1, 1)$, $(x_2, y_2, z_2) = (x_{21} x_{11}^{-1}, x_{22} x_{21}^{-1}, 1)$. $\qquad\square$

Lemma 6.4.5 *Let* $\mathbf{x}_i = (x_{i1}, x_{i2}) \in \overline{\mathbb{Q}}^* \times \overline{\mathbb{Q}}^*$ *with* $x_{i1} + x_{i2} = 1$ *for* $i = 1, 2$. *Then for every positive integer* N *there is* $M \in \{N, N + 1\}$ *such that*

$$h(\mathbf{x}_1) \leq \frac{1}{M+1} h\left(\mathbf{x}_2 \mathbf{x}_1^{-2M-1}\right) + \log 8.$$

Proof. If both x_{11}, x_{12} are roots of unity, then $h(\mathbf{x}_1) = 0$ and the lemma is obviously true. Assume that x_{11}, x_{12} are not both roots of unity. Choose a number field K containing x_{i1}, x_{i2} for $i = 1, 2$. By (6.4.3) we have

$$x_{11}^{2M+1} W_M(x_{12}, -1) + x_{12}^{2M+1} W_M(-1, x_{11}) - W_M(x_{11}, x_{12}) = 0$$

for $M \in \mathbb{Z}_{>0}$, while also

$$x_{11}^{2M+1}\left(x_{21} x_{11}^{-2M-1}\right) + x_{12}^{2M+1}\left(x_{22} x_{12}^{-2M-1}\right) - 1 = 0.$$

Let N be a positive integer. By (6.4.4), and since $\mathbf{x}_1$ does not consist of roots of unity, there is $M \in \{N, N + 1\}$ such that the vectors $(x_{21}, x_{22}, -1)$ and

$$\left(x_{11}^{2M+1} W_M(x_{12}, -1), x_{12}^{2M+1} W_M(-1, x_{11}), -W_M(x_{11}, x_{12})\right)$$

are linearly independent. This implies that the two vectors

$$\left(x_{21} x_{11}^{-2M-1}, x_{22} x_{12}^{-2M-1}, -1\right),$$
$$\left(W_M(x_{12}, -1), W_M(-1, x_{11}), -W_M(x_{11}, x_{12})\right) =: (a, b, c)$$

are linearly independent. So by Lemma 6.4.3,

$$(2M + 1)h(\mathbf{x}_1) \leq h\left(\mathbf{x}_2 \mathbf{x}_1^{-2M-1}\right) + h^{\text{hom}}(a, b, c) + \log 2. \qquad (6.4.5)$$

We estimate $h^{\text{hom}}(a, b, c)$. Choose a number field K containing x_{11}, x_{12}. The binary form W_M has integer coefficients, whose absolute values have sum

$$\sum_{m=0}^{M} \binom{2M - m}{M - m}\binom{M + m}{m} = \binom{3M + 1}{M} \leq 2^{3M}$$

(this can be seen by comparing the coefficients of X^M in the power series identity $(1 - X)^{-M-1} \cdot (1 - X)^{-M-1} = (1 - X)^{-2M-2}$). As a consequence, we have for $v \in M_K$,

$$\max(|a|_v, |b|_v, |c|_v) \leq 2^{3Ms(v)} \max(1, |x_{11}|_v, |x_{12}|_v)^M.$$

By taking the product over $v \in M_K$, then logarithms and then dividing by $[K : \mathbb{Q}]$, we obtain

$$h^{\text{hom}}(a, b, c) \leq M \cdot h(\mathbf{x}_1) + 3M \log 2.$$

Together with (6.4.5) this gives

$$(M + 1)h(\mathbf{x}_1) \leq h\left(\mathbf{x}_2 \mathbf{x}_1^{-2M-1}\right) + (3M + 1)\log 2,$$

which easily implies our lemma. $\qquad\square$

The next result, which is needed to deal with the "small" solutions, is due to Beukers and Zagier.

Lemma 6.4.6 *Let* $\mathbf{x}_0 = (x_{i1}, x_{i2})$ $(i = 0, 1, 2)$ *be three distinct points from* $\overline{\mathbb{Q}}^* \times \overline{\mathbb{Q}}^*$ *with* $x_{i1} + x_{i2} = 1$ *for* $i = 0, 1, 2$. *Then*

$$h\left(\mathbf{x}_1 \mathbf{x}_0^{-1}\right) + h\left(\mathbf{x}_2 \mathbf{x}_0^{-1}\right) > 0.09.$$

Proof. This is a consequence of Corollary 2.4 of Beukers and Zagier (1997). A similar result of this type, with a lower bound $1/2400$ instead of 0.09, was obtained earlier by Schlickewei and Wirsing (1997). We give a sketch of the proof of Beukers and Zagier, referring for certain details to their paper.

Write $\mathbf{y}_i = (y_{i1}, y_{i2}) = \mathbf{x}_i \mathbf{x}_0^{-1}$ for $i = 1, 2$ and $a_j := x_{0j}$ for $j = 1, 2$. Then $(1, 1), \mathbf{y}_1, \mathbf{y}_2$ lie on the line $L : a_1 x_1 + a_2 x_2 = 1$. Hence

$$\begin{vmatrix} 1 & 1 & 1 \\ 1 & y_{11} & y_{12} \\ 1 & y_{21} & y_{22} \end{vmatrix} = 0. \tag{6.4.6}$$

Further,

$$\Delta := \begin{vmatrix} 1 & 1 & 1 \\ y_{11}y_{12} & y_{12} & y_{11} \\ y_{21}y_{22} & y_{22} & y_{21} \end{vmatrix} = y_{11}y_{12}y_{21}y_{22}\begin{vmatrix} 1 & 1 & 1 \\ 1 & y_{11}^{-1} & y_{12}^{-1} \\ 1 & y_{21}^{-1} & y_{22}^{-1} \end{vmatrix} \neq 0. \tag{6.4.7}$$

For assume the contrary. Then the three points $(1, 1)$, $\mathbf{y}_1$, $\mathbf{y}_2$ lie on a conic C : $b_1 x_1 x_2 + b_2 x_2 + b_3 x_1 = 0$, where at least two among the coefficients b_1, b_2, b_3 must be non-zero. It is easy to see that L and C can have no more than two points in common, giving a contradiction.

In what follows, we need functions on nine-dimensional complex space. We write points in $\mathbb{C}^9$ as $\mathbf{z}$, $(z_{00}, z_{01}, \ldots, z_{22})$, or as $(\mathbf{z}_0, \mathbf{z}_1, \mathbf{z}_2)$ where $\mathbf{z}_i = (z_{i0}, z_{i1}, z_{i2})$ for $i = 0, 1, 2$. Define $F : \mathbb{C}^9 \to \mathbb{C}$ by

$$F(\mathbf{z}) := \begin{vmatrix} z_{01}z_{02} & z_{02}z_{00} & z_{00}z_{01} \\ z_{11}z_{12} & z_{12}z_{10} & z_{10}z_{11} \\ z_{21}z_{22} & z_{22}z_{20} & z_{20}z_{21} \end{vmatrix}.$$

Notice that $F(\mathbf{z}) = (\prod_{i,j=0}^{2} z_{ij}) \det(z_{ij}^{-1})$ if all $z_{ij} \neq 0$. Let μ, ν be reals with

$$\mu \geq 0, \quad \nu \geq 0, \quad 2\mu + 3\nu = 1,$$

which will be chosen optimally later, and define $\Phi_{\mu,\nu} : \mathbb{C}^9 \to \mathbb{R}$ by

$$\Phi_{\mu,\nu}(\mathbf{z}) := |F(\mathbf{z})|^{\mu} \prod_{i,j=0}^{2} |z_{ij}|^{\nu}.$$

This is equal to $|\det(z_{ij}^{-1})|^{\mu} \prod_{i,j=0}^{2} |z_{ij}|^{\mu+\nu}$ if all $z_{ij} \neq 0$. Finally, define the set

$$D := \left\{ \mathbf{z} \in \mathbb{C}^9 : \det(z_{ij}) = 0, \ |z_{ij}| \leq 1 \text{ for } i, j = 0, 1, 2 \right\}$$

and put

$$m(\mu, \nu) := \sup_{\mathbf{z} \in D} \Phi_{\mu,\nu}(\mathbf{z}).$$

We first show that

$$h(\mathbf{y}_1) + h(\mathbf{y}_2) \geq -\log m(\mu, \nu). \tag{6.4.8}$$

Choose a number field K containing the coordinates of $\mathbf{y}_1, \mathbf{y}_2$. For $v \in M_K$, $i = 1, 2$, let $\lambda_{iv} := \max(1, |y_{i1}|_v, |y_{i2}|_v)$. First, let v be an infinite place, and choose an embedding $\sigma_v : K \hookrightarrow \mathbb{C}$ such that $|\cdot|_v = |\sigma_v(\cdot)|^{s(v)}$, where as usual, $s(v) = 1$ if v is real, $s(v) = 2$ if v is complex. Let $\mathbf{z} = (\mathbf{z}_0, \mathbf{z}_1, \mathbf{z}_2) = (z_{00}, \ldots, z_{22})$, where $\mathbf{z}_0 = (1, 1, 1)$, and $\mathbf{z}_i$ is a scalar multiple of $(1, \sigma_v(y_{i1}), \sigma_v(y_{i2}))$ such that $\max_{0 \leq j \leq 2} |z_{ij}| = 1$, for $i = 1, 2$. Then

$$\frac{|\Delta|_v^{\mu} \cdot |y_{11}y_{12}y_{21}y_{22}|_v^{\nu}}{(\lambda_{1v}\lambda_{2v})^{2\mu+3\nu}} = \Phi_{\mu,\nu}(\mathbf{z})^{s(v)} \leq m(\mu, \nu)^{s(v)}.$$

If v is finite then we have by the ultrametric inequality,

$$\frac{|\Delta|_v^{\mu} \cdot |y_{11}y_{12}y_{21}y_{22}|_v^{\nu}}{(\lambda_{1v}\lambda_{2v})^{2\mu+3\nu}} \leq 1.$$

By taking the product over $v \in M_K$, using (6.4.7) and the Product Formula, and the condition $2\mu + 3\nu = 1$, inequality (6.4.8) easily follows.

We first derive an upper bound for $m(\mu, \nu)$, and then determine the minimum of this upper bound over the set of $(\mu, \nu) \in \mathbb{R}^2_{\geq 0}$ with $2\mu + 3\nu = 1$. Since the set D is compact, the function $\Phi_{\mu,\nu}$ attains a maximum on D, say at $\mathbf{z}$. We use the fact that $\mathbf{z}$ can be chosen in such a way that at most one of the coordinates of $\mathbf{z}$ has absolute value < 1 (see Beukers and Zagier (1997), Lemma 3.2 for a proof). By symmetry, we may assume that $|z_{00}| \leq 1$ and $|z_{ij}| = 1$ if $(i, j) \neq (0, 0)$. So $z_{ij}^{-1} = \overline{z_{ij}}$ if $(i, j) \neq (0, 0)$. Assume for the moment that $z_{00} \neq 0$. Then

$$
\begin{aligned}
\Phi_{\mu,\nu}(\mathbf{z}) &= |z_{00}|^{\mu+\nu} |\det(z_{ij}^{-1})|^\mu = |z_{00}|^{\mu+\nu} |\det(z_{ij}^{-1}) - \det(\overline{z_{ij}})|^\mu \\
&= |z_{00}|^{\mu+\nu} \big(|z_{00}^{-1} - \overline{z_{00}}| \cdot |\overline{z_{11}} \cdot \overline{z_{22}} - \overline{z_{12}} \cdot \overline{z_{21}}| \big)^\mu \\
&\leq 2^\mu |z_{00}|^\nu (1 - |z_{00}|^2)^\mu.
\end{aligned}
$$

This is also true if $z_{00} = 0$ since in that case one can prove directly that $|F(\mathbf{z})| \leq 2$. Computing the maximum of $f(x) = 2^\mu x^\nu (1 - x^2)^\mu$ on $[0, 1]$, we obtain

$$
m(\mu, \nu) \leq 2^\mu \left(\frac{\nu}{2\mu + \nu} \right)^{\nu/2} \left(\frac{2\mu}{2\mu + \nu} \right)^\mu.
$$

We now let μ, ν vary, and determine the minimum of the right-hand side under the constraints $2\mu + 3\nu = 1$, $\mu \geq 0$, $\nu \geq 0$. Elementary calculus (see Beukers and Zagier (1997), Lemma 3.3) shows that this minimum is equal to the unique root $x_0 \in (0, 1)$ of $\frac{1}{2}x^2 + x^6 = 1$. Inserting this into (6.4.8), we obtain

$$
h(\mathbf{y}_1) + h(\mathbf{y}_2) \geq -\log x_0 > 0.09.
$$

This proves our lemma. □

We proceed further with equation (6.4.1) and assume henceforth that Γ is a finitely generated subgroup of $\overline{\mathbb{Q}}^* \times \overline{\mathbb{Q}}^*$ of rank r. Then there exist an algebraic number field K and a finite set of places S of K containing the infinite places, such that

$$
\Gamma \subseteq O_S^* \times O_S^*.
$$

Let $[K : \mathbb{Q}] = d$, $|S| = s$. We denote elements of $\mathbb{R}^{2s}$ as $\mathbf{u} = (u_{iv} : v \in S, i = 1, 2)$, and define a homomorphism from Γ to the additive group of $\mathbb{R}^{2s}$ by

$$
\varphi : (x_1, x_2) \mapsto \left(\frac{\log |x_i|_v}{d} : v \in S, i = 1, 2 \right).
$$

Let $V \subseteq \mathbb{R}^{2s}$ be the real vector space spanned by $\varphi(\Gamma)$. Then V has dimension r. By (6.3.8) we have

$$\widehat{h}(\mathbf{x}) = \|\varphi(\mathbf{x})\| \quad \text{for } \mathbf{x} \in \Gamma, \tag{6.4.9}$$

where $\| \cdot \|$ is the norm on $\mathbb{R}^{2s}$, given by

$$\|\mathbf{u}\| := \tfrac{1}{2} \sum_{v \in S} \sum_{i=1}^{2} |u_{iv}| \quad \text{for } \mathbf{u} = (u_{iv} : v \in S, \ i = 1, 2) \in \mathbb{R}^{2s}.$$

Denote by $\mathcal{S}$ the image under φ of the set of solutions of (6.4.1). We have collected some properties of $\mathcal{S}$.

Lemma 6.4.7 *For every* $\mathbf{u} \in \mathcal{S}$ *there are at most two solutions* $\mathbf{x}$ *of* (6.4.1) *such that* $\varphi(\mathbf{x}) = \mathbf{u}$.

Proof. Let v be an infinite place of K. Then there is an embedding $\sigma : K \hookrightarrow \mathbb{C}$ such that $|x|_v = |\sigma(x)|^{s(v)}$ for $x \in K$, where $s(v) = 1$ if v is real, $s(v) = 2$ if v is complex. Consider the solutions $\mathbf{x} = (x_1, x_2)$ of (6.4.1) with $\varphi(\mathbf{x}) = \mathbf{u}$. For these solutions, the absolute values $|\sigma(x_1)|$ and $|1 - \sigma(x_1)| = |\sigma(x_2)|$ have prescribed values, depending only on $\mathbf{u}$. In geometric terms, $\sigma(x_1)$ is an intersection point of two given circles in the complex plane that depend on $\mathbf{u}$. This implies that for any given $\mathbf{u}$ there are at most two possibilities for $\sigma(x_1)$, hence for $\mathbf{x}$. $\quad\square$

Lemma 6.4.8 *The set* $\mathcal{S}$ *has the following properties:*

(i) for any two distinct $\mathbf{u}_1, \mathbf{u}_2 \in \mathcal{S}$ *we have*

$$\|\mathbf{u}_1\| \leq 2\|\mathbf{u}_2 - \mathbf{u}_1\| + \log 4;$$

(ii) for any two distinct $\mathbf{u}_1, \mathbf{u}_2 \in \mathcal{S}$ *and any positive integer* N, *there is* $M \in \{N, N+1\}$ *such that*

$$\|\mathbf{u}_1\| \leq \frac{2}{M+1} \|\mathbf{u}_2 - (2M+1)\mathbf{u}_1\| + \log 64;$$

(iii) for any three distinct $\mathbf{u}_0, \mathbf{u}_1, \mathbf{u}_2 \in \mathcal{S}$ *we have*

$$\|\mathbf{u}_1 - \mathbf{u}_0\| + \|\mathbf{u}_2 - \mathbf{u}_0\| > 0.09.$$

Proof. This is simply a combination of Lemmas 6.4.4–6.4.6, the inequality $h(\mathbf{x}) \leq \widehat{h}(\mathbf{x}) \leq 2h(\mathbf{x})$ for $\mathbf{x} \in \overline{\mathbb{Q}}^* \times \overline{\mathbb{Q}}^*$, and (6.4.9). $\quad\square$

Our strategy to prove Theorem 6.4.1 is to estimate the cardinality of $\mathcal{S}$, using (i)–(iii) of Lemma 6.4.8. Then in view of Lemma 6.4.7, we only have to multiply the upper bound for $|\mathcal{S}|$ by 2 to get an upper bound for the number of solutions of (6.4.1).

We cover $\mathcal{S}$ by cones and estimate the number of points in a cone. Let $\theta > 0$ be a parameter whose value will be specified later. By Lemma 6.3.4, there is a set $\mathcal{E} \subset \{\mathbf{e} \in V : \|\mathbf{e}\| = 1\}$, with

$$|\mathcal{E}| \leq (1 + 2\theta^{-1})^r \tag{6.4.10}$$

such that for every non-zero $\mathbf{u} \in V$ there is $\mathbf{e} \in \mathcal{E}$ for which

$$\| \, \|\mathbf{u}\|^{-1}\mathbf{u} - \mathbf{e}\| \leq \theta. \tag{6.4.11}$$

For $\mathbf{e} \in \mathcal{E}$ denote by $\mathcal{S}_\mathbf{e}$ the set of $\mathbf{u} \in \mathcal{S}$ with (6.4.11). Notice that every $\mathbf{u} \in \mathcal{S}_\mathbf{e}$ can be written as

$$\mathbf{u} = \|\mathbf{u}\|\mathbf{e} + \mathbf{u}' \quad \text{with } \|\mathbf{u}'\| \leq \theta\|\mathbf{u}\|. \tag{6.4.12}$$

Lemma 6.4.9 *Let* $\mathbf{e} \in \mathcal{E}$, $0 < \theta < \frac{1}{9}$.

(i) For any two distinct $\mathbf{u}_1, \mathbf{u}_2 \in \mathcal{S}_\mathbf{e}$ *with* $\|\mathbf{u}_2\| \geq \|\mathbf{u}_1\| \geq \frac{\log 16}{1 - 9\theta}$ *we have*

$$\|\mathbf{u}_2\| \geq \tfrac{5}{4}\|\mathbf{u}_1\|.$$

(ii) For any two distinct $\mathbf{u}_1, \mathbf{u}_2 \in \mathcal{S}_\mathbf{e}$ *with* $\|\mathbf{u}_2\| \geq \|\mathbf{u}_1\| \geq \frac{\log 64}{1 - 9\theta}$ *we have*

$$\|\mathbf{u}_2\| < 10\theta^{-1}\|\mathbf{u}_1\|.$$

(iii) The set of $\mathbf{u} \in \mathcal{S}_\mathbf{e}$ *with* $\|\mathbf{u}\| \geq \frac{\log 16}{1 - 9\theta}$ *has cardinality at most*

$$3 + [\log(10\theta^{-1})/\log(5/4)].$$

Proof. Part (i) is an elementary gap principle. Part (ii) is the more involved result, based on the polynomial identities from Lemma 6.4.2. Part (iii) follows from (i) and (ii).

We first prove (i) and (ii). Let $\mathbf{u}_1, \mathbf{u}_2$ be distinct elements of $\mathcal{S}_\mathbf{e}$ with $\|\mathbf{u}_2\| \geq \|\mathbf{u}_1\|$. Put $\lambda_i := \|\mathbf{u}_i\|$ for $i = 1, 2$. By (6.4.12), we have $\mathbf{u}_i = \lambda_i\mathbf{e} + \mathbf{u}_i'$ with $\|\mathbf{u}_i'\| \leq \theta\lambda_i$ for $i = 1, 2$.

Assume that $\lambda_2 < \frac{5}{4}\lambda_1$. Then by property (i) of Lemma 6.4.8,

$$\lambda_1 \leq 2\|(\lambda_2 - \lambda_1)\mathbf{e} + \mathbf{u}_2' - \mathbf{u}_1'\| + \log 4$$

$$\leq 2(\lambda_2 - \lambda_1 + \theta\lambda_2 + \theta\lambda_1) + \log 4 < \left(\tfrac{1}{2} + \tfrac{9}{2}\theta\right)\lambda_1 + \log 4.$$

Hence $\lambda_1 < \frac{\log 16}{1 - 9\theta}$. This implies (i).

Next, assume that $\lambda_2 \geq 10\theta^{-1}\lambda_1$. Let N be the positive integer with $2N + 1 \leq \lambda_2/\lambda_1 < 2N + 3$, and let $M \in \{N, N + 1\}$ be the integer from Lemma 6.4.8 (ii). Thus, $|\lambda_2 - (2M + 1)\lambda_1| \leq 2\lambda_1$, and moreover,

$$M > \frac{4}{\theta}.$$

It follows that

$$\lambda_1 \le \tfrac{2}{M+1} \, \|(\lambda_2 - (2M+1)\lambda_1)\mathbf{e} + \mathbf{u}_2' - (2M+1)\mathbf{u}_1'\| + \log 64$$

$$\le \tfrac{2}{M+1} \, (2\lambda_1 + \lambda_2\theta + (2M+1)\lambda_1\theta) + \log 64$$

$$\le \tfrac{2}{M+1} \, (2 + (2M+3)\theta + (2M+1)\theta)\lambda_1 + \log 64$$

$$= \big(\tfrac{4}{M+1} + 8\theta\big)\lambda_1 + \log 64 < 9\theta\lambda_1 + \log 64,$$

implying $\lambda_1 < \frac{\log 64}{1-9\theta}$. This proves (ii).

We next prove (iii). We first consider the points $\mathbf{u} \in \mathcal{S}_e$ with

$$\frac{\log 16}{1 - 9\theta} \le \|\mathbf{u}\| < \frac{\log 64}{1 - 9\theta}. \tag{6.4.13}$$

Let $\mathbf{u}_1, \mathbf{u}_2, \ldots$ be these points, ordered such that $\|\mathbf{u}_1\| \le \|\mathbf{u}_2\| \le \ldots$. Then by (i), we have for the n-th point in this sequence that $\|\mathbf{u}_n\| \ge (5/4)^{n-1}\|\mathbf{u}_1\|$. Hence $(5/4)^{n-1} < (\log 64)/(\log 16) = 3/2$, implying $n \le 2$. So $\mathcal{S}_e$ has at most two points $\mathbf{u}$ with (6.4.13).

Next, we count the points $\mathbf{u} \in \mathcal{S}_e$ with

$$\|\mathbf{u}\| \ge \frac{\log 64}{1 - 9\theta}. \tag{6.4.14}$$

Similarly as above, we order these points in a sequence $\mathbf{u}_1, \mathbf{u}_2, \ldots$ such that $\|\mathbf{u}_1\| \le \|\mathbf{u}_2\| \le \ldots$. Then again, by (i), we have for the n-th point in this sequence that $\|\mathbf{u}_n\| \ge (5/4)^{n-1}\|\mathbf{u}_1\|$. On the other hand, by (ii) we have $\|\mathbf{u}_n\| < 10\theta^{-1}\|\mathbf{u}_1\|$. Hence $(5/4)^{n-1} < 10\theta^{-1}$. Thus, we obtain an upper bound

$$1 + [\log(10\theta^{-1})/\log(5/4)]$$

for the number of $\mathbf{u} \in \mathcal{S}_e$ with (6.4.14). Combined with the upper bound 2 for the number of points with (6.4.13), this gives (iii). $\qquad\square$

Proof of Theorem 6.4.1. Let $0 < \theta < \frac{1}{9}$. We divide $\mathcal{S}$ into large points, i.e., with $\|\mathbf{u}\| \ge \frac{\log 16}{1-9\theta}$, and small points, i.e., with $\|\mathbf{u}\| < \frac{\log 16}{1-9\theta}$.

Combining the upper bound (6.4.10) for $|\mathcal{E}|$ with (iii) of Lemma 6.4.8, we see that $\mathcal{S}$ has at most

$$\left(3 + \left[\frac{\log(10\theta^{-1})}{\log(5/4)}\right]\right) \cdot (1 + 2\theta^{-1})^r$$

large points.

To estimate the number of small points of $\mathcal{S}$, we observe that by Lemma 6.4.8 (iii), for any $\mathbf{u}_0 \in \mathcal{S}$, the set

$$\{\mathbf{u} \in \mathcal{S} : \|\mathbf{u} - \mathbf{u}_0\| \le 0.045\}$$

has cardinality at most 2. By Lemma 6.3.4, the set of small points of S can be covered by at most

$$\left(1 + \frac{2\log 16}{0.045(1 - 9\theta)}\right)^r$$

such sets. Hence S has at most

$$2\left(1 + \frac{2\log 16}{0.045(1 - 9\theta)}\right)^r$$

small points.

We now choose θ such that $\theta^{-1} = (\log 16)/0.045(1 - 9\theta)$, i.e., $\theta^{-1} = 9 + (\log 16)/0.045 = 70.613\ldots$, add the upper bounds for the number of large points and the number of small points of S obtained above, and finally multiply with 2 to get an upper bound for the number of solutions of (6.4.1). The resulting bound is 68×143^r, which is smaller than the bound stated in Theorem 6.4.1. $\quad\square$

6.5 Proof of Theorem 6.1.6

Let K be a field of characteristic 0, $a_1, a_2 \in K^*$, and Γ a subgroup of $K^* \times K^*$ of finite rank r. We consider equations

$$a_1 x_1 + a_2 x_2 = 1 \quad \text{in } (x_1, x_2) \in \Gamma \qquad (6.1.6)$$

having at least three distinct solutions, and we have to show that there are at most $B(r)$ possibilities for the Γ-equivalence class of (a_1, a_2), where $B(r)$ is given by (6.1.7).

Thus, let (u_1, u_2), (v_1, v_2), (w_1, w_2) be three distinct solutions of (6.1.6). Then

$$\begin{vmatrix} 1 & u_1 & u_2 \\ 1 & v_1 & v_2 \\ 1 & w_1 & w_2 \end{vmatrix} = 0,$$

i.e.

$$v_1 w_2 - v_2 w_1 + u_2 w_1 - u_1 w_2 + u_1 v_2 - u_2 v_1 = 0 \qquad (6.5.1)$$

and

each 2×2 subdeterminant of the above determinant is $\neq 0$. $\qquad (6.5.2)$

A vanishing subsum of the left-hand side of (6.5.1) is called minimal if none of the proper subsums of this subsum is 0. We have to distinguish various cases depending on how (6.5.1) splits into minimal vanishing subsums. Clearly, two possible splittings that can be transformed into each other by permuting (u_1, u_2), (v_1, v_2), (w_1, w_2) or interchanging the indices $(1, 2)$ can be treated in the same manner. Notice that, in this way, one can derive at most 12 splittings from a given splitting. More precisely, after permuting (u_1, u_2), (v_1, v_2), (w_1, w_2) or interchanging $(1, 2)$, we are left with the following cases:

(I) no proper subsum of the left-hand side of (6.5.1) vanishes,

(II) $v_1 w_2 + u_2 w_1 = 0$, $-v_2 w_1 - u_1 w_2 + u_1 v_2 - u_2 v_1 = 0$,

(III) $v_1 w_2 - v_2 w_1 + u_2 w_1 = 0$, $-u_1 w_2 + u_1 v_2 - u_2 v_1 = 0$,

(IV) $v_1 w_2 + u_2 w_1 + u_1 v_2 = 0$, $-v_2 w_1 - u_1 w_2 - u_2 v_1 = 0$,

(V) $v_1 w_2 + u_2 w_1 - u_2 v_1 = 0$, $-v_2 w_1 - u_1 w_2 + u_1 v_2 = 0$.

Other splittings into minimal vanishing subsums are in conflict with (6.5.2).

We define the quantities $y_1 := v_1/u_1$, $y_2 := v_2/u_2$, $z_1 := w_1/u_1$, $z_2 := w_2/u_2$. The pair (y_1, y_2) determines uniquely the Γ-equivalence class of (a_1, a_2), since $(a_1 u_1, a_2 u_2)$ is the unique solution (ξ_1, ξ_2) to $\xi_1 + \xi_2 = 1$, $y_1 \xi_1 + y_2 \xi_2 = 1$. Likewise, (z_1, z_2) determines uniquely the Γ-equivalence class of (a_1, a_2).

Case I. We rewrite (6.5.1) (by dividing by $u_2 v_1$) as

$$z_2 - \frac{y_2 z_1}{y_1} + \frac{z_1}{y_1} - \frac{z_2}{y_1} + \frac{y_2}{y_1} = 1.$$

Let Γ' be the image of $\Gamma \times \Gamma$ under the group homomorphism

$$((y_1, y_2), (z_1, z_2)) \mapsto \left(z_2, \frac{y_2 z_1}{y_1}, \frac{z_1}{y_1}, \frac{z_2}{y_1}, \frac{y_2}{y_1} \right).$$

Then Γ' has rank at most $2r$, and $(z_2, \ldots, \frac{y_2}{y_1})$ is a non-degenerate solution of

$$x_1 - x_2 + x_3 - x_4 + x_5 = 1 \quad \text{in } (x_1, \ldots, x_5) \in \Gamma'.$$

By Theorem 6.1.3 there are at most $A(5, 2r)$ possibilities for $(z_2, \ldots, \frac{y_2}{y_1})$. Such a tuple determines z_2 and $\frac{y_2 z_1}{y_1} (\frac{y_2}{y_1})^{-1} = z_1$, hence the Γ-equivalence class of (a_1, a_2). So case I gives rise to at most

$$A(5, 2r)$$

possible Γ-equivalence classes of pairs (a_1, a_2).

Case II. This implies

$$\frac{y_1 z_2}{z_1} = -1, \quad -\frac{y_2 z_1}{y_1} - \frac{z_2}{y_1} + \frac{y_2}{y_1} = 1.$$

Theorem 6.1.3 implies that we have at most $A(3, 2r)$ possibilities for the triple $(-\frac{y_2 z_1}{y_1}, \frac{z_2}{y_1}, \frac{y_2}{y_1})$ (using again the argument based on a homomorphic image of

$\Gamma \times \Gamma$). In combination with $\frac{y_1 z_2}{z_1} = -1$ this tuple determines $\frac{y_1 z_2}{z_1} \frac{y_2 z_1}{y_1} \frac{y_1}{z_2} = y_1 y_2$, hence $y_1^2 = y_1 y_2 y_1 / y_2$. This leads to two possibilities for (y_1, y_2), hence two possible Γ-equivalence classes for (a_1, a_2). So case II gives rise to at most

$$2A(3, 2r)$$

possible Γ-equivalence classes of pairs (a_1, a_2).

Case III. This implies

$$\frac{y_1 z_2}{z_1} - y_2 = 1, \quad -\frac{z_2}{y_1} + \frac{y_2}{y_1} = 1.$$

By Theorem 6.1.3 we have at most $A(2, 2r)^2$ possibilities for $(\frac{y_1 z_2}{z_1}, y_2, \frac{z_2}{y_1}, \frac{y_2}{y_1})$. Each such tuple determines uniquely the pair (y_1, y_2), hence the Γ-equivalence class of (a_1, a_2). So case III gives rise to at most

$$A(2, 2r)^2$$

possible Γ-equivalence classes of pairs (a_1, a_2).

Case IV. This implies

$$-\frac{y_1 z_2}{y_2} - \frac{z_1}{y_2} = 1, \quad -\frac{y_2 z_1}{y_1} - \frac{z_2}{y_1} = 1.$$

According to Theorem 6.1.3, there are at most $A(2, 2r)^2$ possibilities for the tuple $(\frac{y_1 z_2}{y_2}, \frac{z_1}{y_2}, \frac{y_2 z_1}{y_1}, \frac{z_2}{y_1})$. From this tuple we can compute

$$\frac{y_2}{y_1 z_2} \frac{y_2}{z_1} \frac{y_2 z_1}{y_1} \frac{z_2}{y_1} = \left(\frac{y_2}{y_1}\right)^3.$$

This gives three possibilities for $\frac{y_2}{y_1}$, hence for (z_1, z_2), and hence for the Γ-equivalence class of (a_1, a_2). Thus in case IV there are at most

$$3A(2, 2r)^2$$

possible Γ-equivalence classes of pairs (a_1, a_2).

Case V. This implies

$$z_2 + \frac{z_1}{y_1} = 1, \quad z_1 + \frac{z_2}{y_2} = 1.$$

Theorem 6.1.3 implies that we have at most $A(2, 2r)^2$ possibilities for the triple $(z_2, \frac{z_1}{y_1}, z_1, \frac{z_2}{y_2})$, hence for (z_1, z_2), and hence for the Γ-equivalence class of (a_1, a_2). So in case V we have at most

$$A(2, 2r)^2$$

possibilities for the Γ-equivalence class of (a_1, a_2).

By adding the upper bounds for the numbers of possible Γ-equivalence classes of pairs (a_1, a_2) found in cases I–V, and multiplying this with the number of permutations of (u_1, u_2), (v_1, v_2), (w_1, w_2) and of the indices 1, 2, we obtain that the number of Γ-equivalence classes of pairs $(a_1, a_2) \in K^* \times K^*$ such that equation (6.1.6) has more than two solutions is at most

$$12(A(5, 2r) + 2A(3, 2r) + 5A(2, 2r)^2) = B(r).$$

This proves Theorem 6.1.6. $\qquad\qquad\qquad\qquad\qquad\qquad\qquad\qquad\square$

6.6 Proofs of Theorems 6.1.7 and 6.1.8

Proof of Theorem 6.1.7. We follow Konyagin and Soundararajan (2007). Recall that we are considering the equation

$$x_1 + x_2 = 1 \quad \text{in } x_1, x_2 \in \left\{ \pm p_1^{z_1} \cdots p_t^{z_t} : z_1, \ldots, z_t \in \mathbb{Z} \right\}, \qquad (6.1.8)$$

where $S = \{p_1, \ldots, p_t\}$ is a set of distinct primes. We intend to show that for every $\beta < 2 - \sqrt{2}$ there are sets of primes S of arbitrarily large cardinality t such that the number $N(S)$ of solutions of (6.1.8) is at least $\exp(t^\beta)$.

Let y be a large real number and fix real numbers β, γ with $0 < \beta, \gamma < 1$, which will be chosen optimally later. We introduce two sets $\mathcal{L}, \mathcal{M}$. The set $\mathcal{L}$ is the set of numbers that are the product of exactly $[y^\beta]$ distinct primes from the interval $[y/2, y]$, while $\mathcal{M}$ is the set of numbers that are the product of exactly $[\gamma y^\beta]$ distinct primes from $[y/4, y/2)$. Thus, the integers in $\mathcal{L}$ are coprime to those in $\mathcal{M}$.

Using the Prime Number Theorem and

$$\frac{\log \binom{a}{b}}{b \log(a/b)} \to 1 \quad \text{as } b, \frac{a}{b} \to \infty$$

(which follows from Stirling's Formula) we obtain that the set $\mathcal{L}$ has cardinality

$$|\mathcal{L}| = \binom{\pi(y) - \pi(y/2)}{[y^\beta]} = L^{1-\beta+o(1)}, \qquad (6.6.1)$$

where $L := y^{[y^\beta]}$ and here and below, $o(1)$ is used to denote functions of y that tend to 0 as $y \to \infty$. In a similar manner,

$$|\mathcal{M}| = L^{\gamma(1-\beta)+o(1)}. \qquad (6.6.2)$$

The idea is to find a positive integer u for which there are many triples (l_1, l_2, m) such that $l_1, l_2 \in \mathcal{L}$, $m \in \mathcal{M}$ and $(l_1 - l_2)/m = u$, and then take for S the set of primes in $[y/4, y]$ and those dividing u. Then the pairs $(um/l_1, l_2/l_1)$ yield many solutions to (6.1.8).

We first count the number of triples (l_1, l_2, m) with

$$l_1 \equiv l_2 (\mathrm{mod}\, m), \quad l_1, l_2 \in \mathcal{L}, \quad m \in \mathcal{M}, \quad l_1 > l_2. \qquad (6.6.3)$$

Let $m \in \mathcal{M}$. For $a \in \mathbb{Z}$, denote by $r(\mathcal{L}, a, m)$ the number of elements in $\mathcal{L}$ that lie in the residue class $a \pmod m$. By the Cauchy–Schwarz inequality,

$$\sum_{a=1}^{m} r(\mathcal{L}, a, m)^2 \geq \frac{1}{m} \left(\sum_{a=1}^{m} r(\mathcal{L}, a, m) \right)^2 = \frac{|\mathcal{L}|^2}{m}.$$

The left-hand side counts the pairs (l_1, l_2) in $\mathcal{L}$ that are congruent modulo m. Among these, there are $|\mathcal{L}|$ trivial solutions with $l_1 = l_2$. Note that $m \leq y^{[\gamma y^\beta]} \leq L^\gamma$. We assume that $\gamma < 1 - \beta$. Then in view of (6.6.1), the integer m is of smaller order of magnitude than $|\mathcal{L}|$. Deleting the pairs with $l_1 = l_2$ and using symmetry, we infer that for any fixed $m \in \mathcal{M}$, the number of pairs $l_1, l_2 \in \mathcal{L}$ with $l_1 > l_2$ that are congruent modulo m is bounded below by

$$\frac{|\mathcal{L}|^2}{2m} - |\mathcal{L}| = L^{2(1-\beta)-\gamma+o(1)}.$$

Now, using (6.6.2), and summing over the elements of $\mathcal{M}$, we see that the number of triples (l_1, l_2, m) with (6.6.3) is at least

$$L^{2(1-\beta)-\beta\gamma+o(1)}.$$

The elements of $\mathcal{L}$ are all $\leq y^{[y^\beta]} = L$, and the integers of $\mathcal{M}$ are all $\geq (y/4)^{[\gamma y^\beta]}$, so the integers $(l_1 - l_2)/m$ with l_1, l_2, m satisfying (6.6.3), are all bounded above by $L^{1-\gamma+o(1)}$. If we assume that $2(1 - \beta) - \beta\gamma > 1 - \gamma$, or equivalently,

$$(2 + \gamma)(1 - \beta) > 1,$$

we see that there is a positive integer $u \leq L^{1-\gamma+o(1)}$ with the property that there are at least

$$L^{2(1-\beta)-\beta\gamma-1+\gamma+o(1)} = L^{(2+\gamma)(1-\beta)-1+o(1)}$$

triples (l_1, l_2, m) with $l_1, l_2 \in \mathcal{L}$, $m \in \mathcal{M}$ and $(l_1 - l_2)/m = u$. Notice that $\gcd(l_1, l_2)$ is a divisor of u. By elementary number theory (see, e.g., Hardy and Wright (1980), chapter XVIII, Theorem 317), the number u has at most $u^{O(1/\log\log u)} = L^{o(1)}$ divisors. Hence there are a divisor v of u, and at least $L^{(2+\gamma)(1-\beta)-1+o(1)}$ triples (l_1', l_2', m) such that l_1', l_2' are coprime integers composed of primes from $[y/2, y]$, m is an integer composed of primes from $[y/4, y)$, and $(l_1' - l_2')/m = v$. Let S consist of the primes from $[y/4, y]$ and the primes dividing v. Then these triples (l_1', l_2', m) yield at least $L^{(2+\gamma)(1-\beta)-1+o(1)}$ solutions $(mv/l_1', l_2'/l_1')$ to (6.1.8).

In the course of the proof we assumed that $\gamma < 1 - \beta$ and $(2 + \gamma)(1 - \beta) >$ 1. Such a number γ exists precisely if $\beta < 2 - \sqrt{2}$. Since $v \le u \le L^{1-\gamma+o(1)}$, the cardinality of S is at most $\pi(y) - \pi(y/4) + \log v < y$ for y sufficiently large. Further, for y sufficiently large, we have

$$L^{(2+\gamma)(1-\beta)-1+o(1)} = y^{[y^\beta]((2+\gamma)(1-\beta)-1+o(1))} > \exp(y^\beta).$$

This completes the proof of Theorem 6.1.7.   □

Proof of Theorem 6.1.8. Let $\beta < 2 - \sqrt{2}$ and choose t and a set of primes $S = \{p_1, \ldots, p_t\}$ such that $N(S) \ge \exp(t^\beta) =: A(t)$. We consider

$$x_1 + \cdots + x_n = 1 \quad \text{in } x_1, \ldots, x_n \in \{\pm p_1^{z_1} \cdots p_t^{z_t} : z_1, \ldots, z_t \in \mathbb{Z}\}. \quad (6.1.9)$$

Recall that $g(n, S)$ denotes the minimal integer g such that there exists a nonzero polynomial $P \in \mathbb{C}[X_1, \ldots, X_n]$ of total degree g, which is not divisible by $X_1 + \cdots + X_n - 1$, and which has the property that $P(x_1, \ldots, x_n) = 0$ for every solution $(x_1, \ldots, x_n)$ of (6.1.9). We prove by induction on n that $g(n, S) \ge A(t)$ for $n \ge 2$.

First, let $n = 2$. Let $P \in \mathbb{C}[X_1, X_2]$ be a polynomial of total degree $g(2, S)$, not divisible by $X_1 + X_2 - 1$, such that $P(x_1, x_2) = 0$ for every solution (x_1, x_2) of (6.1.8). Let $Q(X) := P(X, 1 - X)$. Then $Q(x_1) = 0$ for every x_1 for which there exists x_2 such that (x_1, x_2) is a solution of (6.1.8). Hence

$$g(2, S) = \deg P = \deg Q \ge A(t).$$

Suppose now that $n \ge 3$, and that $g(n - 1, S) \ge A(t)$ is known to hold. Let U be the set of tuples

$$(x_1, \ldots, x_n) = (y_1, \ldots, y_{n-2}, y_{n-1}x_1, y_{n-1}x_2),$$

where $(y_1, \ldots, y_{n-1})$ runs through the solutions of

$$y_1 + \cdots + y_{n-1} = 1, \quad y_1, \ldots, y_{n-1} \in \{\pm p_1^{z_1} \cdots p_t^{z_t} : z_1, \ldots, z_t \in \mathbb{Z}\} \quad (6.6.4)$$

and where (x_1, x_2) runs through the solutions of (6.1.8). By construction, the tuples in U satisfy

$$y_1 + \cdots + y_{n-2} + y_{n-1}(x_1 + x_2) = 1$$

and so they are solutions of (6.1.9).

Let $P \in \mathbb{C}[X_1, \ldots, X_n]$ be a polynomial of total degree $g(n, S)$, not divisible by $X_1 + \cdots + X_n - 1$, such that $P(x_1, \ldots, x_{n-1}, x_n) = 0$ for every solution

$(x_1, \ldots, x_n)$ of (6.1.8). Put

$$Q(X_1, \ldots, X_{n-1}) := P(X_1, \ldots, X_{n-1}, 1 - X_1 - \cdots - X_{n-1}).$$

Then Q has total degree $g(n, S)$, and is not identically 0. So we have to prove that Q has total degree at least $A(t)$.

Clearly, we have

$$Q(y_1, \ldots, y_{n-2}, y_{n-1}x_1) = 0 \qquad (6.6.5)$$

for every solution $(y_1, \ldots, y_{n-1})$ of (6.6.4) and every solution (x_1, x_2) of (6.1.8). Define a new polynomial in $n - 1$ variables,

$$Q^*(Y_1, \ldots, Y_{n-2}, Z) := Q(Y_1, \ldots, Y_{n-2}, Z \cdot (1 - Y_1 - \cdots - Y_{n-2})).$$
$$(6.6.6)$$

Then Q^* is not identically zero since Q is not identically zero and since the change of variables

$$(X_1, \ldots, X_{n-1}) \mapsto (Y_1, \ldots, Y_{n-2}, Z \cdot (1 - Y_1 - \cdots - Y_{n-2}))$$

is invertible. Now from (6.6.6), (6.6.5), it follows that

$$Q^*(y_1, \ldots, y_{n-2}, x_1) = 0 \qquad (6.6.7)$$

for every solution $(y_1, \ldots, y_{n-1})$ of (6.6.4) and every solution (x_1, x_2) of (6.1.8). We distinguish two cases.

Case I. There is a solution (x_1, x_2) of (6.1.8) such that the polynomial

$$Q^*_{x_1}(Y_1, \ldots, Y_{n-2}) := Q^*(Y_1, \ldots, Y_{n-2}, x_1)$$

is not identically zero. Then by (6.6.7), $Q^*_{x_1}$ is a non-zero polynomial with $Q^*_{x_1}(y_1, \ldots, y_{n-2}) = 0$ for every solution $(y_1, \ldots, y_{n-1})$ of (6.6.4). Hence $Q^*_{x_1}$ has total degree

$$\geq g(n - 1, S) \geq A(t).$$

Now by (6.6.6) this implies that the total degree of Q is at least $A(t)$.

Case II. The polynomial $Q^*_{x_1}(Y_1, \ldots, Y_{n-2})$ is identically zero for every solution (x_1, x_2) of (6.1.8). Then since (6.1.8) has at least $A(t)$ solutions, the polynomial Q^* must have degree at least $A(t)$ in the variable Z. By (6.6.6) this implies that Q has degree at least $A(t)$ in the variable X_{n-1}. So again we conclude that the total degree of Q is at least $A(t)$. This completes our induction step and our proof. $\square$

6.7 Notes

We recall some history concerning the number of solutions of unit equations and discuss some related results.

- Lewis and Mahler (1961) obtained an explicit upper bound for the number of solutions of the S-unit equation over $\mathbb{Q}$,

$$x_1 + x_2 = 1 \quad \text{in } x_1, x_2 \in \mathbb{Z}_S^*, \tag{6.7.1}$$

where $S = \{\infty, p_1, \ldots, p_t\}$ with distinct primes $p_1, \ldots, p_t$ and $\mathbb{Z}_S^* = \{\pm p_1^{z_1} \cdots p_t^{z_t} : z_i \in \mathbb{Z}\}$ is the corresponding group of S-units. But their bound depends on $p_1, \ldots, p_t$. Lewis and Mahler derived this by applying a general result of theirs on Thue–Mahler equations to equations of the type $|ax^n + by^n| = p_1^{z_1} \cdots p_t^{z_t}$ in $x, y, z_1, \ldots, z_t \in \mathbb{Z}$. In fact, as was unnoticed by Lewis and Mahler, applying their general result instead to $|xy(x+y)| = p_1^{z_1} \cdots p_t^{z_t}$ implies an upper bound c^{t+1} for the number of solutions of (6.7.1), with c an absolute constant independent of $p_1, \ldots, p_t$. A similar result was independently obtained by Silverman around 1984, by a different method (unpublished).

The above result was generalized and improved by Evertse (1984a) as follows. Let K be an algebraic number field of degree d, S a finite set of places of K of cardinality s containing the infinite places, and $a_1, a_2 \in K^*$. Then the equation

$$a_1 x_1 + a_2 x_2 = 1 \quad \text{in } x_1, x_2 \in O_S^*$$

has at most $3 \times 7^{d+2s}$ solutions. We note that earlier Győry (1979), under certain assumptions concerning the S-norms of a_1 and a_2, obtained the better upper bound $4s + 1$ by means of the theory of logarithmic forms. These were the first upper bounds that depend only on d and s, but not on the coefficients a_1, a_2.

- Schlickewei considered the equation

$$a_1 x_1 + a_2 x_1 = 1 \quad \text{in } (x_1, x_2) \in \Gamma, \tag{6.1.6}$$

where a_1, a_2 are non-zero elements of an arbitrary field K of characteristic 0, and Γ is a subgroup of $K^* \times K^*$ of finite rank r. He derived a uniform upper bound for the number of solutions, depending only on r. His unpublished result was later improved by Beukers and Schlickewei (1996) who obtained the upper bound $2^{8(r+2)}$ for the number of solutions.

- First Poe (1997) alone in a special case, and then Poe together with Bombieri and Mueller (Bombieri, Mueller and Poe (1997)) developed a "cluster principle" for the solutions of (6.1.6), in the case that K is a number field and Γ

is again a subgroup of $K^* \times K^*$ of rank r. Here, a cluster is a set of solutions such that for any two solutions $\mathbf{x}_1, \mathbf{x}_2$ in the cluster, the height $h(\mathbf{x}_1\mathbf{x}_2^{-1})$ is small, and the cluster principle gives an upper bound for the number of such clusters. By combining this cluster principle with Baker-type upper bounds for the heights of the solutions of (6.1.6), Bombieri et al. proved that (6.1.6) has at most $d^{9r}e^{86r^2}$ solutions, where $d = [K : \mathbb{Q}]$. Although this bound is much larger than that of Beukers and Schlickewei, the method of proof is very different, and it may be applicable to other situations.

- In the special case when K is a number field and $\Gamma = O_S^* \times O_S^*$, where O_S^* is the group of S-units for some finite set of places S of K containing the infinite places, a weaker but effective version of Theorem 6.1.6 was established in Evertse, Győry, Stewart and Tijdeman (1988a). Using some earlier versions of Theorems 3.2.5, 3.2.7 and Corollary 4.1.5, due to Baker, van der Poorten and Győry, respectively, it was proved that apart from finitely many and effectively determinable O_S^*-equivalence classes of pairs $(a_1, a_2) \in K^* \times K^*$, the equation $a_1x_1 + a_2x_2 = 1$ has at most $s + 1$ solutions $(x_1, x_2) \in O_S^* \times O_S^*$, where s denotes the cardinality of S. Further, in the case when S is the set of infinite places of K and $a_1, a_2 \in \mathbb{Q}^*$, the following more precise result was obtained in Brindza and Győry (1990). For given coprime positive integers a_1, a_2, there are only finitely many and effectively determinable positive integers c such that the equation $a_1x_1 + a_2x_2 = c$ has more than one solution (up to conjugacy) in $x_1, x_2 \in O_K^*$. The proof utilizes a simultaneous variant of Baker's method.

- Corvaja and Zannier (2006), and in a more general extent Levin (2006) considered one-parameter families of S-unit equations

$$a_1(t)x_1 + a_2(t)x_2 = c(t) \quad \text{in } t \in K, \ x_1, x_2 \in O_S^*, \tag{6.7.2}$$

where as before K is a number field, S a finite set of places of K containing all infinite places, and where $a_1, a_2, c \in K[X]$ are given polynomials. In his paper, Levin proved, among other things, that if (a_1, a_2, c) is a general triple of non-constant polynomials with $\deg a_1 + \deg a_2 = \deg c > 2$, then (6.7.2) has only finitely many solutions with $a_1(t)a_2(t)c(t) \neq 0$. Here "general" means that if we view triples (a_1, a_2, c) as points in the affine space $K^{2 \deg c + 3}$, then the set of triples (a_1, a_2, c) for which the above mentioned finiteness result does not hold is contained in a proper Zariski closed subset of $K^{2 \deg c + 3}$. Levin's proof allows us to effectively determine this Zariski closed subset. Notice that Levin's result provides many examples of S-unit equations that have no solutions. In his proof, Levin heavily uses the finiteness results of Corvaja and Zannier (2002a, 2004b) on S-integral points on curves and surfaces, which they derived from the Subspace Theorem. As a consequence, Levin's finiteness result on (6.7.2) is ineffective.

- As was mentioned in Section 4.7, Győry and Pintér (2008) considered over $\mathbb{Q}$ the three-parameter family of S-unit equations

$$u_1^n x_1 + u_2^n x_2 = 1 \quad \text{in } u_1, u_2 \in \mathbb{Z} \setminus \{0\}, n \geq 3, x_1, x_2 \in \mathbb{Z}_S^*$$
$$\text{with } \gcd(u_1 u_2, p_1 \cdots p_t) = 1, \qquad (6.7.3)$$

where $p_1, \ldots, p_t$ are distinct rational primes, $S = \{\infty, p_1, \ldots, p_t\}$ and $\mathbb{Z}_S^*$ denotes the group of S-units in $\mathbb{Q}$. They showed that apart from finitely many and effectively computable pairs (u_1^n, u_2^n), the equations under consideration have no solution in x_1, x_2.

- We now compare the above result with the special case $K = \mathbb{Q}, \Gamma = \mathbb{Z}_S^* \times \mathbb{Z}_S^*$ of Theorem 6.1.6 and with the remark occurring after that theorem. Further, we complete these results with a new one.

For given $a_1, a_2 \in \mathbb{Q}^*$, consider the equation

$$a_1 x_1 + a_2 x_2 = 1 \text{ in } x_1, x_2 \in \mathbb{Z}_S^*. \qquad (6.7.4)$$

We call two pairs $(a_1, a_2), (b_1, b_2) \in \mathbb{Q}^* \times \mathbb{Q}^*$ S-equivalent if they are $\mathbb{Z}_S^* \times \mathbb{Z}_S^*$-equivalent, i.e., if there is $(\varepsilon_1, \varepsilon_2) \in \mathbb{Z}_S^* \times \mathbb{Z}_S^*$ such that $b_i = a_i \varepsilon_i$ for $i = 1, 2$. Then the number of solutions of (6.7.4) does not change if (a_1, a_2) is replaced by an S-equivalent pair.

Theorem 6.7.1 *The following assertions hold.*

 (i) *There are only finitely many S-equivalence classes of pairs (a_1, a_2) in $\mathbb{Q}^* \times \mathbb{Q}^*$ for which equation (6.7.4) has more than two solutions.*

(ii) *For each $N \in \{0, 1, 2\}$, there are infinitely many S-equivalence classes of pairs $(a_1, a_2) \in \mathbb{Q}^* \times \mathbb{Q}^*$ such that equation (6.7.4) has exactly N solutions.*

The assertion (i) is a special case of Theorem 6.1.6, hence is ineffective. The statement (ii) for $N = 2$ has been proved in a more general form after the enunciation of Theorem 6.1.6, while for $N = 0$ is an immediate consequence of the above result concerning equation (6.7.3). We now give a sketch of the proof for $N = 1$. The proof of each case of (ii) is constructive.

Sketch of the proof of the case $N = 1$ of (ii). Let A be a large integer, and $\mathscr{S}$ the set of integers composed of the primes $p_1, \ldots, p_t$. Denote by $H(A)$ the set of pairs (a, b) of relatively prime positive integers a, b with $a, b \leq A$, and by $P(A)$ the set of those triples (a, b, c) of positive integers a, b, c for which $(a, b) \in H(A), \gcd(ab, p_1 \cdots p_t) = 1$ and $a + b = c$. It is known that $H(A)$ has cardinality at least $c_1 A^2$, where c_1 is an effectively computable positive absolute constant. This implies that the cardinality of $P(N)$ is at least $c_2 A^2$. Here c_2 and c_3, c_4, c_5 below are effectively computable positive numbers depending only on $p_1, \ldots, p_t$. If x, y, z is a solution of the equation

$$ax + by = cz \quad \text{in } x, y, z \in \mathscr{S} \text{ with } \gcd(x, y, z) = 1 \qquad (6.7.5)$$

for some (a, b, c) in $P(A)$, then by Corollary 4.1.5

$$\max(|x|, |y|, |z|) \leq c_3 A^{c_4}.$$

Thus the total number of triples (x, y, z) with relatively prime $x, y, z \in \mathscr{S}$ for which there exists (a, b, c) in $P(A)$ satisfying (6.7.5) is at most $c_5(\log A)^{3t}$.

If $(a, b, c) \in P(A)$ so that (6.7.5) holds for some relatively prime x, y, z in $\mathscr{S}$ and (x, y, z) is not $(1, 1, 1)$ or $(-1, -1, -1)$, then $(a, b, c) = (y - z, z - x, y - x)/d$ with $d = \gcd(y - z, z - x, y - x)$. Hence (a, b, c) is uniquely determined by (x, y, z). Consequently, the number of $(a, b, c) \in P(A)$ for which up to proportionality $(1, 1, 1)$ is the only solution of (6.7.5) in $\mathscr{S}$ is at least $c_2 A^2 - c_5(\log A)^{3t}$, which tends to infinity as A tends to infinity. One can inductively construct an infinite sequence of such (a, b, c). For a triple (a, b, c) of this kind, write $c = \sigma c_0$ with positive integers σ, c_0 such that $\sigma \in \mathscr{S}, \gcd(c_0, p_1 \cdots p_t) = 1$. Then $(1/\sigma, 1/\sigma)$ is the only solution of the equation $(a/c_0)x_1 + (b/c_0)x_2 = 1$ in $x_1, x_2 \in \mathbb{Z}_S^*$. Since a, b and c_0 are pairwise relatively prime, the pairs $(a/c_0, b/c_0)$ under consideration are pairwise S-inequivalent. This proves the case $N = 1$ of (ii). $\square$

- We discussed above results that give bounds for the number of solutions of S-unit equations. Here, we consider equations of the form

$$x_1 + x_2 = 1 \quad \text{in } x_1, x_2 \in O_K^*, \qquad (6.7.6)$$

where K is a number field. Recall that Evertse's result mentioned above gives an upper bound $3 \times 7^{2d+3}$ for the number of solutions of this equation. Grant (1996) gave examples of number fields K of arbitrarily large degree d such that (6.7.6) has $\gg d^2$ solutions. In fact, Grant's examples were cyclotomic fields $\mathbb{Q}(e^{2\pi i/p})$ with p a prime, and certain number fields arising from elliptic curves.

   We can get much better upper bounds for the number of solutions of (6.7.6) if we impose some restrictions on x_1, x_2. For instance, Silverman (1995) proved that if ε is a fixed element of O_K^*, then the equation $\varepsilon^m + y = 1$ has at most $d^{1+o(1)}$ solutions $m \in \mathbb{Z}, y \in O_K^*$.

- We now deal with the equation

$$a_1 x_1 + \cdots + a_n x_n = 1 \quad \text{in } (x_1, \ldots, x_n) \in \Gamma, \qquad (6.1.3)$$

where $a_1, \ldots, a_n$ are non-zero elements of a field K of characteristic 0 and Γ is a subgroup of finite rank of the n-fold direct product $(K^*)^n$. Already in the 1970s, Dubois and Rhin (1976) and independently Schlickewei (1977a) obtained finiteness results for (6.1.3) in the special case that $K = \mathbb{Q}$ and $\Gamma = (\mathbb{Z}_S^*)^n$ for some finite set of places S of $\mathbb{Q}$, and also with a condition imposed on the solutions stronger than non-degeneracy. The general result that for arbitrary K of characteristic 0 and Γ of finite rank, equation (6.1.3) has only finitely many non-degenerate solutions, was proved in several steps in the 1980s. Van der Poorten and Schlickewei in their unpublished preprint van der Poorten and Schlickewei (1982) and independently Evertse (1984b) proved that this equation has only finitely many non-degenerate solutions if K is a number field and $\Gamma = (O_S^*)^n$ for some finite set of places S of K. Also in their above mentioned preprint, van der Poorten and Schlickewei claimed a generalization of this to the case that K is an arbitrary field of characteristic 0 and Γ a finitely generated subgroup of $(K^*)^n$, but their proof was incomplete. In van der Poorten and Schlickewei (1991) they published the complete proof of their claim. Meanwhile, Evertse and Győry (1988b) gave a different proof of the claim of van der Poorten and Schlickewei, and showed that the number of non-degenerate solutions can be estimated from above by a (with their method of proof not effectively computable) number depending only on n, K and Γ. Further, Laurent (1984) developed some Kummer theory, which made it possible to extend the finiteness result on (6.1.3) from finitely generated groups Γ to groups of finite rank.

- Schlickewei (1990) was the first to obtain an explicit upper bound for the number of non-degenerate solutions of (6.1.3) in the case that K is a number field and $\Gamma = (O_S^*)^n$, where S is a finite set of places of K, containing all infinite places. His bound was improved in Evertse (1995) to $(2^{35}n^2)^{n^3 s}$, where s is the cardinality of S (see also Subsection 9.5.2 of the present book). In the case where K is a number field and $\Gamma = (O_S^*)^n$ this has not been improved so far.

  Building further on unpublished weaker results of the last two authors, Evertse, Schlickewei and Schmidt (2002) proved that if K is any field of zero characteristic, $a_1, \ldots, a_n \in K^*$, and Γ a subgroup of rank r of $(K^*)^n$, then (6.1.3) has at most $A(n, r) = \exp((6n)^{3n}(r + 1))$ non-degenerate solutions. In Amoroso and Viada (2009) this was improved to $A(n, r) = (8n)^{4n^4(n+r+1)}$.

- We consider the case that Γ has rank 0, i.e., we consider the equation

$$a_1\zeta_1 + \cdots + a_n\zeta_n = 1 \quad \text{in roots of unity } \zeta_1, \ldots, \zeta_n, \qquad (6.7.7)$$

where $a_1, \ldots, a_n$ again lie in a field K of characteristic 0. Results from Mann (1965) and Conway and Jones (1976) imply that if $a_1, \ldots, a_n \in \mathbb{Q}^*$, then for each non-degenerate solution $(\zeta_1, \ldots, \zeta_n)$ of (6.7.7), the lowest common multiple of the orders of $\zeta_1, \ldots, \zeta_n$ is $\le C(n)$ with $C(n)$ effectively computable in terms of n only. Further, results from Schinzel (1988) and Dvornicich and Zannier (2000) imply that if $a_1, \ldots, a_n$ generate a number field K of degree d, then for each non-degenerate solution of (6.7.7) the lowest common multiple of the orders of their components is bounded above by an effectively computable number $C(n, d)$ depending on n and d only. This implies that the non-degenerate solutions of (6.7.7) can be determined effectively, and it implies also that the number of non-degenerate solutions of (6.7.7) is bounded above by a number depending on n and d only. Schlickewei (1996b) considered equations (6.7.7) with coefficients $a_1, \ldots, a_n$ from an arbitrary field K of characteristic 0 and obtained an upper bound $2^{4(n+1)!}$ for the number of non-degenerate solutions. This was improved by Evertse (1999) to $(n + 1)^{3(n+1)^2}$. The proofs of the two last mentioned results use only simple properties of cyclotomic fields.

- Evertse, Győry, Stewart and Tijdeman (1988a) proved the following result, which shows that Theorem 6.1.6 has no obvious generalization to equations in more than two unknowns. Let K be a field of characteristic 0, $n \ge 3$, and Γ a subgroup of $(K^*)^n$ of finite rank. Call two tuples of coefficients $(a_1, \ldots, a_n), (b_1, \ldots, b_n) \in (K^*)^n$ Γ-equivalent if $(a_1 b_1^{-1}, \ldots, a_n b_n^{-1}) \in \Gamma$. Then there are groups Γ of finite rank such that for every $m > 0$ there are infinitely many Γ-equivalence classes of tuples $(a_1, \ldots, a_n) \in (K^*)^n$ with the property that (6.1.3) has at least m non-degenerate solutions.

  We give an easy construction different from that of Evertse et al. Choose m points $(x_{i1}, \ldots, x_{i,n-1}) \in (K^*)^{n-1}$ such that $x_{i1} + \cdots + x_{i,n-1} = 1$ for $i = 1, \ldots, m$ and no proper subsums of the left-hand sides vanish. Let Γ_1 be the multiplicative group generated by x_{ij}, for all $i = 1, \ldots, m, j = 1, \ldots, n - 1$. Then the equation

$$x_1 + \cdots + x_{n-1} = 1 \quad \text{in } (x_1, \ldots, x_{n-1}) \in \Gamma_1^{n-1}$$

has at least m non-degenerate solutions. It follows that for all but finitely many $\alpha \in K \setminus \{0, -1\}$, the equation

$$\frac{1}{1+\alpha} x_1 + \cdots + \frac{1}{1+\alpha} x_{n-1} + \frac{\alpha}{1+\alpha} x_n = 1 \quad \text{in } (x_1, \ldots, x_n) \in \Gamma_1^n$$

has at least m non-degenerate solutions, all with $x_n = 1$. We claim that the tuples $\varphi(\alpha) := (\frac{1}{1+\alpha}, \ldots, \frac{1}{1+\alpha}, \frac{\alpha}{1+\alpha})$ with $\alpha \in K \setminus \{0, -1\}$ lie in infinitely many different Γ_1^n-equivalence classes. Indeed, it is easy to see that the Γ_1^n-equivalence of $\varphi(\alpha), \varphi(\beta)$ implies that $\alpha/\beta \in \Gamma_1$. Now if the tuples $\varphi(\alpha)$ with $\alpha \in K \setminus \{0, -1\}$ lay in finitely many Γ_1^n-equivalence classes, the group K^* would be finitely generated, which is clearly absurd.

Instead, Evertse and Győry (1998b) proved the following result.

Theorem 6.7.2 *Let K be a field of characteristic 0 and Γ a subgroup of $(K^*)^n$ of finite rank. Then for all tuples $(a_1, , \ldots, a_n) \in (K^*)^n$ with the exception of at most finitely many Γ-equivalence classes, the (non-degenerate or degenerate) solutions of (6.1.3) lie in a union of at most $2^{(n+1)!}$ proper linear subspaces of K^n.*

This was improved in Evertse (2004) to 2^{n+1}. This bound is probably not best possible. It is as yet not clear what the optimal bound should be.

- Let K be a number field, S a finite set of places of K containing the infinite places, and consider again equation (6.1.3) in S-units $x_1, \ldots, x_n$. There is as yet no general effective method to find all non-degenerate solutions if the number of unknowns is larger than 2. However, in his thesis, Vojta (1983) gave an effective method to determine all non-degenerate solutions in S-units of (6.1.3) if the number n of unknowns is 3, and cardinality of the set S is at most 3. Recently, Bennett (not published when this book went to press) extended this to $n = 4$, $|S| \leq 3$. Both Vojta and Bennett proved more general effective results for systems of S-unit equations. More recently, Levin (2014) extended Vojta's result to an effective result for S-integral points on certain quasi-projective varieties, where again $|S|$ is small enough. The proofs of Vojta, Bennett and Levin all use Baker-type lower bounds for linear forms in logarithms.

- An effective version of the p-adic Subspace Theorem of Schmidt and Schlick-ewei would yield an effective version of Corollary 6.1.2, stating the finiteness of the number of non-degenerate solutions of the S-unit equation (6.1.2), i.e. (6.1.3). It seems, however, hopeless to make the Subspace Theorem effective by the present methods. As is pointed out in Győry (1992a), an effective variant of the following weaker Diophantine result would also imply an effective version of Corollary 6.1.2, which would be of great importance for its applications.

Let $k, n \geq 1$ be integers, $\alpha_0, \ldots, \alpha_k, \beta_1, \ldots, \beta_n$ non-zero elements of a number field K, and $b_{i1}, \ldots, b_{in}$ $(i = 1, \ldots, k)$ rational integers with absolute values at most B such that

$$\Lambda = \sum_{i=1}^{k} \alpha_i \beta_1^{b_{i1}} \cdots \beta_n^{b_{in}} - \alpha_0$$

has no vanishing subsum containing α_0. Let $| \, . \, |_v$ be a normalized absolute value on K.

Proposition *If*

$$0 < |\Lambda|_v < e^{-\delta B}$$

for some $\delta > 0$, then $B < C$, where C is a number depending only on k, n, K, $\alpha_0, \ldots \alpha_k, \beta_1, \ldots, \beta_n, v$ and δ.

For $k = 1$, this is a non-effective version of Baker's Theorem and its p-adic analogue; see Section 3.2. For $k \geq 1$, the above proposition is a straightforward consequence of Proposition 6.2.1, which was deduced from the Subspace Theorem. Hence the bound C is not effectively computable for $k > 1$ by the method of proof.

In Győry (1992a) it is shown that *an effective version of the above Proposition would imply an effective variant of Corollary 6.1.2 on S-unit equations.*

7

Analogues over function fields

Let **k** be an algebraically closed field of characteristic 0, and K a function field in one variable over **k**, i.e., a finitely generated extension of **k** of transcendence degree 1. Thus, K is a finite extension of the field of rational functions $\mathbf{k}(z)$, where z is any element of $K \setminus \mathbf{k}$. For definitions and more information on function fields we refer to Chapter 2.

We denote by $g_{K/\mathbf{k}}$ the genus of $K/\mathbf{k}$. By a valuation on K we mean a discrete valuation on K with value group $\mathbb{Z}$ such that $v(x) = 0$ for $x \in \mathbf{k}^*$. Let M_K denote the set of valuations of K. We recall that for a finite subset S of M_K a non-zero element u of K is called an S-unit if $v(u) = 0$ for all $v \in M_K \setminus S$.

In this chapter we deal with equations

$$a_1 x_1 + a_2 x_2 = 1 \tag{7.1}$$

and, in a less detailed manner,

$$a_1 x_1 + \cdots + a_n x_n = 1 \tag{7.2}$$

to be solved in S-units $x_1, \ldots, x_n$ and with some generalizations. The coefficients are non-zero elements of K.

In Section 7.1 we state Stothers' and Mason's Theorem, giving a function field analogue of the abc-conjecture, as well as a corollary which states that (7.1) has only finitely many solutions in S-units x_1, x_2 with $a_1 x_1, a_2 x_2 \notin \mathbf{k}^*$, which can be effectively determined in a well-defined sense. The theorem of Stothers and Mason and its corollary are proved in Section 7.2. In Section 7.3 we give a survey without proofs on effective results on S-unit equations (7.2) in an arbitrary number of unknowns, and explain the structure of the set of solutions of such equations, which is somewhat more complicated than that of S-unit equations over number fields.

In Sections 7.4 and 7.5 we consider, among other things, the equation

$$a_1 x_1 + a_2 x_2 = 1 \quad \text{in } (x_1, x_2) \in \Gamma, \qquad (7.3)$$

where again $a_1, a_2 \in K^*$ and where Γ is a multiplicative subgroup of $K^* \times K^*$ containing $\mathbf{k}^* \times \mathbf{k}^*$ such that $\Gamma / (\mathbf{k}^* \times \mathbf{k}^*)$ is a group of finite rank r. We prove a result from Evertse and Zannier (2008), stating that equation (7.3) has at most 3^r solutions with $a_1 x_1, a_2 x_2 \notin \mathbf{k}$. The method of proof we use, which is based on algebraic geometry, was developed by Bombieri, Mueller and Zannier (2001) and Zannier (2004).

In the last section of this chapter, we give a brief overview of recent results on unit equations over fields of positive characteristic.

7.1 Mason's inequality

Recall that for any $\alpha \in K$, the *height* of α relative to K is defined by

$$H_K(\alpha) := - \sum_v \min(0, v(\alpha)).$$

We have $H_K(\alpha) \geq 0$, and equality holds precisely when $\alpha \in \mathbf{k}$. Further, we denote by $|S|$ the cardinality of a set S.

We start with a theorem of Mason (1983, 1984). It is a generalization of an earlier result of Stothers (1981).

Theorem 7.1.1 (abc-theorem for function fields) *Let S be a finite, non-empty subset of M_K, and let x_1, x_2 and x_3 be non-zero elements of K with*

$$x_1 + x_2 + x_3 = 0 \qquad (7.1.1)$$

such that

$$v(x_1) = v(x_2) = v(x_3) \text{ for every } v \text{ in } M_K \setminus S. \qquad (7.1.2)$$

Then either x_1/x_2 lies in $\mathbf{k}$, or

$$H_K(x_1/x_2) \leq |S| + 2g_{K/\mathbf{k}} - 2. \qquad (7.1.3)$$

We note that (7.1.3) is best possible in the sense that for every $g \geq 0$ there is a function field K over $\mathbf{k}$ of genus g such that equality holds for infinitely many values of $|S|$; see Silverman (1984) for $g = 0$ and Brownawell and Masser (1986) for arbitrary g.

Theorem 7.1.1 implies at once that if S is again a finite subset of M_K and x_1, x_2 are S-units with $x_1 + x_2 = 1$ and $x_1, x_2 \notin \mathbf{k}$, then $H_K(x_i) \leq |S| +$

$2g_{K/k} - 2$ for $i = 1, 2$. We state a more general result for the S-unit equation

$$a_1 x_1 + a_2 x_2 = 1 \text{ in } S\text{-units } x_1, x_2, \qquad (7.1.4)$$

where $a_1, a_2 \in K^*$.

Theorem 7.1.2 *Let* (x_1, x_2) *be a solution of* (7.1.4) *with* $a_i x_i \notin \mathbf{k}^*$ *for* $i = 1, 2$. *Then*

$$\max_{i=1,2} H_K(x_i) \leq |S| + 2g_{K/k} - 2 + 5 \max_{i=1,2} H_K(a_i).$$

We observe that equation (7.1.4) may have infinitely many solutions (x_1, x_2) such that one of $a_1 x_1, a_2 x_2$ lies in $\mathbf{k}$. Indeed, suppose (7.1.4) has such a solution $(x_{1,0}, x_{2,0})$. Then both $a_1 x_{1,0}, a_2 x_{2,0} \in \mathbf{k}^*$. Put $a_1 x_{1,0}/a_2 x_{2,0} =: \eta$. Then we obtain infinitely many solutions (x_1, x_2) of (7.1.4) with $a_1 x_1, a_2 x_2 \in \mathbf{k}^*$ by taking $(x_1, x_2) = (x_{1,0}\xi, x_{2,0}(1 + (1 - \xi)\eta))$ for any $\xi \in \mathbf{k}^*$.

From Theorem 7.1.2, we obtain the following effective finiteness result. Here it is necessary to assume that $\mathbf{k}$ is *presented explicitly* in the sense of Fröhlich and Shepherdson (1956). This means that there is an algorithm to determine the zeros of any polynomial with coefficients in $\mathbf{k}$. In particular, in this case we can perform the field operations in $\mathbf{k}$. Further, we assume that K is presented explicitly, that is, K is given in the form $\mathbf{k}(z)(y)$ where z is a variable and y is a primitive element of K over $\mathbf{k}(z)$, with explicitly given minimal polynomial in $\mathbf{k}(z)[X]$.

We say that an element x of K is *explicitly given* if it is given in the form

$$x = \sum_{i=1}^{d} (q_i(z)/q(z)) y^{i-1},$$

where $d = [K : \mathbf{k}(z)]$ and $q_1, \ldots, q_d, q$ are explicitly given polynomials from $\mathbf{k}[z]$. We call $(q_1, \ldots, q_d, q)$ a representation for x. We say that a valuation v of K is explicitly given, if we are given a local parameter z_v and a Laurent series y_v in $\mathbf{k}((z_v))$ such that $y \mapsto y_v$ defines an isomorphic embedding of K into $\mathbf{k}((z_v))$. By a Laurent series being explicitly given we mean that we are given an inductive procedure to compute its coefficients one by one. If a non-zero $x \in K$ and a valuation v are explicitly given, then $v(x)$ can be determined by computing a Laurent series of x in terms of z_v and searching for the first non-zero coefficient.

Finally, an element x of K is said to be *effectively determinable* from certain given input data if there is an algorithm to determine an explicit representation of x from these data. We note that if elements x_1 and x_2 of K are effectively determinable, then so are $x_1 \pm x_2, x_1 x_2$ and x_1/x_2 $(x_2 \neq 0)$.

Corollary 7.1.3 *Equation* (7.1.4) *has only finitely many solutions with* $a_i x_i \notin$ **k*** *for* $i = 1, 2$ *and these can be determined effectively if we assume that* **k**, K *are presented explicitly and* a_1, a_2 *and the valuations in* S *are given explicitly in the sense described above.*

For $a_1 = a_2 = 1$, Mason (1983, 1984) deduced this corollary from his Theorem 7.1.1 stated above and from Propositions 2.4.1 and 2.4.2. Further, in this special case he extended Corollary 7.1.3 to the case of positive characteristic. We deduce Corollary 7.1.3 in a manner similar to Mason's, using Theorem 7.1.2 instead of Theorem 7.1.1.

We mention that in his book, Mason (1984) gave various applications of the results mentioned above, to Thue equations, hyper- and superelliptic equations, and curves of genus 0 and genus 1.

7.2 Proofs

We prove Theorems 7.1.1, 7.1.2 and Corollary 7.1.3.

Proof of Theorem 7.1.1. We assume without loss of generality that S is precisely the set of all $v \in M_K$ such that $v(x_1), v(x_2), v(x_3)$ are distinct. For convenience, we write $u := x_3/x_1$. Thus, (7.1.1) implies that $x_2/x_1 = -(u + 1)$ and our assumption translates into

$$S = \{v \in M_K : v(u) \neq 0 \text{ or } v(u + 1) \neq 0\}.$$

We may assume that u does not lie in **k**. Since $v(u + 1) \geq 0$ if $v(u) = 0$, we can partition S into a disjoint union $S_\infty \cup S_0 \cup S_{-1}$, where

$$S_\infty = \{v \in S : v(u) < 0\}, \quad S_0 = \{v \in S : v(u) > 0\},$$
$$S_{-1} = \{v \in S : v(u + 1) > 0\}.$$

These sets are pairwise disjoint. Notice that by the Sum Formula (see (2.1.3) in Section 2.1), $\sum_{v \in S_\infty \cup S_0} v(u) = 0$.

Now choose for every valuation $v \in M_K$ a local parameter z_v. We compare the order of vanishing at v of u and the local derivative du/dz_v. From (2.3.1) and (2.3.2) in Section 2.3, we infer

$$v\left(\frac{du}{dz_v}\right) = v(u) - 1 \quad \text{for } v \in S_\infty \cup S_0,$$

$$v\left(\frac{du}{dz_v}\right) = v\left(\frac{d(u + 1)}{dz_v}\right) = v(u + 1) - 1 \quad \text{for } v \in S_{-1},$$

$$v\left(\frac{du}{dz_v}\right) \geq 0 \quad \text{for } v \in M_K \setminus S.$$

By combining these with Theorem 2.3.1 and the Sum Formula we obtain

$$2g_{K/\mathbf{k}} - 2 = \sum_v v\left(\frac{du}{dz_v}\right) \geq \sum_{v \in S_\infty \cup S_0} (v(u) - 1) + \sum_{v \in S_{-1}} (v(u + 1) - 1)$$

$$= \sum_{v \in S_{-1}} v(u + 1) - |S| = H_K((u + 1)^{-1}) - |S| = H_K(x_1/x_2) - |S|.$$

This implies Theorem 7.1.1. $\qquad\qquad\square$

Proof of Theorem 7.1.2. Put $H := \max_{i=1,2} H_K(a_i)$. For $i = 1, 2$, let S_i be the set of valuations v outside S with $v(a_i) \neq 0$. By (2.2.8) we have $|S_i| \leq 2H_K(a_i) \leq 2H$ for $i = 1, 2$. Take a solution (x_1, x_2) of (7.1.4) with $a_i x_i \notin \mathbf{k}^*$ for $i = 1, 2$. We have $v(a_1 x_1) = v(a_2 x_2) = v(1) = 0$ for $v \in M_K \setminus (S \cup S_1 \cup S_2)$. Notice that by (2.2.7) and (2.2.6),

$$H_K(x_i) \leq H_K(a_i x_i) + H_K\left(a_i^{-1}\right) = H_K(a_i x_i) + H_K(a_i) \leq H_K(a_i x_i) + H$$

for $i = 1, 2$. Now an application of Theorem 7.1.1 with $a_1 x_1, a_2 x_2, 1$ instead of x_1, x_2, x_3 and $S \cup S_1 \cup S_2$ instead of S gives for $i = 1, 2$,

$$H_K(x_i) \leq H_K(a_i x_i) + H \leq |S| + |S_1| + |S_2| + 2g_{K/\mathbf{k}} - 2 + H$$

$$\leq |S| + 2g_{K/\mathbf{k}} - 2 + 5H. \qquad\qquad\square$$

Proof of Corollary 7.1.3. Let (x_1, x_2) be a solution of (7.1.4) with $a_i x_i \notin \mathbf{k}^*$ for $i = 1, 2$. Pick $i \in \{1, 2\}$. Then $v(x_i) = 0$ for every valuation $v \in M_K \setminus S$. Further, by (2.2.8),

$$\sum_{v \in S} |v(x_i)| = 2H_K(x_i) \leq 2C,$$

where C is the bound from Theorem 7.1.2. As explained in Section 2.4, we can compute a minimal polynomial over $\mathbf{k}[z]$ of each a_i, and then estimate from above the heights of the a_i using (2.2.10). This leads to an effectively computable upper bound for C. We conclude that the tuple of integers $(v(x_i) : i = 1, 2, v \in S)$ has only a finite number of effectively determinable possibilities as (x_1, x_2) runs over the solutions of (7.1.4) with $a_i x_i \notin \mathbf{k}^*$ for $i = 1, 2$.

Applying Proposition 2.4.1 we infer that apart from a non-zero factor in $\mathbf{k}$, x_1, x_2 have only a finite number of possibilities which are effectively determinable. Hence for $i = 1, 2$ we may write $x_i = y_i \xi_i$, where ξ_i is some non-zero element of $\mathbf{k}$ and y_i belongs to a finite computable subset of K. Now equation (7.1.4) transforms into

$$b_1 \xi_1 + b_2 \xi_2 = 1, \tag{7.2.1}$$

where $b_i := a_i y_i \notin \mathbf{k}^*$ for $i = 1, 2$. The pair (b_1, b_2) belongs to a finite, effectively determinable set, and for each such pair we have to determine the solutions $\xi_1, \xi_2 \in \mathbf{k}^*$ of (7.2.1).

Fix b_1, b_2. We have seen that $b_1, b_2 \notin \mathbf{k}$. If b_1, b_2 are linearly dependent over $\mathbf{k}$ then (7.2.1) is unsolvable. Assume that b_1, b_2 are linearly independent over $\mathbf{k}$. Then by Proposition 2.4.2, equation (7.2.1) has precisely one solution which can be determined effectively. This completes the proof of our assertion. $\square$

7.3 Effective results in the more unknowns case

For completeness, we now present without proof some generalizations of the results stated in Section 7.1.

Let $n \geq 2$ be a given integer. For non-zero elements $x_0, x_1, \ldots, x_n$ of K, we define the *homogeneous height* by

$$H_K^{\mathrm{hom}}(x_0, \ldots, x_n) = - \sum_v \min(v(x_0), \ldots, v(x_n)).$$

The Sum Formula on K shows that this is actually a height on the projective space $\mathbb{P}^n(K)$. Further, we have

$$H_K(x_i/x_j) \leq H_K^*(x_0, \ldots, x_n)$$

for each i, j with $0 \leq i, j \leq n$.

Write

$$N_0 := 0, \quad N_n := \frac{1}{2}(n-1)(n-2) \text{ if } n \geq 1.$$

Brownawell and Masser (1986) proved the following general theorem.

Theorem 7.3.1 *Suppose that $x_0, \ldots, x_n$ are non-zero elements of K such that*

$$x_0 + \cdots + x_n = 0, \tag{7.3.1}$$

and no proper subset of $\{x_0, \ldots, x_n\}$ is $\mathbf{k}$-linearly dependent. For each valuation v of K, let

$$m(v) = m(v; x_0, \ldots, x_n) := |\{i : 0 \leq i \leq n, v(x_i) = 0\}|.$$

Then

$$H_K^{\mathrm{hom}}(x_0, \ldots, x_n) \leq N_{n+1}(2g_{K/\mathbf{k}} - 2) + \sum_v (N_{n+1} - N_{m(v)}). \tag{7.3.2}$$

The proof of Brownawell and Masser uses logarithmic Wronskians. Fix $z \in K \setminus \mathbf{k}$, so that K is a finite extension of $\mathbf{k}(z)$. For $f \in K$, define $f^{(k)} :=$ $(d/dz)^k f$. Now we define the logarithmic Wronskian of $f_1, \ldots, f_n \in K^*$ by

$$\lambda(f_1, \ldots, f_n) := \det \left(f_i^{(j-1)}/f_i \right)_{i,j=1,\ldots,n}.$$

Given a solution $(x_0, \ldots, x_n)$ of (7.3.1), let λ_i be the logarithmic Wronskian of $x_0, \ldots, x_n$ with x_i omitted. Then the argument of Brownawell and Masser consists of showing that $H_K(x_0, \ldots, x_n) = H_K(\lambda_0, \ldots, \lambda_n)$, and estimating $v(\lambda_i)$ from below for $i = 0, \ldots, n$ and $v \in M_K$, which leads to an upper bound for $H_K(\lambda_0, \ldots, \lambda_n)$.

The following consequence of Theorem 7.3.1 was obtained independently by Voloch (1985).

Corollary 7.3.2 *Suppose that for some finite subset S of M_K, $x_0, \ldots, x_n$ give rise to a solution of (7.3.1) in S-units, and that no proper subset of $\{x_0, \ldots, x_n\}$ is $\mathbf{k}$-linearly dependent. Then*

$$H_K^{\mathrm{hom}}(x_0, \ldots, x_n) \le \frac{1}{2}n(n-1)(|S| + 2g_{K/\mathbf{k}} - 2).$$

In his proof, Voloch did not use the Wronskian argument of Brownawell and Masser, but instead used properties of Weierstrass points on algebraic curves.

Notice that we obtain Mason's result, Theorem 7.1.1, from Corollary 7.3.2 by taking x_1, x_2, x_3 in K^* with $x_1 + x_2 + x_3 = 0$ and with (7.1.2) and applying Corollary 7.3.2 to $(x_1/x_2) + (x_3/x_2) + 1 = 0$.

Independently, Mason (1986a) proved Corollary 7.3.2 with a larger bound in terms of n. Further, he showed that apart from a common proportional S-unit factor, the full range of possibilities for such $x_0, \ldots, x_n$ is finite, and may be determined effectively whenever $\mathbf{k}$, K are presented explicitly and the valuations in S are given explicitly. A sharpening of Corollary 7.3.2 was given by Zannier (1993). Hsia and Wang (2004) obtained a generalization of the result of Brownawell and Masser to function fields of arbitrary transcendence degree over constant fields of arbitrary characteristic. Their proof uses generalized Wronskians.

A solution $x_0, \ldots, x_n$ of (7.3.1) is called *non-degenerate* if $\sum_{i \in I} x_i \ne 0$ for every non-empty proper subset I of $\{0, \ldots, n\}$, and *degenerate* otherwise.

Brownawell and Masser (1986) proved that the inequality (7.3.2) in Theorem 7.3.1 remains true, in a slightly modified form, if the assumption of linear independence is replaced by the weaker hypothesis of non-degeneracy.

Set

$$G_K := \max(0, 2g_{K/\mathbf{k}} - 2).$$

Theorem 7.3.3   *Suppose that* $x_0, \ldots, x_n$ *is a non-degenerate solution of* (7.3.1). *Then*

$$H_K^{\mathrm{hom}}(x_0, \ldots, x_n) \leq N_{n+1} \cdot G_K + \sum_v (N_{n+1} - N_{m(v)}).$$

This implies the following version of Corollary 7.3.2.

Corollary 7.3.4   *Suppose that for some finite subset S of M_K, $x_0, \ldots, x_n$ is a non-degenerate solution of* (7.3.1) *in S-units. Then*

$$H_K^{\mathrm{hom}}(x_0, \ldots, x_n) \leq \frac{1}{2} n(n-1)(|S| + G_K). \tag{7.3.3}$$

It is likely that for $n > 2$ the factor $\frac{1}{2} n(n-1)$ in (7.3.3) is not the best possible one.

We derive a result on the inhomogeneous equation

$$a_1 x_1 + \cdots + a_n x_n = 1 \text{ in } S\text{-units } x_1, \ldots, x_n, \tag{7.3.4}$$

where, as before, S is a finite subset of M_K and where $a_1, \ldots, a_n$ are non-zero elements of K. A solution $(x_1, \ldots, x_n)$ of this equation is called *non-degenerate* if $\sum_{i \in I} a_i x_i \neq 0$ for each non-empty subset I of $\{1, \ldots, n\}$.

Theorem 7.3.5   *For every non-degenerate solution $(x_1, \ldots, x_n)$ of* (7.3.4) *we have*

$$\max_{1 \leq i \leq n} H_K(x_i) \leq \frac{1}{2} n(n-1)(|S| + G_K) + (n^3 - n^2 + 1) \max_{1 \leq i \leq n} H_K(a_i).$$

Proof. Put $H := \max_{1 \leq i \leq n} H_K(a_i)$. For $i = 1, \ldots, n$, let S_i be the set of valuations v outside S for which $v(a_i) \neq 0$. Then by (2.2.8) we have $|S_i| \leq 2 H_K(a_i) \leq 2H$ for $i = 1, \ldots, n$. Choose a non-degenerate solution $(x_1, \ldots, x_n)$ of (7.3.4). Then, completely similarly as in the proof of Theorem 7.1.2, we have $H_K(x_i) \leq H_K(a_i x_i) + H$ for $i = 1, \ldots, n$. Now by applying Corollary 7.3.4 with $(1, a_1 x_1, \ldots, a_n x_n)$ and $S \cup S_1 \cup \cdots \cup S_n$ instead of $(x_0, x_1, \ldots, x_n)$, S, we obtain for $i = 1, \ldots, n$,

$$\begin{aligned}
H_K(x_i) &\leq H + H_K(a_i x_i) \leq H + H_K(1, a_1 x_1, \ldots, a_n x_n) \\
&\leq H + \tfrac{1}{2} n(n-1)(G_K + |S| + |S_1| + \cdots + |S_n|) \\
&\leq \tfrac{1}{2} n(n-1)(G_K + |S|) + (n^3 - n^2 + 1)H. \qquad \square
\end{aligned}$$

The above result does not imply that (7.3.4) has only finitely many solutions. We say that two solutions $(x_1, \ldots, x_n)$, $(\tilde{x}_1, \ldots, \tilde{x}_n)$ of (7.3.4) are **k**-*proportional*, or lie in the same **k**-proportionality class, if $x_i / \tilde{x}_i \in \mathbf{k}$ for $i = 1, \ldots, n$. In general, a **k**-proportionality class may contain infinitely many non-degenerate solutions.

Corollary 7.3.6 *The set of non-degenerate solutions of (7.3.4) is contained in a union of finitely many* **k**-*proportionality classes, and if we assume that* **k**, *K are presented explicitly and S, $a_1, \ldots, a_n$ are explicitly given, a full system of representatives of these classes can be determined effectively.*

Proof. Let $(x_1, \ldots, x_n)$ be a non-degenerate solution of (7.3.4). Then by (2.2.8) we have

$$\sum_{v \in S} |v(x_i)| = 2H_K(x_i) \leq 2C \quad \text{for } i = 1, \ldots, n,$$

where C is the upper bound from Theorem 7.3.5. By the same method as in the proof of Corollary 7.1.3, one can effectively compute an upper bound for C. This shows that the tuple $(v(x_i) : i = 1, \ldots, n, v \in S)$ runs through a finite, effectively determinable set as $(x_1, \ldots, x_n)$ runs through the non-degenerate solutions of (7.3.4), and the non-degenerate solutions corresponding to a given tuple lie in the same **k**-proportionality class. Hence the non-degenerate solutions of (7.3.4) lie in only finitely many **k**-proportionality classes.

By Proposition 2.4.1, it can be decided effectively whether for a given tuple $(c_{iv} : i = 1, \ldots, n, v \in S)$ there exist $b_1, \ldots, b_n \in O_S^*$ with $v(b_i) = c_{iv}$ for $i = 1, \ldots, n, v \in S$ and if so, such $b_1, \ldots, b_n$ can be determined effectively. The non-degenerate solutions of (7.3.4) that are **k**-proportional to $(b_1, \ldots, b_n)$ are of the shape $(b_1\xi_1, \ldots, b_n\xi_n)$ with

$$a_1 b_1 \xi_1 + \cdots + a_n b_n \xi_n = 1, \quad (\xi_1, \ldots, \xi_n) \in \mathbf{k}^n, \tag{7.3.5}$$

$$\sum_{i \in I} a_i b_i \xi \neq 0 \quad \text{for each non-empty } I \subset \{1, \ldots, n\}. \tag{7.3.6}$$

The tuples $(\xi_1, \ldots, \xi_n)$ with (7.3.5) form a linear subvariety V of $\mathbf{k}^n$, and the elements of V with (7.3.6) lie in the complement of a finite number of linear subvarieties of V, say $V_1, \ldots, V_r$. Thus, the set of non-degenerate solutions of (7.3.4) that are **k**-proportional to $(b_1, \ldots, b_n)$ can be described as $(b_1\xi_1, \ldots, b_n\xi_n)$ with $(\xi_1, \ldots, \xi_n) \in \mathcal{U} := V \setminus (V_1 \cup \cdots \cup V_r)$.

So we have to decide whether or not $\mathcal{U} \neq \emptyset$ and if so, find an element of $\mathcal{U}$. Notice that $\mathcal{U} \neq \emptyset$ precisely if $V \neq \emptyset$ and $V_1, \ldots, V_r$ are proper linear subvarieties of V. This can be checked using Proposition 2.4.2. Further, assuming $\mathcal{U} \neq \emptyset$ one can find an element of $\mathcal{U}$ using the parameter representations of $V, V_1, \ldots, V_r$ that can be computed according to Proposition 2.4.2.

Assume that $\mathcal{U} \neq \emptyset$. Then one easily checks that $\mathcal{U}$ consists of only one element if V has dimension 0, that is, if $a_1 b_1, \ldots, a_n b_n$ are linearly independent over **k**, and $\mathcal{U}$ is infinite if V is positive dimensional, which is the case precisely if $a_1 b_1, \ldots, a_n b_n$ are linearly dependent over **k**. $\qquad \square$

7.4 Results on the number of solutions

Let again $\mathbf{k}$ be an algebraically closed field of characteristic 0 and let now K be an extension field of $\mathbf{k}$ of arbitrary positive transcendence degree over $\mathbf{k}$.

Let $n \geq 2$, and denote by $(K^*)^n$ the n-fold direct product of the multiplicative group K^*, endowed with coordinatewise multiplication. We consider equations with solution vectors from a subgroup Γ of $(K^*)^n$ such that $(\mathbf{k}^*)^n \subset \Gamma$, and $\Gamma/(\mathbf{k}^*)^n$ has finite rank r. If $r = 0$ this means that $\Gamma = (\mathbf{k}^*)^n$, while if $r > 0$, this means that there are multiplicatively independent elements $\mathbf{u}_1, \ldots, \mathbf{u}_r$ of Γ such that every element of Γ can be expressed as $\xi \cdot \mathbf{u}_1^{w_1} \cdots \mathbf{u}_r^{w_r}$ with $\xi \in (\mathbf{k}^*)^n$ and $w_1, \ldots, w_r \in \mathbb{Q}$. (The coordinates of $\mathbf{u}_i^{w_i}$ are determined only up to multiplication by roots of unity, but we just make any choice for them.)

We start with the equation in two unknowns

$$a_1 x_1 + a_2 x_2 = 1 \quad \text{in } (x_1, x_2) \in \Gamma, \tag{7.4.1}$$

where Γ is a subgroup of $(K^*)^2$ and $a_1, a_2 \in K^*$. The following theorem is a generalization of a result of Zannier (2004).

Theorem 7.4.1 *Suppose that $\Gamma \supset (\mathbf{k}^*)^2$ and that $\Gamma/(\mathbf{k}^*)^2$ has finite rank $r \geq 0$. Then (7.4.1) has at most 3^r solutions with $a_j x_j \notin \mathbf{k}^*$ for $j = 1, 2$.*

We now consider equations in an arbitrary number of unknowns, i.e.,

$$a_1 x_1 + \cdots + a_n x_n = 1 \quad \text{in } (x_1, \ldots, x_n) \in \Gamma, \tag{7.4.2}$$

where Γ is a subgroup of $(K^*)^n$ and $a_1, \ldots, a_n \in K^*$. Recall that a solution of (7.4.2) is called non-degenerate if $\sum_{i \in I} a_i x_i \neq 0$ for each non-empty subset I of $\{1, \ldots, n\}$. Further, we say that two solutions $(x_1, \ldots, x_n)$, $(\tilde{x}_1, \ldots, \tilde{x}_n)$ are $\mathbf{k}$-proportional, or belong to the same $\mathbf{k}$-proportionality class, if $x_i/\tilde{x}_i \in \mathbf{k}^*$ for $i = 1, \ldots, n$. The next theorem is the main result from Evertse and Zannier (2008).

Theorem 7.4.2 *Let $n \geq 2$. Suppose that $\Gamma \supset (\mathbf{k}^*)^n$ and that $\Gamma/(\mathbf{k}^*)^n$ has finite rank $r \geq 0$. Then the non-degenerate solutions of (7.4.2) lie in at most*

$$\sum_{i=2}^{n+1} \binom{i}{2}^r - n + 1$$

$\mathbf{k}$-*proportionality classes.*

Theorem 7.4.1 follows at once from 7.4.2. Indeed, all solutions of (7.4.1) are non-degenerate. Further, the solutions with $a_j x_j \notin \mathbf{k}^*$ for $j = 1, 2$ are pairwise $\mathbf{k}$-non-proportional, and by substituting $n = 2$ into the bound of Theorem 7.4.2 we obtain precisely the bound 3^r from Theorem 7.4.1.

We mention that the proof of Theorem 7.4.2, given in Evertse and Zannier (2008) depends heavily on ideas introduced in Bombieri, Mueller and Zannier (2001) and Zannier (2004). Weaker and less general results were obtained earlier in Evertse and Győry (1988b) and Mueller (2000).

To give a flavour of the techniques used in the papers mentioned above, in the next section we prove Theorem 7.4.1 in the special case that K has transcendence degree 1 over $\mathbf{k}$. The general case that K has arbitrary transcendence degree over $\mathbf{k}$ can be reduced to this special case by means of a specialization argument. The proof of Theorem 7.4.2, which is not given here, is based on the same ideas as the proof of Theorem 7.4.1.

7.5 Proof of Theorem 7.4.1

Let $\mathbf{k}$ be an algebraically closed field of characteristic 0, let K be an extension of $\mathbf{k}$ of transcendence degree 1, and let Γ be a subgroup of $(K^*)^2$ which contains $(\mathbf{k}^*)^2$ and such that $\Gamma/(\mathbf{k}^*)^2$ has finite rank r. We would like to define in some way the $\mathbf{k}$-closure of Γ, which is such that if a point (x_1, x_2) belongs to this $\mathbf{k}$-closure, then so does (x_1^w, x_2^w) for any $w \in \mathbf{k}$. Then we would like to consider equation (7.4.1) with solutions from the $\mathbf{k}$-closure of Γ instead of Γ itself. The importance of this is that it will allow us to use techniques from algebraic geometry.

It does not suffice to define the $\mathbf{k}$-closure of Γ formally by taking the tensor product of Γ with $\mathbf{k}$, but we have to embed this $\mathbf{k}$-closure somehow into a ring or field, to make sense of our desired extension of (7.4.1). As it turns out, one can define exponentiation with elements from $\mathbf{k}$ for formal power series. Then the $\mathbf{k}$-closure of Γ can be defined after embedding K into a formal power series ring.

7.5.1 Extension to the k-closure of Γ

We keep the notation introduced in Section 7.4. Let again $\mathbf{k}$ be an algebraically closed field of characteristic 0, K an extension of $\mathbf{k}$ of transcendence degree 1, and Γ a subgroup of $(K^*)^2$ such that $\Gamma \supset (\mathbf{k}^*)^2$ and $\Gamma/(\mathbf{k}^*)^2$ has rank r. Choose pairs (b_{i1}, b_{i2}) $(i = 1, \ldots, r)$ from Γ that are multiplicatively independent over $(\mathbf{k}^*)^2$, i.e., there is no non-zero vector $\mathbf{w} = (w_1, \ldots, w_r) \in \mathbb{Z}^r$ with $\prod_{i=1}^r (b_{i1}, b_{i2})^{w_i} \in (\mathbf{k}^*)^2$. Then every element of Γ can be expressed as

$$(\xi_1, \xi_2) \prod_{i=1}^r (b_{i1}, b_{i2})^{w_i}, \tag{7.5.1}$$

where $(\xi_1, \xi_2) \in (\mathbf{k}^*)^2$, $w_1, \ldots, w_r \in \mathbb{Q}$, and where the powers with rational exponents are defined up to multiplication by points consisting of roots of unity.

Let L be the extension of $\mathbf{k}$ generated by a_1, a_2, b_{ij} ($i = 1, \ldots, r, j = 1, 2$). Notice that L is an algebraic function field in one variable over $\mathbf{k}$. Choose a valuation v of L such that $v(a_j) = 0$, $v(b_{ij}) = 0$ for $j = 1, 2, i = 1, \ldots, r$. Since v has residue class field $\mathbf{k}$, after multiplying a_1, a_2 and the b_{ij} by appropriate elements from $\mathbf{k}^*$, we can achieve that

$$v(a_j - 1) > 0, \quad v(b_{ij} - 1) > 0 \text{ for } j = 1, 2, \ i = 1, \ldots, r. \tag{7.5.2}$$

Choose a local parameter z for v. Then the completion of L at v is the field of formal Laurent series $\mathbf{k}((z))$, and we may assume that L is a subfield of $\mathbf{k}((z))$. The valuation v is extended to $\mathbf{k}((z))$ by setting $v(\sum_{i=i_0}^{\infty} c_i z^i) = i_0$ if $c_i \in \mathbf{k}$ for $i \geq i_0$ and $c_{i_0} \neq 0$. We say that a sequence $\{f_m\}_{m=0}^{\infty}$ in $\mathbf{k}((z))$ converges to f if $v(f_m - f) \to \infty$ as $m \to \infty$. The derivative of $f = \sum_{i=i_0}^{\infty} c_i z^i \in \mathbf{k}((z))$ is defined by $f' = \mathrm{d}f/\mathrm{d}z = \sum_{i=i_0}^{\infty} i c_i z^{i-1}$. If $\lim_{m \to \infty} f_m = f$ for some sequence $\{f_m\}$ in $\mathbf{k}((z))$, then also $\lim_{m \to \infty} f'_m = f'$.

Denote by $\mathbf{k}[[z]]$ the ring of formal power series in z, and denote by $1 + z\mathbf{k}[[z]]$ the set of power series

$$1 + \sum_{i=1}^{\infty} c_i z^i \quad \text{with } c_i \in \mathbf{k} \text{ for } i \geq 1.$$

Notice that $1 + z\mathbf{k}[[z]]$ is a multiplicative group.

By (7.5.2) the elements a_j, b_{ij} ($j = 1, 2, i = 1, \ldots, r$) all belong to the group $1 + z\mathbf{k}[[z]]$. Further, they belong to L, hence are algebraic over $\mathbf{k}(z)$.

We are now ready to define exponentiation with elements from $\mathbf{k}$. For $f \in 1 + z\mathbf{k}[[z]]$, $w \in \mathbf{k}$ we put

$$f^w := \sum_{j=0}^{\infty} \binom{w}{j} (f - 1)^j,$$

where $\binom{w}{j} := w(w - 1) \cdots (w - j + 1)/j!$. In the topology of $\mathbf{k}[[z]]$ defined by v, this infinite series converges to a limit which belongs to $1 + z\mathbf{k}[[z]]$. Obviously, by Newton's binomial formula, $f^w = f \cdots f$ (w times) for any non-negative integer w. This exponentiation has the usual properties:

$$(f^w)' = w f^{w-1} f' \text{ for } f \in 1 + z\mathbf{k}[[z]], \ w \in \mathbf{k}; \tag{7.5.3}$$

$$(fg)^w = f^w g^w \text{ for } f, g \in 1 + z\mathbf{k}[[z]], \ w \in \mathbf{k}; \tag{7.5.4}$$

$$f^{w_1 + w_2} = f^{w_1} f^{w_2} \text{ for } f \in 1 + z\mathbf{k}[[z]], \ w_1, w_2 \in \mathbf{k}; \tag{7.5.5}$$

$$(f^{w_1})^{w_2} = f^{w_1 w_2} \text{ for } f \in 1 + z\mathbf{k}[[z]], \ w_1, w_2 \in \mathbf{k}. \tag{7.5.6}$$

Property (7.5.3) can be proved using the fact that $((f - 1)^j)' = j(f - 1)^{j-1} f'$ for all j and that infinite series can be differentiated sumwise. The other properties can be proved using logarithmic derivatives, where the logarithmic derivative of $f \in \mathbf{k}((z))$ is f'/f. The map $f \mapsto f'/f$ defines an injective homomorphism from the multiplicative group $1 + z\mathbf{k}[[z]]$ to the additive group of $\mathbf{k}[[z]]$, and $(f^w)'/f^w = wf'/f$ for $f \in 1 + z\mathbf{k}[[z]]$, $w \in \mathbf{k}$. For instance, (7.5.4) is proved by showing that both $(fg)^w$ and $f^w g^w$ have logarithmic derivatives $w \cdot ((f'/f) + (g'/g))$. The identities (7.5.5) and (7.5.6) can be proved likewise.

In the usual manner, we put $(x_1, \ldots, x_n)^w := (x_1^w, \ldots, x_n^w)$ for $x_1, \ldots, x_n \in 1 + z\mathbf{k}[[z]]$ and $w \in \mathbf{k}$. We now define the $\mathbf{k}$-closure of Γ by

$$\overline{\Gamma} := \left\{ (\xi_1, \xi_2) \prod_{i=1}^{r} (b_{i1}, b_{i2})^{w_i} : \xi_1, \xi_2 \in \mathbf{k}^*, \ w_1, \ldots, w_r \in \mathbf{k} \right\}. \quad (7.5.7)$$

By (7.5.1), this group indeed contains Γ. In what follows, we write $\mathbf{w}$ for vectors $(w_1, \ldots, w_r) \in \mathbf{k}^r$. Our result for $\overline{\Gamma}$ is as follows.

Theorem 7.5.1 *Let r be a positive integer and let a_j, b_{ij} ($j = 1, 2$, $i = 1, \ldots, r$) be elements of $1 + z\mathbf{k}[[z]]$ such that*

$$a_j, b_{ij} \text{ are algebraic over } \mathbf{k}(z) \text{ for } j = 1, 2, \ i = 1, \ldots, r, \quad (7.5.8)$$

$$\text{there is no } \mathbf{w} \in \mathbb{Z}^r \setminus \{0\} \text{ with } \prod_{i=1}^{r} (b_{i1}, b_{i2})^{w_i} \in (\mathbf{k}^*)^2. \quad (7.5.9)$$

Let $\overline{\Gamma}$ be given by (7.5.7). Then the equation

$$a_1 x_1 + a_2 x_2 = 1 \quad \text{in } (x_1, x_2) \in \overline{\Gamma} \text{ with } a_1 x_1 \notin \mathbf{k}^*, a_2 x_2 \notin \mathbf{k}^* \quad (7.5.10)$$

has at most 3^r solutions.

The idea of the proof is to consider the vectors $\mathbf{w} \in \mathbf{k}^r$ in the representation (7.5.7) for the solutions (x_1, x_2) of (7.5.10), and to estimate the number of these $\mathbf{w}$ using techniques from algebraic geometry. Here it will be crucial that $\mathbf{w}$ can be chosen from $\mathbf{k}^r$ and not just from $\mathbb{Q}^r$, as was the case with the group Γ.

7.5.2 Some algebraic geometry

We have collected some basic facts from algebraic geometry. Our basic reference is Hartshorne (1977), chapter 1. As before, $\mathbf{k}$ is an algebraically closed field of characteristic 0. Let r be a positive integer.

By an *algebraic subset* of $\mathbf{k}^r$ we mean the set of common zeros in $\mathbf{k}^r$ of a collection of polynomials in $\mathbf{k}[X_1, \ldots, X_r]$. The algebraic subset of $\mathbf{k}^r$ given by $f_1, \ldots, f_m \in \mathbf{k}[X_1, \ldots, X_m]$, notation $\mathcal{V}(f_1, \ldots, f_m)$, is defined as the set of common zeros in $\mathbf{k}^r$ of $f_1, \ldots, f_m$. The algebraic subsets of $\mathbf{k}^r$ are the closed sets of the *Zariski topology* on $\mathbf{k}^r$. We say that an algebraic subset of $\mathbf{k}^r$ is defined over a subfield $\mathbf{k}'$ of $\mathbf{k}$ if it is the set of common zeros in $\mathbf{k}^r$ of polynomials with coefficients in $\mathbf{k}'$.

The collection of all polynomials $f \in \mathbf{k}[X_1, \ldots, X_r]$ vanishing identically on a given algebraic set $\mathcal{X} \subset \mathbf{k}^r$ is an ideal of $\mathbf{k}[X_1, \ldots, X_r]$, which is denoted by $I(\mathcal{X})$. Clearly, if $\mathcal{X}, \mathcal{Y}$ are algebraic subsets of $\mathbf{k}^r$, then $\mathcal{X} \subseteq \mathcal{Y}$ if and only if $I(\mathcal{X}) \supseteq I(\mathcal{Y})$. Since $\mathbf{k}[X_1, \ldots, X_r]$ is a Noetherian domain, every ascending chain of ideals of $\mathbf{k}[X_1, \ldots, X_r]$ is eventually constant. By applying this to ideals associated with algebraic sets, we obtain the *descending chain property for algebraic sets*:

if $\mathcal{X}_i$ ($i = 1, 2, \ldots$) are algebraic subsets of $\mathbf{k}^r$ with $\mathcal{X}_1 \supseteq \mathcal{X}_2 \supseteq \mathcal{X}_3 \supseteq \cdots$, then there is j_0 such that $\mathcal{X}_j = \mathcal{X}_{j_0}$ for $j \geq j_0$.

An algebraic subset of $\mathbf{k}^r$ is called *irreducible* if it is not the union of two strictly smaller algebraic subsets of $\mathbf{k}^r$. An irreducible algebraic subset of $\mathbf{k}^r$ is also called an *algebraic subvariety* of $\mathbf{k}^r$. A linear subvariety of $\mathbf{k}^r$ is an algebraic subvariety of $\mathbf{k}^r$ given by linear polynomials.

From the descending chain property for algebraic sets it follows easily that every non-empty algebraic subset $\mathcal{X}$ of $\mathbf{k}^r$ is a union $\mathcal{V}_1 \cup \cdots \cup \mathcal{V}_g$ of finitely many algebraic subvarieties of $\mathbf{k}^r$. If we assume in addition that none of these algebraic subvarieties is contained in the union of the others, they are uniquely determined. In that case, $\mathcal{V}_1, \ldots, \mathcal{V}_g$ are called the *irreducible components* of $\mathcal{X}$. From the definition of irreducibility it follows at once that any algebraic subvariety contained in $\mathcal{X}$ must be contained in an irreducible component of $\mathcal{X}$.

Let $\mathcal{V}$ be an algebraic subvariety of $\mathbf{k}^r$. Then its associated ideal $I(\mathcal{V})$ is a prime ideal of $\mathbf{k}[X_1, \ldots, X_r]$, hence the quotient ring $\mathbf{k}[X_1, \ldots, X_r]/I(\mathcal{V})$ is an integral domain. The quotient field of this domain is called the *function field* of $\mathcal{V}$, notation $\mathbf{k}(\mathcal{V})$. The transcendence degree over $\mathbf{k}$ of this field is called the *dimension* of $\mathcal{V}$, notation $\dim \mathcal{V}$. A zero-dimensional algebraic variety is a point, and a one-dimensional algebraic variety is an algebraic curve. Further, if $\mathcal{V}_1, \mathcal{V}_2$ are algebraic subvarieties of $\mathbf{k}^r$ with $\mathcal{V}_1$ strictly contained in $\mathcal{V}_2$, then $\dim \mathcal{V}_1 < \dim \mathcal{V}_2$.

We recall some results from intersection theory. To state these, we need the notion of degree of an algebraic variety. Let $\mathcal{V}$ be an n-dimensional algebraic subvariety of $\mathbf{k}^r$. Denote by V_m the $\mathbf{k}$-vector space of polynomials in

$\mathbf{k}[X_1, \ldots, X_r]$ of total degree at most m and define $H_{\mathcal{V}}(m)$ to be the dimension of the quotient $\mathbf{k}$-vector space $V_m/(V_m \cap I(\mathcal{V}))$. One can show that there is a polynomial $p_{\mathcal{V}} \in \mathbb{Q}[X]$, called the *Hilbert polynomial* of $\mathcal{V}$, such that $H_{\mathcal{V}}(m) = p_{\mathcal{V}}(m)$ for every sufficiently large integer m. Further, there is a positive integer $\deg \mathcal{V}$, called the *degree* of $\mathcal{V}$, such that $p_{\mathcal{V}}(m) = \deg \mathcal{V} \cdot m^n/n! +$ (lower powers of m). For instance, $\deg \mathcal{V} = 1$ if $\mathcal{V} = \mathbf{k}^r$ or if $\mathcal{V}$ is a point.

Proposition 7.5.2 *Let $\mathcal{V}$ be an algebraic subvariety of $\mathbf{k}^r$ and let $\mathcal{X}$ be an algebraic subset of $\mathbf{k}^r$ given by polynomials of total degree at most d. Then $\mathcal{V} \cap \mathcal{X}$ is an algebraic subset of $\mathbf{k}^r$ with at most $d^{\dim \mathcal{V}} \cdot \deg \mathcal{V}$ irreducible components.*

Proof. We proceed by induction on $\dim \mathcal{V}$. If $\mathcal{V}$ has dimension 0, i.e., is a point, the assertion is obvious.

Suppose that $\dim \mathcal{V} = n > 0$. If $\mathcal{V} \subseteq \mathcal{X}$ we are done since by definition, $\mathcal{V}$ itself is irreducible. Assume that $\mathcal{V} \not\subseteq \mathcal{X}$. Then there is a polynomial $f \in \mathbf{k}[X_1, \ldots, X_r]$ of total degree at most d that vanishes identically on $\mathcal{X}$, but does not vanish identically on $\mathcal{V}$. We now invoke a version of Bézout's Theorem (see Hartshorne (1977), chapter 1, Theorem 7.7 for a more precise version with multiplicities) which states that if $\mathcal{V}_1, \ldots, \mathcal{V}_g$ are the irreducible components of $\mathcal{V} \cap \mathcal{V}(f)$, then

$$\dim \mathcal{V}_i = n - 1 \text{ for } i = 1, \ldots, g, \quad \sum_{i=1}^{g} \deg \mathcal{V}_i \leq d \cdot \deg \mathcal{V}.$$

Now, by the induction hypothesis, we have for $i = 1, \ldots, g$ that $\mathcal{V}_i \cap \mathcal{X}$ has at most $d^{n-1} \cdot \deg \mathcal{V}_i$ irreducible components. Consequently, the number of irreducible components of $\mathcal{V} \cap \mathcal{X}$ is at most

$$\sum_{i=1}^{g} d^{n-1} \cdot \deg \mathcal{V}_i \leq d^n \deg \mathcal{V} = d^{\dim \mathcal{V}} \cdot \deg \mathcal{V}. \qquad \square$$

Corollary 7.5.3 *Let $\mathcal{X}$ be an algebraic subset of $\mathbf{k}^r$ given by polynomials of total degree at most d. Let $\mathcal{Y}$ be an algebraic subset of $\mathcal{X}$ such that $\mathcal{X} \setminus \mathcal{Y}$ is finite. Then $\mathcal{X} \setminus \mathcal{Y}$ has cardinality at most d^r.*

Proof. Assume that $\mathcal{X} \setminus \mathcal{Y} = \{P_1, \ldots, P_g\}$. Notice that the irreducible components of $\mathcal{Y}$, together with $P_1, \ldots, P_g$, form a decomposition of $\mathcal{X}$ into irreducible subsets, none of which is contained in the union of the others. This shows that $\{P_1\}, \ldots, \{P_g\}$ are irreducible components of $\mathcal{X}$. Now our corollary follows at once by applying Proposition 7.5.2 with $\mathcal{V} = \mathbf{k}^r$. $\qquad \square$

7.5.3 Proof of Theorem 7.5.1

We keep the notation and assumptions from Subsection 7.5.1. We denote by E the algebraic closure of $\mathbf{k}(z)$ in the field $\mathbf{k}((z))$. It is important to observe that the differentiation $x \mapsto x' = \mathrm{d}x/\mathrm{d}z$ maps elements from E to elements from E.

Suppose that a_j, b_{ij} ($j = 1, 2$, $i = 1, \ldots, r$) satisfy (7.5.8) and (7.5.9). Denote by $\mathcal{X}$ the set of $\mathbf{w} = (w_1, \ldots, w_r) \in \mathbf{k}^r$ such that the three functions

$$1, \quad a_1 \prod_{i=1}^{r} b_{i1}^{w_i}, \quad a_2 \prod_{i=1}^{r} b_{i2}^{w_i}$$

are $\mathbf{k}$-linearly dependent. Further, denote by $\mathcal{Y}$ the set of $\mathbf{w} \in \mathbf{k}^r$ such that any two among these functions are $\mathbf{k}$-linearly dependent.

Let (x_1, x_2) be a solution of (7.5.10). Representing (x_1, x_2) as in (7.5.7), we obtain

$$\xi_1 a_1 \prod_{i=1}^{r} b_{i1}^{w_i} + \xi_2 a_2 \prod_{i=1}^{r} b_{i1}^{w_i} = 1 \tag{7.5.11}$$

with $\xi_1, \xi_2 \in \mathbf{k}^*$, $\mathbf{w} = (w_1, \ldots, w_r) \in \mathbf{k}^r$. Hence $\mathbf{w} \in \mathcal{X}$. Further, the condition $a_1 x_1 \notin \mathbf{k}^*$, $a_2 x_2 \notin \mathbf{k}^*$ implies that $\mathbf{w} \notin \mathcal{Y}$. This shows that $\mathbf{w} \in \mathcal{X} \setminus \mathcal{Y}$. Conversely, let $\mathbf{w} \in \mathcal{X} \setminus \mathcal{Y}$. Then there are unique $\xi_1, \xi_2 \in \mathbf{k}^*$ with (7.5.11), and this leads to a unique solution $(x_1, x_2) = (\xi_1, \xi_2) \prod_{i=1}^{r} (b_{i1}, b_{i2})^{w_i}$ of (7.5.10).

So in order to prove Theorem 7.5.1 it suffices to prove the following.

Proposition 7.5.4 *The set $\mathcal{X} \setminus \mathcal{Y}$ has cardinality at most 3^r.*

Eventually, we will apply Corollary 7.5.3 from the previous subsection. We first show that $\mathcal{X} \setminus \mathcal{Y}$ is finite, and then that $\mathcal{X}, \mathcal{Y}$ are algebraic sets where $\mathcal{X}$ is given by polynomials of total degree at most 3.

We first prove a number of lemmas.

Lemma 7.5.5 *Let L be a finite extension of $\mathbf{k}(z)$ contained in E, and let $\beta_1, \ldots, \beta_r, \alpha \in L^*$ and $w_1, \ldots, w_r \in \mathbf{k}$ be such that $\beta_1^{w_1} \cdots \beta_r^{w_r} = \alpha$. Then for every valuation v of L we have $\sum_{i=1}^{r} w_i v(\beta_i) = v(\alpha)$.*

Proof. We need some facts on residues. Choose a local parameter z_v of v. Then every $f \in L$ can be expressed as a Laurent series $\sum_{i=i_0}^{\infty} c_i z_v^i$ with $c_i \in \mathbf{k}$. We define the residue of f at z_v by $\mathrm{res}_{z_v}(f) := c_{-1}$; this defines a $\mathbf{k}$-linear map from L to $\mathbf{k}$. One can show that the residue depends only on v, i.e., that it is independent of the choice of z_v, but we do not need this. We need only the easily verifiable fact that for the logarithmic derivative of $f \in L^*$ with respect to z_v we have $\mathrm{res}_{z_v}(f^{-1}\mathrm{d}f/\mathrm{d}z_v) = v(f)$.

Recall that the logarithmic derivative $x \mapsto x^{-1} \mathrm{d}x/\mathrm{d}z$ is defined on the group $1 + z\mathbf{k}[[z]]$ and that it maps products of powers with exponents in $\mathbf{k}$ to linear combinations. Thus, $\prod_{i=1}^{r} \beta_i^{w_i} = \alpha$ maps to

$$\sum_{i=1}^{r} w_i \cdot \frac{\mathrm{d}\beta_i/\mathrm{d}z}{\beta_i} = \frac{\mathrm{d}\alpha/\mathrm{d}z}{\alpha},$$

which is an identity with functions in L. By multiplying with $\mathrm{d}z/\mathrm{d}z_v$, which also belongs to L, we obtain the same identity, but with z_v instead of z. Then by taking residues with respect to z_v, our lemma follows. $\qquad \square$

Lemma 7.5.6 *Let β_{ij} ($i = 1, \ldots, r$, $j = 1, \ldots, m$) be elements of $(1 + z\mathbf{k}[[z]]) \cap E$ such that there is no non-zero $\mathbf{w} = (w_1, \ldots, w_r) \in \mathbb{Z}^r$ with*

$$\prod_{i=1}^{r} \beta_{ij}^{w_i} \in \mathbf{k}^* \text{ for } i = 1, \ldots, m.$$

Then the map

$$\psi : \mathbf{w} \mapsto \left(\prod_{i=1}^{r} \beta_{i1}^{w_i}, \ldots, \prod_{i=1}^{r} \beta_{im}^{w_i} \right)$$

defines an injective homomorphism from $\mathbf{k}^r$ to $(1 + z\mathbf{k}[[z]])^m$ with coordinatewise multiplication.

Proof. By (7.5.5), the map ψ defines a homomorphism on $\mathbf{k}^r$. Denote by H the kernel of ψ. Notice that H is the set of $\mathbf{w} \in \mathbf{k}^r$ such that

$$\prod_{i=1}^{r} \beta_{ij}^{w_i} = 1 \text{ for } j = 1, \ldots, m. \tag{7.5.12}$$

We have to prove that $H = (\mathbf{0})$.

Let L be the extension of K generated by the elements β_{ij} ($i = 1, \ldots, r$, $j = 1, \ldots, m$). Then $L \subset E$ and L is a finite extension of K. By Lemma 7.5.5, for every $\mathbf{w} \in H$ we have

$$\sum_{i=1}^{r} w_i \cdot v(\beta_{ij}) = 0 \text{ for } j = 1, \ldots, m, v \in M_L,$$

where M_L is the set of valuations of L. The latter system defines a proper linear subspace H' of $\mathbf{k}^r$, defined over $\mathbb{Q}$, containing H. Suppose $H' \neq (\mathbf{0})$. Then H' contains a non-zero vector $\mathbf{w} = (w_1, \ldots, w_r) \in \mathbb{Z}^r$. Put $x_j := \prod_{i=1}^{r} \beta_{ij}^{w_i}$ for $j = 1, \ldots, m$. Then $x_j \in L$ and $v(x_j) = 0$ for $j = 1, \ldots, m$, $v \in M_L$. But this implies $x_j \in \mathbf{k}^*$ for $j = 1, \ldots, m$, contradicting our assumption. $\qquad \square$

Lemma 7.5.7 *Let m, r be positive integers, let $R \in E(U_1, \ldots, U_m)$ be a rational function in m variables, and let $\beta_1, \ldots, \beta_r \in (1 + z\mathbf{k}[[z]]) \cap E$. Then there are only finitely many $\alpha \in F$ for which there exist $\mathbf{w} = (w_1, \ldots, w_r) \in \mathbf{k}^r$, $\mathbf{u} \in \mathbf{k}^m$ such that*

$$\prod_{i=1}^{r} \beta_i^{w_i} = R(\mathbf{u}) = \alpha. \tag{7.5.13}$$

Proof. The assertion is obvious if all β_i are equal to 1. Suppose that not all β_i are equal to 1. Then since $(1 + z\mathbf{k}[[z]]) \cap \mathbf{k} = \{1\}$, not all β_i belong to $\mathbf{k}^*$. Write $R = P/Q$, where $P, Q \in E[U_1, \ldots, U_m]$. Let L be the extension of $\mathbf{k}(z)$ generated by $\beta_1, \ldots, \beta_r$ and the coefficients of P, Q. Then L is a finite extension of K with $L \subset E$. There is a valuation v of L such that the integers $v(\beta_i)$ are not all equal to 0.

We claim that there are integers a, b independent of $\mathbf{u}$ such that if $R(\mathbf{u})$ is defined and non-zero, then $a \leq v(R(\mathbf{u})) \leq b$. Choose a local parameter z_v for v. By expressing the coefficients of P as Laurent series in z_v, we obtain

$$P(\mathbf{u}) = \sum_{i=i_0}^{\infty} p_i(\mathbf{u}) z_v^i$$

with $p_i \in \mathbf{k}[U_1, \ldots, U_m]$ not all identically 0. Put $\mathcal{X}_{i_0-1} := \mathbf{k}^r$, and for $j \geq i_0$, denote by $\mathcal{X}_j$ the set of $\mathbf{u} \in \mathbf{k}^r$ such that $p_i(\mathbf{u}) = 0$ for $i_0 \leq i \leq j$. Then $\mathcal{X}_{i_0} \supseteq \mathcal{X}_{i_0+1} \supseteq \cdots$, and by the descending chain property for algebraic sets, there is j_0 such that $\mathcal{X}_j = \mathcal{X}_{j_0}$ for $j \geq j_0$. Let j_0 be the smallest index with this property. If $\mathbf{u} \in \mathbf{k}^r$ is such that $P(\mathbf{u}) \neq 0$ and $v(P(\mathbf{u})) = j$, say, we have $\mathbf{u} \in \mathcal{X}_{j-1} \setminus \mathcal{X}_j$. Hence $v(P(\mathbf{u})) \in \{i_0, \ldots, j_0\}$. By applying the same reasoning to Q, our claim follows.

Let $\alpha \in E$ for which there exist $\mathbf{w} \in \mathbf{k}^r$, $\mathbf{u} \in \mathbf{k}^m$ with (7.5.13). By Lemma 7.5.5 we have

$$j = \sum_{i=1}^{r} w_i v(\alpha_i),$$

where $j = v(R(\mathbf{u})) \in \{a, a+1, \ldots, b\}$. Thus, $\mathbf{w}$ belongs to one of finitely many linear subvarieties of $\mathbf{k}^r$ of dimension $r - 1$, all defined over $\mathbb{Q}$.

We now proceed by induction on r. If $r = 1$, we have only finitely many possibilities for $\mathbf{w}$, and then our lemma follows at once. Suppose that $r \geq 2$, and let $\mathcal{L}$ be one of the linear varieties from above. We have to show that $\mathcal{L}$ gives rise to only finitely many α as in (7.5.13). There are $\mathbf{c}_0 \in \mathbb{Q}^r$ and linearly independent $\mathbf{c}_k \in \mathbb{Z}^r$ ($k = 1, \ldots, r - 1$) such that every $\mathbf{w} \in \mathcal{L}$ can be expressed as $\mathbf{w} = \mathbf{c}_0 + \sum_{k=1}^{r-1} t_k \mathbf{c}_k$ with $t_1, \ldots, t_{r-1} \in \mathbf{k}$. Write $\mathbf{c}_k = (c_{1k}, \ldots, c_{rk})$,

$\gamma_k := \prod_{i=1}^{r} \beta_i^{c_{ik}}$ for $k = 0, \dots, r$. Notice that $\gamma_k \in E$ for $k = 0, \dots, r$. By substituting our expression for $\mathbf{w} \in \mathcal{L}$ into (7.5.13), we obtain

$$\prod_{k=1}^{r-1} \gamma_k^{t_k} = \gamma_0^{-1} R(\mathbf{u}) = \gamma_0^{-1} \alpha.$$

By the induction hypothesis, we have only finitely many possibilities for $\gamma_0^{-1}\alpha$, hence for α. This completes our induction step. $\square$

Lemma 7.5.8 *The set $\mathcal{X} \setminus \mathcal{Y}$ is finite.*

Proof. Let $\mathbf{w} = (w_1, \dots, w_r) \in \mathcal{X} \setminus \mathcal{Y}$. Put $y_j := \prod_{i=1}^{r} b_{ij}^{w_j}$ for $j = 1, 2$. Then the logarithmic derivative of $a_j y_j$ with respect to z is

$$\frac{a_j'}{a_j} + \sum_{i=1}^{r} w_i \cdot \frac{b_{ij}'}{b_{ij}} =: Q_j \quad \text{for } j = 1, 2.$$

Since $\mathbf{w} \in \mathcal{X}$, there are $\xi_1, \xi_2 \in \mathbf{k}^*$ with

$$\xi_1 a_1 y_1 + \xi_2 a_2 y_2 = 1.$$

Upon differentiating this identity with respect to z, we obtain

$$Q_1 \cdot \xi_1 a_1 y_1 + Q_2 \cdot \xi_2 a_2 y_2 = 0.$$

Since $\mathbf{w} \notin \mathcal{Y}$ we have $a_1 y_1 \neq a_2 y_2$, and since logarithmic differentiation $x \mapsto x'/x$ is injective on $1 + z\mathbf{k}[[z]]$, we have $Q_1 \neq Q_2$. Hence the last two equations have a unique solution (y_1, y_2), and on applying Cramer's rule we obtain

$$\prod_{i=1}^{r} b_{ij}^{w_j} = y_j = R_j(\xi_1, \xi_2, w_1, \dots, w_r) \text{ for } j = 1, 2,$$

where R_1, R_2 are certain rational functions in $E(U_1, \dots, U_{r+2})$.

Put $\psi(\mathbf{w}) := (\prod_{i=1}^{r} b_{i1}^{w_i}, \prod_{i=1}^{r} b_{i2}^{w_i})$. Lemma 7.5.7 implies that if $\mathbf{w}$ runs through $\mathcal{X} \setminus \mathcal{Y}$, then $\psi(\mathbf{w})$ runs through a finite set. On the other hand, by condition (7.5.9) and Lemma 7.5.6, ψ defines an injective map. This shows that $\mathcal{X} \setminus \mathcal{Y}$ is finite. $\square$

Lemma 7.5.9 *The set $\mathcal{X}$ is an algebraic subset of $\mathbf{k}^r$, given by polynomials in $\mathbf{k}[X_1, \dots, X_r]$ of total degree at most 3. Further, $\mathcal{Y}$ is an algebraic subset of $\mathcal{X}$.*

Proof. We apply the Wronskian criterion, that functions $1, f_1, f_2 \in \mathbf{k}((z))$ are linearly dependent over $\mathbf{k}$ if and only if $f_1' f_2'' - f_2'' f_1' = 0$, where

$f_j'' := \mathrm{d}^2 f_j / \mathrm{d}z^2$. Let $f_j := a_j \prod_{i=1}^{r} b_{ij}^{w_i}$ for $j = 1, 2$. By a straightforward computation, $f_j' = p_j(\mathbf{w}) f_j$, $f_j'' = q_j(\mathbf{w}) f_j$ for $j = 1, 2$, where p_j are linear polynomials, and q_j are quadratic polynomials with coefficients in E, for $j = 1, 2$. Since $f_1 f_2 \neq 0$ for every $\mathbf{w}$, it follows that $\mathbf{w} \in \mathcal{X}$ if and only if $h(\mathbf{w}) = 0$, where $h := p_1 q_2 - p_2 q_1$. Notice that h has total degree at most 3. From the coefficients of h we select a maximal subset which is linearly independent over $\mathbf{k}$, $\{a_1, \dots, a_s\}$, say. Then the other coefficients of h can be expressed as $\mathbf{k}$-linear combinations of $a_1, \dots, a_s$, and we get $h = \sum_{k=1}^{s} a_s h_s$ with polynomials $h_k \in \mathbf{k}[X_1, \dots, X_r]$ $(k = 1, \dots, s)$ of total degree at most 3. Now, clearly, for $\mathbf{w} \in \mathbf{k}^r$ we have $h(\mathbf{w}) = 0$ if and only if $h_k(\mathbf{w}) = 0$ for $k = 1, \dots, s$. Hence $\mathcal{X}$ is an algebraic set given by $h_1, \dots, h_s$.

To prove that $\mathcal{Y}$ is an algebraic subset of $\mathcal{X}$, we use the Wronskian criterion that two functions f_1, $f_2 \in \mathbf{k}((z))$ are $\mathbf{k}$-linearly dependent if and only if $f_1 f_2' - f_1' f_2 = 0$, and follow the same arguments as above.   □

Now Proposition 7.5.4 follows at once by combining Lemmas 7.5.8 and 7.5.9 with Corollary 7.5.3.   □

7.6 Results in positive characteristic

We give an overview of some results on S-unit equations and generalizations thereof over function fields of positive characteristic.

Let $\mathbf{k}$ be a field of characteristic $p > 0$ and K a finite extension of the rational function field $\mathbf{k}(z)$. We assume that $\mathbf{k}$ is algebraically closed in K. Denote by $g_{K/\mathbf{k}}$ the genus of K over $\mathbf{k}$.

Similarly as in the characteristic 0 case, we can endow K with a set of valuations M_K (i.e., normalized discrete valuations that are trivial on $\mathbf{k}$) satisfying the Sum Formula $\sum_{v \in M_K} v(x) = 0$ for $x \in K^*$. Further, we define for $\mathbf{x} = (x_1, \dots, x_n) \in K^n \setminus \{0\}$,

$$v(\mathbf{x}) := -\min(v(x_1), \dots, v(x_n)) \quad \text{for } v \in M_K,$$
$$H_K^{\mathrm{hom}}(\mathbf{x}) := \sum_{v \in M_K} v(\mathbf{x}).$$

The height of $x \in K$ is given by $H_K(x) := -\sum_{v \in M_K} \min(0, v(x))$. For a finite set of valuations S of M_K, we define the group of S-units

$$O_S^* := \{x \in K : v(x) = 0 \text{ for } v \in M_K \setminus S\}.$$

Recall that the Frobenius map $x \mapsto x^p$ defines an injective field homomorphism on K. As a consequence, the sets $K^{p^m} := \{x^{p^m} : x \in K\}$

$(m = 1, 2, \ldots)$ are subfields of K. We say that $a_1, \ldots, a_m \in K$ are linearly independent over a subset U of K if there are no $c_1, \ldots, c_m \in U \setminus \{0\}$ such that $c_1 a_1 + \cdots + c_m a_m = 0$.

We start with an analogue of Theorem 7.1.1 in the case of characteristic p, also due to Mason (1984), chapter VI, Lemma 10.

Theorem 7.6.1 *Let x_1, x_2, x_3 be non-zero elements of K, and let S be a finite set of valuations of K such that*

$$x_1 + x_2 + x_3 = 0,$$

$$v(x_1) = v(x_2) = v(x_3) \quad \text{for } v \in M_K \setminus S.$$

Then either $x_1/x_2 \in K^p$ or

$$H_K(x_1/x_2) \leq 2g_{K/\mathbf{k}} - 2 + |S|.$$

Of course, if we allow $x_1/x_2 \in K^p$, the result may become false, for instance if x_1, x_2, x_3 are any elements of K with $x_1 + x_2 + x_3 = 0$ and x_1/x_2 not in the constant field $\mathbf{k}$, then we have also $x_1^{p^m} + x_2^{p^m} + x_3^{p^m} = 0$ for every positive integer m and clearly $H_K(x_1^{p^m}/x_2^{p^m})$ may become arbitrarily large.

Mason's proof of Theorem 7.6.1 is similar to his proof of Theorem 7.1.1, based on derivations. Silverman (1984) proved a similar result (stated in another but equivalent form) by a different geometric method, based on the Riemann–Hurwitz formula and properties of Fermat curves $x^N + y^N = 1$.

We now turn to S-unit equations in several unknowns. First Mason (1986b) and later in a sharper form Wang (1996, 1999) proved analogues of Corollary 7.3.4 in the case of positive characteristic. We recall Wang (1996), Corollary. 1.

Theorem 7.6.2 *Suppose that K has genus g. Let S be a finite set of valuations of K, let m be a positive integer, and let $x_0, \ldots, x_n$ be elements of O_S^* such that*

$$x_0 + \cdots + x_n = 0$$

and every proper subset of $\{x_0, \ldots, x_n\}$ is linearly independent over $\mathbf{k} \cdot K^{p^m}$. Then

$$H_K^{\text{hom}}(\mathbf{x}) \leq \frac{n(n-1)}{2} p^{m-1} \max(0, 2g - 2 + |S|).$$

Wang's proof is a positive characteristic analogue of the Wronskian argument of Brownawell and Masser. In Wang (1999) she slightly sharpened her result. Hsia and Wang (2004) proved a further extension to function fields of arbitrary transcendence degree, in the cases of both zero characteristic and positive characteristic. Their proof uses generalized Wronskians.

We now turn to linear equations with unknowns from a finitely generated multiplicative group. Let p be a prime, $\mathbb{F}_p$ the field of p elements, and $\overline{\mathbb{F}}_p$ its algebraic closure. Again, K is a field of characteristic $p > 0$ but we now assume that K is a finitely generated, transcendental extension of $\mathbb{F}_p$. Let $\mathbf{k}$ be the algebraic closure of $\mathbb{F}_p$ in K; so $\mathbf{k}$ is a finite field.

Let us start with the equation

$$a_1 x_1 + a_2 x_2 = 1 \quad \text{in } \mathbf{x} = (x_1, x_2) \in \Gamma, \tag{7.6.1}$$

where Γ is a subgroup of $(K^*)^2$ of finite rank not contained in $(\overline{\mathbb{F}_p}^*)^2$ and $\mathbf{a} = (a_1, a_2) \in (K^*)^2$. For instance, if there is an integer l coprime with p such that $\mathbf{a}^l \in \Gamma$, then (7.6.1) may have infinitely many solutions. Specifically, let q be a power of p such that $l \mid q - 1$ and let $\mathbf{u} = (u_1, u_2)$ be a solution of (7.6.1) with $\mathbf{u} \notin (\overline{\mathbb{F}_p}^*)^2$, then

$$\mathbf{a}^{q^e - 1} \cdot \mathbf{u}^{q^e} \quad (e = 0, 1, 2, \ldots) \tag{7.6.2}$$

yield infinitely many different solutions of (7.6.1). The following nice result is due to Voloch (1998), Theorem 2.

Theorem 7.6.3 *Let Γ have rank $r \geq 0$. Assume there is no positive integer l such that $\mathbf{a}^l \in \Gamma$. Then (7.6.1) has at most*

$$\frac{p^r(p^r + p - 2)}{p - 1}$$

solutions.

The proof uses derivations on K.

Now let $n \geq 2$, Γ a finitely generated subgroup of $(K^*)^n$ (n-fold direct product, not to be confused with the field of p-th powers K^p in case n, p happen to be equal), and $\mathbf{a} = (a_1, \ldots, a_n) \in (K^*)^n$, and consider the equation

$$a_1 x_1 + \cdots + a_n x_n = 1 \quad \text{in } \mathbf{x} = (x_1, \ldots, x_n) \in \Gamma. \tag{7.6.3}$$

The following result is a consequence of Hsia's and Wang's analogue of Theorem 7.6.2 for function fields in several variables mentioned above.

Theorem 7.6.4 *Equation (7.6.3) has only finitely many solutions such that $a_1 x_1, \ldots, a_n x_n$ are linearly independent over K^p.*

We want to weaken the condition that $a_1 x_1, \ldots, a_n x_n$ be linearly independent over K^p to the condition that $(x_1, \ldots, x_n)$ be non-degenerate, that is,

$$\sum_{i \in I} a_i x_i \neq 0 \quad \text{for each proper subset } I \text{ of } \{1, \ldots, n\}.$$

Then the situation becomes much more complicated, due to the Frobenius action on K, and in fact the solutions can be divided into finitely many infinite classes with a particular structure.

In Mason (1986b), Masser (2004), Adamczewski and Bell (2012) and Derksen and Masser (2012) various descriptions are given for the set of non-degenerate solutions of (7.6.3). We discuss here a result from the last paper, which unlike Voloch's result does not give an upper bound for the number of classes of solutions, but instead implies that these classes can be determined effectively.

As an introduction to the result of Derksen and Masser, note that we can write the set of solutions of (7.6.1) given in (7.6.2) as

$$\psi_{\mathbf{a}}^{-1} \varphi_q^e \psi_{\mathbf{a}}(\mathbf{u}) \quad (e = 0, 1, 2, \ldots), \tag{7.6.4}$$

where $\psi_{\mathbf{a}}$, φ_q are the maps given by

$$\psi_{\mathbf{a}}(\mathbf{x}) := \mathbf{a} \cdot \mathbf{x}, \quad \varphi_q(\mathbf{x}) := \mathbf{x}^q.$$

We now proceed to state the result of Derksen and Masser on the non-degenerate solutions of (7.6.3). Denote by Γ_K the group of $\mathbf{u} \in (K^*)^n$ for which there exists $l \in \mathbb{Z}_{>0}$ with $\mathbf{u}^l \in \Gamma$. For a power q of p and for $\mathbf{u} \in \Gamma$, $\mathbf{g}_1, \ldots, \mathbf{g}_h \in \Gamma_K$, define the set

$$[\mathbf{g}_1, \ldots, \mathbf{g}_h]_q(\mathbf{u}) := \left\{ \psi_{\mathbf{g}_1}^{-1} \varphi_q^{e_1} \psi_{\mathbf{g}_1} \cdots \psi_{\mathbf{g}_h}^{-1} \varphi_q^{e_h} \psi_{\mathbf{g}_h}(\mathbf{u}) : e_1, \ldots, e_h \in \mathbb{Z}_{\geq 0} \right\},$$

where φ_q and the $\psi_{\mathbf{g}_i}$ are maps from $(K^*)^n$ to $(K^*)^n$ given by (7.6.4). We call such a set a (Γ_K, q)-*set of order h*. We agree here that a (Γ_K, q)-set of order 0 is a single element. Computing a (Γ_K, q)-set means computing q and the tuple $(\mathbf{g}_1, \ldots, \mathbf{g}_h, \mathbf{u})$ by which it is defined.

The following result is part of Derksen and Masser (2012), Theorem 3.

Theorem 7.6.5 *Let* $a_1, \ldots, a_n \in K^*$. *Assume that* Γ *is not contained in* $(\overline{\mathbb{F}_p}^*)^n$ *and that* Γ_K *is finitely generated. Then there is a power q of p such that the set of non-degenerate solutions of* (7.6.3) *is contained in a finite union of* (G_K, q)-*sets of order at most $n - 1$. Further, if suitable effective representations of K and Γ are given, the prime power q and these* (Γ_K, q)-*sets can be determined effectively.*

In their proof, Derksen and Masser derived a sharpening of Theorem 7.6.4, basically by extending the argument of Brownawell and Masser based on Wronskians. From there, they completed the proof of Theorem 7.6.5 by means of an inductive argument.

We mention that earlier, Masser (2004) obtained a less precise and ineffective version of Theorem 7.6.5. Building further on work from Derksen (2007) on linear recurrence sequences, Adamczewski and Bell (2012), Theorem 3.1 gave a description of the set of non-degenerate solutions of (7.6.3) in terms of finite p-automata, and in fact they proved a much more general result for semi-abelian varieties. Finally, general results on semi-abelian varieties over fields of positive characteristic, implying Theorem 7.6.5, have been proved by means of methods from logic, see Hrushovki (1996), Moosa and Scanlon (2002, 2004) and Ghioca (2008).

We illustrate Theorem 7.6.5 with an example from Masser (2004). Let $K = \mathbb{F}_p(z)$ with z transcendental over $\mathbb{F}_p$ and let $\Gamma = G^3$, where G is the multiplicative subgroup of K^* generated by z and $1 - z$. Consider the equation

$$x_1 + x_2 - x_3 = 1 \quad \text{in } (x_1, x_2, x_3) \in \Gamma. \tag{7.6.5}$$

This equation has (among others) the non-degenerate solutions

$$(z^{(q-1)q'}, (1 - z)^{qq'}, z^{(q-1)q'}(1 - z)^{q'}),$$

where q, q' run independently through the powers of p different from 1. This set of solutions may be described as $[\mathbf{g}_1, \mathbf{g}_2]_p(\mathbf{u})$, where

$$\mathbf{g}_1 = (1, 1, 1), \quad \mathbf{g}_2 = (z, 1, z(1 - z)^{-1}),$$
$$\mathbf{u} = (z^{(p-1)p}, (1 - z)^{p^2}, z^{(p-1)p}(1 - z)^p).$$

Leitner (2012), Theorem 2 gave a complete description of the set of solutions of (7.6.5) as a union of (Γ_K, p)-sets. As it turned out, the set of non-degenerate solutions is contained in a union of 40 (Γ_K, p)-sets if $p \geq 5$, 48 $(\Gamma_K, 3)$-sets if $p = 3$, and 240 $(\Gamma_K, 2)$-sets if $p = 2$. For $p = 2$, his result was obtained earlier by Arenas-Carmona, Berend and Bergelson (2008).

8

Effective results for unit equations in two unknowns over finitely generated domains

In Chapter 4 we established effective finiteness results on unit equations and S-unit equations in two unknowns in an algebraic number field. In this chapter we extend these results to the finitely generated case. More precisely, let $A \supset \mathbb{Z}$ be an integral domain which is finitely generated over $\mathbb{Z}$, i.e., $A = \mathbb{Z}[z_1, \ldots, z_r]$ for certain not necessarily algebraic generators $z_1, \ldots, z_r$, K the quotient field of A, and a_1, a_2, a_3 non-zero elements of K. By a theorem of Roquette (1957), the group A^* of units of A is finitely generated, hence we know from, e.g., the results of Chapter 6 that the equation

$$a_1 x_1 + a_2 x_2 = a_3 \quad \text{in } x_1, x_2 \in A^* \tag{8.1}$$

has only finitely many solutions. This result is, however, ineffective.

In this chapter, we give an effective proof of this finiteness statement, which is valid for any arbitrary integral domain A that is finitely generated over $\mathbb{Z}$. In fact, our main result, Theorem 8.1.1, provides effective upper bounds for the "sizes" of the solutions x_1, x_2 in terms of suitable effective representations for A, a_1, a_2, a_3. This enables one to determine all solutions in principle; see Corollary 8.1.2 below. As a further consequence of Theorem 8.1.1 we deduce an effective finiteness theorem on equation (8.1) in unknowns x_1, x_2 taken from a finitely generated and effectively given multiplicative subgroup Γ of K^*, see Theorem 8.1.3 below.

Our strategy of proof of Theorem 8.1.1 is roughly as follows. We construct an integral domain $B \supset A$ of a special type that can be dealt with more easily, and consider instead of (8.1) the equation

$$a_1 x_1 + a_2 x_2 = a_3 \quad \text{in } x_1, x_2 \in B^*. \tag{8.2}$$

In the construction of B we use ideas from Seidenberg (1974). We reduce (8.2) to the function field case, and using Mason's Theorem 7.1.1 we derive an effective upper bound for the degrees of the solutions. Next, by means of effective specializations, i.e., explicitly given ring homomorphisms $B \to \overline{\mathbb{Q}}$, we reduce (8.2) to various S-unit equations in different algebraic number fields, and apply the results of Chapter 4 to the S-unit equations obtained. This provides enough information to derive an effective upper bound for the heights of the solutions of (8.2). From this, we can effectively determine all solutions of (8.2). The final, crucial step is to go back from (8.2) to (8.1) and to select from the solutions in B^* those that belong to A^*. For this we have developed an effective procedure, based on an effective result of Aschenbrenner (2004) on systems of linear equations over polynomial rings over $\mathbb{Z}$.

The above approach was developed by Győry (1983, 1984). However, at that time Aschenbrenner's result was not yet available. Hence, to select those solutions from B^* of the equations under consideration that belong to A^*, certain restrictions on the integral domain A had to be imposed.

This chapter is organized as follows. In Section 8.1 we give the necessary definitions and state our results. In Sections 8.2–8.6 we prove Theorem 8.1.1. More precisely, in Sections 8.2 and 8.3 we construct the domain B and give the effective procedure to select those elements of B^* that belong to A^*, in Section 8.4 we reduce (8.2) to the function field case and apply Mason's Theorem, in Section 8.5 we develop some effective specialization theory, and in Section 8.6 we reduce (8.2) to S-unit equations over number fields, apply the results from Chapter 4, and complete the proof. In Section 8.7 we prove Theorem 8.1.3. In Section 8.8 we briefly discuss some related results.

The results in this chapter were proved for the first time in Evertse and Győry (2013). We closely follow the exposition of that paper.

8.1 Statements of the results

We introduce the notation used in our theorems. Let again $A \supset \mathbb{Z}$ be an integral domain which is finitely generated over $\mathbb{Z}$, say $A = \mathbb{Z}[z_1, \ldots, z_r]$. Let I be the ideal of polynomials $f \in \mathbb{Z}[X_1, \ldots, X_r]$ such that $f(z_1, \ldots, z_r) = 0$. Then I is finitely generated, hence

$$A \cong \mathbb{Z}[X_1, \ldots, X_r]/I, \quad I = (f_1, \ldots, f_m)$$

for some finite set of polynomials $f_1, \ldots, f_m \in \mathbb{Z}[X_1, \ldots, X_r]$. We observe here that given $f_1, \ldots, f_m$, it can be checked effectively whether A is a domain containing $\mathbb{Z}$. Indeed, this holds if and only if I is a prime ideal of $\mathbb{Z}[X_1, \ldots, X_r]$ with $I \cap \mathbb{Z} = (0)$, and the latter can be checked effectively for instance using Aschenbrenner (2004), Proposition 4.10, Corollary 3.5.

Denote by K the quotient field of A. For $\alpha \in A$, we call f a *representative* for α or say that f represents α if $f \in \mathbb{Z}[X_1, \ldots, X_r]$ and $\alpha = f(z_1, \ldots, z_r)$. Further, for $\alpha \in K$, we call (f, g) a *pair of representatives* for α or say that (f, g) represents α if $f, g \in \mathbb{Z}[X_1, \ldots, X_r]$, $g \notin I$ and $\alpha = f(z_1, \ldots, z_r)/g(z_1, \ldots, z_r)$. We say that $\alpha \in A$ (resp. $\alpha \in K$) is given if a representative (resp. pair of representatives) for α is given.

To do explicit computations in A and K, one needs an *ideal membership algorithm* for $\mathbb{Z}[X_1, \ldots, X_r]$, which decides, for any given polynomial and ideal of $\mathbb{Z}[X_1, \ldots, X_r]$, whether the polynomial belongs to the ideal. In the literature there are various such algorithms; we mention only the algorithm of Simmons (1970), and the more precise algorithm of Aschenbrenner (2004) which plays an important role in this chapter; see Lemma 8.2.5 below for a statement of his result. One can perform arithmetic operations on A and K by using representatives. Further, one can decide effectively whether two polynomials g_1, g_2 represent the same element of A, i.e., $g_1 - g_2 \in I$, or whether two pairs of polynomials $(g_1, h_1), (g_2, h_2)$ represent the same element of K, i.e., $g_1 h_2 - g_2 h_1 \in I$, by using one of the ideal membership algorithms mentioned above.

The *degree* $\deg f$ of a polynomial $f \in \mathbb{Z}[X_1, \ldots, X_r]$ is by definition its total degree. By the *logarithmic height* $h(f)$ of f we mean the logarithm of the maximum of the absolute values of its coefficients. The *size* of f is defined by

$$s(f) := \max(1, \deg f, h(f)).$$

Clearly, there are only finitely many polynomials in $\mathbb{Z}[X_1, \ldots, X_r]$ of size below a given bound, and these can be determined effectively.

We consider equations

$$a_1 x_1 + a_2 x_2 = a_3 \quad \text{in } x_1, x_2 \in A^*, \tag{8.1.1}$$

where a_1, a_2, a_3 are non-zero elements of A.

Theorem 8.1.1 *Assume that $r \geq 1$. Let $\tilde{a}_1, \tilde{a}_2, \tilde{a}_3$ be representatives for a_1, a_2, a_3, respectively. Assume that $f_1, \ldots, f_m$ and $\tilde{a}_1, \tilde{a}_2, \tilde{a}_3$ all have degree at most d and logarithmic height at most h, where $d \geq 1$, $h \geq 1$. Then for each solution (x_1, x_2) of (8.1.1), there are representatives $\tilde{x}_1, \tilde{x}_1{}', \tilde{x}_2, \tilde{x}_2{}'$ of $x_1, x_1^{-1}, x_2, x_2^{-1}$, respectively, such that*

$$s(\tilde{x}_i), \; s(\tilde{x}_i{}') \leq \exp\left((2d)^{c_1}(h+1)\right) \quad \text{for } i = 1, 2,$$

where c_1 is an effectively computable absolute constant > 1.

By a theorem of Roquette (1957), the unit group of an integral domain finitely generated over $\mathbb{Z}$ is finitely generated. In the case that $A = O_S$ is the ring of S-integers of a number field it is possible to determine effectively a system of generators for A^*, and this was used in all effective finiteness proofs for (8.1.1) with $A = O_S$. However, no general algorithm is known to determine a system of generators for the unit group of an arbitrary finitely generated domain A. In our proof of Theorem 8.1.1, we do not need any information on the generators of A^*.

By combining Theorem 8.1.1 with an ideal membership algorithm for the polynomial ring $\mathbb{Z}[X_1, \ldots, X_r]$, one easily deduces the following.

Corollary 8.1.2 *Given $f_1, \ldots, f_m$ such that A is an integral domain containing $\mathbb{Z}$, and given $a_1, a_2, a_3 \in A \setminus \{0\}$, the solutions of (8.1.1) can be determined effectively.*

Proof. Clearly, (x_1, x_2) is a solution of (8.1.1) if and only if for $i = 1, 2$, there are polynomials $\tilde{x}_i, \tilde{x}_i' \in \mathbb{Z}[X_1, \ldots, X_r]$ $(i = 1, 2)$ such that $\tilde{x}_i$ represents x_i for $i = 1, 2$, and

$$\tilde{a}_1 \cdot \tilde{x}_1 + \tilde{a}_2 \cdot \tilde{x}_2 - \tilde{a}_3 \in I, \quad , \quad \tilde{x}_i \cdot \tilde{x}_i' - 1 \in I \quad \text{for } i = 1, 2. \quad (8.1.2)$$

Thus, we obtain all solutions of (8.1.1) by checking, for each quadruple of polynomials $\tilde{x}_1, \tilde{x}_1', \tilde{x}_2, \tilde{x}_2' \in \mathbb{Z}[X_1, \ldots, X_r]$ of size at most $\exp((2d)^{c_i'}(h + 1))$ whether it satisfies (8.1.2). Further, using the ideal membership algorithm, it can be checked effectively whether two different pairs $(\tilde{x}_1, \tilde{x}_2)$ represent the same solution of (8.1.1). Thus, we can make a list of representatives, one for each solution of (8.1.1). $\square$

Let $\gamma_1, \ldots, \gamma_s$ be multiplicatively independent elements of K^*. For given elements $\gamma_1, \ldots, \gamma_s \in K^*$ the multiplicative independence of $\gamma_1, \ldots, \gamma_s$ can be checked effectively, see for instance Lemma 8.7.2 below. Let again a_1, a_2, a_3 be non-zero elements of A and consider the equation

$$a_1 \gamma_1^{v_1} \cdots \gamma_s^{v_s} + a_2 \gamma_1^{w_1} \cdots \gamma_s^{w_s} = a_3 \quad \text{in } v_1, \ldots, v_s, \, w_1, \ldots, w_s \in \mathbb{Z}. \quad (8.1.3)$$

Theorem 8.1.3 *Let $\tilde{a}_1, \tilde{a}_2, \tilde{a}_3$ be representatives for a_1, a_2, a_3 and for $i = 1, \ldots, s$, let (g_{i1}, g_{i2}) be a pair of representatives for γ_i. Suppose that $f_1, \ldots, f_m, \tilde{a}_1, \tilde{a}_2, \tilde{a}_3$, and g_{i1}, g_{i2} $(i = 1, \ldots, s)$ all have degree at most d and logarithmic height at most h, where $d \geq 1$, $h \geq 1$. Then for each solution $(v_1, \ldots, w_s)$ of (8.1.3) we have*

$$\max(|v_1|, \ldots, |v_s|, \, |w_1|, \ldots, |w_s|) \leq \exp\left((2^s d)^{c_2'}(h + 1)\right),$$

where c_2 is an effectively computable absolute constant > 1.

An immediate consequence of Theorem 8.1.3 is that for given $f_1, \ldots, f_m$, a_1, a_2, a_3 and $\gamma_1, \ldots, \gamma_s$, the solutions of (8.1.3) can be determined effectively.

Since every integral domain finitely generated over $\mathbb{Z}$ has a finitely generated unit group, equation (8.1.1) may be viewed as a special case of (8.1.3). But since no general effective algorithm is known to find a finite system of generators for the unit group of a finitely generated integral domain, we cannot deduce an effective result for (8.1.1) from Theorem 8.1.3. In fact, we argue reversely, and prove Theorem 8.1.3 by combining Theorem 8.1.1 with an effective result on Diophantine equations of the type $\gamma_1^{v_1} \cdots \gamma_s^{v_s} = \gamma_0$ in integers $v_1, \ldots, v_s$, where $\gamma_1, \ldots, \gamma_s, \gamma_0 \in K^*$ (see Corollary 8.7.3 below).

8.2 Effective linear algebra over polynomial rings

We have gathered from the literature some effective results for systems of linear equations to be solved in polynomials with coefficients in a field, or with coefficients in $\mathbb{Z}$.

As usual, we write

$$\log^* u := \max(1, \log u) \quad \text{for } u > 0, \log^* 0 := 1.$$

We use the notation $\mathfrak{O}(\cdot)$ as an abbreviation for c times the expression between the parentheses, where c is an effectively computable positive absolute constant (notice that the meaning of the $\mathfrak{O}$-symbol is different from that of the usual O-symbol which means "at most c times the expression between the parentheses"). At each occurrence of $\mathfrak{O}(\cdot)$, the value of c may be different.

Given a ring R, we denote by $R^{m,n}$ the R-module of $m \times n$-matrices with entries in R and by R^n the R-module of n-dimensional column vectors with entries in R. Further, as usual GL(n, R) denotes the group of matrices in $R^{n,n}$ with determinant in the unit group R^*. The degree of a polynomial $f \in R[X_1, \ldots, X_N]$, that is, its total degree, is denoted by $\deg f$.

From matrices U, V with the same number of rows, we form a matrix $[U, V]$ by placing the columns of V after those of U, and from two matrices U, V with the same number of columns we form $\left[\begin{smallmatrix} U \\ V \end{smallmatrix}\right]$ by placing the rows of V below those of U.

The logarithmic height $h(\mathcal{S})$ of a finite set $\mathcal{S} = \{a_1, \ldots, a_t\} \subset \mathbb{Z}$ is defined by $h(\mathcal{S}) := \log \max(|a_1|, \ldots, |a_t|)$. The logarithmic height $h(U)$ of a matrix with entries in $\mathbb{Z}$ is defined by the logarithmic height of the set of entries of U. The logarithmic height $h(f)$ of a polynomial with coefficients in $\mathbb{Z}$ is the logarithmic height of the set of non-zero coefficients of f.

Lemma 8.2.1 *Let $U \in \mathbb{Z}^{m,n}$. Then the $\mathbb{Q}$-vector space of $\mathbf{y} \in \mathbb{Q}^n$ with $U\mathbf{y} = \mathbf{0}$ is generated by vectors in $\mathbb{Z}^n$ of logarithmic height at most $mh(U) + \frac{1}{2}m \log m$.*

Proof. Without loss of generality we may assume that U has rank m, and moreover, that the matrix V consisting of the first m columns of U is invertible. Let $\Delta := \det V$. By multiplying with ΔV^{-1}, we can rewrite $U\mathbf{y} = \mathbf{0}$ as $[\Delta I_m, W]\mathbf{y} = \mathbf{0}$, where I_m is the $m \times m$-unit matrix, and W consists of $m \times m$-subdeterminants of U. The solution space of this system is generated by the columns of $\begin{bmatrix} -W \\ \Delta I_{n-m} \end{bmatrix}$. An application of Hadamard's inequality gives the upper bound from the lemma for the logarithmic heights of these columns.   □

Proposition 8.2.2 *Let F be a field, $N \geq 1$, and $R := F[X_1, \ldots, X_N]$. Further, let U be an $m \times n$-matrix and $\mathbf{b}$ an m-dimensional column vector, both consisting of polynomials from R of degree $\leq d$ where $d \geq 1$.*

(i) *The R-module of $\mathbf{x} \in R^n$ with $U\mathbf{x} = \mathbf{0}$ is generated by vectors $\mathbf{x}$ whose coordinates are polynomials of degree at most $(2md)^{2^N}$.*

(ii) *Suppose that $U\mathbf{x} = \mathbf{b}$ is solvable in $\mathbf{x} \in R^n$. Then it has a solution $\mathbf{x}$ whose coordinates are polynomials of degree at most $(2md)^{2^N}$.*

Proof. See Aschenbrenner (2004), theorems 3.2, 3.4. Part (ii) of Proposition 8.2.2 was obtained earlier by Masser and Wüstholz (1983), Proposition on p.440, estimate on top of p.442, with the slightly smaller bound $(2md)^{2^{N-1}}$. Results of this type, but not with a completely correct proof, were given in Hermann (1926) and Seidenberg (1974).   □

Corollary 8.2.3 *Let $R := \mathbb{Q}[X_1, \ldots, X_N]$. Further, Let U be an $m \times n$-matrix of polynomials in $\mathbb{Z}[X_1, \ldots, X_N]$ of degrees at most d and logarithmic heights at most h where $d \geq 1$, $h \geq 1$. Then the R-module of $\mathbf{x} \in R^n$ with $U\mathbf{x} = \mathbf{0}$ is generated by vectors $\mathbf{x}$, consisting of polynomials in $\mathbb{Z}[X_1, \ldots, X_N]$ of degree at most $(2md)^{2^N}$ and height at most $(2md)^{6^N}(h + 1)$.*

Proof. By Proposition 8.2.2 (i) we have to study $U\mathbf{x} = \mathbf{0}$, restricted to vectors $\mathbf{x} \in R^n$ consisting of polynomials of degree at most $(2d)^{2^N}$. The set of these $\mathbf{x}$ is a finite dimensional $\mathbb{Q}$-vector space, and we have to prove that it is generated by vectors whose coordinates are polynomials in $\mathbb{Z}[X_1, \ldots, X_N]$ of logarithmic height at most $(2md)^{6^N}(h + 1)$.

If $\mathbf{x}$ consists of polynomials of degree at most $(2md)^{2^N}$, then $U\mathbf{x}$ consists of m polynomials with coefficients in $\mathbb{Q}$ of degrees at most $(2md)^{2^N} + d$, all of whose coefficients have to be set to 0. This leads to a system of linear equations $V\mathbf{y} = \mathbf{0}$, where $\mathbf{y}$ consists of the coefficients of the polynomials in $\mathbf{x}$ and V consists of integers of logarithmic heights at most h. Notice that the number m^* of rows of V is m times the number of monomials in N variables of degree at most $(2md)^{2^N} + d$, that is

$$m^* \leq m \binom{(2md)^{2^N} + d + N}{N}.$$

By Lemma 8.2.1 the solution space of $V\mathbf{y} = \mathbf{0}$ is generated by integer vectors of logarithmic height at most

$$m^* h + \tfrac{1}{2} m^* \log m^* \leq (2md)^{6^N} (h + 1).$$

This completes the proof of our corollary. □

Lemma 8.2.4 *Let* $U \in \mathbb{Z}^{m,n}$, $\mathbf{b} \in \mathbb{Z}^m$ *be such that* $U\mathbf{y} = \mathbf{b}$ *is solvable in* $\mathbb{Z}^n$. *Then it has a solution* $\mathbf{y} \in \mathbb{Z}^n$ *with* $h(\mathbf{y}) \leq mh([U, \mathbf{b}]) + \tfrac{1}{2} m \log m$.

Proof. Assume without loss of generality that U and $[U, \mathbf{b}]$ have rank m. By a result of Borosh, Flahive, Rubin and Treybig (1989) (see also Lemma 4.3.5), $U\mathbf{y} = \mathbf{b}$ has a solution $\mathbf{y} \in \mathbb{Z}^n$ such that the absolute values of the entries of $\mathbf{y}$ are bounded above by the maximum of the absolute values of the $m \times m$-subdeterminants of $[U, \mathbf{b}]$. The upper bound for $h(\mathbf{y})$ as in the lemma easily follows from Hadamard's inequality. □

Proposition 8.2.5 *Let* $N \geq 1$ *and let* $f_1, \ldots, f_m, b \in \mathbb{Z}[X_1, \ldots, X_N]$ *be polynomials of degrees at most* d *and logarithmic heights at most* h *where* $d \geq 1$, $h \geq 1$, *such that*

$$f_1 x_1 + \cdots + f_m x_m = b \tag{8.2.1}$$

is solvable in $x_1, \ldots, x_m \in \mathbb{Z}[X_1, \ldots, X_N]$. *Then* (8.2.1) *has a solution in polynomials* $x_1, \ldots, x_m \in \mathbb{Z}[X_1, \ldots, X_N]$ *with*

$$\deg x_i \leq (2d)^{\exp \mathcal{O}(N \log^* N)}(h + 1), \quad h(x_i) \leq (2d)^{\exp \mathcal{O}(N \log^* N)}(h + 1)^{N+1} \tag{8.2.2}$$

for $i = 1, \ldots, m$.

Proof. Aschenbrenner's main theorem (Aschenbrenner (2004), Theorem A) states that equation (8.2.1) has a solution $x_1, \ldots, x_m \in \mathbb{Z}[X_1, \ldots, X_N]$ with $\deg x_i \leq d_0$ for $i = 1, \ldots, m$, where

$$d_0 = (2d)^{\exp \mathcal{O}(N \log^* N)}(h + 1).$$

So it remains to show the existence of a solution with small logarithmic height.

Let us restrict ourselves to solutions $(x_1, \ldots, x_m)$ of (8.2.1) of degree $\leq d_0$, and denote by $\mathbf{y}$ the vector of coefficients of the polynomials $x_1, \ldots, x_m$. Then (8.2.1) translates into a system of linear equations $U\mathbf{y} = \mathbf{b}$ which is solvable over $\mathbb{Z}$. Here, the number of equations, i.e., number of rows of U, is equal to $m^* := \binom{d_0+d+N}{N}$. Further, $h(U, \mathbf{b}) \leq h$. By Lemma 8.2.4, $U\mathbf{y} = \mathbf{b}$ has a solution $\mathbf{y}$ with coordinates in $\mathbb{Z}$ of height at most

$$m^*h + \tfrac{1}{2}m^* \log m^* \leq (2d)^{\exp \mathfrak{O}(N \log^* N)}(h + 1)^{N+1}.$$

It follows that (8.2.1) has a solution $x_1, \ldots, x_m \in \mathbb{Z}[X_1, \ldots, X_N]$ satisfying (8.2.2). □

Remarks

1 Aschenbrenner (2004) gives an example which shows that the upper bound for the degrees of the x_i cannot depend on d and N only.

2 The above lemma gives an effective criterion for ideal membership in the polynomial ring $\mathbb{Z}[X_1, \ldots, X_N]$. Let $b \in \mathbb{Z}[X_1, \ldots, X_N]$ be given. Further, suppose that an ideal I of $\mathbb{Z}[X_1, \ldots, X_N]$ is given by a finite set of generators $f_1, \ldots, f_m$. By the above lemma, if $b \in I$ then there are $x_1, \ldots, x_m \in \mathbb{Z}[X_1, \ldots, X_N]$ with upper bounds for the degrees and heights as in (8.2.2) such that $b = \sum_{i=1}^{m} x_i f_i$. It requires only a finite computation to check whether such x_i exist.

8.3 A reduction

We reduce the general unit equation (8.1.1) to a unit equation over an integral domain B of a special type that can be dealt with more easily.

Let again $A = \mathbb{Z}[z_1, \ldots, z_r]$ be an integral domain finitely generated over $\mathbb{Z}$ and denote by K the quotient field of A. We assume that $r > 0$. We have

$$A \cong \mathbb{Z}[X_1, \ldots, X_r]/I, \tag{8.3.1}$$

where I is the ideal of polynomials $f \in \mathbb{Z}[X_1, \ldots, X_r]$ such that $f(z_1, \ldots, z_r) = 0$. The ideal I is finitely generated. Let $d \geq 1$, $h \geq 1$ and assume that

$$I = (f_1, \ldots, f_m) \quad \text{with } \deg f_i \leq d, \quad h(f_i) \leq h \ (i = 1, \ldots, m). \tag{8.3.2}$$

Suppose that K has transcendence degree $q \geq 0$. In the case that $q > 0$, we assume without loss of generality that $z_1, \ldots, z_q$ form a transcendence basis of $K/\mathbb{Q}$. We write $t := r - q$ and rename $z_{q+1}, \ldots, z_r$ as $y_1, \ldots, y_t$, respectively. In the case that $t = 0$ we have $A = \mathbb{Z}[z_1, \ldots, z_q]$, $A^* = \{\pm 1\}$ and Theorem 8.1.1 is trivial. So we assume henceforth that $t > 0$.

Define

$$A_0 := \mathbb{Z}[z_1, \ldots, z_q], \quad K_0 := \mathbb{Q}(z_1, \ldots, z_q) \quad \text{if } q > 0,$$
$$A_0 := \mathbb{Z}, \quad K_0 := \mathbb{Q} \quad \text{if } q = 0.$$

Then

$$A = A_0[y_1, \ldots, y_t], \quad K = K_0(y_1, \ldots, y_t).$$

Clearly, K is a finite extension of K_0, so in particular is an algebraic number field if $q = 0$. Using standard algebra techniques, worked out in detail below, one can show that there exist $y \in A$ and $f \in A_0$ such that $K = K_0(y)$, y is integral over A_0, and

$$A \subseteq B := A_0[f^{-1}, y], \quad a_1, a_2, a_3 \in B^*,$$

where a_1, a_2, a_3 are the coefficients in (8.1.1). If $x_1, x_2 \in A^*$ is a solution to (8.1.1), then $x_i' := a_i x_i / a_3$ ($i = 1, 2$) satisfy

$$x_1' + x_2' = 1, \quad x_1', x_2' \in B^*. \tag{8.3.3}$$

At the end of this section, we formulate Proposition 8.3.7 which gives an effective result for equations of the type (8.3.3). More precisely, we introduce a different type of degree and height, $\overline{\deg}(\alpha)$ and $\overline{h}(\alpha)$, for elements α of B, and give effective upper bounds for the $\overline{\deg}$ and $\overline{h}$ of x_1', x_2'. Subsequently we deduce Theorem 8.1.1.

The deduction of Theorem 8.1.1 is based on some auxiliary results which are proved first. We start with an explicit construction of y, f, with effective upper bounds in terms of r, d, h and a_1, a_2, a_3 for the degrees and logarithmic heights of f and of the coefficients in A_0 of the monic minimal polynomial of y over A_0. Here we follow more or less Seidenberg (1974). Second, for a given solution x_1, x_2 of (8.1.1), we derive effective upper bounds for the degrees and logarithmic heights of representatives for x_i, x_i^{-1}, ($i = 1, 2$) in terms of $\overline{\deg}(x_i')$, $\overline{h}(x_i')$ ($i = 1, 2$). Here we use Proposition 8.2.5 (Aschenbrenner's result).

We introduce some further notation. First let $q > 0$. Then since $z_1, \ldots, z_q$ are algebraically independent, we may view them as independent variables, and for $\alpha \in A_0$, we denote by $\deg \alpha$, $h(\alpha)$ the total degree and logarithmic height of α, viewed as a polynomial in $z_1, \ldots, z_q$. In the case that $q = 0$, we have $A_0 = \mathbb{Z}$, and we agree that $\deg \alpha = 0$, $h(\alpha) = \log|\alpha|$ for $\alpha \in A_0$.

We write $\mathbf{Y} = (X_{q+1}, \ldots, X_r)$ and $K_0(\mathbf{Y}) = K_0(X_{q+1}, \ldots, X_r)$. Given $f \in \mathbb{Q}(X_1, \ldots, X_r)$ we denote by f^* the rational function of $K_0(\mathbf{Y})$ obtained by substituting z_i for X_i for $i = 1, \ldots, q$ (and $f^* = f$ if $q = 0$). We denote by $\deg_{\mathbf{Y}} f^*$ the (total) degree of $f^* \in K_0[\mathbf{Y}]$ with respect to $\mathbf{Y}$. We recall that the total degree $\deg g$ is defined for elements $g \in A_0$ and is taken with respect to

$z_1, \ldots, z_q$. With this notation, we can rewrite (8.3.1) and (8.3.2) as

$$
\left.
\begin{aligned}
&A \cong A_0[\mathbf{Y}]/(f_1^*, \ldots, f_m^*),\\
&\deg_{\mathbf{Y}} f_i^* \leq d \quad \text{for } i = 1, \ldots, m,\\
&\text{the coefficients of } f_1^*, \ldots, f_m^* \text{ in } A_0 \text{ have degrees at most } d\\
&\text{and logarithmic heights at most } h.
\end{aligned}
\right\} \quad (8.3.4)
$$

Put $D := [K : K_0]$ and denote by $\sigma_1, \ldots, \sigma_D$ the K_0-isomorphic embeddings of K in an algebraic closure $\overline{K_0}$ of K_0.

Lemma 8.3.1

(i) We have $D \leq d^t$.

(ii) There exist integers $a_1, \ldots, a_t$ with $|a_i| \leq D^2$ for $i = 1, \ldots, t$ such that for $w := a_1 y_1 + \cdots + a_t y_t$ we have $K = K_0(w)$.

Proof. (i) The images of $(y_1, \ldots, y_t)$ under $\sigma_1, \ldots, \sigma_D$ all lie in

$$
\mathcal{W} := \{\mathbf{y} \in \overline{K_0}^t : f_1^*(\mathbf{y}) = \cdots = f_m^*(\mathbf{y}) = 0\}.
$$

Conversely, using the fact that $K \cong K_0[\mathbf{Y}]/(f_1^*, \ldots, f_m^*)$, one sees that each assignment $(y_1, \ldots, y_t) \mapsto \mathbf{y}$ with $\mathbf{y} \in \mathcal{W}$ yields a K_0-isomorphic embedding of K in $\overline{K_0}$. Hence $|\mathcal{W}| = D < \infty$. Now Corollary 7.5.3 with $\mathbf{k} = \overline{K_0}$, $\mathcal{X} = \overline{K_0}^t$, $\mathcal{Y} = \emptyset$ implies that $|\mathcal{W}| \leq d^t$. Hence $D \leq d^t$.

(ii) Let $a_1, \ldots, a_t$ be integers. Then $w := \sum_{i=1}^t a_i y_i$ generates K over K_0 if and only if $\sum_{j=1}^t a_j \sigma_i(y_j)$ $(i = 1, \ldots, D)$ are distinct. There are integers a_i with $|a_i| \leq D^2$ for which this holds. $\qquad\square$

Lemma 8.3.2 *There are $\mathcal{G}_0, \ldots, \mathcal{G}_D \in A_0$ such that*

$$
\sum_{i=0}^D \mathcal{G}_i w^{D-i} = 0, \quad \mathcal{G}_0 \mathcal{G}_D \neq 0, \tag{8.3.5}
$$

$$
\deg \mathcal{G}_i \leq (2d)^{\exp \mathfrak{O}(r)}, \quad h(\mathcal{G}_i) \leq (2d)^{\exp \mathfrak{O}(r)}(h+1), \quad (i = 0, \ldots, D). \tag{8.3.6}
$$

Proof. In what follows we write $\mathbf{Y} = (X_{q+1}, \ldots, X_r)$ and $\mathbf{Y}^{\mathbf{u}} := X_{q+1}^{u_1} \cdots X_{q+t}^{u_t}$, $|\mathbf{u}| := u_1 + \cdots + u_t$ for tuples of non-negative integers $\mathbf{u} = (u_1, \ldots, u_t)$. Further, we define $W := \sum_{j=1}^t a_j X_{q+j}$.

$\mathcal{G}_0, \ldots, \mathcal{G}_D$ as in (8.3.5) clearly exist since w has degree D over K_0. By (8.3.4), there are $g_1^*, \ldots, g_m^* \in A_0[\mathbf{Y}]$ such that

$$
\sum_{i=0}^D \mathcal{G}_i W^{D-i} = \sum_{j=1}^m g_j^* f_j^*. \tag{8.3.7}
$$

By Proposition 8.2.2 (ii), applied with the field $F = K_0$, there are polynomials $g_j^* \in K_0[\mathbf{Y}]$ (so with coefficients being rational functions in $\mathbf{z}$) satisfying (8.3.7) of degree at most $(2\max(d, D))^{2^t} \le (2d^t)^{2^t} =: d_0$ in $\mathbf{Y}$. By multiplying $\mathcal{G}_0, \ldots, \mathcal{G}_D$ with an appropriate non-zero factor from A_0 we may assume that the g_j^* are polynomials in $A_0[\mathbf{Y}]$ of degree at most d_0 in $\mathbf{Y}$. By considering (8.3.7) with such polynomials g_j^*, we obtain

$$\sum_{i=0}^{D} \mathcal{G}_i W^{D-i} = \sum_{j=1}^{m} \left(\sum_{|\mathbf{u}| \le d_0} g_{j,\mathbf{u}} \mathbf{Y}^{\mathbf{u}} \right) \cdot \left(\sum_{|\mathbf{v}| \le d} f_{j,\mathbf{v}} \mathbf{Y}^{\mathbf{v}} \right), \qquad (8.3.8)$$

where $g_{j,\mathbf{u}} \in A_0$ and $f_j^* = \sum_{|\mathbf{v}| \le d} f_{j,\mathbf{v}} \mathbf{Y}^{\mathbf{v}}$ with $f_{j,\mathbf{v}} \in A_0$. We view $\mathcal{G}_0, \ldots, \mathcal{G}_D$ and the polynomials $g_{j,\mathbf{u}}$ as the unknowns of (8.3.8). Then (8.3.8) has solutions with $\mathcal{G}_0 \mathcal{G}_D \ne 0$.

We may view (8.3.8) as a system of linear equations $U\mathbf{x} = \mathbf{0}$ over K_0, where $\mathbf{x}$ consists of $\mathcal{G}_i$ ($i = 0, \ldots, D$) and $g_{j,\mathbf{u}}$ ($j = 1, \ldots, m$, $|\mathbf{u}| \le d_0$). By Lemma 8.3.1 and an elementary estimate, the polynomial $W^{D-i} = (\sum_{k=1}^{t} a_k X_{q+k})^{D-i}$ has logarithmic height at most $\mathcal{O}(D \log(2D^2 t)) \le (2d)^{\mathcal{O}(t)}$. By combining this with (8.3.4), it follows that the entries of the matrix U are elements of A_0 of degrees at most d and logarithmic heights at most $h_0 := \max((2d)^{\mathcal{O}(t)}, h)$. Further, the number of rows of U is at most the number of monomials in $\mathbf{Y}$ of degree at most $d_0 + d$ which is bounded above by $m_0 := \binom{d_0 + d + t}{t}$. So, by Corollary 8.2.3, the solution module of (8.3.8) is generated by vectors $\mathbf{x} = (\mathcal{G}_0, \ldots, \mathcal{G}_D, \{g_{i,\mathbf{u}}\})$, consisting of elements from A_0 of degree and height at most

$$(2m_0 d)^{2^q} \le (2d)^{\exp \mathcal{O}(r)}, \quad (2m_0 d)^{6^q}(h_0 + 1) \le (2d)^{\exp \mathcal{O}(r)}(h + 1),$$

respectively.

At least one of these vectors $\mathbf{x}$ must have $\mathcal{G}_0 \mathcal{G}_D \ne 0$ since otherwise (8.3.8) would have no solution with $\mathcal{G}_0 \mathcal{G}_D \ne 0$, contradicting what we already observed about (8.3.5). Thus, there exists a solution $\mathbf{x}$ whose components $\mathcal{G}_0, \ldots, \mathcal{G}_D$ satisfy both (8.3.5) and (8.3.6). This proves our lemma. $\quad\square$

It will be more convenient to work with

$$y := \mathcal{G}_0 w = \mathcal{G}_0 \cdot (a_1 y_1 + \cdots + a_t y_t).$$

In the case $D = 1$ we set $y := 1$. The following properties of y follow at once from Corollary 1.9.5 and Lemmas 8.3.1 and 8.3.2.

Corollary 8.3.3 *We have $K = K_0(y)$, $y \in A$, y is integral over A_0, and y has minimal polynomial $\mathcal{F}(X) = X^D + \mathcal{F}_1 X^{D-1} + \cdots + \mathcal{F}_D$ over K_0 with*

$$\mathcal{F}_i \in A_0, \quad \deg \mathcal{F}_i \le (2d)^{\exp \mathfrak{D}(r)}, \quad h(\mathcal{F}_i) \le (2d)^{\exp \mathfrak{D}(r)}(h+1)$$

for $i = 1, \ldots, D$.

Recall that $A_0 = \mathbb{Z}$ if $q = 0$ and $\mathbb{Z}[z_1, \ldots, z_q]$ if $q > 0$, where in the latter case, $z_1, \ldots, z_q$ are algebraically independent. Hence A_0 is a unique factorization domain, and so the greatest common divisor of a finite set of elements of A_0 is well-defined and up to sign uniquely determined. With every element $\alpha \in K$ we can associate an up to sign unique tuple $P_{\alpha,0}, \ldots, P_{\alpha,D-1}, Q_\alpha$ of elements of A_0 such that

$$\alpha = Q_\alpha^{-1} \sum_{j=0}^{D-1} P_{\alpha,j} y^j \quad \text{with } Q_\alpha \ne 0, \ \gcd(P_{\alpha,0}, \ldots, P_{\alpha,D-1}, Q_\alpha) = 1.$$

$$(8.3.9)$$

Put

$$\overline{\deg}\,\alpha := \max(\deg P_{\alpha,0}, \ldots, \deg P_{\alpha,D-1}, \deg Q_\alpha),$$
$$\overline{h}(\alpha) := \max(h(P_{\alpha,0}), \ldots, h(P_{\alpha,D-1}), h(Q_\alpha)).$$

Then for $q = 0$ we have $\overline{\deg}\,\alpha = 0$, $\overline{h}(\alpha) = \log \max(|P_{\alpha,0}|, \ldots, |P_{\alpha,D-1}|, |Q_\alpha|)$.

Lemma 8.3.4 *Let $\alpha \in K^*$ and let (a, b) be a pair of representatives for α, with $a, b \in \mathbb{Z}[X_1, \ldots, X_r]$, $b \notin I$. Put*

$$d^* := \max(d, \deg a, \deg b), \quad h^* := \max(h, h(a), h(b)).$$

Then

$$\overline{\deg}\,\alpha \le (2d^*)^{\exp \mathfrak{D}(r)}, \quad \overline{h}(\alpha) \le (2d^*)^{\exp \mathfrak{D}(r)}(h^* + 1).$$

Proof. Consider the linear equation

$$Q \cdot \alpha = \sum_{j=0}^{D-1} P_j y^j \qquad (8.3.10)$$

in unknowns $P_0, \ldots, P_{D-1}, Q \in A_0$. This equation has a solution with $Q \ne 0$, since $\alpha \in K = K_0(y)$ and y has degree D over K_0. Write again $\mathbf{Y} = (X_{q+1}, \ldots, X_r)$ and put $Y := \mathcal{G}_0 \cdot (\sum_{j=1}^t a_j X_{q+j})$. Let $a^*, b^* \in A_0[\mathbf{Y}]$ be obtained from a, b by substituting z_i for X_i for $i = 1, \ldots, q$ ($a^* = a$, $b^* = b$

if $q = 0$). By (8.3.4), there are $g_j^* \in A_0[\mathbf{Y}]$ such that

$$Q \cdot a^* - b^* \sum_{j=0}^{D-1} P_j Y^j = \sum_{j=1}^{m} g_j^* f_j^*. \tag{8.3.11}$$

By Proposition 8.2.2 (ii) this identity holds with polynomials $g_j^* \in A_0[\mathbf{Y}]$ of degree in $\mathbf{Y}$ at most $(2 \max(d^*, D))^{2^t} \le (2d^*)^{t2^t}$, where possibly we have to multiply $(P_0, \ldots, P_{D-1}, Q)$ by a non-zero element from A_0. Now completely similarly as in the proof of Lemma 8.3.2, one can rewrite (8.3.11) as a system of linear equations over K_0 and then apply Corollary 8.2.3. It follows that (8.3.10) is satisfied by $P_0, \ldots, P_{D-1}, Q \in A_0$ with $Q \neq 0$ and

$$\deg P_i, \deg Q \le (2d^*)^{\exp \mathfrak{O}(r)},$$
$$h(P_i), h(Q) \le (2d^*)^{\exp \mathfrak{O}(r)}(h^* + 1) \quad (i = 0, \ldots, D - 1).$$

By dividing $P_0, \ldots, P_{D-1}, Q$ by their greatest common divisor and using Corollary 1.9.5 we obtain $P_{\alpha,0}, \ldots, P_{D-1,\alpha}, Q_\alpha \in A_0$ satisfying both (8.3.9) and

$$\deg P_{i,\alpha}, \deg Q_\alpha \le (2d^*)^{\exp \mathfrak{O}(r)},$$
$$h(P_{i,\alpha}), h(Q_\alpha) \le (2d^*)^{\exp \mathfrak{O}(r)}(h^* + 1) \quad (i = 0, \ldots, D - 1). \qquad \square$$

Lemma 8.3.5 *Let* $\alpha_1, \ldots, \alpha_n \in K^*$. *For* $i = 1, \ldots, n$, *let* (a_i, b_i) *be a pair of representatives for* α_i, *with* $a_i, b_i \in \mathbb{Z}[X_1, \ldots, X_r]$, $b_i \notin I$. *Put*

$$d^{**} := \max(d, \deg a_1, \deg b_1, \ldots, \deg a_n, \deg b_n),$$
$$h^{**} := \max(h, h(a_1), h(b_1), \ldots, h(a_n), h(b_n)).$$

Then there is a non-zero $f \in A_0$ *such that*

$$A \subseteq A_0[y, f^{-1}], \quad \alpha_1, \ldots, \alpha_n \in A_0[y, f^{-1}]^*, \tag{8.3.12}$$

$$\left.\begin{array}{l} \deg f \le (n + 1)(2d^{**})^{\exp \mathfrak{O}(r)}, \\ h(f) \le (n + 1)(2d^{**})^{\exp \mathfrak{O}(r)}(h^{**} + 1). \end{array}\right\} \tag{8.3.13}$$

Proof. Take

$$f := \prod_{i=1}^{t} Q_{y_i} \cdot \prod_{j=1}^{n} \left(Q_{\alpha_i} Q_{\alpha_i^{-1}} \right).$$

Since in general, $Q_\beta \beta \in A_0[y]$ for $\beta \in K^*$, we have $f\beta \in A_0[y]$ for $\beta = y_1, \ldots, y_t, \alpha_1, \alpha_1^{-1}, \ldots, \alpha_n, \alpha_n^{-1}$. This implies (8.3.12). The inequalities (8.3.13) follow at once from Lemma 8.3.4 and Corollary 1.9.5. $\qquad \square$

Lemma 8.3.6 *Let* $\lambda \in K^*$ *and let* x *be a non-zero element of* A. *Let* (a, b) *with* $a, b \in \mathbb{Z}[X_1, \ldots, X_r]$ *be a pair of representatives for* λ. *Put*

$$d_0 := \max(\deg f_1, \ldots, \deg f_m, \deg a, \deg b, \overline{\deg} \lambda x),$$
$$h_0 := \max(h(f_1), \ldots, h(f_m), h(a), h(b), \overline{h}(\lambda x)).$$

Then x *has a representative* $\widetilde{x} \in \mathbb{Z}[X_1, \ldots, X_r]$ *such that*

$$\deg \widetilde{x} \leq (2d_0)^{\exp \mathfrak{O}(r \log^* r)}(h_0 + 1), \quad h(\widetilde{x}) \leq (2d_0)^{\exp \mathfrak{O}(r \log^* r)}(h_0 + 1)^{r+1}.$$

If moreover $x \in A^*$, *then* x^{-1} *has a representative* $\widetilde{x}' \in \mathbb{Z}[X_1, \ldots, X_r]$ *with*

$$\deg \widetilde{x}' \leq (2d_0)^{\exp \mathfrak{O}(r \log^* r)}(h_0 + 1), \quad h(\widetilde{x}') \leq (2d_0)^{\exp \mathfrak{O}(r \log^* r)}(h_0 + 1)^{r+1}.$$

Proof. In the case $q > 0$, we identify z_i with X_i and view elements of A_0 as polynomials in $\mathbb{Z}[X_1, \ldots, X_q]$. Put $Y := \mathcal{G}_0 \cdot (\sum_{i=1}^t a_i X_{q+i})$. We have

$$\lambda x = Q^{-1} \sum_{i=0}^{D-1} P_i y^i \tag{8.3.14}$$

with $P_0, \ldots, P_{D-1}, Q \in A_0$ and $\gcd(P_0, \ldots, P_{D-1}, Q) = 1$. According to (8.3.14), $\widetilde{x} \in \mathbb{Z}[X_1, \ldots, X_r]$ is a representative for x if and only if there are $g_1, \ldots, g_m \in \mathbb{Z}[X_1, \ldots, X_r]$ such that

$$\widetilde{x} \cdot (Q \cdot a) + \sum_{i=1}^m g_i f_i = b \sum_{i=0}^{D-1} P_i Y^i. \tag{8.3.15}$$

We may view (8.3.15) as an inhomogeneous linear equation in the unknowns $\widetilde{x}, g_1, \ldots, g_m$. Notice that by Lemmas 8.3.1–8.3.4 the degrees and logarithmic heights of Qa and $b \sum_{i=0}^{D-1} P_i Y^i$ are all bounded above by

$$(2d_0)^{\exp \mathfrak{O}(r)}, \quad (2d_0)^{\exp \mathfrak{O}(r)}(h_0 + 1),$$

respectively. Now Proposition 8.2.5 implies that (8.3.15) has a solution with upper bounds for $\deg \widetilde{x}$, $h(\widetilde{x})$, as stated in the lemma.

Now suppose that $x \in A^*$. Again by (8.3.14), $\widetilde{x}' \in \mathbb{Z}[X_1, \ldots, X_r]$ is a representative for x^{-1} if and only if there are $g_1', \ldots, g_m' \in \mathbb{Z}[X_1, \ldots, X_r]$ such that

$$\widetilde{x}' \cdot b \sum_{i=0}^{D-1} P_i Y^i + \sum_{i=1}^m g_i' f_i = Q \cdot a.$$

Similarly as above, this equation has a solution with upper bounds for $\deg \widetilde{x}'$, $h(\widetilde{x}')$ as stated in the lemma. $\square$

Recall that we have defined $A_0 = \mathbb{Z}[z_1, \ldots, z_q]$, $K_0 = \mathbb{Q}(z_1, \ldots, z_q)$ if $q > 0$ and $A_0 = \mathbb{Z}$, $K_0 = \mathbb{Q}$ if $q = 0$, and that in the case $q = 0$, degrees and $\overline{\deg}$-s are always zero. Theorem 8.1.1 can be deduced from the following proposition, which makes sense also if $q = 0$. The proof of this proposition is given in Sections 8.4–8.6.

Proposition 8.3.7 *Let $f \in A_0$ with $f \neq 0$, and let*

$$\mathcal{F} = X^D + \mathcal{F}_1 X^{D-1} + \cdots + \mathcal{F}_D \in A_0[X] \quad (D \geq 1)$$

be the minimal polynomial of y over K_0. Let $d_1 \geq 1$, $h_1 \geq 1$ and suppose

$$\left.\begin{array}{r} \max(\deg f, \deg \mathcal{F}_1, \ldots, \deg \mathcal{F}_D) \leq d_1, \\ \max(h(f), h(\mathcal{F}_1), \ldots, h(\mathcal{F}_D)) \leq h_1. \end{array}\right\} \tag{8.3.16}$$

Define the domain $B := A_0[y, f^{-1}]$. Then for each pair (x_1, x_2) with

$$x_1 + x_2 = 1, \quad x_1, x_2 \in B^* \tag{8.3.17}$$

we have

$$\overline{\deg} x_1, \overline{\deg} x_2 \leq 4qD^2 \cdot d_1, \tag{8.3.18}$$

$$\overline{h}(x_1), \overline{h}(x_2) \leq \exp \mathfrak{O}\big(2D(q + d_1)(\log^*\{2D(q + d_1)\})^2 + D \log^* Dh_1\big). \tag{8.3.19}$$

Proof of Theorem 8.1.1. Let $a_1, a_2, a_3 \in A \setminus \{0\}$ be the coefficients of (8.1.1), and $\tilde{a}_1, \tilde{a}_2, \tilde{a}_3$ the representatives for a_1, a_2, a_3 from the statement of Theorem 8.1.1. By Lemma 8.3.5, there exists non-zero $f \in A_0$ such that $A \subseteq B := A_0[y, f^{-1}]$, $a_1, a_2, a_3 \in B^*$, and moreover,

$$\deg f \leq (2d)^{\exp \mathfrak{O}(r)}, \quad h(f) \leq (2d)^{\exp \mathfrak{O}(r)}(h + 1).$$

By Corollary 8.3.3 we have the same type of upper bounds for the degrees and logarithmic heights of $\mathcal{F}_1, \ldots, \mathcal{F}_D$. So in Proposition 8.3.7 we may take $d_1 = (2d)^{\exp \mathfrak{O}(r)}$, $h_1 = (2d)^{\exp \mathfrak{O}(r)}(h + 1)$. Finally, by Lemma 8.3.1 we have $D \leq d^t$.

Let (x_1, x_2) be a solution of (8.1.1) and put $x_i' := a_i x_i / a_3$ for $i = 1, 2$. Let $i \in \{1, 2\}$. By Proposition 8.3.7 we have

$$\overline{\deg} x_i' \leq 4qd^{2t}(2d)^{\exp \mathfrak{O}(r)} \leq (2d)^{\exp \mathfrak{O}(r)}, \quad \overline{h}(x_i') \leq \exp((2d)^{\exp \mathfrak{O}(r)}(h + 1)).$$

We apply Lemma 8.3.6 with $\lambda = a_i / a_3$. Notice that λ is represented by $(\tilde{a}_i, \tilde{a}_3)$. By assumption, $\tilde{a}_i$ and $\tilde{a}_3$ have degrees at most d and logarithmic heights at most h. Letting $\tilde{a}_i, \tilde{a}_3$ play the role of a, b in Lemma 8.3.6, we see that in that lemma we may take $h_0 = \exp((2d)^{\exp \mathfrak{O}(r)}(h + 1))$ and $d_0 = (2d)^{\exp \mathfrak{O}(r)}$. It follows that x_i, x_i^{-1} have representatives $\widetilde{x}_i, \widetilde{x}_i' \in \mathbb{Z}[X_1, \dots, X_r]$ such that

$$\deg \widetilde{x}_i, \ \deg \widetilde{x}_i', \ h(\widetilde{x}_i), \ h(\widetilde{x}_i') \leq \exp\left((2d)^{\exp \mathfrak{O}(r)}(h + 1)\right).$$

We observe here that the upper bound for $\overline{h}(x_i)$ dominates by far the other terms in our estimation. This completes the proof of Theorem 8.1.1. □

Proposition 8.3.7 is proved in Sections 8.4–8.6. In Section 8.4 we deduce the degree bound (8.3.18). Here, our main tool is Theorem 7.1.1 (Mason's effective result on S-unit equations over function fields). In Section 8.5 we work out a more precise version of an effective specialization argument of Győry (1983, 1984). In Section 8.6 we prove (8.3.19) by combining the specialization argument from Section 8.5 with a recent effective result for S-unit equations over number fields, due to Győry and Yu (2006).

8.4 Bounding the degree in Proposition 8.3.7

We keep the notation from Proposition 8.3.7. We may assume that $q > 0$ because the case $q = 0$ is trivial. Let as before $K_0 = \mathbb{Q}(z_1, \dots, z_q)$, $K = K_0(y)$, $A_0 = \mathbb{Z}[z_1, \dots, z_q]$, $B = \mathbb{Z}[z_1, \dots, z_q, f^{-1}, y]$. Choose an algebraic closure $\overline{K_0}$ of K_0. Then there are precisely D K_0-isomorphic embeddings of K into $\overline{K_0}$, which we denote by $x \mapsto x^{(i)}$ ($i = 1, \dots, D$).

Fix $i \in \{1, \dots, q\}$. Let $\mathbf{k}_i$ be an algebraic closure of $\mathbb{Q}(z_1, \dots, z_{i-1}, z_{i+1}, \dots, z_q)$, contained in K_0. Thus, A_0 is contained in $\mathbf{k}_i[z_i]$. Define the field

$$M_i := \mathbf{k}_i\left(z_i, y^{(1)}, \dots, y^{(D)}\right).$$

That is, M_i is the splitting field of the polynomial $X^D + \mathcal{F}_1 X^{D-1} + \cdots + \mathcal{F}_D$ over $\mathbf{k}_i(z_i)$. The subring

$$B_i := \mathbf{k}_i\left[z_i, f^{-1}, y^{(1)}, \dots, y^{(D)}\right]$$

of M_i contains $B = \mathbb{Z}[z_1, \dots, z_q, f^{-1}, y]$ as a subring. Put $\Delta_i := [M_i : \mathbf{k}_i(z_i)]$.

We will apply the estimates from Sections 2.2 and 2.3 with $z_i, \mathbf{k}_i, M_i$ instead of $z, \mathbf{k}, K$. Denote by $g_{M_i/\mathbf{k}_i}$ the genus of M_i over $\mathbf{k}_i$. The height H_{M_i} is taken with respect to $M_i/\mathbf{k}_i$. For $G \in A_0$, we denote by $\deg_{z_i} G$ the degree of G in the variable z_i.

Lemma 8.4.1 *For* $\alpha \in K$ *we have*

$$\overline{\deg}\,\alpha \leq qD \cdot d_1 + \sum_{i=1}^{q} \Delta_i^{-1} \sum_{j=1}^{D} H_{M_i}\left(\alpha^{(j)}\right).$$

Proof. We have

$$\alpha = Q^{-1} \sum_{j=0}^{D-1} P_j y^j$$

for certain $P_0, \ldots, P_{D-1}, Q \in A_0$ with $\gcd(Q, P_0, \ldots, P_{D-1}) = 1$. Clearly,

$$\overline{\deg}\,\alpha \le \sum_{i=1}^{q} \mu_i, \quad \text{where } \mu_i := \max(\deg_{z_i} Q, \deg_{z_i} P_0, \ldots, \deg_{z_i} P_{D-1}).$$

(8.4.1)

Below, we estimate $\mu_1, \ldots, \mu_q$ from above. We heavily use the height properties listed in Section 2.2. We fix $i \in \{1, \ldots, q\}$ and use the notation introduced above.

By taking conjugates, we obtain

$$\alpha^{(k)} = Q^{-1} \sum_{j=0}^{D-1} P_j \cdot (y^{(k)})^j \quad \text{for } k = 1, \ldots, D.$$

Let Ω be the $D \times D$-matrix with rows

$$(1, \ldots, 1), \left(y^{(1)}, \ldots, y^{(D)}\right), \ldots, \left(\left(y^{(1)}\right)^{D-1}, \ldots, \left(y^{(D)}\right)^{D-1}\right).$$

By Cramer's rule, $P_j/Q = \delta_j/\delta$, where $\delta = \det \Omega$, and δ_j is the determinant of the matrix obtained by replacing the j-th row of Ω by $(\alpha^{(1)}, \ldots, \alpha^{(D)})$.

Gauss' Lemma implies that $\gcd(P_0, \ldots, P_{D-1}, Q) = 1$ in the ring $\mathbf{k}_i[z_i]$. So by (2.2.2) (with z_i in place of z) we have

$$\mu_i = H_{\mathbf{k}_i(z_i)}^{\mathrm{hom}}(Q, P_0, \ldots, P_{D-1}).$$

Notice that $(\delta, \delta_1, \ldots, \delta_D)$ is a scalar multiple of $(Q, P_0, \ldots, P_{D-1})$. By combining (2.2.3), (2.2.1) and inserting $[M_i : \mathbf{k}_i(z_i)] = \Delta_i$, we obtain

$$\mu_i = \Delta_i^{-1} H_{M_i}(Q, P_0, \ldots, P_{D-1}) = H_{M_i}(\delta, \delta_1, \ldots, \delta_D). \qquad (8.4.2)$$

We bound from above the right-hand side. A straightforward estimate yields that for every valuation v of $M_i/\mathbf{k}_i$,

$$-\min(v(\delta), v(\delta_1), \ldots, v(\delta_D)) \le -D \sum_{j=1}^{D} \min(0, v(y^{(j)})) - \sum_{j=1}^{D} \min(0, v(\alpha^{(j)})),$$

and then summation over v gives

$$H_{M_i}(\delta, \delta_1, \ldots, \delta_D) \le D \sum_{j=1}^{D} H_{M_i}(y^{(j)}) + \sum_{j=1}^{D} H_{M_i}(\alpha^{(j)}). \qquad (8.4.3)$$

A combination of (2.2.10), (2.2.3), (2.2.2) and assumption (8.3.16) gives

$$\sum_{j=1}^{D} H_{M_i}\left(y^{(j)}\right) = H_{M_i}(\mathcal{F}) = \Delta_i H_{\mathbf{k}_i(z_i)}(\mathcal{F})$$

$$= \Delta_i \max(0, \deg_{z_i} \mathcal{F}_1, \ldots, \deg_{z_i} \mathcal{F}_D) \le \Delta_i \cdot d_1.$$

Together with (8.4.2) and (8.4.3) this leads to

$$\mu_i \le D d_1 + \Delta_i^{-1} \sum_{j=1}^{D} H_{M_i}\left(\alpha^{(j)}\right).$$

Now these bounds for $i = 1, \ldots, q$ together with (8.4.1) imply our lemma. $\quad\square$

Proof of (8.3.18). We fix again $i \in \{1, \ldots, q\}$ and use the notation introduced above. By Proposition 2.3.2, applied with $\mathbf{k}_i$, z_i, M_i instead of $\mathbf{k}$, z, K and with $F = \mathcal{F} = X^D + \mathcal{F}_1 X^{D-1} + \cdots + \mathcal{F}_D$, we have

$$g_{M_i/\mathbf{k}_i} \le (\Delta_i - 1)D \max_j \deg_{z_i}(\mathcal{F}_j) \le (\Delta_i - 1) \cdot D d_1. \qquad (8.4.4)$$

Let S denote the subset of valuations v of $M_i/\mathbf{k}_i$ such that $v(z_i) < 0$ or $v(f) > 0$. Each valuation of $\mathbf{k}_i(z_i)$ can be extended to at most $[M_i : \overline{\mathbf{k}}_i(z_i)] = \Delta_i$ valuations of M_i. Hence M_i has at most Δ_i valuations v with $v(z_i) < 0$ and at most $\Delta_i \deg f$ valuations with $v(f) > 0$. Thus,

$$|S| \le \Delta_i + \Delta_i \deg_{z_i} f \le \Delta_i(1 + \deg f) \le \Delta_i(1 + d_1). \qquad (8.4.5)$$

Define the ring of S-integers in M_i,

$$O_S = \{x \in M_i : v(x) \ge 0 \text{ for } v \in M_{M_i} \setminus S\}.$$

This ring contains $\mathbf{k}_i$, z_i, f and is integrally closed. As a consequence, $A_0 = \mathbb{Z}[z_1, \ldots, z_q] \subset O_S$. The elements $y^{(1)}, \ldots, y^{(D)}$ belong to M_i and are integral over A_0 so they certainly belong to O_S. As a consequence, the elements of B and their conjugates over $\mathbb{Q}(z_1, \ldots, z_q)$ belong to O_S. In particular, if $x_1, x_2 \in B^*$ and $x_1 + x_2 = 1$, then

$$x_1^{(j)} + x_2^{(j)} = 1, \ x_1^{(j)}, x_2^{(j)} \in O_S^* \quad \text{for } j = 1, \ldots, D.$$

We apply Mason's inequality, Theorem 7.1.1, and insert the upper bounds (8.4.4) and (8.4.5). It follows that for $j = 1, \ldots, D$ we have either $x_1^{(j)} \in \overline{\mathbf{k}}_i$ or

$$H_{M_i}\left(x_1^{(j)}\right) \le |S| + 2g_{M_i/\mathbf{k}_i} - 2 \le 3\Delta_i \cdot D d_1;$$

in fact the last upper bound is valid also if $x_1^{(j)} \in \overline{\mathbf{k}}_i$. Together with Lemma 8.4.1 this gives

$$\overline{\deg}\, x_1 \le q D d_1 + q D \cdot 3 D d_1 \le 4 q D^2 d_1.$$

For $\overline{\deg}\, x_2$ we derive the same estimate. This proves (8.3.18). $\quad\square$

8.5 Specializations

In this section we prove some results about specialization homomorphisms from B to $\overline{\mathbb{Q}}$, where B is the integral domain B from Proposition 8.3.7. We start with three auxiliary results that are used in the construction of our specializations.

Lemma 8.5.1 *Let $m \geq 1$, let $\alpha_1, \ldots, \alpha_m \in \overline{\mathbb{Q}}$ be distinct and suppose that $G(X) := \prod_{i=1}^{m}(X - \alpha_i) \in \mathbb{Z}[X]$. Let $q, p_0, \ldots, p_{m-1}$ be integers with*

$$\gcd(q, p_0, \ldots, p_{m-1}) = 1,$$

and put

$$\beta_i := \frac{1}{q} \sum_{j=0}^{m-1} p_j \alpha_i^j \quad (i = 1, \ldots, m).$$

Then

$$\log \max(|q|, |p_0|, \ldots, |p_{m-1}|) \leq 2m^2 + (m-1)h(G) + \sum_{j=1}^{m} h(\beta_j).$$

Proof. We use the height estimates from Section 1.9. For $m = 1$ the assertion is obvious, so we assume $m \geq 2$. Let $L = \mathbb{Q}(\alpha_1, \ldots, \alpha_m)$. Let Ω be the $m \times m$ matrix with rows $(\alpha_1^i, \ldots, \alpha_m^i)$ $(i = 0, \ldots, m-1)$. By Cramer's rule we have $p_i/q = \delta_i/\delta$ $(i = 0, \ldots, m-1)$, where $\delta = \det \Omega$ and δ_i is the determinant of the matrix, obtained by replacing the i-th row of Ω by $(\beta_1, \ldots, \beta_m)$. Put

$$\mu := \log \max(|q|, |p_0|, \ldots, |p_{m-1}|).$$

Then since $(\delta, \delta_1, \ldots, \delta_{m-1})$ is a scalar multiple of $(q, p_1 \cdots p_{m-1})$ we have, by (1.9.4) and (1.9.8),

$$\mu = h^{\mathrm{hom}}(q, p_1, \ldots, p_{m-1}) = h^{\mathrm{hom}}(\delta, \delta_1, \ldots, \delta_{m-1})$$

$$= \frac{1}{d} \sum_{v \in M_L} \log \max(|\delta|_v, |\delta_1|_v, \ldots, |\delta_{m-1}|_v). \tag{8.5.1}$$

Estimating the determinants using Hadamard's inequality for the infinite places and the ultrametric inequality for the finite places, we get

$$\max(|\delta|_v, |\delta_1|_v, \ldots, |\delta_m|_v) \leq m^{ms(v)/2} \prod_{i=1}^{m} \max(1, |\alpha_i|_v)^{m-1} \max(1, |\beta_i|_v).$$

for $v \in M_L$, where $s(v) = 1$ if v is real, $s(v) = 2$ if v is complex, and $s(v) = 0$ if v is finite. Together with (8.5.1) this implies

$$\mu \leq \tfrac{1}{2} m \log m + \sum_{i=1}^{m} ((m-1)h(\alpha_i) + h(\beta_i)).$$

A combination with Corollary 1.9.6 implies Lemma 8.5.1. □

Lemma 8.5.2 *Let $g \in \mathbb{Z}[z_1, \ldots, z_q]$ be a non-zero polynomial of degree d and $\mathcal{N}$ a subset of $\mathbb{Z}$ of cardinality $> d$. Then*

$$|\{\mathbf{u} \in \mathcal{N}^q : g(\mathbf{u}) = 0\}| \leq d |\mathcal{N}|^{q-1}.$$

Proof. We proceed by induction on q. For $q = 1$ the assertion is clear. Let $q \geq 2$. Write $g = \sum_{i=0}^{d_0} g_i(z_1, \ldots, z_{q-1}) z_q^i$ with $g_i \in \mathbb{Z}[z_1, \ldots, z_{q-1}]$ and $g_{d_0} \neq 0$. Then $\deg g_{d_0} \leq d - d_0$. The induction hypothesis implies that there are at most $(d - d_0)|\mathcal{N}|^{q-2} \cdot |\mathcal{N}|$ tuples $(u_1, \ldots, u_q) \in \mathcal{N}^q$ with $g_{d_0}(u_1, \ldots, u_{q-1}) = 0$. Further, there are at most $|\mathcal{N}|^{q-1} \cdot d_0$ tuples $\mathbf{u} \in \mathcal{N}^q$ with $g_{d_0}(u_1, \ldots, u_{q-1}) \neq 0$ and $g(u_1, \ldots, u_q) = 0$. Summing these two quantities implies that g has at most $d|\mathcal{N}|^{q-1}$ zeros in $\mathcal{N}^q$. □

Lemma 8.5.3 *Let $g_1, g_2 \in \mathbb{Z}[z_1, \ldots, z_q]$ be two non-zero polynomials of degrees D_1, D_2, respectively, and let N be an integer $\geq \max(D_1, D_2)$. Define*

$$\mathcal{S} := \{\mathbf{u} \in \mathbb{Z}^q : |\mathbf{u}| \leq N, \, g_2(\mathbf{u}) \neq 0\}.$$

Then $\mathcal{S}$ is non-empty, and

$$|g_1|_p \leq (4N)^{q D_1 (D_1+1)/2} \max\{|g_1(\mathbf{u})|_p : \mathbf{u} \in \mathcal{S}\}$$

$$\text{for } p \in M_{\mathbb{Q}} = \{\infty\} \cup \{\text{primes}\}. \qquad (8.5.2)$$

Proof. Put $C_p := \max\{|g_1(\mathbf{u})|_p : \mathbf{u} \in \mathcal{S}\}$ for $p \in M_{\mathbb{Q}}$. We proceed by induction on q, starting with $q = 0$. In the case $q = 0$ we interpret g_1, g_2 as non-zero constants with $|g_1|_p = C_p$ for $p \in M_{\mathbb{Q}}$. Then the lemma is trivial. Let $q \geq 1$. Write

$$g_1 = \sum_{j=0}^{D_1'} g_{1j}(z_1, \ldots, z_{q-1}) z_q^j, \quad g_2 = \sum_{j=0}^{D_2'} g_{2j}(z_1, \ldots, z_{q-1}) z_q^j,$$

where $g_{1,D_1'}, g_{2,D_2'} \neq 0$. By the induction hypothesis, the set

$$\mathcal{S}' := \{\mathbf{u}' \in \mathbb{Z}^{q-1} : |\mathbf{u}'| \leq N, \, g_{2,D_2'}(\mathbf{u}') \neq 0\}$$

is non-empty and moreover,

$$\max_{0 \leq j \leq D_1'} |g_{1j}|_p \leq (4N)^{(q-1)D_1(D_1+1)/2} C_p' \quad \text{for } p \in M_{\mathbb{Q}}, \qquad (8.5.3)$$

where

$$C'_p := \max\{|g_{1j}(\mathbf{u}')|_p : \mathbf{u}' \in \mathcal{S}', \ j = 0, \ldots, D'_1\}.$$

We estimate C'_p from above in terms of C_p. Fix $\mathbf{u}' \in \mathcal{S}'$. There are at least $2N + 1 - D'_2 \geq D'_1 + 1$ integers u_q with $|u_q| \leq N$ such that $g_2(\mathbf{u}', u_q) \neq 0$. Let $a_0, \ldots, a_{D'_1}$ be distinct integers from this set. By Lagrange's interpolation formula,

$$g_1(\mathbf{u}', X) = \sum_{j=0}^{D'_1} g_{1j}(\mathbf{u}')X^j$$

$$= \sum_{j=0}^{D'_1} g_1(\mathbf{u}', a_j) \prod_{\substack{i=0 \\ i \neq j}}^{D'_1} \frac{X - a_i}{a_j - a_i}.$$

Since, in general, the coefficients of a polynomial $\prod_{k=1}^{m}(X - c_k)$ with $c_1, \ldots, c_m \in \mathbb{C}$ have absolute values at most $\prod_{k=1}^{m}(1 + |c_k|)$, we deduce

$$\max_{0 \leq j \leq D'_1} |g_{1j}(\mathbf{u}')| \leq C_\infty \sum_{j=0}^{D'_1} \prod_{\substack{i=0 \\ i \neq j}}^{D'_1} \frac{1 + |a_i|}{|a_j - a_i|}$$

$$\leq C_\infty(D'_1 + 1)(N + 1)^{D'_1} \leq (4N)^{D'_1(D'_1+1)/2}C_\infty.$$

Now let p be a prime and put $\Delta := \prod_{1 \leq i < j \leq D'_1} |a_j - a_i|$. Then

$$\max_{0 \leq j \leq D'_1} |g_{1j}(\mathbf{u}')|_p \leq C_p|\Delta|_p^{-1} \leq \Delta C_p \leq (4N)^{D'_1(D'_1+1)/2}C_p.$$

It follows that $C'_p \leq (4N)^{D'_1(D'_1+1)/2}C_p$ for $p \in M_{\mathbb{Q}}$. A combination with (8.5.3) gives (8.5.2). $\qquad\square$

We now introduce our specializations $B \to \overline{\mathbb{Q}}$ and prove some properties. We assume $q > 0$ and apart from that keep the notation and assumptions from Proposition 8.3.7. In particular, $A_0 = \mathbb{Z}[z_1, \ldots, z_q]$, $K_0 = \mathbb{Q}(z_1, \ldots, z_q)$ and

$$K = \mathbb{Q}(z_1, \ldots, z_q, y), \quad B = \mathbb{Z}[z_1, \ldots, z_q, f^{-1}, y],$$

where f is a non-zero element of A_0, y is integral over A_0, and y has minimal polynomial

$$\mathcal{F} := X^D + \mathcal{F}_1 X^{D-1} + \cdots + \mathcal{F}_D \in A_0[X]$$

over K_0. In the case $D = 1$, we take $y = 1$, $\mathcal{F} = X - 1$.

To allow for other applications (e.g., Lemma 8.7.2 below), we consider a more general situation than what is needed for the proof of Proposition 8.3.7.

Let $d_1 \geq d_0 \geq 1$, $h_1 \geq h_0 \geq 1$ and assume that

$$\max(\deg \mathcal{F}_1, \ldots, \deg \mathcal{F}_D) \leq d_0, \quad \max(d_0, \deg f) \leq d_1,$$
$$\max(h(\mathcal{F}_1), \ldots, h(\mathcal{F}_D)) \leq h_0, \quad \max(h_0, h(f)) \leq h_1.$$

Let $\mathbf{u} = (u_1, \ldots, u_q) \in \mathbb{Z}^q$. Then the substitution $z_1 \mapsto u_1, \ldots, z_q \mapsto u_q$ defines a ring homomorphism (specialization)

$$\varphi_{\mathbf{u}} : \alpha \mapsto \alpha(\mathbf{u}) : \{g_1/g_2 : g_1, g_2 \in A_0, \; g_2(\mathbf{u}) \neq 0\} \to \mathbb{Q}.$$

We want to extend this to a ring homomorphism from B to $\overline{\mathbb{Q}}$ and for this, we have to impose some restrictions on $\mathbf{u}$. Denote by $\Delta_{\mathcal{F}}$ the discriminant of $\mathcal{F}$ (with $\Delta_{\mathcal{F}} := 1$ if $D = \deg \mathcal{F} = 1$), and let

$$\mathcal{H} := \Delta_{\mathcal{F}} \mathcal{F}_D \cdot f. \tag{8.5.4}$$

Then $\mathcal{H} \in A_0$. Using the fact that $\Delta_{\mathcal{F}}$ is a polynomial of degree $2D - 2$ with integer coefficients in $\mathcal{F}_1, \ldots, \mathcal{F}_D$, it follows easily that

$$\deg \mathcal{H} \leq (2D - 1)d_0 + d_1 \leq 2Dd_1. \tag{8.5.5}$$

Now assume that

$$\mathcal{H}(\mathbf{u}) \neq 0.$$

Then $f(\mathbf{u}) \neq 0$ and, moreover, the polynomial

$$\mathcal{F}_{\mathbf{u}} := X^D + \mathcal{F}_1(\mathbf{u})X^{D-1} + \cdots + \mathcal{F}_D(\mathbf{u})$$

has D distinct zeros which are all different from 0, say $y_1(\mathbf{u}), \ldots, y_D(\mathbf{u})$ (these numbers should not be confused with the algebraic functions $y_1, \ldots, y_t$ from Section 8.3). Thus, for $j = 1, \ldots, D$ the assignment

$$z_1 \mapsto u_1, \ldots, z_q \mapsto u_q, \quad y \mapsto y_j(\mathbf{u})$$

defines a ring homomorphism $\varphi_{\mathbf{u},j}$ from B to $\overline{\mathbb{Q}}$; in the case $D = 1$ it is just $\varphi_{\mathbf{u}}$. The image of $\alpha \in B$ under $\varphi_{\mathbf{u},j}$ is denoted by $\alpha_j(\mathbf{u})$. Recall that we may express elements α of B as

$$\alpha = \sum_{i=0}^{D-1} (P_i/Q)y^i \tag{8.5.6}$$

with $P_0, \ldots, P_{D-1}, Q \in A_0$, $\gcd(P_0, \ldots, P_{D-1}, Q) = 1$.

Since $\alpha \in B$, the denominator Q must divide a power of f, hence $Q(\mathbf{u}) \neq 0$. So we have

$$\alpha_j(\mathbf{u}) = \sum_{i=0}^{D-1} (P_i(\mathbf{u})/Q(\mathbf{u}))y_j(\mathbf{u})^i \quad (j = 1, \ldots, D). \tag{8.5.7}$$

It is obvious that $\varphi_{\mathbf{u},j}$ is the identity on $B \cap \mathbb{Q}$. Thus, if $\alpha \in B \cap \overline{\mathbb{Q}}$, then $\varphi_{\mathbf{u},j}(\alpha)$ has the same minimal polynomial as α and so it is conjugate to α.

For $\mathbf{u} = (u_1, \ldots, u_q) \in \mathbb{Z}^q$, we put $|\mathbf{u}| := \max(|u_1|, \ldots, |u_q|)$. It is easy to verify that for any $g \in A_0, \mathbf{u} \in \mathbb{Z}^q$,

$$\log |g(\mathbf{u})| \leq q \log \deg g + h(g) + \deg g \log \max(1, |\mathbf{u}|). \tag{8.5.8}$$

In particular,

$$h(\mathcal{F}_{\mathbf{u}}) \leq q \log d_0 + h_0 + d_0 \log \max(1, |\mathbf{u}|). \tag{8.5.9}$$

Now an application of Corollary 1.9.6 gives

$$\sum_{j=1}^{D} h(y_j(\mathbf{u})) \leq D + 1 + q \log d_0 + h_0 + d_0 \log \max(1, |\mathbf{u}|). \tag{8.5.10}$$

Define the algebraic number fields $K_{\mathbf{u},j} := \mathbb{Q}(y_j(\mathbf{u}))$ $(j = 1, \ldots, D)$. We derive an upper bound for the discriminant $D_{K_{\mathbf{u},j}}$ of $K_{\mathbf{u},j}$.

Lemma 8.5.4 *Let* $\mathbf{u} \in \mathbb{Z}^q$ *with* $\mathcal{H}(\mathbf{u}) \neq 0$. *Then for* $j = 1, \ldots, D$ *we have* $[K_{\mathbf{u},j} : \mathbb{Q}] \leq D$ *and*

$$|D_{K_{\mathbf{u},j}}| \leq D^{2D-1} \left(d_0^q \cdot e^{h_0} \max(1, |\mathbf{u}|)^{d_0} \right)^{2D-2}.$$

Proof. Let $j \in \{1, \ldots, D\}$. The estimate for the degree is obvious. By Lemma 1.5.1 we have

$$|D_{K_{\mathbf{u},j}}| \leq D^{2D-1} H(\mathcal{F}_{\mathbf{u}})^{2D-2},$$

where $H(\mathcal{F}_{\mathbf{u}})$ denotes the maximum of the absolute values of the coefficients of $\mathcal{F}_{\mathbf{u}}$. Now our lemma follows by combining this with (8.5.9). $\qquad\square$

We finish with two lemmas, which relate $\overline{h}(\alpha)$ to the heights of $\alpha_j(\mathbf{u})$ for $\alpha \in B, \mathbf{u} \in \mathbb{Z}^q$.

Lemma 8.5.5 *Let* $\mathbf{u} \in \mathbb{Z}^q$ *with* $\mathcal{H}(\mathbf{u}) \neq 0$. *Let* $\alpha \in B$. *Then for* $j = 1, \ldots, D$,

$$h(\alpha_j(\mathbf{u})) \leq D^2 + q(D \log d_0 + \log \overline{\deg} \alpha) + Dh_0 + \overline{h}(\alpha)$$
$$+ (Dd_0 + \overline{\deg} \alpha) \log \max(1, |\mathbf{u}|).$$

Proof. Let $P_0, \ldots, P_{D-1}, Q$ be as in (8.5.6) and write $\alpha_j(\mathbf{u})$ as in (8.5.7). Let $L = \mathbb{Q}(y_j(\mathbf{u}))$. Then for $v \in M_L$, we have

$$|\alpha_j(\mathbf{u})|_v \leq D^{s(v)} A_v \max(1, |y_j(\mathbf{u})|)_v^{D-1},$$

where $s(v) = 1$ if v is real, $s(v) = 2$ if v is complex, $s(v) = 0$ if v is finite, and

$$A_v = \max(1, |P_0(\mathbf{u})/Q(\mathbf{u})|_v, \ldots, |P_{D-1}(\mathbf{u})/Q(\mathbf{u})|_v).$$

Hence

$$h(\alpha_j(\mathbf{u})) \le \log D + \frac{1}{[L:\mathbb{Q}]} \sum_{v \in M_L} \log A_v + (D-1)h(y_j(\mathbf{u})). \qquad (8.5.11)$$

From (1.9.8), (1.9.4) and (8.5.8) we infer

$$\frac{1}{[L:\mathbb{Q}]} \sum_{v \in M_L} \log A_v = h(P_0(\mathbf{u})/Q(\mathbf{u}), \dots, P_{D-1}(\mathbf{u})/Q(\mathbf{u}))$$

$$= h^{\mathrm{hom}}(Q(\mathbf{u}), P_0(\mathbf{u}), \dots, P_{D-1}(\mathbf{u}))$$

$$\le \log \max(|Q(\mathbf{u})|, |P_0(\mathbf{u})|, \dots, |P_{D-1}(\mathbf{u})|)$$

$$\le q \log \overline{\deg} \alpha + \overline{h}(\alpha) + \overline{\deg} \alpha \cdot \log \max(1, |\mathbf{u}|).$$

By combining (8.5.11) with this inequality and with (8.5.10), our lemma easily follows. $\qquad \square$

Lemma 8.5.6 *Let $\alpha \in B$, $\alpha \neq 0$, and let N be an integer with*

$$N \ge \max(\overline{\deg} \alpha, \, 2Dd_0 + 2(q+1)(d_1+1)).$$

Then the set

$$\mathcal{S} := \{\mathbf{u} \in \mathbb{Z}^q : |\mathbf{u}| \le N, \, \mathcal{H}(\mathbf{u}) \neq 0\}$$

is non-empty, and

$$\overline{h}(\alpha) \le 5N^4(h_1+1)^2 + 2D(h_1+1)H$$

where $H := \max\{h(\alpha_j(\mathbf{u})) : \mathbf{u} \in \mathcal{S}, \, j = 1, \dots, D\}$.

Proof. It follows from our assumption on N, (8.5.5) and Lemma 8.5.3 that $\mathcal{S}$ is non-empty. We proceed with estimating $\overline{h}(\alpha)$.

Let $P_0, \dots, P_{D-1}, Q \in A_0$ be as in (8.5.6). We analyse Q more closely. Let

$$f = \pm p_1^{k_1} \cdots p_m^{k_m} g_1^{l_1} \cdots g_n^{l_n}$$

be the unique factorization of f in A_0, where $p_1, \dots, p_m$ are distinct prime numbers, and $\pm g_1, \dots, \pm g_n$ distinct irreducible elements of A_0 of positive degree. Notice that

$$m \le h(f)/\log 2 \le h_1/\log 2, \qquad (8.5.12)$$

$$\sum_{i=1}^n l_i h(g_i) \le qd_1 + h_1, \qquad (8.5.13)$$

where the last inequality is a consequence of Corollary 1.9.5. Since $\alpha \in B$, the polynomial Q is also composed of $p_1, \dots, p_m, g_1, \dots, g_n$. Hence

$$Q = a\widetilde{Q} \quad \text{with } a = \pm p_1^{k_1'} \cdots p_m^{k_m'}, \quad \widetilde{Q} = g_1^{l_1'} \cdots g_n^{l_n'} \qquad (8.5.14)$$

for certain non-negative integers $k'_1, \ldots, l'_n$. Clearly,

$$l'_1 + \cdots + l'_n \le \deg Q \le \overline{\deg} \, \alpha \le N,$$

and by Corollary 1.9.5 and (8.5.13),

$$h(\widetilde{Q}) \le q \deg Q + \sum_{i=1}^{n} l'_i h(g_i) \le N(q + qd_1 + h_1) \le N^2(h_1 + 1). \quad (8.5.15)$$

In view of (8.5.8), we have for $\mathbf{u} \in \mathcal{S}$,

$$\log |\widetilde{Q}(\mathbf{u})| \le q \log d_1 + h(\widetilde{Q}) + \deg Q \log N$$
$$\le \tfrac{3}{2} N \log N + N^2(h_1 + 1) \le N^2(h_1 + 2).$$

Hence

$$h(\widetilde{Q}(\mathbf{u})\alpha_j(\mathbf{u})) \le N^2(h_1 + 2) + H$$

for $\mathbf{u} \in \mathcal{S}$, $j = 1, \ldots, D$. Further, by (8.5.7) and (8.5.13) we have

$$\widetilde{Q}(\mathbf{u})\alpha_j(\mathbf{u}) = \sum_{i=0}^{D-1} (P_i(\mathbf{u})/a) y_j(\mathbf{u})^i.$$

Put

$$\delta(\mathbf{u}) := \gcd(a, P_0(\mathbf{u}), \ldots, P_{D-1}(\mathbf{u})).$$

Then by applying Lemma 8.5.1 and then (8.5.9) we obtain

$$\log\left(\frac{\max(|a|, |P_0(\mathbf{u})|, \ldots, |P_{D-1}(\mathbf{u})|)}{\delta(\mathbf{u})}\right)$$
$$\le 2D^2 + (D-1)h(\mathcal{F}_{\mathbf{u}}) + D\big(N^2(h_1 + 2) + H\big)$$
$$\le 2D^2 + (D-1)(q \log d_1 + h_1 + d_1 \log N) + D\big(N^2(h_1 + 2) + H\big)$$
$$\le N^3(h_1 + 2) + DH. \quad (8.5.16)$$

Our assumption that $\gcd(Q, P_0, \ldots, P_{D-1}) = 1$ implies that the greatest common divisor of a and the coefficients of $P_0, \ldots, P_{D-1}$ is 1. Let $p \in \{p_1, \ldots, p_m\}$ be one of the prime factors of a. There is $j \in \{0, \ldots, D-1\}$ such that $|P_j|_p = 1$. Our assumption on N and (8.5.5) imply that $N \ge \max(\deg \mathcal{H}, \deg P_j)$. This means that Lemma 8.5.3 is applicable with $g_1 = P_j$ and $g_2 = \mathcal{H}$. It follows that

$$\max\{|P_j(\mathbf{u})|_p : \mathbf{u} \in \mathcal{S}\} \ge (4N)^{-qN(N+1)/2}.$$

That is, there is $\mathbf{u}_0 \in \mathcal{S}$ with $|P_j(\mathbf{u}_0)|_p \ge (4N)^{-qN(N+1)/2}$. Hence

$$|\delta(\mathbf{u}_0)|_p \ge (4N)^{-qN(N+1)/2}.$$

Together with (8.5.16), this implies

$$\log |a|_p^{-1} \leq \log |a/\delta(\mathbf{u}_0)| + \log |\delta(\mathbf{u}_0)|_p^{-1}$$
$$\leq N^3(h_1 + 2) + DH + \tfrac{1}{2}N^3 \log 4N \leq N^4(h_1 + 3) + DH.$$

Combining this with the upper bound (8.5.12) for the number of prime factors of a, we obtain

$$\log |a| \leq 2N^4 h_1(h_1 + 3) + 2Dh_1 \cdot H. \tag{8.5.17}$$

Together with (8.5.14) and (8.5.15), this implies

$$h(Q) \leq 2N^4 h_1(h_1 + 3) + 2Dh_1 \cdot H + N^2(h_1 + 1)$$
$$\leq 3N^4(h_1 + 1)^2 + 2Dh_1 \cdot H. \tag{8.5.18}$$

Further, the right-hand side of (8.5.17) is also an upper bound for $\log \delta(\mathbf{u})$, for $\mathbf{u} \in \mathcal{S}$. Combining this with (8.5.16) gives

$$\log \max\{|P_j(\mathbf{u})| : \mathbf{u} \in \mathcal{S}, \ j = 0, \ldots, D - 1\}$$
$$\leq N^3(h_1 + 2) + DH + 3N^4(h_1 + 1)^2 + 2Dh_1 \cdot H$$
$$\leq 4N^4(h_1 + 1)^2 + 2D(h_1 + 1) \cdot H.$$

Another application of Lemma 8.5.3 yields

$$h(P_j) \leq \tfrac{1}{2}qN(N + 1)\log 4N + 4N^4(h_1 + 1)^2 + 2D(h_1 + 1) \cdot H$$
$$\leq 5N^4(h_1 + 1)^2 + 2D(h_1 + 1) \cdot H$$

for $j = 0, \ldots, D - 1$. Together with (8.5.18) this gives the upper bound for $\overline{h}(\alpha)$ from our lemma. $\qquad\square$

8.6 Bounding the height in Proposition 8.3.7

It remains to prove the height bound in (8.3.19). As before, we use $\mathfrak{O}(\cdot)$ to denote a quantity which is c times the expression between the parentheses, where c is an effectively computable positive absolute constant which may be different at each occurrence of the $\mathfrak{O}$-symbol.

We first consider the case $q > 0$. Pick $\mathbf{u} \in \mathbb{Z}^q$ with $\mathcal{H}(\mathbf{u}) \neq 0$, pick $j \in \{1, \ldots, D\}$ and put $L := K_{\mathbf{u}, j}$. Further, let the set of places S consist of all infinite places of L, and all finite places of L lying above the rational prime divisors of $f(\mathbf{u})$. Let $\mathfrak{p}_1, \ldots, \mathfrak{p}_t$ be the prime ideals in S and define, in the usual manner,

$$s := |S|, \quad P := \max(2, N_K(\mathfrak{p}_1), \ldots, N_K(\mathfrak{p}_t)), \quad Q := \prod_{i=1}^{t} N_K(\mathfrak{p}_i).$$

Further, denote by R_S the S-regulator (see (1.8.2)). Note that $y_j(\mathbf{u})$ is an algebraic integer, and $f(\mathbf{u}) \in O_S^*$. Hence $\varphi_{\mathbf{u},j}(B) \subseteq O_S$ and $\varphi_{\mathbf{u},j}(B^*) \subseteq O_S^*$. Let x_1, x_2 be a solution of (8.3.17). So

$$x_{1,j}(\mathbf{u}) + x_{2,j}(\mathbf{u}) = 1, \quad x_{1,j}(\mathbf{u}), \, x_{2,j}(\mathbf{u}) \in O_S^*,$$

where $x_{1,j}(\mathbf{u}), x_{2,j}(\mathbf{u})$ are the images of x_1, x_2 under $\varphi_{\mathbf{u},j}$. We apply Corollary 4.1.5. In a slightly less precise form, this result gives

$$\max(h(x_{1,j}(\mathbf{u})), \, h(x_{2,j}(\mathbf{u})))$$
$$\leq \exp(\mathfrak{O}(s \log s)) \frac{P}{\log P} \cdot R_S \cdot \max(\log P, \log^* R_S). \qquad (8.6.1)$$

We estimate this bound from above. By assumption, f has degree at most d_1 and logarithmic height at most h_1, hence

$$|f(\mathbf{u})| \leq d_1^q e^{h_1} \max(1, |\mathbf{u}|)^{d_1} =: R(\mathbf{u}). \qquad (8.6.2)$$

Since the degree of L is at most D, the cardinality s of S is at most $s \leq D(1 + \omega)$, where ω is the number of prime divisors of $f(\mathbf{u})$. Using the inequality from elementary number theory, $\omega \leq \mathfrak{O}(\log |f(\mathbf{u})| / \log \log |f(\mathbf{u})|)$, we obtain

$$s \leq \mathfrak{O}\left(\frac{D \log^* R(\mathbf{u})}{\log^* \log^* R(\mathbf{u})}\right). \qquad (8.6.3)$$

Next, we estimate P and R_S. By (8.6.2), we have

$$P \leq Q \leq |f(\mathbf{u})|^D \leq \exp \mathfrak{O}(D \log^* R(\mathbf{u})). \qquad (8.6.4)$$

By inequality (1.8.4) we have

$$R_S \leq |D_L|^{1/2} (\log^* |D_L|)^{D-1} \prod_{i=1}^{t} \log N_K(\mathfrak{p}_i) \leq |D_L|^{1/2} (\log^* |D_L|)^{D-1} (\log Q)^s.$$

In view of Lemma 8.5.4 (using $d_0 \leq d_1$) we have

$$|D_L| \leq D^{2D-1} \left(d_1^q e^{h_1} \max(1, |\mathbf{u}|)^{d_1}\right)^{2D-2} \leq \exp \mathfrak{O}(D \log^* D R(\mathbf{u})),$$

and this easily implies

$$|\Delta_L|^{1/2} (\log^* \Delta_L)^{D-1} \leq \exp \mathfrak{O}(D \log^* D R(\mathbf{u})).$$

Together with the estimates (8.6.3) and (8.6.4) for s and Q, this leads to

$$R_S \leq \exp \mathfrak{O}(D \log^* D R(\mathbf{u}) + s \log^* \log^* Q) \leq \exp \mathfrak{O}(D \log^* D R(\mathbf{u})).$$
(8.6.5)

Now by inserting (8.6.3)–(8.6.5) into the upper bound (8.6.1) we obtain

$$h(x_{1,j}(\mathbf{u})), \; h(x_{2,j}(\mathbf{u})) \leq \exp \mathfrak{O}(D \log^* D \log^* R(\mathbf{u})).$$

We apply Lemma 8.5.6 with $N := 4D^2(q + d_1 + 1)^2$. From the already established (8.3.18) it follows that $\overline{\deg} \, x_1$, $\overline{\deg} \, x_2 \leq N$. Further, since $d_1 \geq d_0$ we have $N \geq 2Dd_0 + 2(d_1 + 1)(q + 1)$. So indeed, Lemma 8.5.6 is applicable with this value of N. It follows that the set $\mathcal{S} := \{\mathbf{u} \in \mathbb{Z}^q : |\mathbf{u}| \leq N, \; \mathcal{H}(\mathbf{u}) \neq 0\}$ is not empty. Further, for $\mathbf{u} \in \mathcal{S}$, $j = 1, \ldots, D$, we have

$$h(x_{1,j}(\mathbf{u})) \leq \exp \mathfrak{O}(D \log^* D(q \log d_1 + h_1 + d_1 \log^* N))$$
$$\leq \exp \mathfrak{O}(N^{1/2}(\log^* N)^2 + (D \log^* D)h_1),$$

and so by Lemma 8.5.6,

$$\overline{h}(x_1) \leq \exp \mathfrak{O}(N^{1/2}(\log^* N)^2 + (D \log^* D)h_1).$$

For $\overline{h}(x_2)$ we obtain the same upper bound. This easily implies (8.3.19) in the case $q > 0$.

Now assume $q = 0$. In this case, $K_0 = \mathbb{Q}$, $A_0 = \mathbb{Z}$ and $B = \mathbb{Z}[f^{-1}, y]$, where y is an algebraic integer with minimal polynomial $\mathcal{F} = X^D + \mathcal{F}_1 X^{D-1} + \cdots + \mathcal{F}_D \in \mathbb{Z}[X]$ over $\mathbb{Q}$, and f is a non-zero rational integer. By assumption, $\log|f| \leq h_1$, $\log|\mathcal{F}_i| \leq h_1$ for $i = 1, \ldots, D$. Denote by $y_1, \ldots, y_D$ the conjugates of y, and let $L = \mathbb{Q}(y_j)$ for some j. By Lemma 1.5.1 we have $|\Delta_L| \leq D^{2D-1} e^{(2D-2)h_1}$. The isomorphism given by $y \mapsto y_j$ maps K to L and B to O_S, where S consists of the infinite places of L and of the prime ideals of O_L that divide f. The estimates (8.6.2)–(8.6.5) remain valid if we replace $R(\mathbf{u})$ by e^{h_1}. Hence for any solution (x_1, x_2) of (8.3.17),

$$h(x_{1,j}), \; h(x_{2,j}) \leq \exp \mathfrak{O}((D \log^* D)h_1),$$

where $x_{1,j}$, $x_{2,j}$ are the j-th conjugates of x_1, x_2, respectively. Now an application of Lemma 8.5.1 with $g = \mathcal{F}$, $m = D$, $\beta_j = x_{1,j}$ gives

$$\overline{h}(x_1) \leq \exp \mathfrak{O}((D \log^* D)h_1).$$

Again we derive the same upper bound for $\overline{h}(x_2)$, and deduce (8.3.19). This completes the proof of Proposition 8.3.7. $\qquad\square$

8.7 Proof of Theorem 8.1.3

We start with some results on multiplicative (in)dependence. We first recall a result that was published by Loher and Masser, but attributed by them to Yu Kunrui. Another result of this type was obtained earlier in Loxton and van der Poorten (1983).

Lemma 8.7.1 *Let L be an algebraic number field of degree d, and $\gamma_0, \ldots, \gamma_s$ non-zero elements of L such that $\gamma_0, \ldots, \gamma_s$ are multiplicatively dependent, but any s elements among $\gamma_0, \ldots, \gamma_s$ are multiplicatively independent. Then there are non-zero integers $k_0, \ldots, k_s$ such that*

$$\gamma_0^{k_0} \cdots \gamma_s^{k_s} = 1,$$

$$|k_i| \leq 58(s! e^s / s^s) \cdot d^{s+1} (\log^* d) h(\gamma_0) \cdots h(\gamma_s) / h(\gamma_i) \quad \text{for } i = 0, \ldots, s.$$

Proof. See Loher and Masser (2004), Corollary 3.2. □

We prove a generalization for arbitrary finitely generated integral domains. As before, let $A = \mathbb{Z}[z_1, \ldots, z_r] \supseteq \mathbb{Z}$ be an integral domain finitely generated over $\mathbb{Z}$, and suppose that the ideal I of polynomials $f \in \mathbb{Z}[X_1, \ldots, X_r]$ with $f(z_1, \ldots, z_r) = 0$ is generated by $f_1, \ldots, f_m$. Let K be the quotient field of A.

Let $\gamma_0, \ldots, \gamma_s$ be non-zero elements of K, and for $i = 1, \ldots, s$, let (g_{i1}, g_{i2}) be a pair of representatives for γ_i, i.e., elements of $\mathbb{Z}[X_1, \ldots, X_r]$ such that

$$\gamma_i = \frac{g_{i1}(z_1, \ldots, z_r)}{g_{i2}(z_1, \ldots, z_r)}.$$

Lemma 8.7.2 *Assume that $f_1, \ldots, f_m$ and g_{i1}, g_{i2} ($i = 0, \ldots, s$) have degrees at most d and logarithmic heights at most h, where $d \geq 1$, $h \geq 1$. Further, assume that $\gamma_0, \ldots, \gamma_s$ are multiplicatively dependent. Then there are integers $k_0, \ldots, k_s$, not all equal to 0, such that*

$$\gamma_0^{k_0} \cdots \gamma_s^{k_s} = 1,$$

$$|k_i| \leq (2d)^{\exp \mathfrak{O}(r+s)} (h+1)^s \quad \text{for } i = 0, \ldots, s.$$

Proof. We assume without loss of generality that any s numbers among $\gamma_0, \ldots, \gamma_s$ are multiplicatively independent (if this is not the case, take a minimal multiplicatively dependent subset of $\{\gamma_0, \ldots, \gamma_s\}$ and proceed further with this subset). We first assume that $q > 0$. We use an argument of van der Poorten and Schlickewei (1991). We keep the notation and assumptions from Sections 8.3–8.5. In particular, we assume that $z_1, \ldots, z_q$ is a transcendence basis of K, and rename $z_{q+1}, \ldots, z_r$ as $y_1, \ldots, y_t$, respectively. For brevity, we have included the case $t = 0$ as well in our proof. But it should be possible to prove in this case a sharper result by means of a more elementary method. In the case $t > 0$, y and $\mathcal{F} = X^D + \mathcal{F}_1 X^{D-1} + \cdots + \mathcal{F}_D$ will be as in Corollary 8.3.3. In the case $t = 0$ we take $m = 1$, $f_1 = 0$, $d = h = 1$, $y = 1$, $\mathcal{F} = X - 1$, $D = 1$. We construct a specialization such that among the images of $\gamma_0, \ldots, \gamma_s$ no s elements are multiplicatively dependent, and then apply Lemma 8.7.1.

Let $V \geq 2d$ be a positive integer. Later we shall make our choice of V more precise. Let

$$\mathcal{V} := \left\{ \begin{array}{l} \mathbf{v} = (v_0, \ldots, v_s) \in \mathbb{Z}^{s+1} \setminus \{\mathbf{0}\} : \\ |v_i| \leq V \quad \text{for } i = 0, \ldots, s, \quad v_i = 0 \text{ for some } i \end{array} \right\}. \tag{8.7.1}$$

Then

$$\gamma_{\mathbf{v}} := \left(\prod_{i=0}^{s} \gamma_i^{v_i} \right) - 1 \quad (\mathbf{v} \in \mathcal{V})$$

are non-zero elements of K, since each proper subset of $\{\gamma_0, \ldots, \gamma_s\}$ is multiplicatively independent. It is not difficult to show that for $\mathbf{v} \in \mathcal{V}$, $\gamma_{\mathbf{v}}$ has a pair of representatives $(g_{1,\mathbf{v}}, g_{2,\mathbf{v}})$ such that

$$\deg g_{1,\mathbf{v}}, \quad \deg g_{2,\mathbf{v}} \leq sdV.$$

In the case $t > 0$, there exists by Lemma 8.3.5 a non-zero $f \in A_0$ such that

$$A \subseteq B := A_0[y, f^{-1}], \quad \gamma_{\mathbf{v}} \in B^* \quad \text{for } \mathbf{v} \in \mathcal{V}$$

and

$$\deg f \leq V^{s+1}(2sdV)^{\exp \mathfrak{O}(r)} \leq V^{\exp \mathfrak{O}(r+s)}.$$

In the case $t = 0$ this holds true as well, with $y = 1$ and $f = \prod_{\mathbf{v} \in \mathcal{V}} (g_{1,\mathbf{v}} \cdot g_{2,\mathbf{v}})$. We apply the theory on specializations explained in Section 8.5 with this f. We put $\mathcal{H} := \Delta_{\mathcal{F}} \mathcal{F}_D f$, where $\Delta_{\mathcal{F}}$ is the discriminant of $\mathcal{F}$. Using Corollary 8.3.3 and inserting the bound $D \leq d^t$ from Lemma 8.3.1 we get for $t > 0$,

$$d_0 \leq (2d)^{\exp \mathfrak{O}(r)}, \quad h_0 \leq (2d)^{\exp \mathfrak{O}(r)}(h + 1), \tag{8.7.2}$$

where

$$d_0 := \max(\deg f_1, \ldots, \deg f_m, \deg \mathcal{F}_1, \ldots, \deg \mathcal{F}_D),$$
$$h_0 := \max(h(f_1), \ldots, h(f_m), h(\mathcal{F}_1), \ldots, h(\mathcal{F}_D)).$$

With the provision $\deg 0 = h(0) = -\infty$, the inequalities (8.7.2) hold true also if $t = 0$. Combining this with Lemma 8.3.4, we obtain

$$\deg \mathcal{H} \leq (2D - 1)d_0 + \deg f \leq V^{\exp \mathfrak{O}(r+s)}.$$

By Lemma 8.5.2 there exists $\mathbf{u} \in \mathbb{Z}^q$ with

$$\mathcal{H}(\mathbf{u}) \neq 0, \quad |\mathbf{u}| \leq V^{\exp \mathfrak{O}(r+s)}. \tag{8.7.3}$$

We proceed further with this $\mathbf{u}$.

As we have seen before, $\gamma_\mathbf{v} \in B^*$ for $\mathbf{v} \in \mathcal{V}$. By our choice of $\mathbf{u}$, there are D distinct specialization maps $\varphi_{\mathbf{u},j}$ ($j = 1, \ldots, D$) from B to $\overline{\mathbb{Q}}$. We fix one of these specializations, say $\varphi_\mathbf{u}$. Given $\alpha \in B$, we write $\alpha(\mathbf{u})$ for $\varphi_\mathbf{u}(\alpha)$. As the elements $\gamma_\mathbf{v}$ are all units in B, their images under $\varphi_\mathbf{u}$ are non-zero. So we have

$$\prod_{i=0}^{s} \gamma_i(\mathbf{u})^{v_i} \neq 1 \quad \text{for } \mathbf{v} \in \mathcal{V}, \tag{8.7.4}$$

where $\mathcal{V}$ is defined by (8.7.1).

We use Lemma 8.5.5 to estimate the heights $h(\gamma_i(\mathbf{u}))$ for $i = 0, \ldots, s$. Recall that by Lemma 8.3.4 we have

$$\overline{\deg} \, \gamma_i \leq (2d)^{\exp \mathfrak{O}(r)}, \quad \overline{h}(\gamma_i) \leq (2d)^{\exp \mathfrak{O}(r)}(h + 1)$$

for $i = 0, \ldots, s$. By inserting these bounds, together with the bound $D \leq d^t$ from Lemma 8.3.1, those for d_0, h_0 from (8.7.2) and that for $\mathbf{u}$ from (8.7.3) into the bound from Lemma 8.5.5, we obtain for $i = 0, \ldots, s$,

$$h(\gamma_i(\mathbf{u})) \leq (2d)^{\exp \mathfrak{O}(r)}(1 + h + \log \max(1, |\mathbf{u}|)) \tag{8.7.5}$$
$$\leq (2d)^{\exp \mathfrak{O}(r+s)}(1 + h + \log V).$$

Assume that among $\gamma_0(\mathbf{u}), \ldots, \gamma_s(\mathbf{u})$ there are s numbers that are multiplicatively dependent. By Lemma 8.7.1 there are integers $k_0, \ldots, k_s$, at least one of which is non-zero and at least one of which is 0, such that

$$\prod_{i=0}^{s} \gamma_i(\mathbf{u})^{k_i} = 0,$$
$$|k_i| \leq (2d)^{\exp \mathfrak{O}(r+s)}(1 + h + \log V)^{s-1} \quad \text{for } i = 0, \ldots, s.$$

Now for

$$V = (2d)^{\exp \mathfrak{O}(r+s)}(h + 1)^{s-1} \tag{8.7.6}$$

(with a sufficiently large constant in the $\mathfrak{O}$-symbol), the upper bound for the numbers $|k_i|$ is smaller than V. But this would imply that $\prod_{i=0}^{s} \gamma_i(\mathbf{u})^{v_i} = 1$ for some $\mathbf{v} \in \mathcal{V}$, contrary to (8.7.4). Thus we conclude that with the choice (8.7.6) for V, there exists $\mathbf{u} \in \mathbb{Z}^q$ with (8.7.3), such that any s numbers among $\gamma_0(\mathbf{u}), \ldots, \gamma_s(\mathbf{u})$ are multiplicatively independent. The numbers $\gamma_0(\mathbf{u}), \ldots, \gamma_s(\mathbf{u})$

are multiplicatively dependent, since they are the images under $\varphi_{\mathbf{u}}$ of $\gamma_0, \ldots, \gamma_s$, which are multiplicatively dependent. Substituting (8.7.6) into (8.7.5) we obtain

$$h(\gamma_i(\mathbf{u})) \leq (2d)^{\exp \mathfrak{D}(r+s)}(h+1) \quad \text{for } i = 0, \ldots, s. \qquad (8.7.7)$$

Now Lemma 8.7.1 implies that there are non-zero integers $k_0, \ldots, k_s$ such that

$$\prod_{i=0}^{s} \gamma_i(\mathbf{u})^{k_i} = 1, \qquad (8.7.8)$$

$$|k_i| \leq (2d)^{\exp \mathfrak{D}(r+s)}(h+1)^s \quad \text{for } i = 0, \ldots, s. \qquad (8.7.9)$$

Our assumption on $\gamma_0, \ldots, \gamma_s$ implies that there are non-zero integers $l_0, \ldots, l_s$ such that $\prod_{i=0}^{s} \gamma_i^{l_i} = 1$. Hence $\prod_{i=0}^{s} \gamma_i(\mathbf{u})^{l_i} = 1$. Together with (8.7.8) this implies

$$\prod_{i=1}^{s} \gamma_i(\mathbf{u})^{l_0 k_i - l_i k_0} = 1.$$

But $\gamma_1(\mathbf{u}), \ldots, \gamma_s(\mathbf{u})$ are multiplicatively independent, hence $l_i k_0 = k_i l_0$ for $i = 1, \ldots, s$. Therefore,

$$\left(\gamma_0^{k_0} \cdots \gamma_s^{k_s}\right)^{l_0} = \left(\gamma_0^{l_0} \cdots \gamma_s^{l_s}\right)^{k_0} = 1,$$

implying that $\prod_{i=0}^{s} \gamma_i^{k_i} = \rho$ for some root of unity ρ. But $\varphi_{\mathbf{u}}(\rho) = 1$ and it is conjugate to ρ. Hence $\rho = 1$. So in fact we have $\prod_{i=0}^{s} \gamma_i^{k_i} = 1$ with non-zero integers k_i satisfying (8.7.9). This proves our lemma, but under the assumption $q > 0$. If $q = 0$ then a much simpler argument, without specializations, gives $h(\gamma_i) \leq (2d)^{\exp \mathfrak{D}(r+s)}(h+1)$ for $i = 0, \ldots, s$ instead of (8.7.7). Then the proof is finished in the same way as in the case $q > 0$. $\qquad \square$

Corollary 8.7.3 *Assume that $f_1, \ldots, f_m$ and g_{i1}, g_{i2} $(i = 0, \ldots, s)$ have degrees at most d and logarithmic heights at most h, where $d \geq 1$, $h \geq 1$. Further, assume that $\gamma_1, \ldots, \gamma_s$ are multiplicatively independent and*

$$\gamma_0 = \gamma_1^{k_1} \cdots \gamma_s^{k_s}$$

for certain integers $k_1, \ldots, k_s$. Then

$$|k_i| \leq (2d)^{\exp \mathfrak{D}(r+s)}(h+1)^s \quad \text{for } i = 1, \ldots, s.$$

Proof. By Lemma 8.7.2, and by the multiplicative independence of $\gamma_1, \ldots, \gamma_s$, there are integers $l_0, \ldots, l_m$ such that

$$\prod_{i=0}^{m} \gamma_i^{l_i} = 1,$$

$$l_0 \neq 0, \quad |l_i| \leq (2d)^{\exp \mathfrak{D}(r+s)}(h+1)^s \quad \text{for } i = 0, \ldots, s.$$

Now, clearly, we have also

$$\prod_{i=1}^{s} \gamma_i^{l_0 k_i - l_i} = 1,$$

hence $l_0 k_i - l_i = 0$ for $i = 1, \ldots, s$. It follows that

$$|k_i| = |l_i / l_0| \le (2d)^{\exp \mathfrak{O}(r+s)} (h+1)^s \quad \text{for } i = 1, \ldots, s.$$

This implies our corollary. □

Proof of Theorem 8.1.3. We keep the notation and assumptions from the statement of Theorem 8.1.3. For $i = 1, \ldots, s$, $j = 1, 2$, let $\alpha_{ij} := g_{ij}(z_1, \ldots, z_r)$. Then $\alpha_{i1}, \alpha_{i2} \in A$ and $\gamma_i = \alpha_{i1}/\alpha_{i2}$ for $i = 1, \ldots, s$. Further, let

$$g := \prod_{i=1}^{s} (g_{i1} g_{i2}), \quad \gamma := \prod_{i=1}^{s} (\alpha_{i1} \alpha_{i2})$$

and define the ring

$$\widetilde{A} := A[\gamma^{-1}].$$

Then

$$\widetilde{A} \cong \mathbb{Z}[X_1, \ldots, X_r, X_{r+1}]/\widetilde{I} \quad \text{with } \widetilde{I} = (f_1, \ldots, f_m, gX_{r+1} - 1).$$

Clearly, $\gamma \in \widetilde{A}^*$, therefore also $\alpha_{i1}, \alpha_{i2} \in \widetilde{A}^*$, and hence $\gamma_i \in \widetilde{A}^*$ for $i = 1, \ldots, s$. Further, g has total degree at most $\mathfrak{O}(sd)$ and logarithmic height at most $\mathfrak{O}(sh)$. As a consequence, $\widetilde{I}$ is generated by polynomials of total degrees at most $\mathfrak{O}(sd)$ and logarithmic heights at most $\mathfrak{O}(sh)$.

Let $(v_1, \ldots, w_s)$ be a solution of (8.1.3), and put

$$x_1 := \prod_{i=1}^{s} \gamma_i^{v_i}, \quad x_2 := \prod_{i=1}^{s} \gamma_i^{w_i}.$$

Then

$$a_1 x_1 + a_2 x_2 = a_3, \quad x_1, x_2 \in \widetilde{A}^*.$$

By Theorem 8.1.1, x_1 has a representative $\widetilde{x_1} \in \mathbb{Z}[X_1, \ldots, X_{r+1}]$ of degree and logarithmic height both bounded above by

$$\exp \left((2sd)^{\exp \mathfrak{O}(r)} (h+1) \right).$$

Now Corollary 8.7.3 implies

$$|v_i| \le \exp \left((2^s d)^{\exp \mathfrak{O}(r)} (h+1) \right) \quad \text{for } i = 1, \ldots, s.$$

For $|w_i|$ ($i = 1, \ldots, s$) we derive a similar upper bound. This completes the proof of Theorem 8.1.3. □

8.8 Notes

- It is well-known that if A is a finitely generated integral domain over $\mathbb{Z}$, then its quotient field K can be represented in the form $K = \mathbb{Q}(z_1, \ldots, z_q, y)$, where $\{z_1, \ldots, z_q\}$ is a transcendence basis of K over $\mathbb{Q}$, y is an integral element over $A_0 := \mathbb{Z}[z_1, \ldots, z_q]$, and A is contained in the integral domain $B := \mathbb{Z}[z_1, \ldots, z_q, f^{-1}, y]$, for some non-zero $f \in A_0$. As was seen in Sections 8.4–8.6, such an overring B has the advantage that it is easier to deal with its elements. The generating sets $\{z_1, \ldots, z_q, y\}$ of the above type proved to be useful in several other applications, among others in transcendental number theory; see e.g. Waldschmidt (1973, 1974) where an appropriate size is introduced for the elements of K with respect to a generating set $\{z_1, \ldots, z_q, y\}$.

- Following a method of Győry (1983, 1984), analogues of some results of this chapter can be established over integral domains finitely generated over a field of characteristic zero instead of $\mathbb{Z}$. However, in this case finiteness results cannot be obtained, upper bounds can be derived only for the degrees of the solutions.

- Theorems 8.1.1–8.1.3 as well as the method of their proofs have several applications. For instance, Theorems 8.1.1–8.1.3 are applied in our book on discriminant equations. Further, their methods of proof are used in several papers to obtain general effective finiteness results for various classes of Diophantine equations over finitely generated domains over $\mathbb{Z}$, namely for Thue equations and superelliptic equations in Bérczes, Evertse and Győry (2014), for polynomial equations $f(x, y) = 0$ in solutions x, y from a finitely generated multiplicative group, and even from the division group of the latter, in Bérczes (2015a, 2015b), and Koymans (2015) for Catalan's equation.

9

Decomposable form equations

Let $F \in \mathbb{Z}[X, Y]$ be a binary form, i.e., a homogeneous polynomial of degree $n \geq 3$, that is irreducible over $\mathbb{Q}$, and δ a non-zero integer. Thue (1909) proved that the equation

$$F(x, y) = \delta \quad \text{in } x, y \in \mathbb{Z}$$

has only finitely many solutions. This was extended by Mahler (1933a) as follows. Let $p_1, \ldots, p_t$ be distinct prime numbers. Then the equation

$$F(x, y) = \pm \delta p_1^{z_1} \cdots p_t^{z_t} \quad \text{in } x, y, z_1, \ldots, z_t \in \mathbb{Z}$$
$$\text{with } \gcd(x, y, p_1 \cdots p_t) = 1$$

has only finitely many solutions. Mahler's result can be reformulated as follows. In accordance with terminology introduced before, let $S = \{\infty, p_1, \ldots, p_t\}$, where ∞ is the infinite place of $\mathbb{Q}$, and let $\mathbb{Z}_S = \mathbb{Z}[(p_1 \cdots p_t)^{-1}]$ be the ring of S-integers. Then the set of solutions of the equation

$$F(x, y) \in \delta \mathbb{Z}_S^* \quad \text{in } x, y \in \mathbb{Z}_S$$

is a union of only finitely many $\mathbb{Z}_S^*$-cosets, i.e., sets of the form $\{\varepsilon(x_0, y_0) : \varepsilon \in \mathbb{Z}_S^*\}$, where (x_0, y_0) is a solution of the equation.

In this chapter, we deal with generalizations, where instead of equations over $\mathbb{Z}$ or $\mathbb{Z}_S$ we consider equations over integral domains that are finitely generated over $\mathbb{Z}$, and where instead of a binary form F we take a *decomposable form* in an arbitrary number of variables, that is, a homogeneous polynomial that factors into linear forms over an extension of its field of definition.

More precisely, let K be a finitely generated (but not necessarily algebraic) extension field of $\mathbb{Q}$, and $F \in K[X_1, \ldots, X_m]$ a decomposable form in $m \geq 2$ variables, which factors into linear forms over a finite extension of K. Further, let $\delta \in K^*$ and let A be a subring of K that is finitely generated over $\mathbb{Z}$. We

231

consider the equations

$$F(\mathbf{x}) = \delta \quad \text{in } \mathbf{x} = (x_1, \ldots, x_m) \in A^m \tag{9.1}$$

and

$$F(\mathbf{x}) \in \delta A^* \quad \text{in } \mathbf{x} = (x_1, \ldots, x_m) \in A^m, \tag{9.2}$$

where A^* denotes the unit group of A. Equations of the type (9.1) and (9.2) are called *decomposable form equations*. The set of solutions of (9.2) can be divided into A^*-cosets $\mathbf{x}_0 A^* = \{\varepsilon \mathbf{x}_0 : \varepsilon \in A^*\}$ where $\mathbf{x}_0$ is a solution of (9.2). By Roquette's Theorem (Roquette (1957), p.3), the unit group A^* is finitely generated. Hence it is easy to see that (9.2) can be reduced to finitely many equations of the form (9.1).

Clearly, every binary form is a decomposable form in two variables. Equations of type (9.1) and (9.2) are called *Thue equations* and *Thue–Mahler equations*, respectively in the case that F is a binary form. Unlike in the results of Thue and Mahler mentioned above, we do not require that the binary form is irreducible over its field of definition. Other important special cases of decomposable form equations are norm form equations, discriminant form equations and index form equations. As we shall see, decomposable form equations are in a certain sense equivalent to unit equations and in particular, Thue equations are equivalent to unit equations in two unknowns. Decomposable form equations have many number-theoretic applications.

This chapter is basically an extensive survey, in which for some of the stated results we give a complete proof, whereas for the proofs of others we refer to the literature. Below, we give a brief overview.

In Section 9.1 we present a general finiteness criterion for equations (9.1) and (9.2), and in particular for Thue equations when F is a binary form. For convenience for applications, we establish our criterion for slightly more general equations. This criterion gives effectively decidable necessary and sufficient conditions in terms of the linear factors of F such that equations (9.1) and (9.2) have only finitely many (A^*-cosets of) solutions.

In Section 9.2 we explain how our finiteness criterion for decomposable form equations implies the finiteness result for unit equations established in Chapter 6. In Section 9.3 we deduce our finiteness criterion for equations (9.1) and (9.2) from the finiteness result for unit equations. This shows that the finiteness results for unit equations and decomposable form equations are equivalent.

In Section 9.4 we give, without proof, a complete description of the set of solutions of (9.1) and (9.2) in the case when this set is infinite. More precisely, in this case, the set of solutions can be divided in a natural way into infinite families, and the number of these families is finite.

In Section 9.5, we present, without proofs, explicit upper bounds for the number of families and, in the case of finitely many solutions, for the number of S-integral solutions for decomposable form equations over the ring of S-integers of a number field. In Section 9.6 we derive, with full proofs, effective bounds for the heights of the S-integral solutions of Thue equations, and of decomposable form equations in an arbitrary number of unknowns from a restricted class, including discriminant form equations and certain norm form equations. The proofs are based on effective results from Chapter 4 concerning S-unit equations.

In our next book, *Discriminant Equations in Diophantine Number Theory*, we work out various applications of unit equations to discriminant form equations, index form equations, decomposable form equations of discriminant type and related problems.

There is an extensive literature on decomposable form equations. Almost all books and survey papers listed in the Preface of this book that deal with unit equations and their applications are also concerned with decomposable form equations. We refer also to Borevich and Shafarevich (1967), Evertse and Győry (1988d), Feldman and Nesterenko (1998) and Győry (1999) on the subject. Some further references can be found in the Notes (Section 9.7).

9.1 A finiteness criterion for decomposable form equations

We present a general finiteness criterion which guarantees the finiteness of the number of solutions of equation (9.1), and the finiteness of the number of A^*-cosets of solutions of equation (9.2) for every $\delta \in K^*$ and every subring A of K which is finitely generated over $\mathbb{Z}$.

Let again K be a field which is finitely generated over $\mathbb{Q}$. We fix an algebraic closure $\overline{K}$ of K. Let $A \subset K$ be a ring finitely generated over $\mathbb{Z}$. Further, let $F \in K[X_1, \ldots, X_m]$ be a non-zero decomposable form in $m \geq 2$ variables and let $\delta \in K^*$. For applications it is convenient to extend (9.1) and (9.2) mentioned in the introduction and to consider the equations

$$F(\mathbf{x}) = \delta \text{ in } \mathbf{x} \in \mathcal{M} \quad \text{with } l(\mathbf{x}) \neq 0 \quad \text{for } l \in \mathcal{L}, \quad (9.1.1)$$

and

$$F(\mathbf{x}) \in \delta A^* \text{ in } \mathbf{x} \in \mathcal{M} \quad \text{with } l(\mathbf{x}) \neq 0 \quad \text{for } l \in \mathcal{L}, \quad (9.1.2)$$

where $\mathcal{M}$ is a finitely generated A-module with $\mathcal{M} \subset K^m$, and $\mathcal{L}$ is a finite set of non-zero linear forms from $\overline{K}[X_1, \ldots, X_m]$. In the special case where $\mathcal{M} = A^m$ and $\mathcal{L}$ consists of the linear factors of F, equations (9.1.1) and (9.1.2) give (9.1) and (9.2), respectively.

We give necessary and sufficient conditions, in terms of $\mathcal{L}$ and the linear factors of F, such that (9.1.1) and (9.1.2) have only finitely many (A^*-cosets of) solutions.

We introduce some convenient notation. In what follows we let G be a finite, normal extension of K over which F factorizes into linear factors. For a linear form $l = \alpha_1 X_1 + \cdots + \alpha_m X_m \in G[X_1, \ldots, X_m]$ and for $\sigma \in \mathrm{Gal}(G/K)$ we define $\sigma(l) := \sigma(\alpha_1)X_1 + \cdots + \sigma(\alpha_m)X_m$. For a subset $\mathcal{L}$ of linear forms from $G[X_1, \ldots, X_m]$, we define the following:

$\sigma(\mathcal{L}) = \{\sigma(l) : l \in \mathcal{L}\}$ for $\sigma \in \mathrm{Gal}(G/K)$;

$[\mathcal{L}]$ is the G-vector space generated by $\mathcal{L}$;

$\mathcal{L}$ is called $\mathrm{Gal}(G/K)$-*stable* if $\sigma(\mathcal{L}) = \mathcal{L}$ for each $\sigma \in \mathrm{Gal}(G/K)$;

$\mathcal{L}$ is called $\mathrm{Gal}(G/K)$-*proper* if for each $\sigma \in \mathrm{Gal}(G/K)$ we have either $\sigma(\mathcal{L}) = \mathcal{L}$ or $\sigma(\mathcal{L}) \cap \mathcal{L} = \emptyset$.

Given G-linear subspaces $V_1, \ldots, V_t$ of the space of linear forms from $G[X_1, \ldots, X_m]$, we denote by $V_1 + \cdots + V_t$ the smallest G-vector space containing them.

We have

$$F = c l_1^{e_1} \cdots l_n^{e_n}, \tag{9.1.3}$$

where

$$\mathcal{L}_0 := \{l_1, \ldots, l_n\} \subset G[X_1, \ldots, X_m]$$

is a $\mathrm{Gal}(G/K)$-stable set of pairwise non-proportional linear forms, $c \in K^*$ and $e_1, \ldots, e_n$ are positive integers. Clearly, $e_i = e_j$ if $l_i = \sigma(l_j)$ for some $\sigma \in \mathrm{Gal}(G/K)$.

Let $\mathcal{L}$ be a finite set of pairwise non-proportional linear forms with

$$\mathcal{L} \supseteq \mathcal{L}_0, \quad \mathcal{L} \subset G[X_1, \ldots, X_m].$$

The main result of this section is as follows.

Theorem 9.1.1 *Let m, K, F, G, $\mathcal{L}_0$, $\mathcal{L}$ be as above. Then the following three assertions are equivalent:*

(i) $\mathrm{rank}_G \mathcal{L}_0 = m$, and for each non-empty subset $\mathcal{L}_1 \subsetneqq \mathcal{L}_0$ such that $\mathcal{L}_1$ is $\mathrm{Gal}(G/K)$-proper, we have

$$\mathcal{L} \cap \left(\sum_{\sigma \in \mathrm{Gal}(G/K)} [\sigma(\mathcal{L}_1)] \cap [\mathcal{L}_0 \setminus \sigma(\mathcal{L}_1)] \right) \neq \emptyset; \tag{9.1.4}$$

(ii) for every subring A of K which is finitely generated over $\mathbb{Z}$, *every finitely generated A-module* $\mathcal{M} \subset K^m$, *and every* $\delta \in K^*$, *equation (9.1.1) has only finitely many solutions;*

(iii) for every A, $\mathcal{M}$, δ as in (ii), equation (9.1.2) has only finitely many A^*-cosets of solutions.

Theorem 9.1.1 is new. We shall prove it in Section 9.3. An important feature of this theorem is that it relates statements (i.e., (ii), (iii)) about Diophantine equations to a statement (i.e., (i)) in linear algebra. Furthermore, assertion (i) is effectively decidable provided K, G and the coefficients of the linear forms in $\mathcal{L}_0$ and $\mathcal{L}$ are effectively given in some sense.

The following corollary is an immediate consequence of Theorem 9.1.1.

Corollary 9.1.2 *Let m, K, F, G, $\mathcal{L}_0$ and $\mathcal{L}$ be as in Theorem 9.1.1. If*

(i') $\text{rank}_G \mathcal{L}_0 = m$ and $\mathcal{L} \cap ([\mathcal{L}_1] \cap [\mathcal{L}_0 \setminus \mathcal{L}_1]) \neq \emptyset$ for every proper, non-empty subset $\mathcal{L}_1$ of $\mathcal{L}_0$,

then (ii), (iii) hold. Moreover, if $G = K$, then (i'), (ii), (iii) are equivalent.

Similar finiteness criteria were established in Evertse and Győry (1988c); see also Evertse, Gaál and Győry (1989).

We deduce some further consequences.

Corollary 9.1.3 *Let m, K, F, G, $\mathcal{L}_0$ and $\mathcal{L}$ be as in Theorem 9.1.1, and let $\mathcal{L} = \mathcal{L}_0$. Assume that $|\mathcal{L}_0| \geq 2m - 1$ and that $\mathcal{L}_0$ is in general position, i.e., each subset of m linear forms from $\mathcal{L}_0$ is linearly independent. Then equation (9.1.1) has only finitely many solutions.*

For $\mathcal{M} = A^m$, this was proved in Győry (1993b).

Proof of Corollary 9.1.3. We apply Corollary 9.1.2. Notice that for each proper, non-empty subset $\mathcal{L}_1$ of $\mathcal{L}_0$ we have $|\mathcal{L}_1| \geq m$ or $|\mathcal{L}_0 \setminus \mathcal{L}_1| \geq m$, i.e., $\text{rank}_G[\mathcal{L}_1] = m$ or $\text{rank}_G[\mathcal{L}_0 \setminus \mathcal{L}_1] = m$. Hence $[\mathcal{L}_1] \cap [\mathcal{L}_0 \setminus \mathcal{L}_1]$ contains $\mathcal{L}_1$ or $\mathcal{L}_0 \setminus \mathcal{L}_1$. This implies (i') with $\mathcal{L} = \mathcal{L}_0$, and, by Corollary 9.1.2, (ii) follows. $\square$

Corollary 9.1.4 *Let $F \in K[X, Y]$ be a non-zero binary form. Then the following two assertions are equivalent:*

(iv) F is divisible by at least three pairwise non-proportional linear forms from $\overline{K}[X, Y]$;

(v) for every subring A of K which is finitely generated over $\mathbb{Z}$ and every $\delta \in K^*$, the equation

$$F(x, y) = \delta \quad \text{in } x, y \in A \tag{9.1.5}$$

has only finitely many solutions.

The implication (iv) $\Rightarrow$ (v) follows from work in Thue (1909) for $K = \mathbb{Q}$, $A = \mathbb{Z}$, Siegel (1921) for K an arbitrary algebraic number field and A its ring of integers, Mahler (1933a) for $K = \mathbb{Q}$ and $A = \mathbb{Z}_S$ for some finite set of places S of $\mathbb{Q}$, Parry (1950) for K an arbitrary algebraic number field and $A = O_S$ for some finite set of places S of K, and Lang (1960) in the most general case. As was mentioned before, equation (9.1.5) is usually called a *Thue equation*.

Proof of Corollary 9.1.4. Let G be the splitting field of F over K. Then G is a finite, normal extension of K. Let $\mathcal{L}_0$ be a maximal $\mathrm{Gal}(G/K)$-stable set of pairwise non-proportional linear forms from $G[X, Y]$ that divide F. We have to show that (i) with $m = 2$, $\mathcal{L} = \mathcal{L}_0$ is equivalent to $|\mathcal{L}_0| \geq 3$. First assume that $|\mathcal{L}_0| \geq 3$. Then $\mathrm{rank}_G \mathcal{L}_0 = 2$. Next, for each proper, non-empty subset $\mathcal{L}_1$ of $\mathcal{L}_0$ we have $|\mathcal{L}_1| \geq 2$ or $|\mathcal{L}_0 \setminus \mathcal{L}_1| \geq 2$, i.e., $\mathrm{rank}_G[\mathcal{L}_1] = 2$ or $\mathrm{rank}_G[\mathcal{L}_0 \setminus \mathcal{L}_1] = 2$ and this implies that $[\mathcal{L}_1] \cap [\mathcal{L}_0 \setminus \mathcal{L}_1]$ contains $\mathcal{L}_1$ or $\mathcal{L}_0 \setminus \mathcal{L}_1$. This gives (9.1.4). Conversely, assume that $|\mathcal{L}_0| = 2$. Then each proper, non-empty subset $\mathcal{L}_1$ of $\mathcal{L}_0$ has $|\mathcal{L}_1| = 1$ and so $[\sigma(\mathcal{L}_1)] \cap [\mathcal{L}_0 \setminus \sigma(\mathcal{L}_1)] = (0)$ for every $\sigma \in \mathrm{Gal}(G/K)$. Hence (9.1.4) cannot hold. □

9.2 Reduction of unit equations to decomposable form equations

It can be shown that unit equations and decomposable form equations are equivalent in the sense that every unit equation leads to a decomposable form equation (over a suitable ring which is finitely generated over $\mathbb{Z}$), and every decomposable form equation can be reduced to finitely many unit equations (in an appropriate finite field extension). Consequently, general finiteness results for unit equations imply general finiteness results for decomposable form equations and vice versa. In the two unknowns case (i.e. for unit equations in two unknowns and for Thue equations) this equivalence was (implicitly) pointed out by Siegel (1926, 1929), while the general case was worked out by Evertse and Győry (1988c).

More precisely, we show that Theorem 9.1.1 is equivalent to the following.

Theorem 9.2.1 *Let K be a field of characteristc 0, Γ a finitely generated multiplicative subgroup of K^* and let $a_1, \ldots, a_m \in K^*$. Then the equation*

$$a_1 x_1 + \cdots + a_m x_m = 1 \quad in \ x_1, \ldots, x_m \in \Gamma \tag{9.2.1}$$

has at most finitely many non-degenerate solutions, i.e., with

$$\sum_{i \in I} a_i x_i \neq 0 \quad for \ each \ non-empty \ subset \ I \ of \ \{1, \ldots, m\}.$$

This theorem is proved in Chapter 6 in a more precise quantitative form, see Theorem 6.1.3. For further historical comments, see Section 6.7. All known proofs of Theorem 9.2.1 are ineffective.

In the next section, we prove Theorem 9.1.1, taking Theorem 9.2.1 as a starting point. In the present section we show that Theorem 9.1.1 implies Theorem 9.2.1.

Proof of the implication Theorem 9.1.1 $\Rightarrow$ Theorem 9.2.1. Let K be a field of characteristic 0, Γ a finitely generated subgroup of K^*, and $a_1, \ldots, a_m \in K^*$ with $m \geq 2$. Define the decomposable form

$$F := X_1 \cdots X_m(a_1 X_1 + \cdots + a_m X_m).$$

Let $\mathcal{L}_0 = \{a_1 X_1, \ldots, a_m X_m, a_1 X_1 + \ldots + a_m X_m\}$, and $\mathcal{L}$ be the set of all linear forms of the form $a_{i_1} X_{i_1} + \cdots + a_{i_s} X_{i_s}$, where $\{i_1, \ldots, i_s\}$ is a non-empty subset of $\{1, \ldots, m\}$. Then we have $\mathcal{L} \supset \mathcal{L}_0$. It is easy to check that these $\mathcal{L}_0$ and $\mathcal{L}$ satisfy statement (i) in Theorem 9.1.1 with $G = K$ (and even (i′) in Corollary 9.1.2). Let A be the subring of K generated by $a_1, \ldots, a_m$ and the elements of Γ. Then A is finitely generated over $\mathbb{Z}$, and Γ is a subgroup of A^*.

It is now clear that every non-degenerate solution $\mathbf{x} = (x_1, \ldots, x_m)$ of (9.2.1) satisfies

$$F(\mathbf{x}) \in A^*, \quad \mathbf{x} \in A^m, \quad l(\mathbf{x}) \neq 0 \quad \text{for } l \in \mathcal{L}. \tag{9.2.2}$$

Theorem 9.1.1 (or in this case Corollary 9.1.2) implies that there are at most finitely many pairwise linearly independent $\mathbf{x}$ with (9.2.2). This implies that (9.2.1) has only finitely many pairwise linearly independent non-degenerate solutions. But obviously, any two linearly dependent solutions of (9.2.1) have to be equal. This implies Theorem 9.2.1. $\qquad\square$

9.3 Reduction of decomposable form equations to unit equations

In this section we prove the equivalence of assertions (i), (ii) and (iii) of Theorem 9.1.1, taking Theorem 9.2.1 as starting point. This section has been divided into three subsections: the first contains the proof of the equivalence of (ii) and (iii), which is elementary and is independent of unit equations, the second contains the proof of the implication (i)$\Rightarrow$(iii) and the last the proof of the implication (iii)$\Rightarrow$(i). In both the second and third subsections we have used Theorem 9.2.1. We keep the notation and definitions from Section 9.1.

We need a few facts on Noetherian rings and modules; for a proof of these, we refer to Lang (1984), chapter 6. A commutative ring is called Noetherian if all its ideals are finitely generated. Let A be a Noetherian commutative ring. Then for any ideal I of A the residue class ring A/I is Noetherian. Further, for any integer $r \geq 1$ the polynomial ring in r variables $A[X_1, \ldots, X_r]$ is Noetherian. Any finitely generated A-module is Noetherian, i.e., all its A-submodules are finitely generated. Any integral domain finitely generated over $\mathbb{Z}$ is isomorphic to $\mathbb{Z}[X_1, \ldots, X_r]/I$ for some ideal I of $\mathbb{Z}[X_1, \ldots, X_r]$, hence it is Noetherian.

9.3.1 Proof of the equivalence (ii) $\Longleftrightarrow$ (iii) in Theorem 9.1.1

We need the following result of Roquette.

Proposition 9.3.1 *Let A be an integral domain that is finitely generated over $\mathbb{Z}$. Then A^* is a finitely generated group.*

Proof. See Roquette (1957). □

(ii)$\Longleftrightarrow$(iii). First assume that (ii) holds. Let A, $\mathcal{M}$, δ be as in the statement of (iii). Proposition 9.3.1 implies that there are a finite set $\mathscr{S} \subset A^*$ such that every $\varepsilon \in A^*$ can be expressed as $\eta \zeta^n$ with $\eta \in \mathscr{S}$, $\zeta \in A^*$, where $n := \deg F$. Now if $\mathbf{x} \in \mathcal{M}$ is a solution of (9.1.2), then $F(\mathbf{x}) = \delta \varepsilon$ with $\varepsilon \in A^*$. Hence there are $\eta \in \mathscr{S}$, $\zeta \in A^*$ such that $F(\zeta^{-1}\mathbf{x}) = \delta \eta$. By (ii), each equation $F(\mathbf{y}) = \delta \eta$ ($\eta \in \mathscr{S}$) in $\mathbf{y} \in \mathcal{M}$ with $l(\mathbf{y}) \neq 0$ for $l \in \mathcal{L}$ has only finitely many solutions. This implies (iii).

Conversely, assume (iii), and take again A, $\mathcal{M}$, δ as in (ii). Then the solutions of (9.1.1) lie in finitely many A^*-cosets. If $\mathbf{x}_1$, $\mathbf{x}_2$ are two solutions in the same A^*-coset then $\mathbf{x}_2 = \varepsilon \mathbf{x}_1$ for some $\varepsilon \in A^*$, and $\varepsilon^n = F(\mathbf{x}_2)/F(\mathbf{x}_1) = 1$. So each A^*-coset contains at most n solutions of (9.1.1). This proves (ii). □

9.3.2 Proof of the implication (i) $\Rightarrow$ (iii) in Theorem 9.1.1

We need the following consequence of Theorem 9.2.1.

Proposition 9.3.2 *The solutions of (9.2.1) lie in a union of finitely many proper linear subspaces of K^m.*

Proof. The degenerate solutions, i.e., with $\sum_{i \in I} a_i x_i = 0$ for some non-empty subset I of $\{1, \ldots, m\}$, lie in finitely many subspaces and so do the non-degenerate solutions. □

Remark It was pointed out in Evertse and Győry (1988b), and it is also implicit in the proof of Theorem 6.1.3, that Theorem 9.2.1 and Proposition 9.3.2 are equivalent.

(i) ⇒ (iii). We assume assertion (i) of Theorem 9.1.1 and deduce (iii). Let A, $\mathcal{M}$, F, $\mathcal{L}_0$, $\mathcal{L}$, δ be as in (i), (iii) and let $V := K\mathcal{M}$. We proceed by induction on $\dim_K V$. If $\dim_K V = 1$, assertion (iii) is trivially true. Assume that $\dim_K V =: d \geq 2$, and that the implication (i) $\Rightarrow$ (iii) is true for finitely generated A-modules in K^m that generate a K-vector space of dimension smaller than d. Without loss of generality we assume that none of the linear forms $l \in \mathcal{L}$ vanishes identically on V.

We say that a set of linear forms $\{l_1, \ldots, l_t\} \subset G[X_1, \ldots, X_m]$ is V-linearly dependent if there are $c_1, \ldots, c_t \in G$, not all 0, such that $c_1 l_1 + \cdots + c_t l_t$ vanishes identically on V; otherwise, $\{l_1, \ldots, l_t\}$ is called V-linearly independent. Further, $\{l_1, \ldots, l_t\}$ is said to be minimally V-linearly dependent if the set itself is linearly dependent on V, but each of its non-empty proper subsets is linearly independent on V.

We first show that there is a subset of $\mathcal{L}_0$ of cardinality ≥ 3 that is minimally V-linearly dependent. Assume the contrary. We divide $\mathcal{L}_0$ into classes such that two linear forms belong to the same class if and only if they are V-linearly dependent. Let $\{l_1, \ldots, l_s\}$ be a full set of representatives for these classes. Then by our assumption, $\{l_1, \ldots, l_s\}$ is V-linearly independent. Let $\mathcal{L}_1$ be the class of l_1. As is easily seen, $\mathcal{L}_1$ is $\mathrm{Gal}(G/K)$-proper.

We show that all linear forms in $W := [\mathcal{L}_1] \cap [\mathcal{L}_0 \setminus \mathcal{L}_1]$ vanish identically on V. Let $l \in W$. Since all linear forms in $\mathcal{L}_1$ are V-linearly dependent on l_1 and since each linear form in $\mathcal{L}_0 \setminus \mathcal{L}_1$ is V-linearly dependent on one of $l_2, \ldots, l_s$, there are $c_1, \ldots, c_s \in G$ such that

$$l(\mathbf{x}) = c_1 l_1(\mathbf{x}) = -\sum_{i=2}^{s} c_i l_i(\mathbf{x}) \quad \text{for } \mathbf{x} \in V.$$

But then, $\sum_{i=1}^{s} c_i l_i$ vanishes identically on V, implying that $c_1 = \cdots = c_s = 0$. So l vanishes identically on V.

In the same way it follows that for each $\sigma \in \mathrm{Gal}(G/K)$, the linear forms in $\sigma(W) := [\sigma(\mathcal{L}_1)] \cap [\mathcal{L}_0 \setminus \sigma(\mathcal{L}_1)]$ vanish identically on V, hence so do the linear forms in $\sum_{\sigma \in \mathrm{Gal}(G/K)} \sigma(W)$. But then the latter vector space cannot contain elements of $\mathcal{L}$ since we assumed that these do not vanish identically on V. This violates assumption (i).

So $\mathcal{L}_0$ has a minimal V-linearly dependent subset, say $\{l_0, \ldots, l_t\}$ with $t \geq 2$. This implies that there are $a_1, \ldots, a_t \in G^*$ such that

$$l_0(\mathbf{x}) = a_1 l_1(\mathbf{x}) + \cdots + a_t l_t(\mathbf{x}) \quad \text{for } \mathbf{x} \in V.$$

Let the set $\mathcal{S}$ consist of the coefficients of $l_1, \ldots, l_n$ (i.e., the linear factors of F in $\mathcal{L}_0$), a finite set of generators for $\mathcal{M}$, and $c, c^{-1}, \delta, \delta^{-1}$, where c, δ are as in (9.1.3). Let $B := A[\mathcal{S}]$. Then for any solution $\mathbf{x} \in \mathcal{M}$ of (9.1.2) we have

$$l_1(\mathbf{x}) \in B^*, \ldots, l_n(\mathbf{x}) \in B^*.$$

This shows that if $\mathbf{x} \in \mathcal{M}$ is a solution to (9.1.2), then the tuple

$$\left(\frac{l_1(\mathbf{x})}{l_0(\mathbf{x})}, \ldots, \frac{l_t(\mathbf{x})}{l_0(\mathbf{x})} \right)$$

is a solution to

$$a_1 y_1 + \cdots + a_t y_t = 1 \quad \text{in } y_1, \ldots, y_t \in B^*. \tag{9.3.1}$$

The domain B is finitely generated over $\mathbb{Z}$, so by Proposition 9.3.1, the group B^* is finitely generated. By Proposition 9.3.2, there are at most finitely many non-zero tuples $(b_1, \ldots, b_t) \in G^t$ such that every solution $\mathbf{y} = (y_1, \ldots, y_t)$ of (9.3.1) satisfies one of the relations $b_1 y_1 + \cdots + b_t y_t = 0$. As a consequence, every solution $\mathbf{x} \in \mathcal{M}$ of (9.1.2) satisfies one of the relations

$$b_1 l_1(\mathbf{x}) + \cdots + b_t l_t(\mathbf{x}) = 0.$$

Since $\{l_0, \ldots, l_t\}$ is minimally V-linearly dependent, each of these relations defines a proper linear subspace of V. Hence the solutions $\mathbf{x} \in \mathcal{M}$ of (9.1.2) lie in a finite union of proper linear subspaces of V. By applying the induction hypothesis to the intersection of $\mathcal{M}$ with any of these subspaces (which is a finitely generated A-module since A is a Noetherian ring and $\mathcal{M}$ is a finitely generated A-module), we infer that (9.1.2) has only finitely many solutions. $\square$

9.3.3 Proof of the implication (iii) $\Rightarrow$ (i) in Theorem 9.1.1

We need another consequence of Theorem 9.2.1. It is in fact a special case of the Skolem–Mahler–Lech Theorem on the zero multiplicity of linear recurrence sequences, see Theorem 10.11.1 below.

Proposition 9.3.3 *Let* $a_1, \ldots, a_m, b_1, \ldots, b_m \in K^*$ *and suppose that none of the quotients* b_i / b_j $(1 \le i < j \le m)$ *is a root of unity. Then there are only finitely many* $z \in \mathbb{Z}$ *with*

$$a_1 b_1^z + \cdots + a_m b_m^z = 0. \tag{9.3.2}$$

Proof. We proceed by induction on m. For $m = 2$ the assertion is clear. Let $m \ge 3$. Apply Proposition 9.3.2 with Γ the group generated by $b_1, \ldots, b_m$. By that Proposition, there are a finite number of tuples $(c_1, \ldots, c_{m-1}) \ne \mathbf{0}$ such that each solution of (9.3.2) satisfies one of the relations

$$\sum_{i=1}^{m-1} c_i (b_i/b_m)^z = 0.$$

By the induction hypothesis, each of these relations is satisfied by at most finitely many integers z. $\qquad\square$

(iii) $\Rightarrow$ *(i)*. We assume that assertion (i) of Theorem 9.1.1 does not hold and deduce that (iii) does not hold, that is, there are A, $\mathcal{M}$, δ as in (iii), such that equation (9.1.2) has infinitely many A^*-cosets of solutions.

First assume that $\mathrm{rank}_G \mathcal{L}_0 < m$. Then the vector space of $\mathbf{x} \in G^m$ with $l(\mathbf{x}) = 0$ for $l \in \mathcal{L}$ is non-zero. By Lemma 1.1.1 and since $\mathcal{L}_0$ is $\mathrm{Gal}(G/K)$-stable this vector space has a basis from K^m. So we can choose $\mathbf{x}_1 \in K^m \setminus \{\mathbf{0}\}$ with $l(\mathbf{x}_1) = 0$ for $l \in \mathcal{L}_0$. Choose $\mathbf{x}_0 \in K^m$ with $l(\mathbf{x}_0) \neq 0$ for $l \in \mathcal{L}$. Let A be any subring of K which is finitely generated over $\mathbb{Z}$, $\mathcal{M}$ the A-module generated by $\mathbf{x}_0$, $\mathbf{x}_1$ and $\delta = F(\mathbf{x}_0)$. Consider the vectors $\mathbf{x}_0 + k\mathbf{x}_1$ ($k \in \mathbb{Z}$). These vectors lie in different A^*-cosets since $\mathbf{x}_0$, $\mathbf{x}_1$ are K-linearly independent. For all but finitely many k we have $l(\mathbf{x}_0 + k\mathbf{x}_1) \neq 0$ for $l \in \mathcal{L}$, and by (9.1.3),

$$F(\mathbf{x}_0 + k\mathbf{x}_1) = c \prod_{i=1}^{n} l_i(\mathbf{x}_0 + k\mathbf{x}_1)^{e_i} = F(\mathbf{x}_0) = \delta.$$

Hence (9.1.2) has infinitely many A^*-cosets of solutions.

Now assume that $\mathrm{rank}_G \mathcal{L}_0 = m$. Since by assumption, assertion (i) of Theorem 9.1.1 does not hold, there is a $\mathrm{Gal}(G/K)$-proper subset $\mathcal{L}_1$ of $\mathcal{L}_0$ with $\emptyset \subsetneq \mathcal{L}_1 \subsetneq \mathcal{L}_0$ such that

$$\mathcal{L} \cap W = \emptyset, \quad \text{with } W := \sum_{\sigma \in \mathrm{Gal}(G/K)} [\sigma(\mathcal{L}_1)] \cap [\mathcal{L}_0 \setminus \sigma(\mathcal{L}_1)].$$

Then $\dim_G W < m$. Hence the G-vector space

$$W^* := \{\mathbf{x} \in G^m : l(\mathbf{x}) = 0 \text{ for all } l \in W\}$$

has dimension $m - \dim_G W > 0$. Since also W is $\mathrm{Gal}(G/K)$-stable, we infer from Lemma 1.1.1 that W^* is generated by vectors from K^m. Moreover, none of the linear forms in $\mathcal{L}$ vanishes identically on W^* and so neither do they on $W^* \cap K^m$. Thus, there is $\mathbf{x}_0 \in K^m$ with

$$l(\mathbf{x}_0) = 0 \quad \text{for } l \in W, \quad l(\mathbf{x}_0) \neq 0 \quad \text{for } l \in \mathcal{L}. \tag{9.3.3}$$

We make a partition $\{\mathcal{L}_1, \ldots, \mathcal{L}_t\}$ of $\mathcal{L}_0$ as follows. Take the distinct sets among $\sigma(\mathcal{L}_1)$ ($\sigma \in \mathrm{Gal}(G/K)$). Since $\mathcal{L}_1$ is $\mathrm{Gal}(G/K)$-proper, these sets are pairwise disjoint. Let

$$\mathcal{L}_0^* := \bigcup_{\sigma \in \mathrm{Gal}(G/K)} \sigma(\mathcal{L}_1).$$

If $\mathcal{L}_0^* = \mathcal{L}_0$, let $\mathcal{L}_1, \ldots, \mathcal{L}_t$ be the distinct sets among $\sigma(\mathcal{L}_1)$ ($\sigma \in \mathrm{Gal}(G/K)$). If $\mathcal{L}_0^* \subsetneqq \mathcal{L}_0$, let $\mathcal{L}_1, \ldots, \mathcal{L}_{t-1}$ be the distinct sets among $\sigma(\mathcal{L}_1)$ ($\sigma \in \mathrm{Gal}(G/K)$), and take $\mathcal{L}_t := \mathcal{L}_0 \setminus \mathcal{L}_0^*$. Then $\sigma(\mathcal{L}_t) = \mathcal{L}_t$ for all $\sigma \in \mathrm{Gal}(G/K)$.

Let

$$U := \left\{ \mathbf{u} = (u_l : l \in \mathcal{L}_0) \in G^n : \sum_{l \in \mathcal{L}_0} u_l l = 0 \right\}.$$

We show that

$$\sum_{l \in \mathcal{L}_i} u_l l(\mathbf{x}) = 0 \quad \text{for } \mathbf{x} \in W^*, \mathbf{u} \in U, i = 1, \ldots, t. \tag{9.3.4}$$

Let $\mathbf{u} \in U$, $\mathbf{x} \in W^*$. For $\sigma \in \mathrm{Gal}(G/K)$ we have

$$\sum_{l \in \sigma(\mathcal{L}_1)} u_l l = - \sum_{l \in \mathcal{L}_0 \setminus \sigma(\mathcal{L}_1)} u_l l \in [\sigma(\mathcal{L}_1)] \cap [\mathcal{L}_0 \setminus \sigma(\mathcal{L}_1)] \subseteq W.$$

If $\mathcal{L}_0^* = \mathcal{L}_0$ then (9.3.4) follows at once. If $\mathcal{L}_0^* \subsetneqq \mathcal{L}_0$, then (9.3.4) holds for $i = 1, \ldots, t - 1$. But since $\sum_{l \in \mathcal{L}_0} u_l l(\mathbf{x}) = 0$, it must hold for $i = t$ as well.

We now construct numbers $\theta_l \in G^*$ ($l \in \mathcal{L}_0$) with the following properties:

$$\theta_l = \theta_i \quad \text{for } l \in \mathcal{L}_i, i = 1, \ldots, t, \quad \text{with } \theta_i \text{ independent of } l; \tag{9.3.5}$$

$$\theta_{\sigma(l)} = \sigma(\theta_l) \quad \text{for } l \in \mathcal{L}_0, \ \sigma \in \mathrm{Gal}(G/K); \tag{9.3.6}$$

$$\theta_i / \theta_j \quad \text{is not a root of unity for } 1 \leq i < j \leq t. \tag{9.3.7}$$

The construction is as follows. Define the field M by

$$\mathrm{Gal}(G/M) := \{\sigma \in \mathrm{Gal}(G/K) : \sigma(\mathcal{L}_1) = \mathcal{L}_1\}.$$

We first show that there is θ_1 such that $M = K(\theta_1)$ and no quotient of any two distinct conjugates of θ_1 over K is a root of unity. We start by taking θ with $M = K(\theta)$. Let $\theta^{(1)}, \ldots, \theta^{(d)}$ be the conjugates of θ over K in G. Since the field G is finitely generated, its group of roots of unity is finite, say of order D. Now we may take $\theta_1 := \theta + a$, where $a \in \mathbb{Z}$ is such that the numbers $(\theta^{(i)} + a)^D$ ($i = 1, \ldots, d$) are distinct.

Let $\theta_1 \in M$ be as above and put $\theta_i := \sigma_i(\theta_1)$, where $\sigma_i \in \mathrm{Gal}(G/K)$ is such that $\sigma_i(\mathcal{L}_1) = \mathcal{L}_i$. This does not depend on the choice of σ_i. In the case $\mathcal{L}_0^* \subsetneqq \mathcal{L}_0$, choose $\theta_t \in K^*$ such that θ_t / θ_i is not a root of unity for $i = 1, \ldots, t - 1$. Finally, put $\theta_l := \theta_i$ for $l \in \mathcal{L}_i, i = 1, \ldots, t$. If $\sigma \in \mathrm{Gal}(G/K)$ is such that $\sigma(\mathcal{L}_i) = \mathcal{L}_j$, with $1 \leq i < j \leq t$ if $\mathcal{L}_0^* = \mathcal{L}_0$, and with $1 \leq i < j \leq t - 1$ if $\mathcal{L}_0^* \subsetneqq \mathcal{L}_0$, then

$\sigma_j^{-1}\sigma\sigma_i \in \text{Gal}(G/M)$, hence $\theta_j = \sigma(\theta_i)$. Further, if $\mathcal{L}_0^* \subsetneqq \mathcal{L}_0$, then $\sigma(\mathcal{L}_t) = \mathcal{L}_t$ and $\sigma(\theta_t) = \theta_t$ for $\sigma \in \text{Gal}(G/K)$. Thus, (9.3.5)–(9.3.7) follow.

We now construct A, $\mathcal{M}$, δ such that (9.1.2) has infinitely many A^*-cosets of solutions. Pick $\mathbf{x}_0 \in K^m$ with (9.3.3). We claim that for every $k \in \mathbb{Z}_{\geq 0}$ there is a unique $\mathbf{x}_k \in K^m$ such that

$$l(\mathbf{x}_k) = l(\mathbf{x}_0)\theta_l^k \quad \text{for } l \in \mathcal{L}_0, \tag{9.3.8}$$

and that, moreover, these vectors $\mathbf{x}_k$ are pairwise non-proportional. Indeed, by (9.3.3), (9.3.4) and (9.3.5) we have for any $\mathbf{u} \in U$,

$$\sum_{l \in \mathcal{L}_0} u_l l(\mathbf{x}_0)\theta_l^k = \sum_{i=1}^{t} \left(\sum_{l \in \mathcal{L}_i} u_l l(\mathbf{x}_0) \right) \theta_i^k = 0.$$

Hence there is $\mathbf{x}_k \in G^m$ with (9.3.8). Further, since $\text{rank}_G\,\mathcal{L}_0 = m$, it is uniquely determined. By (9.3.6) we have

$$\sigma(l)(\sigma(\mathbf{x}_k)) = \sigma(l)(\mathbf{x}_0)\theta_{\sigma(l)}^k \quad \text{for } l \in \mathcal{L}_0,\ \sigma \in \text{Gal}(G/K),$$

and then $\sigma(\mathbf{x}_k)$ satisfies (9.3.8) since $\mathcal{L}_0$ is $\text{Gal}(G/K)$-stable. Now since (9.3.8) has only one solution, we must have $\sigma(\mathbf{x}_k) = \mathbf{x}_k$ for $\sigma \in \text{Gal}(G/K)$, hence $\mathbf{x}_k \in K^m$. Finally, by (9.3.7), the tuples $(\theta_l^k : l \in \mathcal{L}_0)$ $(k \in \mathbb{Z}_{\geq 0})$ are pairwise non-proportional. Hence the vectors $\mathbf{x}_k$ $(k \in \mathbb{Z}_{\geq 0})$ are pairwise non-proportional.

Notice that by (9.1.3), (9.3.8) and (9.3.6),

$$F(\mathbf{x}_k) = c \prod_{l \in \mathcal{L}_0} l(\mathbf{x}_k)^{e_l} = F(\mathbf{x}_0)u^k, \quad \text{where } u := \prod_{l \in \mathcal{L}_0} \theta_l^{e_l} \in K^*.$$

We show that $l(\mathbf{x}_k) \neq 0$ for $l \in \mathcal{L}$ and for all but finitely many k. Let $l^* \in \mathcal{L}$. Then since $\text{rank}_G\,\mathcal{L}_0 = m$, we have

$$l^* = \sum_{l \in \mathcal{L}_0} \eta_l l \quad \text{with } \eta_l \in G \quad \text{for } l \in \mathcal{L}_0.$$

So by (9.3.5),

$$l^*(\mathbf{x}_k) = \sum_{i=1}^{t} \left(\sum_{l \in \mathcal{L}_i} \eta_l l(\mathbf{x}_0) \right) \theta_i^k.$$

By $l^*(\mathbf{x}_0) \neq 0$ and Proposition 9.3.3, we have $l^*(\mathbf{x}_k) = 0$ for at most finitely many k. Putting all this together, we infer for all but finitely many $k \in \mathbb{Z}_{\geq 0}$,

$$F(\mathbf{x}_k) = F(\mathbf{x}_0)u^k, \quad l(\mathbf{x}_k) \neq 0 \text{ for } l \in \mathcal{L}. \tag{9.3.9}$$

We finish by constructing A, $\mathcal{M}$, δ. Let $\delta := F(\mathbf{x}_0)$. Then $\delta \neq 0$ since $\mathcal{L}_0 \subsetneq \mathcal{L}$. Further, let $f(X) = X^s + c_{s-1}X^{s-1} + \cdots + c_0 \in K[X]$ be a monic polynomial such that θ_l ($l \in \mathcal{L}_0$) are all zeros of f. Let

$$A := \mathbb{Z}[u, u^{-1}, c_0, \ldots, c_{s-1}].$$

Then $u \in A^*$. Clearly, for $k \in \mathbb{Z}$, $k \geq s$, $l \in \mathcal{L}_0$, we have

$$\theta_l^k = -c_{s-1}\theta_l^{k-1} - \cdots - c_0\theta_l^{k-s},$$

and so, by the fact that $\mathbf{x}_k \in K^m$ is the only solution of (9.3.8),

$$\mathbf{x}_k = -c_{s-1}\mathbf{x}_{k-1} - \cdots - c_0\mathbf{x}_{k-s} \quad \text{for } k \in \mathbb{Z}, k \geq s.$$

Now let $\mathcal{M}$ be the A-module generated by $\mathbf{x}_0, \ldots, \mathbf{x}_{s-1}$. Then $\mathbf{x}_k \in \mathcal{M}$ for $k \in \mathbb{Z}_{\geq 0}$. Invoking (9.3.9), we infer that for all but finitely many k the vector $\mathbf{x}_k$ is a solution to (9.1.2). Moreover, the vectors $\mathbf{x}_k$ are pairwise non-proportional. Hence (9.1.2) has infinitely many distinct A^*-cosets of solutions, i.e., assertion (iii) of Theorem 9.1.1 does not hold. $\qquad\square$

9.4 Finiteness of the number of families of solutions

In this section we describe the structure of the set of solutions of the decomposable form equations (9.1) and (9.2).

Let K be a finitely generated extension field of $\mathbb{Q}$, L a finite extension of K of degree $n \geq 2$ and G a finite, normal extension of K containing L. There are n distinct K-isomorphisms of L in G, $\sigma_1, \ldots, \sigma_n$ say. Let $\alpha_1, \ldots, \alpha_m$ ($m \geq 2$) be elements of L and consider the linear form $l = \alpha_1 X_1 + \cdots + \alpha_m X_m$. Define the conjugates of l, $l^{(i)} = \sigma_i(l) = \sum_{j=1}^{m} \sigma_i(\alpha_j)X_j$ ($i = 1, \ldots, n$). Then

$$N_{L/K}(l) := \prod_{i=1}^{n} l^{(i)} = \prod_{i=1}^{n} (\sigma_i(\alpha_1)$$

is a decomposable form of degree n in $K[X_1, \ldots, X_m]$, called a *norm form*, and the equation

$$N_{L/K}(l(\mathbf{x})) = \delta \quad \text{in } \mathbf{x} = (x_1, \ldots, x_m) \in A^m \tag{9.4.1}$$

is called a *norm form equation* over K, where $\delta \in K^*$ and A is a subring of K which is finitely generated over $\mathbb{Z}$.

In what follows, it will be more convenient to consider equation (9.4.1) in the form

$$N_{L/K}(\mu) = \delta \quad \text{in } \mu \in \mathcal{M}, \tag{9.4.2}$$

where $\mathcal{M} := \{\mu = l(\mathbf{x}) : \mathbf{x} \in A^m\}$. Notice that $\mathcal{M}$ is a finitely generated A-submodule of L. If we assume that $\alpha_1, \ldots, \alpha_m$ are linearly independent over K, there is a one-to-one correspondence between the solutions of (9.4.1) and (9.4.2).

Using the Subspace Theorem and its p-adic generalization, Schmidt (1971, 1972) and Schlickewei (1977c) established very important finiteness theorems on these equations over $\mathbb{Q}$. The results of Schmidt and Schlickewei were later extended in Laurent (1984) to the case where the ground field K is a finitely generated extension of $\mathbb{Q}$. These will be presented later as special cases of more general results concerning decomposable form equations stated below.

Equation (9.4.1) is a special decomposable form equation. Let now $F \in K[X_1, \ldots, X_m]$ be an arbitrary decomposable form of degree $n \geq 2$ and let G be a finite, normal extension of K over which F factorizes into linear factors. Consider the decomposable form equation

$$F(\mathbf{x}) = \delta \quad \text{in } \mathbf{x} = (x_1, \ldots, x_m) \in A^m, \tag{9.1}$$

where A is a subring of K which is finitely generated over $\mathbb{Z}$. We can reformulate this in a shape similar to (9.4.2) as follows. First observe that F can be expressed as

$$F = c \prod_{j=1}^{q} N_{L_j/K}(l_j), \tag{9.4.3}$$

where $L_1, \ldots, L_q$ are finite extensions of K, l_j is a linear form from $L_j[X_1, \ldots, X_m]$ for $j = 1, \ldots, q$, and $c \in K^*$. Indeed, we may write F as

$$F = cl_1 \cdots l_n, \tag{9.4.4}$$

where $c \in K^*$ and $l_j = X_{n_j} + \alpha_{n_j+1,j}X_{n_j+1} + \cdots + \alpha_{mj}X_m$ with $\alpha_{ij} \in G$ for $j = 1, \ldots, n$, $i \in \{n_j + 1, \ldots, m\}$. For each $\sigma \in \text{Gal}(G/K)$ we have $\sigma(F) = F \in K[X_1, \ldots, X_m]$. Since $G[X_1, \ldots, X_m]$ is a unique factorization domain, (9.4.4) implies that there is a permutation $(\sigma(1), \ldots, \sigma(n))$ of $(1, \ldots, n)$ such that $\sigma(l_j) = l_{\sigma(j)}$ for $j = 1, \ldots, n$. The index set $\{1, \ldots, n\}$ can be partitioned into subsets $C_1, \ldots, C_q$ such that i, j belong to the same subset if and only if $\sigma(i) = j$ for some $\sigma \in \text{Gal}(G/K)$. Assume without loss of generality that $j \in C_j$ for $j = 1, \ldots, q$, and let $L_j = K(\alpha_{n_j+1,j}, \ldots, \alpha_{mj})$ for $j = 1, \ldots, q$. Then

$$\prod_{i \in C_j} l_i = N_{L_j/K}(l_j) \quad \text{for } j = 1, \ldots, q,$$

and (9.4.3) follows.

Define the K-algebra

$$\Omega := L_1 \times \cdots \times L_q$$

which is endowed with coordinatewise addition and multiplication. Recall that any K-algebra isomorphic to a direct product of finite field extensions of K is called a *finite étale K-algebra*. Let $\mathbf{1} = (1, \ldots, 1)$ denote the unit element of Ω. We agree that K-subalgebras of Ω contain by default $\mathbf{1}$. It can be shown that any K-subalgebra of Ω is itself a finite étale K-algebra. We view K as a subalgebra of Ω by identifying $a \in K$ with $a \cdot \mathbf{1}$.

We define the norm $N_{\Omega/K}(\boldsymbol{\alpha})$ of $\boldsymbol{\alpha} = (\alpha_1, \ldots, \alpha_q) \in \Omega$ by

$$N_{\Omega/K}(\boldsymbol{\alpha}) = N_{L_1/K}(\alpha_1) \cdots N_{L_q/K}(\alpha_q). \tag{9.4.5}$$

It can be shown that this is the determinant of the K-linear map $\mathbf{x} \mapsto \boldsymbol{\alpha}\mathbf{x}$ from Ω to itself.

The A-module

$$\mathcal{M} := \left\{ \boldsymbol{\mu} = (l_1(\mathbf{x}), \ldots, l_q(\mathbf{x})) : \mathbf{x} \in A^m \right\}$$

is contained in Ω. Replacing δ/c by δ in (9.1), the identities (9.4.3) and (9.4.5) imply that every solution $\mathbf{x}$ of the equation (9.1) yields a solution of the equation

$$N_{\Omega/K}(\boldsymbol{\mu}) = \delta \quad \text{in } \boldsymbol{\mu} \in \mathcal{M}. \tag{9.4.6}$$

Further, if F is of maximal rank, that is, if F has m linearly independent linear factors in its factorization over G, then there is a one-to-one correspondence between the solutions of (9.1) and (9.4.6). For $q = 1$, (9.4.6) reduces to a norm form equation.

In what follows we consider (9.4.6) where we allow $\mathcal{M}$ to be any finitely generated non-zero A-module in Ω. Denote by $K\mathcal{M}$ the vector space generated by $\mathcal{M}$ in Ω. For each K-subalgebra Υ of Ω, denote by A_Υ the integral closure of A in Υ, and by E_Υ the multiplicative subgroup of A_Υ^*, consisting of all elements $\boldsymbol{\varepsilon} \in A_\Upsilon^*$ with $N_{\Omega/K}(\boldsymbol{\varepsilon}) = 1$. The group E_Υ is finitely generated. For every solution $\boldsymbol{\mu}$ of (9.4.6) and every K-subalgebra Υ of Ω for which $\boldsymbol{\mu}\Upsilon \subseteq K\mathcal{M}$, all elements of $(\boldsymbol{\mu}E_\Upsilon^*) \cap \mathcal{M}$ are solutions of (9.4.6). Such a subset of solutions $(\boldsymbol{\mu}E_\Upsilon^*) \cap \mathcal{M}$ is called a *wide $(\mathcal{M}, \Upsilon)$-family of solutions* of (9.4.6).

We state some results of Győry without proof.

Theorem 9.4.1 *The set of solutions of* (9.4.6) *is a union of at most finitely many wide families of solutions of* (9.4.6).

Proof. See Győry (1993a). $\qquad\qquad\qquad\qquad\qquad\qquad\qquad\qquad\qquad\qquad\square$

Consider now the equation

$$N_{\Omega/K}(\mu) \in \delta A^* \quad \text{in } \mu \in \mathcal{M}. \qquad (9.4.7)$$

For any K-subalgebra Υ of Ω, denote by U_Υ the subgroup of A^*_Υ consisting of all elements ε with $N_{\Omega/K}(\varepsilon) \in A^*$. The group U_Υ is finitely generated. Further, for every solution μ of (9.4.7) with $\mu \Upsilon \subseteq K\mathcal{M}$, all elements of $(\mu U_\Upsilon) \cap \mathcal{M}$ are also solutions of (9.4.7). Such a set of solutions is called a *wide* $(\mathcal{M}, \Upsilon)$-*family of solutions* of (9.4.7).

Theorem 9.4.1 easily follows from the following.

Theorem 9.4.2 *The set of solutions of (9.4.7) is a union of finitely many wide families of solutions of (9.4.7).*

Proof. See Győry (1993a). □

The proof of Theorem 9.4.2 depends again on Proposition 9.3.2 concerning the unit equation (9.2.1).

Let V be a non-zero K-linear subspace of Ω. For a K-subalgebra Υ of Ω define

$$V^\Upsilon := \{\mu \in V : \mu \Upsilon \subseteq V\}.$$

We call V *non-degenerate* if $V^\Upsilon = (0)$ for every K-subalgebra Υ of K different from K, and *degenerate* otherwise. Let $\mathcal{M}$ be a finitely generated A-module in Ω with $K\mathcal{M} = V$. If V is non-degenerate, then by Theorem 9.4.1, all solutions of (9.4.6) are contained in a union of finitely many sets of the form $(\mu E_K) \cap \mathcal{M}$. But E_K is finite, hence this proves the implication (i) $\Rightarrow$ (ii) of the following.

Theorem 9.4.3 *Let V be a fixed, non-zero K-linear subspace of Ω. Then the following three statements are equivalent:*

(i) V is non-degenerate;

(ii) for every ring A with quotient field K which is finitely generated over $\mathbb{Z}$, every finitely generated A-module $\mathcal{M}$ with $K\mathcal{M} = V$ and every $\delta \in K^*$, equation (9.4.6) has only finitely many solutions;

(iii) for every A, $\mathcal{M}$, δ as in (ii), equation (9.4.7) has only finitely many A^*-cosets of solutions.

Proof. See Győry (1993a). □

This theorem is equivalent to Theorem 9.1.1 with $\mathcal{L} = \mathcal{L}_0$.

We now specialize the above results to the **norm form equations** (9.4.1) and (9.4.2). Then, in the classical case $K = \mathbb{Q}$, $A = \mathbb{Z}$, Schmidt (1971) proved a fundamental theorem, which states that the norm form equation (9.4.2) has finitely many solutions for all $\delta \in \mathbb{Q}^*$ if and only if the $\mathbb{Q}$-vector space $\mathbb{Q}\mathcal{M}$ has no subspace of the form $\mu L'$, where $\mu \in L^*$ and L' is a subfield of L different from $\mathbb{Q}$ and the imaginary quadratic number fields. The result of Schmidt was generalized by Schlickewei (1977c) for the case $K = \mathbb{Q}$ and A a ring of S-integers. In case of norm form equations, Theorem 9.4.3 as well as Theorem 9.4.1 and Theorem 9.4.2 were proved in Laurent (1984).

In the **number field case**, when in (9.4.1) and (9.4.2) K is an algebraic number field, Schmidt (1972) for $K = \mathbb{Q}$, $A = \mathbb{Z}$, Schlickewei (1977c) for $K = \mathbb{Q}$ and Laurent (1984) for an arbitrary number field K gave a more precise description of the set of solutions of the norm form equations (9.4.1) and (9.4.2), in which the solutions are divided into more restrictive *families of solutions* instead of the wide families of Theorems 9.4.1 and 9.4.2. The next theorem is a generalization of these results to arbitrary decomposable form equations.

Let K be an algebraic number field, S a finite set of places on K containing all infinite places, O_S the ring of S-integers in K, and Ω a finite étale K-algebra. Let $\delta \in K^*$, $\mathcal{M}$ a finitely generated O_S-module contained in Ω, and consider the equation

$$N_{\Omega/K}(\boldsymbol{\mu}) \in \delta O_S^* \quad \text{in } \boldsymbol{\mu} \in \mathcal{M}. \tag{9.4.8}$$

For each K-subalgebra Υ of Ω, denote by $O_{S,\Upsilon}$ the integral closure of O_S in Υ. Further, we define the sets

$$(K\mathcal{M})^\Upsilon := \{\boldsymbol{\mu} \in K\mathcal{M} : \boldsymbol{\mu}\Upsilon \subseteq K\mathcal{M}\}, \quad \mathcal{M}^\Upsilon := (K\mathcal{M})^\Upsilon \cap \mathcal{M},$$

where $K\mathcal{M}$ is the K-vector space in Ω generated by $\mathcal{M}$. Consider the subgroup

$$U_{\mathcal{M},\Upsilon} := \left\{ \boldsymbol{\varepsilon} \in O_{S,\Upsilon}^* : \boldsymbol{\varepsilon}\mathcal{M}^\Upsilon = \mathcal{M}^\Upsilon \right\}$$

of the unit group of $O_{S,\Upsilon}$. The group $O_{S,\Upsilon}^*$ is finitely generated, hence its rank is finite. One can show that $U_{\mathcal{M},\Upsilon}$ is of finite index in $O_{S,\Upsilon}^*$ thus has the same rank as $O_{S,\Upsilon}^*$. An $(\mathcal{M}, \Upsilon)$-*family of solutions* of (9.4.8) is a coset $\mu U_{\mathcal{M},\Upsilon}$, where Υ is a K-subalgebra of Ω and $\mu \in \mathcal{M}^\Upsilon$ is a solution of (9.4.8). Every element of $\mu U_{\mathcal{M},\Upsilon}$ is a solution of (9.4.8).

Theorem 9.4.4 *The set of solution of* (9.4.8) *is a union of finitely many families.*

Proof. See Győry (1993a). □

As was mentioned above, in the case of norm form equations, Theorem 9.4.4 is due to Schmidt (1972) for $K = \mathbb{Q}$, $O_S = \mathbb{Z}$, Schlickewei (1977c) for $K = \mathbb{Q}$ and Laurent (1984) for arbitrary number fields K. Theorem 9.4.4 was deduced from Theorem 9.4.2 by showing that every wide family of solutions splits into finitely many families of solutions. It follows from an observation of Laurent (1984) that Theorem 9.4.4 cannot be extended to the case of an arbitrary finitely generated ground field K.

The proofs of Theorems 9.4.1–9.4.4 in Győry (1993a) are based on Theorem 9.2.1 on unit equations. See also Bombieri and Gubler (2006) where, in the case of norm form equations over $\mathbb{Z}$, the proof of the above Theorem 9.4.4 involves also Theorem 9.2.1. As was explained in Chapter 6, the proof of Theorem 9.2.1 depends on the p-adic Subspace Theorem. We note that in contrast, Schmidt and Schlickewei deduced their results concerning norm form equations directly from the Subspace Theorem and its p-adic generalization.

9.5 Upper bounds for the number of solutions

In this section, we consider decomposable form equations over the ring of S-integers in an algebraic number field. We give an overview of quantitative results, giving explicit upper bounds for the number of solutions. We first recall some history. In Subsection 9.5.1 we recall from Evertse (1995) a general result on systems of S-unit equations with a Galois action, and in Subsection 9.5.2 we deduce, among other things, a quantitative version of Theorem 9.1.1.

Let K be an algebraic number field, S a finite set of places of K containing the infinite places, δ a non-zero element of O_S, and $F \in O_S[X, Y]$ a binary form of degree $n \geq 3$ with at least three pairwise non-proportional linear factors over $\overline{K}$. Consider the equation

$$F(x, y) \in \delta O_S^* \quad \text{in } (x, y) \in (O_S^*)^2. \tag{9.5.1}$$

Denote by s the cardinality of S and by $\omega_S(\delta)$ the number of prime ideals outside S occurring in the factorization of δ. The solutions of (9.5.1) are divided into O_S^*-cosets in the usual manner. Lewis and Mahler (1961) were the first to give, in the case $K = \mathbb{Q}$, a completely explicit upper bound for the number of O_S^*-cosets of solutions of (9.5.1), depending on s, $\omega_S(\delta)$, n, and also on the heights of the coefficients of F. In chapter 6 of his PhD thesis Evertse (1983) extended the result of Lewis and Mahler to arbitrary number fields and sets of places S, and derived an explicit upper bound for the number of O_S^*-cosets of solutions of (9.5.1) that depends only on n, s, $\omega_S(\delta)$, and so is independent of the coefficients of F. On the other hand, Evertse's bound had a much worse dependence on the degree n of F than that of Lewis and Mahler. Later, Evertse's bound was reduced substantially in the case that F is irreducible over K. Bombieri and Schmidt (1987) proved that if $F \in \mathbb{Z}[X, Y]$ is an irreducible binary form of degree $n \geq 3$, then the equation

$$F(x, y) = 1 \quad \text{in } x, y \in \mathbb{Z}$$

has at most cn solutions with c an absolute constant. For n sufficiently large, c can be taken equal to 430. The example $(x - a_1 y) \cdots (x - a_n y) + y^n = 1$ with $a_1, \ldots, a_n$ distinct integers shows that the bound of Bombieri and Schmidt is best possible in terms of n. Bombieri considered more generally (9.5.1) with arbitrary K, S but with F irreducible over K and of degree $n \geq 6$. In Bombieri (1994) he obtained the upper bound $(12n)^{12(s + \omega_S(\delta))}$ for the number of O_S^*-cosets of solutions of (9.5.1). For binary forms $F \in O_S[X, Y]$ of degree $n \geq 3$ irreducible over K, this was improved in Evertse (1997) to $(10^5 n)^{s + \omega_S(\delta)}$. This is still the best bound for general Thue–Mahler equations.

Schmidt generalized the above mentioned results on Thue equations to norm form equations over $\mathbb{Z}$ in more than two unknowns. Let L be a number field of degree n and $l = \alpha_1 X_1 + \cdots + \alpha_m X_m$, where $L = \mathbb{Q}(\alpha_1, \ldots, \alpha_m)$ and $\alpha_1, \ldots, \alpha_m$ are linearly independent over $\mathbb{Q}$. Let c be a non-zero integer such that

$$F := c N_{L/\mathbb{Q}}(l) = c \prod_\sigma (\sigma(\alpha_1) X_1 + \cdots + \sigma(\alpha_n) X_n) \in \mathbb{Z}[X_1, \ldots, X_m],$$

where the product is over all embeddings $\sigma : L \hookrightarrow \overline{\mathbb{Q}}$. Recall that F is called non-degenerate if the $\mathbb{Q}$-vector space $V := \{l(\mathbf{x}) : \mathbf{x} \in \mathbb{Q}^m\}$ does not contain $\mu L'$ for some $\mu \in L^*$ and some subfield L' of L that is not equal to $\mathbb{Q}$ or an imaginary quadratic field. Under this hypothesis, Schmidt (1990), Theorem 1 proved that the equation

$$|F(\mathbf{x})| = 1 \quad \text{in } \mathbf{x} \in \mathbb{Z}^m$$

has at most

$$c_1(m, n) = \min\left(n^{2^{30m} n^2}, n^{c_2(m)}\right) \quad \text{with } c_2(m) = (2m)^{m \times 2^{m+4}}$$

solutions. In the same paper, Schmidt proved also that if δ is any positive integer, then the equation

$$|F(\mathbf{x})| = \delta$$

has at most

$$c_1(m, n) \binom{n}{m-1}^{\omega(\delta)} d_{m-1}(\delta^n)$$

primitive solutions, i.e., with coordinates having greatest common divisor 1, where $\omega(\delta)$ denotes the number of distinct primes dividing δ, and $d_{m-1}(\delta^n)$ denotes the number of ways that δ^n can be expressed as a product of $m - 1$ positive integers. Schmidt's main tool was his quantitative version of the Subspace Theorem that he had established shortly before in Schmidt (1989).

Győry (1993a) gave explicit upper bounds for the number of solutions of arbitrary decomposable form equations over the ring of S-integers of a number field K, in the case that this number is finite. More generally, in the case that the number of solutions is infinite, he gave an explicit upper bound for the number of families of solutions. He derived his bounds by making a reduction to S-unit equations over the splitting field over K of the decomposable form involved and this led to bounds that are exponential in both the cardinality of S and the degree of the splitting field. Notice that if the decomposable form involved has degree n, then in the worst case, its splitting field has degree $n!$ and then Győry's bound is exponential in $n!$.

Evertse (1995) proved a general quantitative result on "Galois symmetric S-unit vectors", and this enabled him to prove much sharper upper bounds for the number of solutions (if finite) of decomposable form equations over O_S. This was extended in Evertse and Győry (1997) to estimates for the number of families of solutions in the case when the number of solutions is infinite.

In the next subsection, we recall, without proof, Evertse's result on Galois symmetric S-unit vectors. In the subsequent subsection we discuss some consequences for decomposable form equations and S-unit equations.

9.5.1 Galois symmetric S-unit vectors

Let K be an algebraic number field, S a finite set of places of K, and G a finite normal extension of K. Denote by $O_{S,G}$ the integral closure of O_S in G.

Let $n \geq 3$ be an integer and Σ an action of $\mathrm{Gal}(G/K)$ on $\{1, \ldots, n\}$, i.e., a homomorphism from $\mathrm{Gal}(G/K)$ to the permutation group of $\{1, \ldots, n\}$. That is, Σ maps $\sigma \in \mathrm{Gal}(G/K)$ to a permutation $(\sigma(1), \ldots, \sigma(n))$ of $(1, \ldots, n)$. We define the K-algebra

$$\Lambda_\Sigma := \left\{ \begin{array}{l} \mathbf{u} = (u_1, \ldots, u_n) \in G^n : \\ \sigma(u_i) = u_{\sigma(i)} \text{ for } \sigma \in \mathrm{Gal}(G/K), i = 1, \ldots, n \end{array} \right\}$$

with coordinatewise addition, multiplication, and scalar multiplication with K. The unit element of Λ_Σ is $\mathbf{1} := (1, \ldots, 1)$. We embed K into Λ_Σ via $\iota : a \mapsto a \cdot \mathbf{1}$.

A Σ-*symmetric* partition is a collection of non-empty, pairwise disjoint sets $\mathcal{P} = \{P_1, \ldots, P_t\}$ such that

$$\bigcup_{i=1}^{t} P_i = \{1, \ldots, n\}, \quad \sigma(P_i) \in \mathcal{P} \text{ for } \sigma \in \mathrm{Gal}(G/K), \ i = 1, \ldots, t.$$

In particular we have the trivial Σ-symmetric partition $\mathcal{P}_0 := \{\{1, \ldots, n\}\}$.

A pair $i \overset{\mathcal{P}}{\sim} j$ is a pair $i, j \in \{1, \ldots, n\}$ belonging to the same set of $\mathcal{P}$. With a Σ-symmetric partition $\mathcal{P}$ we associate the sets

$$\Lambda_{\mathcal{P}} := \left\{ \mathbf{u} \in \Lambda_{\Sigma} : u_i = u_j \text{ for each pair } i \overset{\mathcal{P}}{\sim} j \right\},$$

$$O_{S,\mathcal{P}} := O^n_{S,G} \cap \Lambda_{\mathcal{P}}.$$

The set $\Lambda_{\mathcal{P}}$ is a K-subalgebra of Λ_{Σ}, and $O_{S,\mathcal{P}}$ is the integral closure of O_S in $\Lambda_{\mathcal{P}}$. For instance, $\Lambda_{\mathcal{P}_0} = \iota(K)$, and $\Lambda_{\mathcal{P}} = \Lambda$ for $\mathcal{P} = \{\{1\}, \ldots, \{n\}\}$.

Let W be a K-linear subspace of Λ_{Σ}. Define

$$W^{\perp} := \left\{ \mathbf{y} = (y_1, \ldots, y_n) \in G^n : \sum_{i=1}^n y_i u_i = 0 \text{ for all } \mathbf{u} \in W \right\}.$$

For a Σ-symmetric partition $\mathcal{P} = \{P_1, \ldots, P_t\}$, we define the subspace of W,

$$W_{\mathcal{P}} := \left\{ \mathbf{u} \in W : \sum_{j \in P_i} y_j u_j = 0 \text{ for all } \mathbf{y} \in W^{\perp}, \ i = 1, \ldots, t \right\}.$$

One can show that

$$W_{\mathcal{P}} = \{ \mathbf{u} \in W : \mathbf{u}\Lambda_{\mathcal{P}} \subseteq W \}$$

(see Evertse (1995), Lemma 10). As a consequence, $\Lambda_{\mathcal{P}} W_{\mathcal{P}} \subseteq W_{\mathcal{P}}$.

A K^*-coset is a set $\{ a \cdot \mathbf{u} : a \in K^* \}$ with some fixed $\mathbf{u} \in \Lambda_{\Sigma}$.

We are now ready to state our result.

Theorem 9.5.1 *Let K be a number field, G a finite normal extension of K, $n \geq 3$, Σ a $\mathrm{Gal}(G/K)$-action on $\{1, \ldots, n\}$, W a K-linear subspace of Λ_{Σ} of dimension m and S a finite set of places of K of cardinality s, containing all infinite places. Then the set of $\mathbf{u} = (u_1, \ldots, u_n)$ with*

$$\mathbf{u} \in W, \quad u_1 \cdots u_n \neq 0, \quad u_i / u_j \in O^*_{S,G} \text{ for } i, j = 1, \ldots, n, \quad (9.5.2)$$

$$\mathbf{u} \notin W_{\mathcal{P}} \quad \text{for each } \Sigma\text{-symmetric partition } \mathcal{P}$$
$$\text{such that } O^*_{S,\mathcal{P}} / \iota(O^*_S) \text{ is infinite} \quad (9.5.3)$$

is a union of at most $(2^{33} n^2)^{m^3 s}$ K^*-cosets.

Proof. See Evertse (1995), Theorem 4, Lemma 10. The proof is based on a quantitative version of the Subspace Theorem, proved in Evertse (1996). $\square$

9.5.2 Consequences for decomposable form equations and S-unit equations

Let K be a number field and S a finite set of places of K containing all infinite places. Suppose $|S| = s$. For non-zero $\delta \in O_S$, we denote by $\omega_S(\delta)$ the number of prime ideals outside S occurring in the prime ideal factorization of δ.

Let $F \in O_S[X_1, \ldots, X_m]$ be a decomposable form, and denote by G its splitting field over K. Recall that there exists a $\mathrm{Gal}(G/K)$-stable set $\mathcal{L}_0 = \{l_1, \ldots, l_n\} \subset G[X_1, \ldots, X_m]$ of pairwise non-proportional linear forms, $c \in K^*$ and positive integers $e_1, \ldots, e_n$, such that

$$F = c l_1^{e_1} \cdots l_n^{e_n}.$$

Let $\mathcal{L}$ be a finite set of pairwise non-proportional linear forms from $G[X_1, \ldots, X_m]$ with $\mathcal{L} \supseteq \mathcal{L}_0$.

We deduce a quantitative version of the implication (i) $\Rightarrow$ (iii) of Theorem 9.1.1 from Theorem 9.5.1.

Theorem 9.5.2 *Let m, K, S, F, G, $\mathcal{L}_0$, $\mathcal{L}$ be as above. Assume that*

$$\mathrm{rank}_G \, \mathcal{L}_0 = m, \tag{9.5.4}$$

$$\mathcal{L} \cap \left(\sum_{\sigma \in \mathrm{Gal}(G/K)} [\sigma(\mathcal{L}_1)] \cap [\mathcal{L}_0 \setminus \sigma(\mathcal{L}_1)] \right) \neq \emptyset \tag{9.5.5}$$

for each non-empty $\mathrm{Gal}(G/K)$-proper subset $\mathcal{L}_1 \subsetneqq \mathcal{L}_0$. Then the solutions of

$$F(\mathbf{x}) \in \delta O_S^* \quad in \ \mathbf{x} \in O_S^m \quad with \ l(\mathbf{x}) \neq 0 \ for \ l \in \mathcal{L} \tag{9.5.6}$$

lie in at most $(2^{33} n^2)^{m^3(s + \omega_S(\delta))} \, O_S^*$-cosets.

Proof. Let S' consist of the places in S and of the prime ideals in the factorization of δ. Then $|S'| = s + \omega_S(\delta)$. Assume that (9.5.6) is solvable (if not we are done) and choose a solution $\mathbf{x}_0 \in O_S^m$. After multiplying $l_1, \ldots, l_n$ by suitable scalars, which does not affect the above assumptions on $\mathcal{L}_0$, $\mathcal{L}$, we may assume that $l_i(\mathbf{x}_0) = 1$ for $i = 1, \ldots, n$ and $c \in O_{S'}^*$.

Denote by $O_{S', G}$ the integral closure of $O_{S'}$ in G. We first show that for every solution $\mathbf{x} \in O_S^m$ of (9.5.6) we have

$$l_i(\mathbf{x}) \in O_{S', G}^* \quad \text{for } i = 1, \ldots, n. \tag{9.5.7}$$

This is equivalent to the assertion that $|l_i(\mathbf{x})|_V = 1$ for $i = 1, \ldots, n$ and every place V of G not lying above a place from S'. To prove this, take such a place V. For a polynomial H with coefficients in G denote by $|H|_V$ the maximum of

the $|\cdot|_V$-values of the coefficients of H. By our assumption on $l_1, \ldots, l_n$ we
have $|l_i|_V \geq 1$ for $i = 1, \ldots, n$ while on the other hand, by Proposition 1.9.4
and our assumption $F \in O_S[X_1, \ldots, X_n]$,

$$\prod_{i=1}^{n} |l_i|_V^{e_i} = |F|_V \leq 1.$$

Hence $|l_i|_V = 1$ for $i = 1, \ldots, n$. So if $\mathbf{x} \in O_S^m$ is a solution of (9.5.6) then
$|l_i(\mathbf{x})|_V \leq 1$ for $i = 1, \ldots, n$ and $\prod_{i=1}^{n} |l_i(\mathbf{x})|_V^{e_i} = |F(\mathbf{x})|_V = 1$. This implies
$|l_i(\mathbf{x})|_V = 1$ for $i = 1, \ldots, n$, as required.

Define the K-linear map

$$\varphi : \mathbf{x} \mapsto (l_1(\mathbf{x}), \ldots, l_n(\mathbf{x})) : K^m \to G^n.$$

By (9.5.4), it is injective. Let $\varphi(K^m) =: W$. Since $\{l_1, \ldots, l_n\}$ is $\mathrm{Gal}(G/K)$-
stable, there is an action Σ of $\mathrm{Gal}(G/K)$ on $\{1, \ldots, n\}$ such that $\sigma(l_i) = l_{\sigma(i)}$
for $i = 1, \ldots, n$, $\sigma \in \mathrm{Gal}(G/K)$. This implies that W is an m-dimensional,
K-linear subspace of Λ_Σ.

In view of (9.5.7) we have for every solution $\mathbf{x} \in O_S^m$ of (9.5.6) that

$$\varphi(\mathbf{x}) \in W, \quad \varphi(\mathbf{x}) \in (O_{S',G}^*)^n, \tag{9.5.8}$$

so certainly, $\mathbf{u} := \varphi(\mathbf{x})$ satisfies (9.5.2).

We next show that if $\mathbf{x} \in O_S^m$ is a solution of (9.5.6), then

$$\varphi(\mathbf{x}) \notin W_{\mathcal{P}} \quad \text{for each } \Sigma\text{-symmetric partition } \mathcal{P} \neq \{\{1, \ldots, n\}\}, \tag{9.5.9}$$

which is stronger than (9.5.3). Let $\mathbf{x} \in O_S^m$ be a solution of (9.5.6), and $\mathcal{P} = \{P_1, \ldots, P_t\}$ a Σ-symmetric partition different from $\{\{1, \ldots, n\}\}$. Further, let
$\mathcal{L}_1 = \{l_i : i \in P_1\}$. Then $\mathcal{L}_1 \subsetneqq \mathcal{L}_0$ and $\mathcal{L}_1$ is $\mathrm{Gal}(G/K)$-proper. By assumption
(9.5.5), there is a linear form in

$$\sum_{\sigma \in \mathrm{Gal}(G/K)} [\sigma(\mathcal{L}_1)] \cap [\mathcal{L}_0 \setminus \sigma(\mathcal{L}_1)]$$

that does not vanish at $\mathbf{x}$. This implies that there are $\sigma \in \mathrm{Gal}(G/K)$ and $l \in [\sigma(\mathcal{L}_1)] \cap [\mathcal{L}_0 \setminus \sigma(\mathcal{L}_1)]$ such that $l(\mathbf{x}) \neq 0$. There is a set $P_i \in \mathcal{P}$ such that
$\sigma(\mathcal{L}_1) = \{l_j : j \in P_i\}$. Now there are $c_j \in G$ for $j = 1, \ldots, n$ such that

$$l = \sum_{j \in P_i} c_j l_j = -\sum_{j \in P_i^c} c_j l_j,$$

where $P_i^c = \{1, \ldots, n\} \setminus P_i$. The vector $(c_1, \ldots, c_n)$ belongs to $W^\perp$, and our
observation $l(\mathbf{x}) \neq 0$ implies that for the vector $\mathbf{u} = \varphi(\mathbf{x})$ we have $\sum_{j \in P_i} c_j u_j \neq 0$. So indeed, $\varphi(\mathbf{x}) \notin W_{\mathcal{P}}$.

We conclude that if $\mathbf{x} \in O_S^m$ is a solution of (9.5.6), then $\varphi(\mathbf{x})$ satisfies (9.5.8), (9.5.9), hence (9.5.2), (9.5.3) with S' instead of S. Now Theorem 9.5.1 with $S', s' = s + \omega_S(\delta)$, instead of S, s, implies that the vectors $\varphi(\mathbf{x})$, with $\mathbf{x} \in O_S^m$ a solution of (9.5.2), lie in at most $N := (2^{33} n^2)^{m^3(s+\omega_S(\delta))} K^*$-cosets. Since φ is an injective, K-linear map, this implies that the solutions $\mathbf{x}$ themselves lie in at most N K^*-cosets. But, clearly, any two solutions of (9.5.6) in the same K^*-coset lie in fact in the same O_S^*-coset. Theorem 9.5.2 follows. $\qquad\square$

The next consequence is an improvement of Theorem 6.1.3 in the case $\Gamma = (O_S^*)^m$.

Theorem 9.5.3 *Let K, S be as above, and let $a_1, \ldots, a_m \in K^*$. Then the equation*

$$a_1 u_1 + \cdots + a_m u_m = 1 \quad in \; u_1, \ldots, u_m \in O_S^* \qquad (9.5.10)$$

has at most $(2^{35} m^2)^{m^3 s}$ solutions with

$$\sum_{i \in I} a_i u_i \neq 0 \quad for \; each \; non\text{-}empty \; I \subseteq \{1, \ldots, m\}. \qquad (9.5.11)$$

Proof. We apply Theorem 9.5.1 with $n = m + 1$, $G = K$, and

$$W = \{(u_1, \ldots, u_m, u_{m+1}) \in K^{m+1} : a_1 u_1 + \cdots + a_m u_m = u_{m+1}\}.$$

Then the points $(u_1, \ldots, u_m, 1)$ with (9.5.10) and (9.5.11) satisfy (9.5.2) and (9.5.3). Since these points lie in different K^*-cosets, Theorem 9.5.3 follows. $\qquad\square$

We state without proof a consequence of Theorem 9.5.1 for the number of families of solutions of decomposable form equations, giving a quantitative version of Theorem 9.4.4. We keep the notation from Section 9.4.

Let as before K be an algebraic number field, S a finite set of places on K containing all infinite places and Ω a finite étale K-algebra. Let c, δ be non-zero elements of O_S, $\mathcal{M}$ a finitely generated O_S-module contained in Ω, and consider the equation

$$c N_{\Omega/K}(\boldsymbol{\mu}) \in \delta O_S^* \quad in \; \boldsymbol{\mu} \in \mathcal{M}. \qquad (9.5.12)$$

We assume that for some O_S-module generating set $\{\boldsymbol{\alpha}_1, \ldots, \boldsymbol{\alpha}_t\}$ of $\mathcal{M}$, the polynomial $c N_{\Omega/K}(X_1 \boldsymbol{\alpha}_1 + \cdots + X_t \boldsymbol{\alpha}_t)$ has its coefficients in O_S. In fact, this does not depend on the choice of the generating set.

For the definition of the submodules $\mathcal{M}^\Upsilon$ and the groups $U_{\mathcal{M},\Upsilon}$ (for Υ a K-subalgebra of Ω) and that of a family of solutions of (9.5.12) we refer to Section 9.4. Recall that $U_{\mathcal{M},\Upsilon}$ is a subgroup of $O_{S,\Upsilon}^*$ of finite index if $\mathcal{M}^\Upsilon \neq (0)$. A family of solutions of (9.5.12) is called *irreducible* if it is not a union of finitely many strictly smaller families of solutions.

Let $n := [\Omega : K]$, $m := \dim_K K\mathcal{M}$, $s := |S|$, and put

$$\psi(\delta) := \binom{n}{m-1}^{\omega_S(\delta)} \prod_{v \notin S} \binom{\mathrm{ord}_v(\delta) + m - 1}{m - 1},$$

where $\mathrm{ord}_v(\delta)$ is the exponent on the prime ideal corresponding to v in the factorization of δ. Consider the K-subalgebras Υ of Ω such that (9.5.12) has irreducible $(\mathcal{M}, \Upsilon)$-families of solutions, and denote by $I_\mathcal{M}$ the maximum of the indices $[O^*_{S,\Upsilon} : U_{\mathcal{M},\Upsilon}]$, taken over all such algebras Υ.

We state without proof the following quantitative result, which can be deduced from Theorem 9.5.1.

Theorem 9.5.4 *The set of solutions of* (9.5.12) *is a union of at most*

$$(2^{33} n^2)^{m^3 s} \psi(\delta) \cdot I_\mathcal{M}$$

irreducible families.

Proof. This is a simplified version of Evertse and Győry (1997), Theorem 1. □

Notice that by taking for Ω a finite extension field of K, we obtain from Theorem 9.5.4 an upper bound for the number of families of solutions of a norm form equation.

By an O^*_S-coset of solutions, we mean a coset μO^*_S, where μ is a solution of (9.5.12).

Corollary 9.5.5 *Assume that* (9.5.12) *has only finitely many O^*_S-cosets of solutions. Then the number of these is at most*

$$(2^{33} n^2)^{m^3 s} \psi(\delta).$$

Proof. By assumption, (9.5.12) cannot have irreducible families of solutions that are the union of infinitely many O^*_S-cosets. So it has only irreducible families that are the union of only finitely many O^*_S-cosets, and such families must be O^*_S-cosets themselves. In this situation, $I_\mathcal{M} = 1$. Corollary 9.5.5 follows. □

We can express the set of solutions of (9.5.12) as a minimal finite union of irreducible families $\mathcal{F}_1 \cup \cdots \cup \mathcal{F}_t$, i.e., none of the families in this union is contained in the union of the others. Evertse and Győry (1997) showed that this way of expressing the set of solutions is unique, and moreover, that $\mathcal{F}_1, \ldots, \mathcal{F}_t$ are precisely the *maximal* irreducible families of solutions of (9.5.2), that is, if $\mathcal{F}$ is any other irreducible family of solutions of (9.5.12), then $\mathcal{F} \subseteq \mathcal{F}_i$ for some $i \in \{1, \ldots, t\}$.

Voutier (2014) showed that if L is an algebraic number field of degree $n > 3$ and $\mathcal{M}$ a free $\mathbb{Z}$-module of rank 3 contained in O_L, then the norm form equation

$$N_{L/\mathbb{Q}}(\boldsymbol{\mu}) = 1 \quad \text{in } \boldsymbol{\mu} \in \mathcal{M} \tag{9.5.13}$$

has at most $10^{969}n^{10}$ families of solutions. On the other hand, in his paper, Voutier showed that for every number field L of degree $n \geq 3$ and every integer $N > 0$, there exists a full module $\mathcal{M} \subseteq O_L$, i.e., of rank equal to n, such that (9.5.13) has at least N families of solutions. This implies that the bound in Theorem 9.5.4 cannot be replaced by one depending only on m, n, s, δ and independent of $\mathcal{M}$.

9.6 Effective results

In this section, effective results are presented for some important classes of decomposable form equations of the form

$$F(\mathbf{x}) = \delta \quad \text{in } \mathbf{x} = (x_1, \ldots, x_m) \in O_S^m \text{ with } l(\mathbf{x}) \neq 0 \text{ for } l \in \mathcal{L} \tag{9.6.1}$$

and

$$F(\mathbf{x}) \in \delta O_S^* \quad \text{in } \mathbf{x} = (x_1, \ldots, x_m) \in O_S^m \text{ with } l(\mathbf{x}) \neq 0 \text{ for } l \in \mathcal{L}, \tag{9.6.2}$$

where O_S is the ring of S-integers of a number field K, $\delta \in O_S \setminus \{0\}$, $F(\mathbf{X})$ is a decomposable form of degree $n \geq 3$ with coefficients in O_S and $\mathcal{L}$ is a finite set of non-zero linear forms from $\overline{K}[X_1, \ldots, X_m]$. Using the effective results of Section 4.1 on S-unit equations, we derive effective bounds for the S-integral solutions of Thue equations, discriminant equations, certain norm form equations and decomposable form equations of an arbitrary number of unknowns. In the case of equation (9.6.1), these imply the finiteness of the number of solutions, and make it possible, at least in principle, to determine the solutions, provided that K, S, δ, n and the coefficients of F are given effectively in the sense described in Section 1.10. The results presented in this section have many important applications in number theory.

As was already mentioned, equation (9.6.2) can be reduced to finitely many equations of the form (9.6.1). This can be carried out in an effective way. Indeed, let $\mathbf{x}$ be a solution of (9.6.2). Then $F(\mathbf{x}) = \delta\eta$ with some $\eta \in O_S^*$. By Proposition 4.3.12 there is an $\varepsilon \in O_S^*$ for which $h(\eta\varepsilon^n)$ and hence $h(\delta\eta\varepsilon^n)$ are effectively bounded. Further, $\varepsilon\mathbf{x}$ is a solution of equation (9.6.1) with δ replaced by $\delta\eta\varepsilon^n$. In what follows, we deal only with equation (9.6.1).

Further effective applications of S-unit equations to discriminant form and index form equations and related Diophantine problems are given in our next book on discriminant equations.

9.6.1 Thue equations

Let K be an algebraic number field and S a finite set of places of K, containing all infinite places. Let $F \in O_S[X, Y]$ be a binary form of degree $n \geq 3$ having at least three pairwise non-proportional linear factors over $\overline{K}$, and let $\delta \in O_S \setminus \{0\}$. Consider the Thue equation

$$F(x, y) = \delta \quad \text{in } x, y \in O_S. \tag{9.6.3}$$

In the classical case when $K = \mathbb{Q}$, $S = \{\infty\}$ and $F(X, Y)$ is irreducible over $\mathbb{Q}$, the first explicit upper bound for the solutions of this equation was obtained in Baker (1968a). His bound depends only on δ, n, and the maximum of the absolute values of the coefficients of F. Baker's proof is based on his effective estimates for linear forms in logarithms of algebraic numbers. Baker's result was extended in Coates (1969) to the case when $K = \mathbb{Q}$ and S is arbitrary and in Kotov and Sprindžuk (1973) to the case of equation (9.6.3). Later, several improvements and generalizations have been established; for references see the Notes (Section 9.7). We note that better bounds can be obtained for the solutions if certain parameters of the number field generated by one or more zeros of $F(x, 1)$ are also involved.

For applications, we give completely explicit upper bounds for the solutions of equation (9.6.3). Let d, h_K and R_K denote the degree, class number and regulator of K. Further, let $s = |S|$, R_S the S-regulator of K (see (1.8.2)),

$$P_K := \max_{\mathfrak{p}} N(\mathfrak{p}) \text{ if } S \supsetneqq M_K^\infty \quad \text{and} \quad P_K := 2 \text{ if } S = M_K^\infty,$$

and

$$Q_K := \prod_{\mathfrak{p}} N(\mathfrak{p}) \text{ if } S \supsetneqq M_K^\infty \quad \text{and} \quad Q_K := 1 \text{ if } S = M_K^\infty,$$

where the maximum and product are taken over all prime ideals $\mathfrak{p}$ from S, and $N(\mathfrak{p}) := |O_K/\mathfrak{p}|$ denotes the norm of $\mathfrak{p}$. The case $s = 1$ being trivial, we assume that $s \geq 2$. Finally, let $H(\geq 2)$ be an upper bound for the maximum of the logarithmic heights of the coefficients of $F(X, Y)$.

The next theorem is a slightly weaker version of Corollary 3 of Győry and Yu (2006).

Theorem 9.6.1 *Suppose that the binary form $F(X, Y)$ in (9.6.3) factorizes over K into linear factors and that at least three of these factors are pairwise non-proportional. Then all solutions x, y of (9.6.3) satisfy*

$$\max(h(x), h(y)) < \frac{1}{n} h(\delta) + 7n(64eds)^{2s+5} P_K \mathcal{N}_K R_S(\log^* R_S), \tag{9.6.4}$$

where

$$\mathcal{N}_K = n^5 H + \frac{1}{d} \log N_S(\delta) + d^d R_K + \frac{h_K}{d} \log Q_K.$$

The proof is based on Corollary 4.1.5 on S-unit equations.

Consider now the case when $F(X, Y)$ does not factorize over K into linear forms. For later convenience, we assume that $F(1, 0) \neq 0$. Then we may assume that three zeros, say $\alpha_1, \alpha_2, \alpha_3$, of $F(X, 1)$ are distinct, and α_1 is not contained in K. Let $L = K(\alpha_1)$, h_L and R_L be the class number and regulator of L, T the set of places of L lying above those of S, and R_T the T-regulator of L. Further, let $M = K(\alpha_1, \alpha_2, \alpha_3)$ and $P_M = P_K^{[M:K]}$ if $S \supsetneq M_K^\infty$ and $P_M = 2$ if $S = M_K^\infty$.

Theorem 9.6.2 *Let $F(X, Y)$ be a binary form as in (9.6.3). Suppose that $F(1, 0) \neq 0$, that $\alpha_1, \alpha_2, \alpha_3$ are distinct zeros of $F(X, 1)$, and that α_1 is not contained in K. Then, with the above notation, all solutions of (9.6.3) satisfy*

$$\max(h(x), h(y)) < \frac{1}{n} h(\delta) + 17(64edn^2s)^{2ns+5} P_M \mathcal{N}_L R_T (\log^* R_T), \quad (9.6.5)$$

where

$$\mathcal{N}_L = n^2 H + \frac{1}{d} \log N_S(\delta) + (nd)^{nd} R_L + \frac{h_L}{d} \log Q_K.$$

In particular, if $K = \mathbb{Q}$, $S = \{\infty\}$ and $F(X, Y)$ is irreducible over $\mathbb{Q}$, then

$$\max(|x|, |y|) < \exp\left\{c(H + \log|\delta| + n^n R_L) R_L (\log^* R_L)\right\}$$

where $c = 34(64en^2)^{2n+6}$.

Apart from the value of c, this latter bound was established in Bugeaud and Győry (1996b). Combining this bound with (1.5.2) and Lemma 1.5.1, one gets at once an upper bound that depends only on δ, n and H.

Using their methods mentioned in Section 4.5, Bombieri (1993) in the case $S = M_K^\infty$, $F(X, 1)$ monic, and Bugeaud (1998) in the case $F(X, 1)$ monic and irreducible over K derived similar bounds for the solutions of equation (9.6.3). Theorem 9.6.2 is a generalization and, apart from the factor $\log^* R_T$ in (9.6.5), is an improvement of these results of Bombieri and Bugeaud.

Remark The restriction $F(1, 0) \neq 0$ is not an essential one. Indeed, there is an $a \in \mathbb{Z}$ with $1 \leq a \leq n$ such that $F(1, a) \neq 0$. Then one may take the binary form $G(X, Y) = F(X, aX + Y)$ instead of $F(X, Y)$, in which the coefficient of X^n is $F(1, a) \neq 0$ and the logarithmic heights of the coefficients of G do not exceed $(n + 1)(H + n \log n) + \log(n + 1)$.

For convenience, we give a common proof for Theorems 9.6.1 and 9.6.2. We first prove Theorem 9.6.1 by means of Corollary 4.1.5. Proving a version of

Theorem 9.6.2, we could also use this corollary in the field $M = K(\alpha_1, \alpha_2, \alpha_3)$ with the set of places V consisting of the set of places of M lying above the places of S. However, we get a better bound by applying Theorem 4.1.3, where one of the unknowns of the V-unit equation involved belongs to a finitely generated subgroup Γ of M^* which is much smaller than the group of V-units in M.

Proof of Theorems 9.6.1 and 9.6.2. We shall use some basic facts from Chapter 1 without any further mention.

In view of the above remark we may assume that $F(1, 0) \neq 0$ holds in Theorem 9.6.1, too. Then, in the proof below of Theorem 9.6.1, one has to work with $H_1 = (n + 1)(H + n \log n) + \log(n + 1)$ instead of H. Further, we may assume that in both cases $\alpha_1, \alpha_2, \alpha_3$ are distinct zeros of $F(X, 1)$ (in the latter case, not necessarily in K). For $i = 1, 2, 3$, let $L_i := K(\alpha_i)$ with $L_1 = L$, h_{L_i}, R_{L_i} the class number and regulator of L_i, T_i with $T_1 = T$ the set of places of L_i lying above those in S, $O_{T_i}, O_{T_i}^*$ the ring of T_i-integers and the group of T_i-units in L_i, and

$$Q_{L_i} = \prod_{\mathfrak{P}} N_{L_i}(\mathfrak{P}) \text{ if } S \supsetneq M_K^\infty \quad \text{and} \quad Q_{L_i} = 1 \text{ if } S = M_K^\infty,$$

where the product is taken over all prime ideals $\mathfrak{P}$ from T_i and $N_{L_i}(\mathfrak{P}) := |O_{L_i}/\mathfrak{P}|$ is the absolute norm of $\mathfrak{P}$.

Let x, y be a solution of (9.6.3), and let $a_0 = F(1, 0)$. The number $a_0 \alpha_i$ is integral over O_S, and so it is in O_{T_i} for $i = 1, 2, 3$. Thus $a_0(x - \alpha_i y)$ is also in O_{T_i}, it divides $a_0^{n-1} F(x, y)$ and hence $a_0^{n-1} \delta$ in O_{T_i}, $i = 1, 2, 3$. By Proposition 4.3.12 there is an ε_i in $O_{T_i}^*$ such that, putting $\delta_i = \varepsilon_i a_0(x - \alpha_i y)$ and using the fact that $N_S(a_0) \leq dh(a_0) \leq dH$, we have

$$h(\delta_i) \leq \frac{1}{[L_i : \mathbb{Q}]} \log N_{T_i}(a_0(x - \alpha_i y)) + 300 R_{L_i} \left(\frac{nd}{2}\right)^{nd} + \frac{h_{L_i}}{[L_i : \mathbb{Q}]} \log Q_{L_i}$$

$$\leq (n - 1)H + \frac{1}{d} \log N_S(\delta) + 300 R_{L_i} \left(\frac{nd}{2}\right)^{nd} + \frac{h_{L_i}}{d} \log Q_K$$

$$=: A_i \quad \text{for } i = 1, 2, 3. \tag{9.6.6}$$

Substituting $x - \alpha_i y = \delta_i/\varepsilon_i$ into the identity

$$(\alpha_3 - \alpha_2)(x - \alpha_1 y) + (\alpha_2 - \alpha_1)(x - \alpha_3 y) + (\alpha_1 - \alpha_3)(x - \alpha_2 y) = 0,$$

we infer that

$$\tau \frac{\varepsilon_2}{\varepsilon_1} + \rho \frac{\varepsilon_2}{\varepsilon_3} = 1, \tag{9.6.7}$$

where

$$\tau = \frac{\alpha_3 - \alpha_2}{\alpha_3 - \alpha_1} \cdot \frac{\delta_1}{\delta_2}, \quad \rho = \frac{\alpha_2 - \alpha_1}{\alpha_3 - \alpha_1} \cdot \frac{\delta_3}{\delta_2}. \tag{9.6.8}$$

We shall give an upper bound for $h(\varepsilon_2/\varepsilon_1)$. First we must derive an upper bound for $h(\tau)$ and $h(\rho)$. The numbers $a_0\alpha_i$ are zeros of the monic polynomial $F'(X) := a_0^{n-1} F(X/a_0, 1)$. The maximum of the logarithmic heights of the coefficients of F' is at most nH. Then Corollary 1.9.6 and (1.9.6) give

$$h(a_0\alpha_i) \le n^2 H + n \log 2 =: A_4 \quad \text{for } i = 1, 2, 3,$$

whence, using (9.6.6) and (9.6.8), it follows that

$$\max(h(\tau), h(\rho)) < 4A_4 + 2\log 2 + 2 \max_{1 \le i \le 3} A_i =: A_5. \tag{9.6.9}$$

We first prove Theorem 9.6.1 when $\alpha_1, \alpha_2, \alpha_3$ are in K. Then we must take H_1 in place of H. Further, $\delta_i \in O_S$, $\varepsilon_i \in O_S^*$ and, instead of (9.6.6), we get

$$h(\delta_i) < (n-1)H_1 + \frac{1}{d} \log N_S(\delta) + 300 R_K \left(\frac{d}{2}\right)^d + \frac{h_K}{d} \log Q_K$$
$$=: A_6 \text{ for } i = 1, 2, 3. \tag{9.6.10}$$

In this case it follows as in (9.6.9) that

$$\max(h(\tau), h(\rho)) < 4A_4 + 2\log 2 + 2A_6 =: A_7$$

with H_1 instead of H in A_4. Since $\varepsilon_2/\varepsilon_1$, $\varepsilon_2/\varepsilon_3$ are S-units in K, we can apply Corollary 4.1.5 to the S-unit equation (9.6.7) and we get

$$h(\varepsilon_2/\varepsilon_1) < 6.5 c_1 c_2 (P_K / \log P_K) A_7 R_S \max(\log(c_1 P_K), \log^*(c_2 R_S)),$$

where $c_1 = 11\lambda s^2 (\log^* s)(16ed)^{3s+2}$ with $\lambda = 12$ if $s = 2$, $\lambda = 1$ if $s \ge 3$, and $c_2 = ((s-1)!)^2/(2^{s-2}d^{s-1})$. But we have $m!e^m/m^m \le e\sqrt{m}$ for any integer $m \ge 1$. Hence after some computation and simplification we obtain

$$h(\varepsilon_2/\varepsilon_1) < 1.9(64eds)^{2s+5}(P_K / \log P_K)\mathcal{N}_K R_S \max(\log P_K, \log^* R_S)$$
$$=: A_8, \tag{9.6.11}$$

where

$$\mathcal{N}_K := n^5 H + \frac{1}{d} \log N_S(\delta) + d^d R_K + \frac{h_K}{d} \log Q_K.$$

We now give an upper bound for $h(x/y)$. Put $\kappa := (x - \alpha_1 y)/(x - \alpha_2 y)$. Then $\kappa = (\delta_1/\delta_2)(\varepsilon_2/\varepsilon_1)$ and, by (9.6.10) and (9.6.11), we infer that $h(\kappa) < 2A_6 + A_8 \le 1.1A_8$. But $x/y = (\kappa\alpha_2 - \alpha_1)/(\kappa - 1)$, hence we get $h(x/y) < 3.3A_8$. Finally, using $y^n F(x/y, 1) = \delta$, we get (9.6.4) for $h(y)$. The bound for $h(x)$ follows in the same way.

Next we prove Theorem 9.6.2. Then α_1 is not contained in K and we may assume that $\alpha_2 = \sigma(\alpha_1)$ with some K-isomorphism σ. We recall that $M = K(\alpha_1, \alpha_2, \alpha_3)$. Let V be the set of places of M lying above the places of S, and O_V, O_V^* the ring of V-integers and group of V-units in M. By Proposition 4.3.9 there exists in L a fundamental system $\{\xi_1, \ldots, \xi_{t-1}\}$ of T-units such that

$$\prod_{j=1}^{t-1} h(\xi_j) \le c_3 R_T,$$

where $t = |T| \le sn$ and $c_3 = ((t-1)!)^2/2^{t-2}[L : \mathbb{Q}]^{t-1}$. Denote by Γ the subgroup of O_V^*, generated by $\sigma(\xi_1)/\xi_1, \ldots, \sigma(\xi_{t-1})/\xi_{t-1}$. In this situation ε_2, δ_2 above can be chosen so that $\varepsilon_2 = \sigma(\varepsilon_1)$ and $\delta_2 = \sigma(\delta_1)$. Then, in the equation (9.6.7), $\varepsilon_2/\varepsilon_1 \in \Gamma$ and $\varepsilon_2/\varepsilon_3 \in O_V^*$. We apply now Theorem 4.1.3 to the equation (9.6.7) under these conditions.

Set

$$\Theta := \prod_{j=1}^{t-1} h(\sigma(\xi_j)/\xi_j).$$

Then by Theorem 4.1.3 we have

$$h(\varepsilon_2/\varepsilon_1) < 6.5c_4 v(P_M/\log P_M)\Theta A_5 \max(\log(c_4 v P_M), \log^* \Theta),$$

where $v = |V|$ and

$$c_4 = 11\lambda t(\log^* t)(16e[M : \mathbb{Q}])^{3t+2} \text{ with } \lambda = 12 \text{ if } t = 1, \lambda = 1 \text{ if } t \ge 2.$$

Further,

$$[M : \mathbb{Q}] \le dn(n-1)(n-2), \quad v \le sn(n-1)(n-2),$$

$t \le sn$, and it follows that

$$\Theta \le 2^{t-1} \prod_{j=1}^{t-1} h(\xi_j) \le 2^{sn} c_3 R_T.$$

Using these inequalities and simplifying the bound so obtained for $h(\varepsilon_2/\varepsilon_1)$ we infer as above in the proof of Theorem 9.6.1 that

$$h(\varepsilon_2/\varepsilon_1) < 5.1(64edn^2s)^{2sn+4.5} P_M \mathcal{N}_L R_T(\log^* R_T),$$

where

$$\mathcal{N}_L = n^2 H + \frac{1}{d} \log N_S(\delta) + (nd)^{nd} R_L + \frac{h_L}{d} \log Q_K.$$

Finally, we can derive the bound in (9.6.5) $h(x)$ and $h(y)$ as at the end of the proof of Theorem 9.6.1. $\square$

9.6.2 Decomposable form equations in an arbitrary number of unknowns

Let again K be an algebraic number field, and S a finite set of places of K containing all infinite places. Consider now the general decomposable form equation

$$F(\mathbf{x}) = \delta \text{ in } \mathbf{x} = (x_1, \ldots, x_m) \in O_S^m \text{ with } l(\mathbf{x}) \neq 0 \text{ for } l \in \mathcal{L}, \quad (9.6.1)$$

where O_S is the ring of S-integers of K, $\delta \in O_S \setminus \{0\}$, $F \in O_S[X_1, \ldots, X_m]$ is a decomposable form of degree $n \geq 3$ and $\mathcal{L}$ is a finite set of non-zero linear forms from $\overline{K}[X_1, \ldots, X_m]$. In this subsection we prove effective finiteness results for some important classes of equations of the form (9.6.1), including discriminant form equations and certain norm form equations. In case of discriminant form equations and norm form equations in an arbitrary number of unknowns the first effective results were established in Győry (1976) and Győry and Papp (1978), respectively.

The arguments of Section 9.3.2 show that equation (9.6.1) leads to systems of unit equations in some finite extension of K. There are no general effective results for unit equations in more than two unknowns, hence one cannot obtain for (9.6.1) effective theorems in full generality. However, it will be seen that if the linear factors of F possess appropriate connectedness properties, then one can arrive at systems of unit equations consisting of equations in two unknowns in which the equations have similar connectedness properties. Then one can apply the effective results from Chapter 4 to the solutions of the arising unit equations, and using the connectedness properties of these equations, one can derive an effective upper bound for the heights of the solutions of (9.6.1). For simplicity, we shall give the bounds explicitly in terms of S only. For completely explicit bounds, we shall refer to some original papers.

Extending the ground field K if necessary, we may assume that in (9.6.1) F factorizes into linear forms over K. These linear factors of F are uniquely determined over K up to proportional factors from K^*. Fix a factorization of F into linear forms $l_1, \ldots, l_n$, and denote by $\mathcal{L}_0$ a maximal subset of pairwise linearly independent linear factors of F. To obtain effective finiteness results on equation (9.6.1), we make some assumptions on $\mathcal{L}_0$.

We denote by $\mathcal{G}(\mathcal{L}_0)$ the graph with vertex set $\mathcal{L}_0$ in which the edges are the unordered pairs $\{l, l'\}$, where l, l' are distinct elements of $\mathcal{L}_0$ with the property that there exists a third linear form $l'' \in \mathcal{L}_0$ that is a K-linear combination of l, l'. If $\mathcal{L}_0$ has at least three elements and $\mathcal{G}(\mathcal{L}_0)$ is connected, then F is said to be *triangularly connected*. In this case one can reduce equation (9.6.1) to a so-called triangularly connected system of unit equations in two unknowns, and, as a consequence, can give an effective upper bound for the heights of the solutions of (9.6.1). The first effective result of this type was obtained in Győry and Papp (1978) for $S = M_K^\infty$, and in Győry (1978/1979, 1980a) for arbitrary S. When $\mathcal{G}(\mathcal{L}_0)$ is not connected, let $\mathcal{L}_{01}, \ldots, \mathcal{L}_{0k}$ denote the vertex sets of the connected components of $\mathcal{G}(\mathcal{L}_0)$. If $k > 1$, we introduce the graph $\mathcal{H}(\mathcal{L}_{01}, \ldots, \mathcal{L}_{0k})$ with vertex set $\{\mathcal{L}_{01}, \ldots, \mathcal{L}_{0k}\}$, in which the pair $\{\mathcal{L}_{0i}, \mathcal{L}_{0j}\}$ is an edge if there exists a non-zero linear form l_{ij} which can be expressed simultaneously as a K-linear combination of the forms in $\mathcal{L}_{0i}$ and in $\mathcal{L}_{0j}$. In this case l_{ij} can be chosen so that the total number of non-zero terms in both representations $l_{ij} = \sum_{l \in \mathcal{L}_{0i}} \lambda_l \cdot l = \sum_{l \in \mathcal{L}_{0j}} \lambda'_l \cdot l$ is minimal. We pick for each edge $\{\mathcal{L}_{0i}, \mathcal{L}_{0j}\}$ such an l_{ij}, and we denote by $\mathcal{L}$ the union of $\mathcal{L}_0$ and the set of the l_{ij} so chosen.

The following generalization was proved in Győry (1998) with an explicit but weaker upper bound in terms of S. The improvement in S given below is due to the use of the recent Theorem 4.1.7 in which the upper bound is better in terms of S than in the other effective results concerning S-unit equations.

In the formulation of the next theorem we keep the notation of Subsection 9.6.1. Namely, s denotes the cardinality of S, R_S the S-regulator of K, P_K the maximal norm and Q_K the product of the norms of the prime ideals in S if $S \supsetneq M_K^\infty$, and $P_K = 2$, $Q_K = 1$ if $S = M_K^\infty$. Further, let d and D_K be the degree and discriminant of K, and H an upper bound for the logarithmic heights of the coefficients of F.

Theorem 9.6.3 *Let $F \in O_S[X_1, \ldots, X_m]$ be a decomposable form of degree n that factors into linear forms over K and satisfies the following conditions:*

(i) the set $\mathcal{L}_0$ has rank m,
(ii) either $k = 1$ or $k > 1$ and the graph $\mathcal{H}(\mathcal{L}_{01}, \ldots, \mathcal{L}_{0k})$ is connected.

Then every solution $\mathbf{x} = (x_1, \ldots, x_m) \in O_S^m$ of (9.6.1) with $l(\mathbf{x}) \neq 0$ for all $l \in \mathcal{L}$ if $k > 1$, satisfies

$$\max_{1 \le i \le m} h(x_i) < c_5^s P_K (\log^* Q_K) R_S, \qquad (9.6.12)$$

where c_5 is an effectively computable positive number which depends only on d, D_K, H, m, n and $h(\delta)$.

The improved dependence on S has applications in Corollaries 9.6.4 and 9.6.5; see also the Notes (Section 9.7). A completely explicit version of Theorem 9.6.3 can be found in Győry and Yu (2006). We mention that Theorem 9.6.3 is also applicable if F does not factor into linear forms over K but over a finite extension, G, say of K: by applying the above theorem with G, T instead of

K, S where T is the set of places of G lying above those in S one obtains an upper bound for $h(x_i)$ $(i = 1, \ldots, m)$ like (9.6.12) but with K, S replaced by G, T. In Győry (1998), another effective result on (9.6.1) has been derived for decomposable forms F satisfying conditions (i) and (ii) and with splitting field $G \supsetneq K$, which gives much better bounds if G is large. This is important for applications to norm form equations and discriminant form equations, see Corollaries 9.6.6–9.6.8 below.

Remark Theorem 9.6.3 implies that under the assumptions (i) and (ii), equation (9.6.1) has only finitely many solutions, and all of them can be determined effectively, at least in principle. We note that the finiteness of the number of solutions in Theorem 9.6.3, and hence in Corollaries 9.6.6–9.6.8 below, follows already from Corollary 9.1.2 in the more general case as well, when K is replaced by a finitely generated extension of $\mathbb{Q}$ and O_S by a finitely generated subring A of K over $\mathbb{Z}$. More precisely, the finiteness condition

(i'') $\mathcal{L} \cap ([\mathcal{L}_1] \cap [\mathcal{L}_0 \setminus \mathcal{L}_1]) \neq \emptyset$ *for every proper, non-empty subset* $\mathcal{L}_1$ *of* $\mathcal{L}_0$
   with $\mathcal{L} = \mathcal{L}_0$ *if* $k = 1$

of Corollary 9.1.2 is a consequence of the condition (ii) of Theorem 9.6.3. Indeed, let $\mathcal{L}_1$ be a proper, non-empty subset of $\mathcal{L}_0$. First consider the case when, in Theorem 9.6.3, $k = 1$. Since $\mathcal{G}(\mathcal{L}_0)$ is connected, there are $l \in \mathcal{L}_1$ and $l' \in \mathcal{L}_0 \setminus \mathcal{L}_1$ such that l, l' are connected by an edge in $\mathcal{G}(\mathcal{L}_0)$, i.e., $\lambda l + \lambda' l' + \lambda'' l'' = 0$ for some $l'' \in \mathcal{L}_0$ and non-zero $\lambda, \lambda', \lambda'' \in K$ which proves (i''). Next assume that $k > 1$. If there is an $\mathcal{L}_{0i}$ with $1 \leq i \leq k$ such that $\mathcal{L}_{0i} \cap \mathcal{L}_1 \neq \emptyset$ and $\mathcal{L}_{0i} \cap (\mathcal{L}_0 \setminus \mathcal{L}_1) \neq \emptyset$, then (i'') follows as in the case $k = 1$. Suppose that any $\mathcal{L}_{0i}$ is either in $\mathcal{L}_1$ or in $\mathcal{L}_0 \setminus \mathcal{L}_1$. Since by assumption $\mathcal{H}(\mathcal{L}_{01}, \ldots, \mathcal{L}_{0k})$ is connected, there is a pair $\mathcal{L}_{0i}, \mathcal{L}_{0j}$ which is an edge in $\mathcal{H}$ and $\mathcal{L}_{0i}$ is in $\mathcal{L}_1$ and $\mathcal{L}_{0j}$ in $\mathcal{L}_0 \setminus \mathcal{L}_1$ or conversely. But there is a non-zero linear form l_{ij} contained in $[\mathcal{L}_{0i}]$ and $[\mathcal{L}_{0j}]$ and hence in $[\mathcal{L}_1]$ and $[\mathcal{L}_0 \setminus \mathcal{L}_1]$ which yields (i'').

We now present some consequences of Theorem 9.6.3. We start with another version of Theorem 9.6.1 which gives a better bound for the solutions of equation (9.6.3) in terms of S. Consider again the equation

$$F(x, y) = \delta \text{ in } x, y \in O_S, \qquad (9.6.3)$$

where $F \in O_S[X, Y]$ is a binary form of degree $n \geq 3$ which factorizes into linear factors over K and at least three of these factors are pairwise non-proportional. Further, let $\delta \in O_S \setminus \{0\}$ and H an upper bound for the logarithmic heights of the coefficients of F. It is easy to check that in this case F satisfies the conditions (i), (ii) of Theorem 9.6.3 with $m = 2$, $k = 1$. Hence Theorem 9.6.3 implies the following.

Corollary 9.6.4 *Under the above assumptions and notation, all solutions x, y of (9.6.3) satisfy*

$$\max(h(x), h(y)) < c_6^s P_K (\log^* Q_K) R_S,$$

where c_6 is an effectively computable positive number depending only on d, D_K, H, n and $h(\delta)$.

For an explicit value of c_6, we refer to Győry and Yu (2006).

The following consequence of Theorem 9.6.3 provides some information about the arithmetical properties of decomposable forms at integral points with coordinates in O_K, where O_K denotes the ring of integers of K. We denote by $\omega(\alpha)$ the number of distinct prime ideal divisors of $\alpha \in O_K \setminus \{0\}$, and by $P(\alpha)$ the greatest of the norms of these prime ideals, with the convention that $P(\alpha) = 1$ if $\alpha \in O_K^*$.

Corollary 9.6.5 *Let $F \in O_K[X_1, \ldots, X_m]$ be a decomposable form as in Theorem 9.6.3, and let N_0 be a positive integer. Further, let $\mathbf{x} = (x_1, \ldots, x_m) \in O_K^m$ be such that*

$$N_K((x_1, \ldots, x_m)) \leq N_0, \quad F(\mathbf{x}) \neq 0, \quad l(\mathbf{x}) \neq 0 \text{ for } l \in \mathcal{L} \text{ if } k > 1.$$

Then

$$P(\log P)^\omega > c_7 (\log N)^{c_8}$$

and

$$P > \begin{cases} c_9 (\log N)^{c_{10}} & \text{if } \omega \leq \log P / \log_2 P, \\ c_{11} (\log_2 N)(\log_3 N)/(\log_4 N) & \text{otherwise,} \end{cases}$$

provided that $N = \max_{1 \leq i \leq m} |N_{K/\mathbb{Q}}(x_i)| \geq N_1$, where $P = P(F(\mathbf{x}))$ and $\omega = \omega(F(\mathbf{x}))$. Here $c_7, \ldots, c_{11}$ and N_1 are effectively computable positive numbers which depend at most on K, F and N_0.

The deduction of this corollary from Theorem 9.6.3 is straightforward, for this we refer to Győry and Yu (2006).

An important special case of Corollary 9.6.5 is $m = 2$, $k = 1$ when F is a binary form with splitting field K and with at least three pairwise non-proportional linear factors. In this special case the corollary implies a similar result for polynomials $F(X) \in O_K[X]$. Corollary 9.6.5 is a generalization and improvement of the corresponding results of Győry (1978/1979, 1981a), Haristoy (2003) and many earlier special lower estimates. It motivates the following.

Conjecture (Győry and Yu (2006)) *Under the assumptions and notation of Corollary 9.6.5,*

$$P > c_{12}(\log N)^{c_{13}} \quad \textit{if } N \geq N_1$$

holds, where c_{12}, c_{13} and N_1 are effectively computable positive numbers depending at most on K, F and N_0.

Let now L be an extension of K of degree $n \geq 3$ and $\alpha_1 = 1, \alpha_2, \ldots, \alpha_m$ K-linearly independent elements of L over K with $m \geq 2$ which are integral over O_S. Consider the norm form equation

$$N_{L/K}(\alpha_1 x_1 + \cdots + \alpha_m x_m) = \delta \text{ in } x_1, \ldots, x_m \in O_S, \qquad (9.6.13)$$

where $\delta \in O_S \setminus \{0\}$. For $m = 2$, this is a Thue equation over O_S.

Corollary 9.6.6 *Suppose that α_m is of degree ≥ 3 over $K(\alpha_1, \ldots, \alpha_{m-1})$. Then all solutions $(x_1, \ldots, x_m)$ of (9.6.13) with $x_m \neq 0$ satisfy*

$$\max_{1 \leq i \leq m} h(x_i) \leq C_1,$$

where C_1 is an effectively computable positive number which depends only on K, L, S, m, n, $\alpha_1, \ldots, \alpha_m$ and δ.

This implies the following.

Corollary 9.6.7 *Suppose that α_{i+1} is of degree ≥ 3 over $K(\alpha_1, \ldots, \alpha_i)$ for $i = 1, \ldots, m - 1$. Then every solution $(x_1, \ldots, x_m)$ of (9.6.13) satisfies*

$$\max_{1 \leq i \leq m} h(x_i) \leq C_2,$$

where C_2 is an effectively computable positive number which depends only on K, L, S, m, n, $\alpha_1, \ldots, \alpha_m$ and δ.

For $S = M_K^\infty$, the first version of Corollary 9.6.7 was obtained in Győry and Papp (1978). In case of arbitrary S, Corollaries 9.6.6 and 9.6.7 were first proved by Győry (1981a, 1981b) and independently by Kotov (1981). The best known, completely explicit upper bounds for the solutions of equation (9.6.13) are given in Bugeaud and Győry (1996b) and Győry (1998).

Remark In Corollaries 9.6.6, 9.6.7 and hence in Theorem 9.6.3, the respective assumptions $x_m \neq 0$ and $l(\mathbf{x}) \neq 0$ for $l \in \mathcal{L}$ cannot be dropped, and the lower bound 3 for the degrees of α_i cannot be diminished in general. Indeed, let $\alpha \in \overline{\mathbb{Q}}$ of degree ≥ 3 over $L_1 = \mathbb{Q}(\sqrt{2})$ and let $L_2 = \mathbb{Q}(\sqrt{2}, \alpha)$. Then the equations

$$N_{L_1/\mathbb{Q}}(x_1 + \sqrt{2}x_2) = \pm 1 \text{ and } N_{L_2/\mathbb{Q}}(x_1 + \sqrt{2}x_2 + \alpha x_3) = \pm 1$$

have infinitely many integral solutions (x_1, x_2, x_3) with $x_3 = 0$.

Let again L be an extension of degree $n \geq 3$ of K and $1, \alpha_1, \ldots, \alpha_m$ K-linearly independent elements of L, integral over O_S, such that $L = K(\alpha_1, \ldots, \alpha_m)$, and let $l = X_0 + \alpha_1 X_1 + \cdots + \alpha_m X_m$. Denote by $\sigma_1, \ldots, \sigma_n$ the K-isomorphic embeddings of L in $\overline{\mathbb{Q}}$, and define $l^{(i)} := X_0 + \sigma_i(\alpha_1)X_1 + \cdots + \sigma_i(\alpha_n)X_n$ for $i = 1, \ldots, n$. Put

$$D_{L/K}(\alpha_1 X_1 + \cdots + \alpha_m X_m) := \prod_{1 \leq i < j \leq n} \left(l^{(i)} - l^{(j)}\right)^2.$$

This is a decomposable form in $O_S[X_1, \ldots, X_m]$ of degree $n(n-1)$, independent of X_0. It is called a *discriminant form*. Consider now the *discriminant form equation*

$$D_{L/K}(\alpha_1 x_1 + \cdots + \alpha_m x_m) = \delta \text{ in } (x_1, \ldots, x_m) \in O_S^m, \qquad (9.6.14)$$

where $\delta \in O_S \setminus \{0\}$.

Corollary 9.6.8 *Under the above assumptions, all solutions $(x_1, \ldots, x_m)$ of (9.6.14) satisfy*

$$\max_{1 \leq i \leq m} h(x_i) < C_3,$$

where C_3 is an effectively computable positive number which depends only on $K, S, m, n, \alpha_1, \ldots, \alpha_m$ and δ.

For $K = \mathbb{Q}$, $S = \{\infty\}$, the first version of this corollary was proved in Győry (1976), and for arbitrary K and S, in Győry and Papp (1977) and Győry (1981b). The best known, explicit version of Corollary 9.6.8 can be found in Győry (1998).

Corollary 9.6.8 and its other versions have several applications, among others to index form equations, algebraic integers of given discriminant or of given index and power integral bases. Such results are treated in detail in our next book on discriminant equations. Some related results are also briefly discussed in the Notes (Section 9.7) and in Section 10.6.

We note that from Theorem 9.6.3 one could easily deduce in Corollaries 9.6.6–9.6.8 explicit bounds in terms of S. Moreover, combining the explicit version of Theorem 9.6.3 from Győry and Yu (2006) with some arguments from Győry (1998), one can give completely explicit version of Corollaries 9.6.6–9.6.8 with slightly better upper bounds than those in Bugeaud and Győry (1996b) and Győry (1998).

Finally, we observe that Corollary 9.6.5 is in particular applicable in the case that F is a discriminant form, or a norm form like in Corollaries 9.6.6 and 9.6.7.

We give only a sketch of the proof of Theorem 9.6.3. For a detailed proof we refer to Győry and Yu (2006).

Proof of Theorem 9.6.3 (sketch). We keep the above notation of Subsection 9.6.2. We shall denote by $c_{14}, c_{15}, \ldots, c_{44}$ effectively computable positive numbers which depend at most on d, the class number h_K and regulator R_K of K, and on H, m, n and $h(\delta)$. But by (1.5.2) and (1.5.3) h_K, R_K can be estimated from above in terms of d and the discriminant D_K of K. Hence we can replace the dependence on h_K and R_K by D_K.

We make some preliminary remarks. We show in two steps that equation (9.6.1) can be written in the form

$$l_1(\mathbf{x}) \cdots l_n(\mathbf{x}) = \delta \text{ in } \mathbf{x} \in O_S^m \text{ with } l(\mathbf{x}) \neq 0 \text{ for } l \in \mathcal{L}, \tag{9.6.15}$$

where, up to a proportional factor, $l_1 \cdots l_n$ is a factorization of F into linear forms in $X_1, \ldots, X_m$ with coefficients in O_K, the logarithmic heights of the coefficients of $l_1, \ldots, l_n$ do not exceed c_{14} and the new $\delta \in O_S \setminus \{0\}$ has height $h(\delta) \leq c_{15} \log^* Q_K$.

First we recall that as in (9.4.4), F can be written as $cl_1' \cdots l_n'$, where $c \in O_S$, $c \neq 0$ and $l_i' = X_{n_i} + \alpha_{n_i+1,i} X_{n_i+1} + \cdots + \alpha_{mi} X_m$ with $\alpha_{ji} \in K$ for $i = 1, \ldots, n$, $j \in \{n_i + 1, \ldots, m\}$. Then by (1.9.5) and (1.9.6) we have $h^{\text{hom}}(F) \leq c_{16}$ and, by Corollary 1.9.5, $h(l_i') = h^{\text{hom}}(l_i') \leq c_{17}$ for $i = 1, \ldots, n$. Thus the maximum of the logarithmic heights of the coefficients of l_i' is at most c_{18}. Since c is a coefficient of F, we have $h(c) \leq c_{19}$ which implies that the coefficients of cl_1' have logarithmic heights at most c_{20}.

In the next step we multiply $cl_1', l_2', \ldots, l_n'$ and δ by the product of the denominators of the coefficients of $cl_1', l_2', \ldots, l_n'$. Then the logarithmic heights of the new δ and the coefficients of the new linear factors, for simplicity denoted again by $l_1, \ldots, l_n$, are at most c_{21}. Therefore our claim is proved.

Let now $\mathbf{x} \in O_S^m$ be a solution of equation (9.6.15) with $l(\mathbf{x}) \neq 0$ for $l \in \mathcal{L}$ if $k > 1$, and write

$$l_i(\mathbf{x}) = \delta_i, \quad i = 1, \ldots, n. \tag{9.6.16}$$

Then δ_i is a divisor of δ in O_S and so, by (1.9.2) and the above upper bound for $h(\delta)$, we have $\log N_S(\delta_i) \leq \log N_S(\delta) \leq c_{22} h(\delta) \leq c_{23}$. By Proposition 4.3.12 there is an $\varepsilon_i \in O_S^*$ such that

$$h(\delta_i/\varepsilon_i) \leq c_{24} \log^* Q_K, \quad i = 1, \ldots, n. \tag{9.6.17}$$

Let $\mathcal{L}_0$ be a maximal subset of pairwise linearly independent linear forms in the set of new linear forms $l_1, \ldots, l_n$. Then the new $\mathcal{L}_0$ and its associated graph $\mathcal{G}(\mathcal{L}_0)$ also satisfy the assumptions (i) and (ii) of the theorem. Let $\mathcal{L}_{01}, \ldots, \mathcal{L}_{0k}$

denote the vertex sets of the connected components of $\mathcal{G}(\mathcal{L}_0)$. First assume that $k = 1$. Then by assumption (i), $\mathcal{G}(\mathcal{L}_0)$ is of order at least 3. If $\{l_i, l_j\}$ is an edge in $\mathcal{G}(\mathcal{L}_0)$, then $\lambda_i l_i + \lambda_j l_j + \lambda l = 0$ for some $l \in \mathcal{L}_0$ and some non-zero λ_i, λ_j, λ in K with logarithmic heights not exceeding c_{25}. Together with (9.6.17) this yields an S-unit equation

$$\tau_i \varepsilon_i + \tau_j \varepsilon_j + \tau \varepsilon = 0 \quad \text{in } \varepsilon_i, \varepsilon_j, \varepsilon \in O_S^* \tag{9.6.18}$$

where the coefficients τ_i, τ_j, τ are non-zero elements of K with logarithmic height $\leq c_{25} \log^* Q_K$. Now applying Theorem 4.1.7 to equation (9.6.18), we infer that

$$\max(h(\varepsilon_i/\varepsilon), h(\varepsilon_j/\varepsilon)) \leq c_{26}^s P_K (\log^* Q_K) R_S$$

and so, by (9.6.16) and (9.6.17),

$$\max(h(\delta_i/\varepsilon), h(\delta_j/\varepsilon)) \leq c_{27}^s P_K (\log^* Q_K) R_S =: A. \tag{9.6.19}$$

If now $\{l_i, l_q\}$ is an edge in $\mathcal{G}(\mathcal{L}_0)$ then we deduce in the same way that there is an $\varepsilon' \in O_S^*$ such that

$$\max(h(\delta_j/\varepsilon'), h(\delta_q/\varepsilon')) \leq A.$$

Together with (9.6.19) this implies $h(\varepsilon'/\varepsilon) \leq 2A$, whence $h(\delta_q/\varepsilon) \leq 3A$. Using the assumption that $\mathcal{G}(\mathcal{L}_0)$ is connected and repeating the above procedure with the shortest path connecting two vertices, we infer that $h(\delta_i/\varepsilon) \leq c_{28} A$ for each i with $l_i \in \mathcal{L}_0$. Further, if $l_{i'} \in \mathcal{L} \setminus \mathcal{L}_0$ is proportional to a linear form $l_i \in \mathcal{L}_0$, then $l_{i'} = \rho l_i$ with some non-zero $\rho \in K$ with $h(\rho) \leq c_{29}$, hence $h(\delta_i/\varepsilon) \leq c_{30} A$ for $i = 1, \ldots, n$. Together with (9.6.15) this gives $h(\delta/\varepsilon^n) \leq c_{31} A$. Thus $h(\varepsilon) \leq c_{32} A$, and so $h(\delta_i) \leq c_{33} A$ for $i = 1, \ldots, n$. Considering (9.6.16) as a system of linear equations in $\mathbf{x} = (x_1, \ldots, x_m)$ and using the assumption (i), we infer by Cramer's Rule that

$$h(x_t) \leq c_{34} A \quad \text{for } t = 1, \ldots, n. \tag{9.6.20}$$

Next consider the case when $k > 1$ and the graph $\mathcal{H}(\mathcal{L}_{01}, \ldots, \mathcal{L}_{0k})$ is connected. For $j = 1, \ldots, k$, let $\mathfrak{J}_j$ denote the set of indices i with $l_i \in \mathcal{L}_{0j}$. We may assume without loss of generality that $\{\mathcal{L}_{01}, \mathcal{L}_{02}\}$ is an edge in this graph. Then by assumption there is a non-zero $l_{1,2} \in \mathcal{L}$ which can be represented in the form

$$\sum_{i \in \mathfrak{J}_1} \lambda_i l_i = \sum_{i \in \mathfrak{J}_2} \lambda_i l_i \tag{9.6.21}$$

such that the total number of non-zero $\lambda_i \in K$ in both sides of (9.6.21) is minimal. Then, up to a proportional factor, these λ_i provide a uniquely determined solution of (9.6.21) as a system of linear equations in λ_i with $i \in \mathfrak{J}_1 \cup \mathfrak{J}_2$. One can prove that there is a non-zero $\lambda_{1,2}$ in K such that $\lambda_{1,2} l_{1,2}$ can be expressed in the form (9.6.21) with non-zero $\lambda_i \in K$ for which $h(\lambda_i) \leq c_{35}$.

As was seen above in the case $k = 1$, we have

$$h(\delta_i/\varepsilon_1) \leq c_{36} A \quad \text{for } i \in \mathfrak{J}_1 \quad \text{and} \quad h(\delta_i/\varepsilon_2) \leq c_{37} A \quad \text{for } i \in \mathfrak{J}_2 \quad (9.6.22)$$

with some $\varepsilon_1, \varepsilon_2 \in O_S^*$. By (9.6.17), this also holds if $\mathfrak{J}_1$ or $\mathfrak{J}_2$ consists of a single element. For the solution $\mathbf{x}$ considered above we deduce from (9.6.21) and (9.6.22) that

$$h(\lambda_{1,2} l_{1,2}(\mathbf{x})/\varepsilon_q) \leq c_{38} A \quad \text{for } q = 1, 2.$$

But $l_{1,2}(\mathbf{x}) \neq 0$, hence it follows that $h(\varepsilon_2/\varepsilon_1) \leq c_{39} A$, whence, by (9.6.22),

$$h(\delta_i/\varepsilon_1) \leq c_{40} A \quad \text{for } i \in \mathfrak{J}_1 \cup \mathfrak{J}_2.$$

Using the fact that the graph $\mathcal{H}(\mathcal{L}_{01}, \ldots, \mathcal{L}_{0k})$ is connected and repeating this process with the shortest path connecting two vertices, we infer that $h(\delta_i/\varepsilon_1) \leq c_{41} A$ for each i in $\mathfrak{J}_1 \cup \cdots \cup \mathfrak{J}_k$. It follows as above in the case $k = 1$ that $h(\delta_i/\varepsilon_1) \leq c_{42} A$ and so, in view of (9.6.15), $h(\delta_i) \leq c_{43} A$ for $i = 1, \ldots, n$. We now infer as in the case $k = 1$ that (9.6.20) holds with a c_{44} in place of c_{34} for $t = 1, \ldots, n$, whence (9.6.12) follows. $\qquad \square$

Proof of Corollary 9.6.6. Put $M = K(\alpha_1, \ldots, \alpha_m)$, and denote by $\mathcal{L}_0$ the set of the conjugates of the linear form $l = \alpha_1 X_1 + \cdots + \alpha_m X_m$ with respect to M/K. By assumption $\alpha_1 = 1$, hence the forms in $\mathcal{L}_0$ are pairwise non-proportional. They form a maximal subset of such forms in the set of linear forms of $N_{L/K}(l)$. Partition the linear forms in $\mathcal{L}_0$ into subsets so that l', l'' belong to the same subset if the coefficients of $X_1, \ldots, X_{m-1}$ in l', l'' coincide. Then we get a partition $\mathcal{L}_{01}, \ldots, \mathcal{L}_{0k}$ with k denoting the degree of $K(\alpha_1, \ldots, \alpha_{m-1})$ over K, and it is easily seen that each of the graphs $\mathcal{G}(\mathcal{L}_{01}), \ldots, \mathcal{G}(\mathcal{L}_{0k})$ defined above is connected. Further, $\mathcal{L}_0$ has the properties (i), (ii) from Theorem 9.6.3 with $\mathcal{L} = \mathcal{L}_0 \cup \{X_m\}$. Considering now equation (9.6.13) over the normal closure, say G, of L over K, Theorem 9.6.3 applies to equation (9.6.13) and gives an effective upper bound for the solutions in terms of $H, m, n, h(\delta)$, the degree g and discriminant D_G of G, and the parameters involved of S_G, that is the set of places of G lying above those of S. But H can be effectively bounded in terms of m, n and $h(\alpha_1), \ldots, h(\alpha_m)$. Further, using explicit estimates form Sections 1.4, 1.5 and 1.8, $g, |D_G|$ and the parameters mentioned can be effectively estimated from above in terms of S and the degrees and discriminants of K and L. This completes the proof. $\qquad \square$

Proof of Corollary 9.6.7. Let $(x_1, \ldots, x_m)$ be a solution of (9.6.13), and denote by m' the greatest integer with $x_{m'} \neq 0$. If $m' \geq 2$, Corollary 9.6.6 applies with m' instead of m, while for $m' = 1$ the assertion is trivial. $\qquad \square$

Proof of Corollary 9.6.8. Using the notation and assumptions of the corollary, $L = K(\alpha_1, \ldots, \alpha_m)$ implies that the linear forms $l^{(1)}, \ldots, l^{(n)}$ are pairwise non-proportional. Further, it follows from the linear independence of $1, \alpha_1, \ldots, \alpha_m$ over K that there are indices $i_1, \ldots, i_{m+1}$ such that $\mathrm{rank}\{l^{(i_1)}, \ldots, l^{(i_{m+1})}\} = m + 1$. Notice that the linear forms $l_{ij} := l^{(i)} - l^{(j)}$ $(1 \leq i, j \leq n)$ depend only on $X_1, \ldots, X_m$, and that

$$D_{L/K}(\alpha_1 X_1 + \cdots + \alpha_m X_m) = (-1)^{n(n-1)/2} \prod_{\substack{1 \leq i, j \leq n \\ i \neq j}} l_{ij}.$$

Further, $\mathrm{rank}\{l_{i_1, i_{m+1}}, \ldots, l_{i_m, i_{m+1}}\} = m$. This means that rank $\mathcal{L}_0 = m$, where $\mathcal{L}_0$ denotes a maximal set of pairwise non-proportional linear factors of the left-hand side. For distinct $u, v, w \in \{1, \ldots, n\}$ we have $l_{uv} + l_{vw} + l_{wu} = 0$. It is easy to check that the graph $\mathcal{G}(\mathcal{L}_0)$ is connected, and hence Theorem 9.6.3 combined with some arguments from the end of the proof of Corollary 9.6.6 yields Corollary 9.6.8. $\qquad \square$

9.7 Notes

We make some *historical notes* and mention some *refinements, applications* and *generalizations* of the results presented in this chapter.

- There are many papers on *effective results* for decomposable form equations. The first effective upper bound for the solutions of Thue equations over $\mathbb{Z}$ was established by Baker (1968b) by means of his effective estimates for linear forms in logarithms. In the case of discriminant form and index form equations, the first effective bounds for the solutions were given in Győry (1976), and for the case of certain norm form and decomposable form equations, in Győry and Papp (1978). Their proofs also involved Baker's method but via Győry's effective results on unit equations in two unknowns. Later, a number of various effective results with explicit bounds and generalizations were obtained on the equations mentioned; for results and references, see the books and survey papers Győry (1980b, 2002), Shorey and Tijdeman (1986), Evertse, Győry, Stewart and Tijdeman (1988b), Evertse and Győry (1988d), Sprindžuk (1993) and Feldman and Nesterenko (1998). Practical algorithms for solving concrete equations of these types were also worked out; see de Weger (1989), Tzanakis and de Weger (1989), Bilu and Hanrot (1996, 1999), Smart (1998), Gaál (2002), our book on discriminant equations and the references given there. All these results were established by Baker's method, many of them via unit equations. As was mentioned in Subsection 9.6.1, another method was developed and used in Bombieri (1993), Bombieri and Cohen (1997, 2003) and Bugeaud (1998) to obtain effective bounds for the solutions of Thue equations.

For the solutions of decomposable form equations, the best effective upper bounds to date are given in Bugeaud and Győry (1996b), Bugeaud (1998), Győry (1998), Győry and Yu (2006) and in Section 9.6 above. The effective results concerning decomposable form equations have many applications in Diophantine number theory and algebraic number theory. Several such applications are treated in detail in our book on discriminant equations.

• Thue equations have many applications. We present here a classical application. Let $f \in \mathbb{Z}[X]$ be a non-linear polynomial of degree n and m a given integer ≥ 2 and consider the equation

$$f(x) = y^m \text{ in } x, y \in \mathbb{Z}. \tag{9.7.1}$$

An important special case is Mordell's equation $x^3 + k = y^2$, where k is a non-zero integer. Equation (9.7.2) is called an *elliptic equation* if $m = 2$, $\deg f = 3$, a *hyperelliptic equation* if $m = 2$ and $\deg f \geq 3$ and a *superelliptic equation* if $m \geq 3$ and $\deg f \geq 2$. The example of the Pell equation $dx^2 + 1 = y^2$ shows that (9.7.1) may have infinitely many solutions if $m = 2$ and $\deg f = 2$.

Mordell (1922b, 1923) in the elliptic case and later Siegel (1926) in the case that f has degree ≥ 3 and no multiple zeros, proved that (9.7.1) has only finitely many solutions. LeVeque (1964) gave a general finiteness criterion for equation (9.7.1). Their proofs are based on Thue's and Siegel's ineffective finiteness theorems on Thue equations over $\mathbb{Q}$ resp. over number fields, hence they are also ineffective.

Baker (1968b, 1968c, 1969) was the first to give effective upper bounds for the solutions of (9.7.1) in the case when f has at least 3 simple zeros if $m = 2$ and at least 2 simple zeros if $m \geq 3$. We sketch the main steps of his proof. Assume, for simplicity, that f is monic, and that α_1, α_2 and, in the case $m = 2$, α_3 are simple zeros of f. Put $K_i = \mathbb{Q}(\alpha_i)$ for $i = 1, 2, 3$. If (x, y) is a solution of (9.7.1), then following Siegel's argument one deduces that $x - \alpha_i = \beta_i \sigma_i^m$, where β_i is a non-zero element of K with bounded height, and σ_i is an unknown integer in K_i for $i = 1, 2, 3$. This implies that

$$\beta_1 \sigma_1^m - \beta_2 \sigma_2^m = \alpha_2 - \alpha_1.$$

For $m \geq 3$, this is a Thue equation over $K_1 K_2$. In this case, Baker applied his effective result concerning Thue equations over number fields to give an effective upper bound for the heights of σ_1, σ_2 and thereby for x and y. If $m = 2$, we have a system of three equations

$$\beta_i \sigma_i^2 - \beta_j \sigma_j^2 = \alpha_j - \alpha_i \quad (1 \leq i < j \leq 3).$$

Following Siegel (1926), Baker reduced this system to a single Thue equation over an appropriate finite extension of $K_1 K_2 K_3$ and applied his effective result on Thue equations to the latter. We mention that alternatively one can combine the above system into a single equation over $K_1 K_2 K_3$ in three unknowns $\sigma_1, \sigma_2, \sigma_3$,

$$\prod_{1 \leq i < j \leq 3} \left(\beta_i \sigma_i^2 - \beta_j \sigma_j^2 \right) = \prod_{1 \leq i < j \leq 3} (\alpha_j - \alpha_i),$$

where the left-hand side is a triangularly connected decomposable form in $\sigma_1, \sigma_2, \sigma_3$, whose linear factors form a system of rank 3, and then apply Theorem 9.6.3. We note that using here Theorem 9.6.1 or 9.6.2, or the explicit version of Theorem 9.6.3 from Győry and Yu (2006) one can get better bounds for the solutions x, y of equation (9.7.1).

Quantitative improvements and generalizations of Baker's theorems were later obtained by many authors, including Brindza (1984), who gave an effective upper bound for the solutions x, y of (9.7.1) under LeVeque's general criterion. For practical methods for complete resolution of elliptic and superelliptic equations, we refer to Gebel, Pethő and Zimmer (1994), Stroeker and Tzanakis (1994), Bilu and Hanrot (1998) and Tzanakis (2013). All these results and methods are based on the theory of logarithmic forms or its elliptic analogue. Recently, Bérczes, Evertse and Győry (2014) proved an effective finiteness result for hyper- and superelliptic equations over finitely generated domains.

There are a couple of results on the number of solutions of hyper- and superelliptic equations. We recall without proof the following special case of a result of Evertse and Silverman (1986). Its proof takes as starting point Evertse's quantitative results on S-unit equations in two unknowns and the Thue–Mahler equation Evertse (1984a) and follows the same lines as the argument sketched above. Let m be an integer ≥ 2 and $f \in \mathbb{Z}[X]$ a polynomial of degree n and discriminant $D(f) \neq 0$. Let $\omega(D(f))$ denote the number of primes dividing $D(f)$. Assume that $n \geq 3$ if $m = 2$ and $n \geq 2$ if $m \geq 3$. Let K be a number field, containing three zeros of f if $m = 2$ and two zeros of f if $m \geq 3$, and denote by $h_m(K)$ the number of ideal classes of O_K whose

m-th power is the principal ideal class. Then the number of solutions of

$$f(x) = y^m \quad \text{in } x, y \in \mathbb{Z} \tag{9.7.1}$$

is at most

$$\begin{cases} 7^{17n^3(\omega(D(f))+1)} h_2(K)^2 & \text{if } m = 2, \\ (17^{13}m^2)^{n^2(\omega(D(f))+1)} h_m(K) & \text{if } m \geq 3. \end{cases}$$

A folklore conjecture asserts that $h_m(K) \ll_{m,[K:\mathbb{Q}],\varepsilon} |D_K|^\varepsilon$ for every $\varepsilon > 0$, where D_K denotes the discriminant of K and the implied constant depends only on m, $[K : \mathbb{Q}]$ and ε. By elementary estimates, one can estimate $|D_K|$ from above by a power of $|D(f)|$. This leads to the conjecture that equation (9.7.1) has $\ll_{m,n,\varepsilon} |D(f)|^\varepsilon$ solutions, for every $\varepsilon > 0$.

- The simplest discriminant equation is

$$D_m(\mathbf{x}) = \prod_{1 \leq i < j \leq m} (x_i - x_j)^2 \in A^* \text{ in } \mathbf{x} = (x_1, \ldots, x_m) \in A^m, \tag{9.7.2}$$

where A is a subring of a field K of characteristic 0 which is integrally closed and finitely generated over $\mathbb{Z}$. The form

$$D_m := \prod_{1 \leq i < j \leq m} (X_i - X_j)^2,$$

called a *decomposable form of discriminant type*, is just the discriminant of the polynomial $f(X) = (X - X_1) \cdots (X - X_m)$. Two solutions $\mathbf{x} = (x_1, \ldots, x_m)$, $\mathbf{x}' = (x_1', \ldots, x_m')$ of (9.7.2) are called *A-equivalent* if there are $u \in A^*$, $a \in A$ such that $x_i' = ux_i + a$ for $i = 1, \ldots, m$.

It is easily seen that the decomposable form D_m is triangularly connected. Hence, in the number field case when $A = O_S$, the ring of S-integers of a number field K, Theorem 9.6.3 gives that the set of solution of (9.7.2) is the union of finitely many O_S-equivalence classes of solutions which can be effectively determined. A generalization for the finitely generated case and other related results will be treated in detail in our next book on discriminant equations.

- For applications, *discriminant form equations* and *index form equations* belong to the most important classes of decomposable form equations. For a detailed treatment of these equations and their applications we refer again to our next book on discriminant equations. We mention here some basic facts only about these equations. See also Section 10.6.

Let K be a field of characteristic 0, L an extension of K of degree $n \geq 2$, A a domain with quotient field K which is integrally closed in K and O an A-*order*, that is a subring of L containing A that as an A-module is free of rank n.

Let $\{\omega_1 = 1, \omega_2, \ldots, \omega_n\}$ be an A-module basis of O. Define the linear form
$l := X_1 + \omega_2 X_2 + \cdots + \omega_n X_n$, let $l^{(1)} = l, l^{(2)}, \ldots, l^{(n)}$ be the conjugates of
l over K, and define the *discriminant form* $D_{L/K}(\omega_2 X_2 + \cdots + \omega_n X_n)$ as in
Section 9.6, i.e., $\prod_{1 \le i < j \le n}(l^{(i)} - l^{(j)})^2$. Then

$$D_{L/K}(\omega_2 X_2 + \cdots + \omega_n X_n) = I(\omega_2 X_2 + \cdots + \omega_n X_n)^2 \cdot \Delta,$$

where $I = I(\omega_2 X_2 + \cdots + \omega_n X_n)$ is a decomposable form in $A[X_2, \ldots, X_n]$
of degree $n(n-1)/2$ and $\Delta = D_{L/K}(1, \omega_2, \ldots, \omega_n)$ is the discriminant of
the basis $\{1, \omega_2, \ldots, \omega_n\}$. Using the finiteness result of Lang (1960) on unit
equations in two unknowns, it was proved in Győry (1982a, 1982b) that apart
from a proportional factor from A^* and a translation of the form $\alpha \to \alpha + a$,
$a \in A$, the equations

$$(i) \quad O = A[\alpha],$$

$$(ii) \quad D_{L/K}(\alpha) \in \delta A^*,$$

$$(iii) \quad D_{L/K}(\omega_2 x_2 + \cdots + \omega_n x_n) \in \delta A^*,$$

$$(iv) \quad I(\omega_2 x_2 + \cdots + \omega_n x_n) \in \delta A^*,$$

where $\delta \in A \setminus \{0\}$, have only finitely many solutions in $\alpha \in O$ resp. in
$x_2, \ldots, x_n \in A$. Moreover, putting $\alpha = \sum_{i=1}^{n} \omega_i x_i$ with $x_1, \ldots, x_n \in A$, one
can show that these equations are equivalent. In the special case $K = \mathbb{Q}$,
$A = \mathbb{Z}$, the quantity $|I(\omega_2 x_2 + \cdots + \omega_n x_n)|$ is just the index of the additive
group $\mathbb{Z}[\alpha]^+$ in O^+, therefore, the form I is called an *index form*. Effective
versions of the above finiteness assertions are given in Győry (1976, 1978b,
1981b) and Győry and Papp (1977) over number fields, and, in our next book
on discriminant equations, over finitely generated domains.

• Let K be a number field with ring of integers O_K, S a finite set of places
of K containing all infinite places, $\mathfrak{p}_1, \ldots, \mathfrak{p}_s$ the prime ideals in S, and
$\mathfrak{p}_i^{h_K} = (\pi_i)$, where h_K denotes the class number of K and $\pi_i \in O_K \setminus \{0\}$
which, by Proposition 4.3.12, can be chosen so that $h(\pi_i)$ is effectively
bounded, for $i = 1, \ldots, s$. Let $F \in K[X_1, \ldots, X_m]$ be a decomposable form,
and let $\delta \in K \setminus \{0\}$. It is easy to see that the equation

$$F(\mathbf{x}) = \delta \text{ in } \mathbf{x} = (x_1, \ldots, x_m) \in O_S^m \tag{9.7.3}$$

leads to finitely many equations of the form

$$F(\mathbf{x}) = \delta' \pi_1^{z_1} \cdots \pi_s^{z_s} \text{ in } \mathbf{x} = (x_1, \ldots, x_m) \in O_K^m \text{ and } z_1, \ldots, z_s \in \mathbb{Z}_{\ge 0},$$
$$\tag{9.7.4}$$

where δ' can take only finitely many and effectively determinable values
from $K \setminus \{0\}$. Conversely, any equation of the shape (9.7.4) can be reduced

to finitely many and effectively determinable equations of the form (9.7.3). In our book, we considered equations in the form (9.7.3), but in the earlier literature many results were formulated and proved for equations (9.7.4).

- Combining the proof of Theorem 9.6.3 with the effective results of Chapter 8 on unit equations, Theorem 9.6.3 and its corollaries formulated in Section 9.6 can be generalized for the case where the ground ring is an arbitrary finitely generated integral domain over $\mathbb{Z}$. Using the method presented in Chapter 8, in Bérczes, Evertse and Győry (2014) and in our next book on discriminant equations *effective finiteness results* are obtained in a more direct way concerning the solutions of Thue equations and discriminant equations over *finitely generated domains*.

- Schmidt (1971) gave a finiteness criterion for decomposable form equations over $\mathbb{Z}$ in m unknowns, $F(\mathbf{x}) = \delta$ in $\mathbf{x} \in \mathbb{Z}^m$, but he restricted himself to decomposable forms F with the property that $F(\mathbf{x}) \neq 0$ for all non-zero $\mathbf{x} \in \mathbb{Z}^m$. Schmidt's result was later extended by Schlickewei (1977d) to the case of decomposable form equations over rings of S-integers in $\mathbb{Q}$. We note that the condition $F(\mathbf{x}) \neq 0$ for $\mathbf{x} \in \mathbb{Z}^m \setminus \{\mathbf{0}\}$ is independent of condition (i) of Theorem 9.1.1. For instance, the decomposable form $F = X_1 \cdots X_m(a_1 X_1 + \cdots + a_m X_m)$ with $a_1, \ldots, a_m$ non-zero integers satisfies (i) with $\mathcal{L}$ consisting of all subsums of $a_1 X_1 + \cdots + a_m X_m$, but it certainly vanishes at non-zero integral points.

- Let K be a number field, S a finite set of place of K of cardinality s containing all infinite places, δ a non-zero element of O_S, and $F \in O_S[X, Y]$ a binary form of degree $n \geq 3$. It was proved in Evertse (1997) that the Thue–Mahler equation

$$F(x, y) \in \delta O_S^* \text{ in } x, y \in O_S \qquad (9.7.5)$$

has at most $(5 \cdot 10^6 n)^{s + \omega_S(\delta)}$ O_S^*-cosets of solutions.

Erdős, Stewart and Tijdeman (1988) proved the following result, which implies that Evertse's bound cannot be replaced by a bound polynomial in s, say. Let $p_1 = 2$, $p_2 = 3, \ldots$ be the sequence of primes and n an integer ≥ 2. Then for every $\epsilon > 0$, there exists $t_0(n, \epsilon)$ such that for every $t \geq t_0(n, \epsilon)$ there is a polynomial $f \in \mathbb{Z}[X]$ of degree n with n distinct zeros in $\mathbb{Q}$ for which the equation

$$f(x) = p_1^{z_1} \cdots p_t^{z_t}$$

has at least $\exp((n^2 - \epsilon)t^{1/n}(\log t)^{-(n-1)/n})$ solutions in $x, z_1, \ldots, z_t \in \mathbb{Z}$. Moree and Stewart (1990) proved a similar result with f irreducible over $\mathbb{Q}$.

On the other hand it turned out that, in a certain sense, most of the equations of type (9.7.5) have much fewer solutions. Let K, S be as above. Two binary forms F, $G \in O_S[X, Y]$ are said to be GL(2, O_S)-*equivalent* if

$$G(X, Y) = \varepsilon F(aX + bY, cX + dY)$$

for some $\varepsilon \in O_S^*$ and $a, b, c, d \in O_S$ with $ad - bc \in O_S^*$. Obviously, the number of O_S^*-cosets of solutions of (9.7.5) does not change when F is replaced by a GL(2, O_S)-equivalent form. Using the number field case of Theorem 6.1.6 (see Evertse, Győry, Stewart and Tijdeman (1988a)), Evertse and Győry (1989) proved the following: for every finite extension L of K and every integer $n \geq 3$ there are up to GL(2, O_S)-equivalence only finitely many binary forms $F \in O_S[X, Y]$ of degree n with non-zero discriminant that factorize into linear factors over L and for which equation (9.7.5) has more than two O_S^*-cosets of solutions. Here the bound 2 is already best possible. Further, the assertion does not remain valid without fixing the splitting field. The proof of Evertse and Győry is ineffective in the sense that it does not allow us to determine effectively a full system of representatives for the exceptional equivalence classes. In Evertse and Győry (1989) the authors established also an effective version, but with the bound $1 + s \cdot \min(m, n(n-1)(n-2))$ instead of 2 where $m := [L : K]$. For a connection with inequalities involving resultants of binary forms, see Section 10.9.

- Mahler (1933b) gave asymptotic formulas for the number of solutions of Thue and Thue–Mahler inequalities, and these were later generalized to decomposable form inequalities. We give an overview of the recent results.

  We need the following notation. Let $S = \{\infty, p_1, \ldots, p_t\}$ be a finite set of places of $\mathbb{Q}$. Call a point $\mathbf{x} = (x_1, \ldots, x_m) \in \mathbb{Z}^m$ S-*primitive* if

$$\gcd(x_1, \ldots, x_m, p_1 \cdots p_t) = 1$$

(in the case $S = \{\infty\}$ this condition is void). Denote by $\mu = \mu_\infty$ the Lebesgue measure on $\mathbb{R}$ normalized such that $\mu_\infty([0, 1]) = 1$. For a prime number p, denote by μ_p the Haar measure on $\mathbb{Q}_p$, normalized such that $\mu_p(\mathbb{Z}_p) = 1$. Further, denote by μ_S the product measure $\prod_{p \in S} \mu_p$ on $\prod_{p \in S} \mathbb{Q}_p$ and for positive integers m, by μ_S^m the product measure on $\prod_{p \in S} \mathbb{Q}_p^m$. We call a point $(\mathbf{x}_p : p \in S) \in \prod_{p \in S} \mathbb{Q}_p^m$ S-primitive if $|\mathbf{x}_p|_p = 1$ for $p \in S \setminus \{\infty\}$, where as usual we define $|\mathbf{x}|_p := \max_i |x_i|_p$ for $\mathbf{x} = (x_1, \ldots, x_m)$.

  For a decomposable form $F \in \mathbb{Z}[X_1, \ldots, X_m]$, we denote by $N_{F,S}(k)$ the number of solutions of the decomposable form inequality

$$\prod_{p \in S} |F(\mathbf{x})|_p \leq k \quad \text{in } S\text{-primitive } \mathbf{x} \in \mathbb{Z}^m$$

and we write $N_F(k)$ for $N_{F,S}(k)$ if $S = \{\infty\}$. Further, we define the set

$$A_{F,S}(k) := \left\{ (\mathbf{x}_p : p \in S) \in \prod_{p \in S} \mathbb{Q}_p^m : \begin{array}{l} \prod_{p \in S} |F(\mathbf{x}_p)|_p \leq k, \\ (\mathbf{x}_p : p \in S) \text{ } S\text{-primitive} \end{array} \right\}.$$

Then for $k > 0$ we have $\mu_S^m(A_{F,S}(k)) = \mu_{F,S} k^{m/n}$, where $\mu_{F,S} := \mu_S^m(A_{F,S}(1))$. In the case $S = \{\infty\}$ we write μ_F for $\mu_{F,S}$.

It is an obvious problem to compare $N_{F,S}(k)$ with $\mu_{F,S} k^{m/n}$. In the 1930s, Mahler (1933b) proved that if $F \in \mathbb{Z}[X, Y]$ is an irreducible binary form of degree $n \geq 3$, then

$$|N_{F,S}(k) - \mu_{F,S} k^{2/n}| \ll_{F,S} k^{1/(n-1)} (\log k)^t \quad \text{as } k \to \infty,$$

where the implied constant depends on F and S. In his master's thesis, de Jong (1999) proved analogues of Mahler's results for certain classes of norm form inequalities. In the case $S = \{\infty\}$, Thunder proved a more substantial generalization of Mahler's result to decomposable form inequalities, and made Mahler's result more precise. For the simplicity of our presentation, we give slightly weaker versions of Thunder's results.

Let $F \in \mathbb{Z}[X_1, \ldots, X_m]$ be a decomposable form satisfying condition (i) of Theorem 9.1.1 with $\mathcal{L} = \mathcal{L}_0$ and also $F(\mathbf{x}) \neq 0$ for all $\mathbf{x} \in \mathbb{Z}^m \setminus \{\mathbf{0}\}$. This condition is slightly stronger than the one imposed by Thunder. In his paper Thunder (2001) proved that $\mu_F \ll_{m,n} 1$ and

$$N_F(k) \ll_{m,n} k^{m/n} \quad \text{as } k \to \infty,$$

where the implied constants are effectively computable and depend only on m, n and moreover,

$$|N_F(k) - \mu_F k^{n/d}| \ll_F k^{n/(d+n^{-2})} \quad \text{as } k \to \infty, \tag{9.7.6}$$

where the implied constant is effectively computable and depends on F. In the special case that $\gcd(m, n) = 1$, Thunder (2005) obtained an estimate similar to (9.7.6) with an effectively computable implied constant depending only on m and n. Thunder's arguments consisted of an application of the Quantitative Subspace Theorem and geometry of numbers. It is still open to prove an estimate like (9.7.6), with implicit constant depending only on m, n, without the constraint $\gcd(m, n) = 1$.

J. Liu (2015) obtained in his PhD-thesis generalizations of Thunder's results for arbitrary finite sets of places S. More precisely, he proved that $\mu_{F,S} \ll_{m,n,S} 1$, $N_{F,S}(k) \ll_{m,n,S} k^{m/n}$ as $k \to \infty$, and

$$|N_{F,S}(k) - \mu_{F,S} k^{n/d}| \ll_{F,S} k^{n/(d+n^{-2})} \quad \text{as } k \to \infty,$$

where now all implied constants depend also on the primes in S. Further, in the case that m and n are coprime, he obtained a similar estimate with implicit constant depending only on m, n and the primes in S. Here again, all implicit constants are effectively computable.

- Let A be an integral domain that is finitely generated over $\mathbb{Z}$ and K its quotient field. Further, let $P \in A[X]$ be a polynomial of degree n without multiple zeros. Consider the *resultant equation*

$$R(P, Q) = \delta \text{ in } Q \in A[X] \text{ with } \deg Q = m, \qquad (9.7.7)$$

where $R(P, Q)$ denotes the resultant of P and Q and where $\delta \in K \setminus \{0\}$. Writing $P = a_0(X - \alpha_1) \cdots (X - \alpha_n)$ with distinct $\alpha_1, \ldots, \alpha_n$ from a finite extension of K and $Q = x_0 X^m + x_1 X^{m-1} + \cdots + x_m \in A[X]$, we have

$$R(P, Q) = a_0^m \prod_{i=1}^{n} \left(x_0\alpha_i^m + x_1\alpha_i^{m-1} + \cdots + x_m\right).$$

Thus, (9.7.7) can be regarded as a decomposable form equation. By means of an earlier version of Corollary 9.1.2 it was proved in Győry (1993b) that this equation has only finitely many solutions if $m < n/2$ and this bound $n/2$ is in general sharp. This improved and generalized results of Wirsing (1971), Schmidt (1973) and Schlickewei (1977e) obtained in the case $K = \mathbb{Q}$, $A = \mathbb{Z}$ or $\mathbb{Z}_S$, a ring of S-integers in $\mathbb{Q}$. In the case when K is a number field and $A = O_S$ is a ring of S-integers in K, a quantitative finiteness result from Evertse (1995) on decomposable form equations was used in Győry (1994) to derive the upper bound $(2^{34}n^2)^{m^3 s}$ for the number of solutions of (9.7.7), where $s = |S|$.

We note that in Sections 10.8, 10.9 other versions of (9.7.7) are considered, where both P and Q are unknowns, but the splitting field of $P \cdot Q$ is fixed.

- The next application is concerned with *irreducible polynomials*. It gave an affirmative answer to a problem of M. Szegedy. Let $P \in \mathbb{Z}[X]$ be a monic polynomial of degree n without multiple zeros. Further, let $p_1, \ldots, p_s$ be distinct primes and denote by $\mathscr{S}$ the set of integers not divisible by primes different from $p_1, \ldots, p_s$. It was proved in Győry (1994) that there are at most $(2^{17}n)^{n^3(s+1)/3}$ values $a \in \mathscr{S}$ for which $P(X) + a$ is reducible over $\mathbb{Q}$. Indeed, if for some $a \in \mathscr{S}$, $Q(X) = X^m + x_1 X^{m-1} + \cdots + x_m$ is a divisor of degree $m \leq n/2$ of $P(X) + a$ in $\mathbb{Z}[X]$ and if $\alpha_1, \ldots, \alpha_n$ denote the zeros of $P(X)$, then $(1, x_1, \ldots, x_m)$ is a solution of the equation

$$F(x_0, x_1, \ldots, x_m) \in \mathscr{S} \text{ in } (x_0, x_1, \ldots, x_m) \in \mathbb{Z}^{m+1}, \qquad (9.7.8)$$

where $F = X_0 \prod_{i=1}^{n}(\alpha_i^m X_0 + \alpha_i^{m-1} X_1 + \cdots + X_m)$ is a decomposable form with coefficients in $\mathbb{Z}$. Using an earlier version of Corollary 9.5.5 from

Evertse (1995), one can get an upper bound for the number of solutions of (9.7.8), i.e. for the number of polynomials Q under consideration and the assertion follows. From this it is easy to deduce that for any monic $P \in \mathbb{Z}[X]$ of degree n there is an $a \in \mathbb{Z}$ with $|a| \leq \exp\{(2^{17}n)^{n^3}\}$ for which $P(X) + a$ is irreducible over $\mathbb{Q}$. It is an important feature of this bound that it depends only on n.

- In the number field case when in (9.2) K is a number field and A is a ring of S-integers in K, Győry (1993a) gave a criterion for (9.2) to have only finitely many A^*-cosets of solutions. Also in the number field case when Γ is a group of S-units in K, Theorem 9.2.1 on unit equations is equivalent not only to Theorem 9.1.1 on decomposable form equations, but also to the following assertion. *For any set of $n + 2$ distinct hyperplanes $H_0, \ldots, H_{n+1}$ in $\mathbb{P}^n(K)$, the set of S-integral points of $\mathbb{P}^n(K) \setminus (H_0 \cup \cdots \cup H_{n+1})$ is contained in a finite union of hyperplanes of $\mathbb{P}^n(K)$;* see LeVesque and Waldschmidt (2011), Ru and Wong (1991), Győry (1993b) and, for more general results, Vojta (1987, 1996) and Levin (2008). Some *refinements* of Theorems 9.4.1 to 9.4.4 can be found in Győry (1993a) and Evertse and Győry (1997).

- Consider a decomposable form equation $F(\mathbf{x}) = \pm\delta$ over $\mathbb{Z}$ and its reformulation of the form (9.4.8) with $K = \mathbb{Q}$, $O_S = \mathbb{Z}$. Assume that this equation has infinitely many solutions in $\mathbf{x} \in \mathbb{Z}^m$. Then the maximal rank r of its families of solutions satisfies $1 \leq r < \infty$. Denote by $P(N)$ the *number of solutions* $\mathbf{x} = (x_1, \ldots, x_m)$ with $\max_{1 \leq i \leq m} |x_i| \leq N$. In Everest and Győry (1997) it was deduced from the case $K = \mathbb{Q}$, $O_S = \mathbb{Z}$ of Theorem 9.4.4 that

$$P(N) = c_1(\log N)^r + O((\log N)^{r-1}) \text{ as } N \to \infty,$$

where c_1 is a positive number which depends only upon F and δ. See also Győry and Pethő (1980) and Evertse and Győry (1997).

- Let G be a finite abelian group, and let $\mathbb{Z}[G]$ denote the *integral group ring* which consists of all formal expressions $\sum_{g \in G} x_g \cdot g$ with $x_g \in \mathbb{Z}$. Then $\mathbb{Z}[G]^*$, the unit group of $\mathbb{Z}[G]$, is finitely generated. There is a considerable interest in the units of $\mathbb{Z}[G]$; see e.g. Karpilovsky (1988) and Sehgal (1978). For $x = \sum_{g \in G} x_g \cdot g \in \mathbb{Z}[G]$, let $|x| := \max_{g \in G} |x_g|$, and let

$$U_G(N) := |\{x \in \mathbb{Z}[G]^* : |x| \leq N\}|.$$

Suppose that $\mathbb{Z}[G]^*$ has rank $r > 0$. It was proved in Everest and Győry (1997) as a special case of the above result concerning $P(N)$ that

$$U_G(N) = c_2(\log N)^r + O((\log N)^{r-1}) \text{ as } N \to \infty,$$

where c_2 is a positive number which depends only on G.

• Let $F \in \mathbb{Z}[X_1, \ldots, X_m]$ be a decomposable form of degree n in $m \geq 2$ variables. For given $c > 0$, $\nu \geq 0$ consider the *decomposable form inequality*

$$0 < |F(\mathbf{x})| \leq c|\mathbf{x}|^{\nu} \text{ in } \mathbf{x} = (x_1, \ldots, x_m) \in \mathbb{Z}^m, \qquad (9.7.9)$$

where $|\mathbf{x}| := \max_{1 \leq i \leq m} |x_i|$. By means of his Subspace Theorem Schmidt (1973, 1980) proved that (9.7.9) has only finitely many solutions, provided that

(i) $n > 2(m - 1)$, $\nu < n - 2(m - 1)$ and the linear factors of F are in general position (i.e. any m of them are linearly independent over $\overline{\mathbb{Q}}$),

(ii) F is not divisible in $\mathbb{Q}[X_1, \ldots, X_m]$ by any form of degree less than m.

This was extended by Schlickewei (1977e) to the case when the ground ring is an arbitrary finitely generated subring of $\mathbb{Q}$. These results have obvious applications to decomposable form equations of the form

$$F(\mathbf{x}) = G(\mathbf{x}) \neq 0,$$

where $G \in \mathbb{Z}[X_1, \ldots, X_m]$ is a non-zero polynomial of degree $\nu < n - 2(m - 1)$.

The above results were generalized in Győry and Ru (1998) for the number field case, without assuming (ii). The proof involves Schmidt's Subspace Theorem with moving targets proved by Ru and Vojta (1997).

• As a generalization of decomposable form equations, several people studied *decomposable polynomial equations* of the form

$$F(\mathbf{x}) = \delta \text{ in } \mathbf{x} = (x_1, \ldots, x_m) \in A^m, \qquad (9.7.10)$$

where A is a subring of a finitely generated extension K of $\mathbb{Q}$ which is finitely generated over $\mathbb{Z}$, $\delta \in K \setminus \{0\}$ and

$$F \in K[X_1, \ldots, X_m]$$

is a *decomposable polynomial*, i.e., it factorizes into not necessarily homogeneous linear polynomials over a finite extension of K. In Evertse, Gaál and Győry (1989) a finiteness criterion was given for equation (9.7.10). Later, in the case $K = \mathbb{Q}$, explicit upper bounds were derived in Bérczes and Győry (2002) for the number of solutions, provided that this number is finite. Over number fields, effective bounds were derived for the solutions in Sprindžuk (1974) and Bilu (1995) for $m = 2$, and in Gaál (1984, 1985, 1986) for certain norm polynomial and discriminant polynomial equations.

- Let $f_1, \ldots, f_n$, G be non-zero polynomials in $K[X_1, \ldots, X_m]$ $(m \geq 2)$, where K is a number field. Let $F = f_1 \cdots f_n$, and assume that

$$\deg F > m \max_{1 \leq i \leq n} (\deg f_i) + \deg G.$$

Further, let O_S be a ring of S-integers in K, and as a generalization of decomposable polynomial equations, consider the equation

$$F(\mathbf{x}) = G(\mathbf{x}) \text{ in } \mathbf{x} \in O_S^m.$$

Let $\mathcal{X}$ be the hypersurface defined by $F = G$. It is proved in Corvaja and Zannier (2004a) that under certain additional assumptions, $\mathcal{X} \cap O_S^m$ is not Zariski dense in $\mathcal{X}$.

- Finally, we note that Győry (1983), Mason (1986a, 1986b, 1987, 1988) and Gaál (1988a, 1988b) established effective results for various decomposable form equations over function fields. Their proofs are based on some earlier variants of results from Chapter 7 concerning unit equations. Gaál and Pohst (2006a, 2006b, 2010) gave the complete resolution of some norm form equations over certain function fields over a finite field.

10

Further applications

In the previous chapters several applications of unit equations were presented or mentioned. Moreover, in Chapter 9 we showed that unit equations and decomposable form equations are in a certain sense equivalent, and using results concerning unit equations we proved several general results for decomposable form equations. Unit equations have, however, a great variety of other applications. In this chapter we briefly present some of these applications in their simplest form, without aiming at completeness. We note that numerous further applications to discriminant equations are treated in our subsequent book *Discriminant Equations in Diophantine Number Theory*.

The following topics are discussed: prime factors of sums of integers in Section 10.1, representations of elements of integral domains as sums of units in Section 10.2, lengths of finite orbits of polynomial maps on integral domains in Section 10.3, divisibility properties of polynomials with few non-zero coefficients in Section 10.4, arithmetic graphs with applications to irreducibility problems for polynomials in Section 10.5, discriminant equations and power integral bases in number fields in Section 10.6, finiteness results for binary forms of given discriminant in Section 10.7, equations involving resultants of monic polynomials in Section 10.8, equations and inequalities involving resultants of binary forms in Section 10.9, Lang's conjecture for tori in Section 10.10, linear recurrence sequences and exponential-polynomial equations in Section 10.11, and finally algebraic independence results for values of lacunary power series in Section 10.12.

10.1 Prime factors of sums of integers

We start with a simple application. Denote by $\omega(n)$ the number of distinct prime factors of a positive integer n, and by $P(n)$ the greatest prime factor of n.

Erdős and Turán (1934) proved that for any finite subset A of $\mathbb{Z}_{>0}$ with $|A| \geq 2$,

$$\omega\left(\prod_{a,a' \in A} (a + a') \right) > c_1 \log |A|,$$

where c_1 denotes an effectively computable positive number. Further, they conjectured, see Erdős (1976) that for every t there is a number $c(t)$ so that if A and B are finite subsets of $\mathbb{Z}_{>0}$ with $|A| = |B| \geq c(t)$ then

$$\omega\left(\prod_{a \in A, b \in B} (a + b) \right) > t.$$

Using the result from Evertse (1984a) on the number of solutions of S-unit equations in two unknowns, see also the Notes in Section 6.7, Győry, Stewart and Tijdeman (1986) proved the conjecture in the following more general and more precise form.

Theorem 10.1.1 *There exists an effectively computable positive absolute constant c_2 such that if A and B are any finite subsets of $\mathbb{Z}_{>0}$ with $|A| \geq |B| \geq 2$, then*

$$\omega\left(\prod_{a \in A, b \in B} (a + b) \right) > c_2 \log |A|. \tag{10.1.1}$$

Since the n-th prime can be estimated from below by a constant times $n \log n$, Theorem 10.1.1 implies the following result.

Corollary 10.1.2 *There exists an effectively computable positive absolute constant c_3 such that if A and B are any finite subsets of $\mathbb{Z}_{>0}$ with $|A| \geq |B| \geq 2$, then there exist integers $a \in A$ and $b \in B$ for which*

$$P(a + b) > c_3 \log |A| \log \log |A|. \tag{10.1.2}$$

Erdős, Stewart and Tijdeman (1988) proved that (10.1.1) and (10.1.2) are not far from being best possible. More precisely, they showed that there is a positive number c_4 such that for each integer k, with $k \geq 3$, there exist subsets A and B of $\mathbb{Z}_{>0}$ with $k = |A| \geq |B| \geq 2$ such that

$$\omega\left(\prod_{a \in A, b \in B} (a + b) \right) < c_4 (\log |A|)^2 \log \log |A|.$$

Further, they obtained a similar result for $P(\prod_{a \in A, b \in B}(a + b))$ as well.

Proof of Theorem 10.1.1. We deduce Theorem 10.1.1 from Corollary 6.1.5 concerning unit equations. It is enough to prove this theorem in the case when

$|B| = 2$. Let $a_1, \ldots, a_k$ denote the elements of A and let b_1, b_2 be the elements of B. Let $p_1, \ldots, p_t$ be the primes which divide

$$\prod_{i=1}^{k} \prod_{j=1}^{2} (a_i + b_j).$$

Each a_i yields a solution $x = a_i + b_1$, $y = a_i + b_2$ of the equation

$$x - y = b_1 - b_2.$$

By Corollary 6.1.5, there are at most $2^{8(2t+2)}$ such pairs $(a_i + b_1, a_i + b_2)$. Hence $k \leq 2^{16(t+1)}$, which gives $t > c_5 \log k$ for some effectively computable positive absolute constant c_5. $\qquad\square$

Győry, Sárközy and Stewart (1996) proved a multiplicative analogue of (10.1.1) by showing that there exists an effectively computable positive number c_6, such that if A and B are any finite subsets of $\mathbb{Z}_{>0}$ with $|A| \geq |B| \geq 2$, then

$$\omega\left(\prod_{a \in A, b \in B} (ab + 1) \right) > c_6 \log |A|. \qquad (10.1.3)$$

This implies a similar multiplicative analogue of (10.1.2). Further, they obtained the following common generalization of (10.1.1) and (10.1.3).

Theorem 10.1.3 *Let $n \geq 2$ be an integer, and let A and B be ordered finite subsets of $\mathbb{Z}_{>0}^n$ with $|A| \geq |B| \geq 2(n - 1)$ and with the following properties: the n-th coordinate of each vector in A is equal to 1 and any n vectors in $B \cup \{(0, \ldots, 0, 1)\}$ are linearly independent. Then*

$$\omega\left(\prod_{\substack{(a_1, \ldots, a_n) \in A \\ (b_1, \ldots, b_n) \in B}} (a_1 b_1 + \cdots + a_n b_n) \right) > c_7 \log |A|$$

with an effectively computable positive number c_7 depending only on n.

Note that (10.1.1) follows from Theorem 10.1.3 by taking $n = 2$ and $b_1 = 1$ for all (b_1, b_2) in B. Further, for $n = 2$, Theorem 10.1.3 gives (10.1.3) if $b_2 = 1$ for each (b_1, b_2) in B.

In Theorem 10.1.3, all assumptions are necessary. The proof of Theorem 10.1.3 depends on some finiteness results of Evertse and Győry (1988c) and Evertse (1995) on decomposable form equations.

In their above-mentioned paper, Győry, Sárközy and Stewart formulated the conjecture that if a, b and c denote distinct positive integers and max $(a, b, c) \to \infty$, then

$$P((ab + 1)(bc + 1)(ca + 1)) \to \infty.$$

The conjecture was confirmed in stronger forms by Corvaja and Zannier (2003) and, independently, by Hernández and Luca (2003). For further related results, see Bugeaud and Luca (2004), Luca (2005) and Zannier (2012).

10.2 Additive unit representations in finitely generated integral domains

Many people have investigated additive unit representations of elements in various rings. A central problem is whether as a $\mathbb{Z}$-module the ring of integers of a number field or, more generally, a finitely generated integral domain of characteristic 0 can be generated by its units. Further, if the answer is yes, how many units are needed to represent the elements of the ring? Ashrafi and Vámos (2005) proved that if K is a quadratic, a complex cubic or a cyclotomic number field generated by a primitive 2^m-th root of unity then there is no integer $n \geq 1$ such that every integer in K can be represented as the sum of not more than n units. Further, they conjectured that it holds true for all algebraic number fields K. Jarden and Narkiewicz (2007) proved the conjecture in the following more general situation.

Theorem 10.2.1 *If A is a finitely generated integral domain of characteristic 0, then there is no integer n such that every element of A is a sum of at most n units.*

In particular, this holds for the rings of integers and the rings of S-integers of number fields.

Theorem 10.2.1 is a consequence of the next theorem from Jarden and Narkiewicz (2007) and a classical result of van der Waerden.

Theorem 10.2.2 *If A is a finitely generated integral domain of characteristic 0 and $n \geq 1$ is an integer then there exists a constant $C_1(A, n)$, depending only on A and n, such that every non-constant arithmetic progression in A having more than $C_1(A, n)$ elements contains an element which is not a sum of n units.*

Theorem 10.2.2 is a special case of the following theorem which was established independently by Hajdu (2007). Let K be a field of characteristic 0 and Γ a multiplicative subgroup of K^* of finite rank r. Further, let $n \geq 1$ be an integer, and $\mathcal{A}$ a non-empty finite subset of K^n of cardinality t.

Theorem 10.2.3 *There exists a constant* $C_2(r, n, t)$ *depending only on* r, n *and* t *such that the length of any non-constant arithmetic progression in the set*

$$\left\{ \sum_{i=1}^{n} a_i s_i \ : \ (a_1, \ldots, a_n) \in \mathcal{A}, \ (s_1, \ldots, s_n) \in \Gamma^n \right\}$$

is at most $C_2(r, n, t)$.

In the special case when A is a finitely generated integral domain of zero characteristic, $\Gamma = A^*$, $t = 1$ and $\mathcal{A} = (1, \ldots, 1)$, Theorem 10.2.3 gives Theorem 10.2.2. We recall that A^*, i.e., the unit group of A, is finitely generated, and hence of finite rank.

The proofs of Theorems 10.2.2 and 10.2.3 are both based on earlier versions of Theorem 6.1.3 on unit equations and a result of van der Waerden (1927) from Ramsey theory.

Later, Hajdu and Luca (2010) proved Theorem 10.2.3 with a completely explicit value of $C_2(r, n, t)$. Its proof depends only on Theorem 6.1.3, and avoids the use of van der Waerden's Theorem.

Below we sketch the proofs of Theorems 10.2.2 and 10.2.1. We shall use the following version of van der Waerden's Theorem.

Theorem 10.2.4 *Let* r, s *be fixed positive integers. Then for any integer* N *sufficiently large in terms of* r, s *the following holds: for any arithmetic progression* P *of length* N *of rational integers, and any splitting of* P *into* r *subsets, at least one of these subsets contains an arithmetic progression of length* s.

Proof. See van der Waerden (1927).　　　　　　　　　　　　　□

Proof of Theorem 10.2.2. We proceed by induction on n. Let first $n = 1$. Let $a_j = a_0 + (j - 1)\delta$, $j = 1, \ldots, N$, be an arithmetic progression consisting of units of A, where δ is a non-zero element of A. We have $a_{j+1} - a_j = \delta$ for $j = 1, \ldots, N - 1$, hence an earlier version, due to Evertse and Győry (1988b), of Theorem 6.1.3 implies that N is bounded by a number depending only on A.

Next let $n \geq 1$, and assume that the assertion holds with a constant $C_1(A, k)$ for each positive integer k not exceeding n. For $\delta \in A \setminus \{0\}$, consider now a finite arithmetic progression $a_j = a_0 + (j - 1)\delta$ in A, $j = 1, \ldots, N$, each term of which is a sum of $n + 1$ units from A. We show that N can be bounded above by a number which depends only on A and n.

Denote by $\Omega(\delta)$ the set of all units u in A which appear in a proper representation of the form $\delta = u_1 + \cdots + u_m$ with $m = 1, 2, \ldots, 2n + 2$, that is the unit sum $u_1 + \cdots + u_m$ has no vanishing subsum. Put $\Omega(\delta) = \{x_1, \ldots, x_M\}$. It follows from the above-mentioned result from Evertse and Győry (1988b) that M is bounded above by a number $C_3(A, n)$ which depends only on A and n. We have

$$a_j = \sum_{k=1}^{n+1} u_{k,j} \quad \text{for some } u_{k,j} \in A^*, \ j = 1, \ldots, N,$$

whence

$$\delta = a_{j+1} - a_j = \sum_{k=1}^{n+1} u_{k,j+1} - \sum_{k=1}^{n+1} u_{k,j}, \ j = 1, \ldots, N. \tag{10.2.1}$$

Cancel the possible vanishing subsums at the right-hand side of (10.2.1). Then, for each j, at least one of the units in (10.2.1) belong to $\Omega(\delta)$. We may assume without loss of generality that for every j either $u_{1,j}$ or $u_{1,j+1}$ belongs to $\Omega(\delta)$. For $t = 1, 2, \ldots, M$, put

$$X_t = \left\{ 1 \leq j \leq N : u_{1,j} = x_t \right\}, \quad Y_t = \left\{ 1 \leq j \leq N : u_{1,j+1} = x_t \right\}.$$

Then the set $\{1, 2, \ldots, N\}$ is the union of the sets $X_t, Y_t, t = 1, \ldots, M$. It follows from van der Waerden's Theorem stated above that at least one of the sets $X_1, \ldots, X_M, Y_1, \ldots, Y_M$ contains an arithmetic progression P of length $T > C_1(A, n)$ if N is sufficiently large with respect to $C_1(A, n)$ and $C_3(A, n)$. We may assume that X_1 has this property. Let d be the difference of P, and put $P = \{n_1, \ldots, n_T\}$, where $n_i = n_1 + (i - 1)d$ for $i = 1, \ldots, T$. Then one can easily verify that $a_{n_i} - x_1 = a_{n_1} - x_1 + (i - 1)d\delta$ for $i = 1, \ldots, T$, and hence $a_{n_1} - x_1, \ldots, a_{n_T} - x_1$ is an arithmetic progression of length $> C_1(A, n)$ in A, each term of which is a sum of n units. This contradicts the induction hypothesis. Thus $N \leq C_4(A, n)$ where $C_4(A, n)$ depends only on A and n. $\square$

Proof of Theorem 10.2.1. Assume that every non-zero element of A can be represented as the sum of at most n units, and let n be the smallest positive integer with this property. Consider a sufficiently long arithmetic progression

$$a_j = a_0 + (j - 1)\delta, \ j = 1, \ldots, N$$

in A, where δ is a non-zero element of A. We follow the above argument. Let X_i, $1 \leq i \leq n$, be the set of those indices $j \in \{1, 2 \ldots, N\}$ for which a_j is a proper sum of i units from A. Then the set $\{1, 2 \ldots, N\}$ is the union of $X_1, \ldots, X_n$. If N is large enough then van der Waerden's Theorem implies that one of the X_i, say X_k, contains a long arithmetic progression $P = \{n_1, \ldots, n_T\}$. Then one can see similarly as above that if N is sufficiently large then $a_{n_1}, \ldots, a_{n_T}$ is a long arithmetic progression each term of which is the sum of k units, contradicting Theorem 10.2.2. This proves Theorem 10.2.1. $\square$

We present some consequences of Theorems 10.2.2 and 10.2.3 as well as some related results. For a finite set S of primes, denote by $\mathbb{Z}_S$ the ring of S-integers, and by $\mathbb{Z}_S^*$ the group of S-units. Jarden and Narkiewicz proved the following consequence of their Theorem 10.2.2.

Corollary 10.2.5 *Let $n \geq 1$ be an integer, and S a finite set of primes. Then the set of positive integers which are sums of at most n elements of $\mathbb{Z}_S^*$ has density zero.*

This follows from Theorem 10.2.2 applied with $A = \mathbb{Z}_S$ and from Szemerédi's Theorem (Szemerédi (1975)) on arithmetic progressions.

In his above-mentioned paper, Hajdu deduced from his Theorem 10.2.3 and from the theorem of Green and Tao (2008) about arithmetic progressions of primes the following.

Corollary 10.2.6 *Let $n \geq 1$, $S = \{p_1, \ldots, p_t\}$ be a finite set of primes, U_S the set of integers of the shape $\pm p_1^{z_1} \cdots p_t^{z_t}$ with $z_1, \ldots, z_t \in \mathbb{Z}_{\geq 0}$, and A a non-empty finite subset of $\mathbb{Z}^n$. Then there are infinitely many primes outside the set*

$$\left\{ \sum_{i=1}^{n} a_i s_i : (a_1, \ldots, a_n) \in \mathcal{A}, \ (s_1, \ldots, s_n) \in U_S^n \right\}.$$

For $\mathcal{A} = (1, \ldots, 1)$ this gives the following

Corollary 10.2.7 *Let $n \geq 1$, S and U_S be as in Corollary 10.2.6. There are infinitely many primes which are not the sum of n elements of A_S.*

For $n = 2$, $S = \{2, 3\}$, this provided a negative answer to a question of Pohst (oral communication) who asked whether every prime can be written in the form $2^u \pm 3^v$ with some non-negative integers u, v.

It is easy to see that there are number fields whose rings of integers cannot be generated by their units. Such number fields are, for example, the imaginary quadratic fields $\mathbb{Q}(\sqrt{d})$ with squarefree integers $d < -3$. Jarden and Narkiewicz (2007) formulated the following problem.

Problem *Give a criterion for an algebraic extension of $\mathbb{Q}$ to have the property that its ring of integers is generated by its units.*

Jarden and Narkiewicz provided some examples of infinite algebraic extensions of $\mathbb{Q}$ having this property. For example, the fields of all algebraic numbers and all real algebraic numbers are such fields. Further, by the Kronecker–Weber Theorem the maximal abelian extension of $\mathbb{Q}$ also has the property mentioned. In particular, the ring of integers of an abelian number field is generated by its units. Besides these results the above problem has been solved for quadratic number fields in Belcher (1974) and later in Ashrafi and Vámos (2005), for pure cubic fields in Tichy and Ziegler (2007) and for pure quartic complex fields in Filipin, Tichy and Ziegler (2008).

Answering affirmatively another problem of Jarden and Narkiewicz, Frei (2012) proved that for any number field K, there exists a finite extension L of K such that the ring of integers of L is generated by its units.

For further related results, we refer to Bertók (2013), Dombek, Hajdu and Pethő (2014), the survey paper Barroero, Frei, Tichy (2011) and the references given there.

10.3 Orbits of polynomial and rational maps

We start with some generalities. Let for the moment X be any non-empty set and $\phi : X \to X$ any map from X to itself (usually called *self-map* of X). We denote by $\phi^{(i)}$ the i-th iterate of ϕ (ϕ applied i times) where we agree that $\phi^{(0)}$ is the identity.

An *orbit* of ϕ is a sequence $O_\phi(a_0) := \{\phi^{(i)}(a_0)\}_{i=0}^{\infty}$, where $a_0 \in X$.

A *cycle* of ϕ is a sequence $(a_0, \ldots, a_{m-1})$ in X, where $a_0, \ldots, a_{m-1}$ are distinct, $a_i = \phi(a_{i-1})$ for $i = 1, \ldots, m - 1$, and $a_0 = \phi(a_{m-1})$. We call m the *length* of the cycle. In this case the orbit $O_\phi(a_0)$ is periodic with period m. Any $a_0 \in X$ that is the starting point of a cycle of ϕ of length m is called a *periodic point* of ϕ of period m.

An orbit $O_\phi(a_0)$ of ϕ is called finite if there are only finitely many distinct elements among $\phi^{(i)}(a_0)$ ($i = 0, 1, 2, \ldots$). Suppose this is the case. Write $a_i := \phi^{(i)}(a_0)$ for $i \geq 0$. Then there exists $l > 0$ such that there is k with $0 \leq k < l$ and $a_k = a_l$. Take l with this property minimal and put $m := l - k$. Then $a_0, \ldots, a_{k+m-1}$ are distinct, and $a_{i+m} = a_i$ for $i \geq k$. We express the orbit $O_\phi(a_0)$ conveniently as

$$O_\phi(a_0) = (a_0, \ldots, a_{k-1}, \overline{a_k, \ldots, a_{k+m-1}}),$$

where the overline indicates that $(a_k, \ldots, a_{k+m-1})$ is the recurring cycle of the orbit. We call $k + m$ the *length* of $O_\phi(a_0)$, $(a_0, \ldots, a_{k-1})$ the *tail* of $O_\phi(a_0)$, and $(a_k, \ldots, a_{k+m-1})$ the *cycle* of $O_\phi(a_0)$. Any $a_0 \in X$ for which $O_\phi(a_0)$ is finite is called a *preperiodic point* of ϕ. In particular, every periodic point is preperiodic.

There is a vast literature on orbits of maps defined by polynomials on rings (see for instance Narkiewicz (1995)) or of morphisms of algebraic varieties (see, e.g., Silverman (2007)). Here, we restrict ourselves to certain aspects that are closely related to unit equations. These concern bounding the lengths of cycles and finite orbits in the cases that $X = A$ is an integral domain and ϕ is defined by a polynomial, or X is the one-dimensional projective line $\mathbb{P}^1(K)$ over a field K and ϕ is a rational self-map of $\mathbb{P}^1(K)$.

Let A be an integral domain. A *polynomial cycle*, resp. *finite polynomial orbit* in A is a cycle, resp. finite orbit of a map of the type $x \mapsto f(x) : A \to A$ where $f \in A[X]$. We sloppily say that it is a cycle or finite orbit of f. We denote by $f^{(i)}$ the i-th iterate of the map $x \mapsto f(x)$.

Notice that linear polynomials $cX + d$ with $c, d \in A$, $c \neq 0$ do not give rise to finite orbits or cycles in A, unless c is a root of unity different from 1.

Narkiewicz (1989) proved that every polynomial cycle in $\mathbb{Z}$ has length at most 2, and in Narkiewicz and Pezda (1997) it is shown that every finite polynomial orbit in $\mathbb{Z}$ has length at most 4. Results from Narkiewicz (1989), Pezda (1994) and Narkiewicz and Pezda (1997) imply that for a large class of integral domains A the lengths of polynomial cycles and finite polynomial orbits in A are uniformly bounded in terms of A. We define the following quantities:

$$N_1(A, b) := |\{(x_1, x_2) \in A^* \times A^* : x_1 + x_2 = b\}| \quad (b \in A \setminus \{0\}),$$

$$N_1(A) := \sup \{N_1(A, b) : b \in A \setminus \{0\}\}.$$

The following result is part of Pezda (2014), Theorem 1. Its proof is based on ideas from Narkiewicz (1989).

Theorem 10.3.1 *Let A be an integral domain for which $N_1(A)$ is finite. Then every polynomial cycle in A has length at most $6(N_1(A) + 2)^2$.*

By Roquette's Theorem (Roquette (1957)) (see also Proposition 9.3.1 in this book), if A is an integral domain of characteristic 0 that is finitely generated over $\mathbb{Z}$, then its unit group A^* is finitely generated. We consider more generally integral domains of which the unit group has finite rank. If A is such a domain, and A^* has rank r, then $N_1(A) \leq 2^{16r+16}$ by Corollary 6.1.5. This leads at once to the following corollary.

Corollary 10.3.2 *Let A be an integral domain of characteristic 0 such that A^* has finite rank r. Then every polynomial cycle of A has length at most 2^{32r+35}.*

In the proof of Theorem 10.3.1 we need the following simple lemma.

Lemma 10.3.3 *Let $g \in A[X]$ and $a, b \in A$ with $a \neq b$. Then $a - b$ divides $g(a) - g(b)$.*

Proof. Use the fact that $\frac{g(X) - g(a)}{X - a} \in A[X]$. $\qquad\qquad\square$

Proof of Theorem 10.3.1. Notice that if $(a_0, a_1, \ldots, a_{m-1})$ (with $m \geq 2$) is a cycle in A of $f \in A[X]$, then $(0, 1, \frac{a_2 - a_0}{a_1 - a_0}, \ldots, \frac{a_{m-1} - a_0}{a_1 - a_0})$ is a cycle of $g(X) :=$ $(a_1 - a_0)^{-1}(f((a_1 - a_0)X + a_0) - a_0)$, which is a polynomial in $A[X]$. So there is no loss of generality to consider only polynomial cycles starting with $0, 1$.

Let $(a_0 = 0, a_1 = 1, a_2, \ldots, a_{m-1})$ be such a cycle, say of $f \in A[X]$, and assume without loss of generality that $m \geq 6$.

Let $\mathcal{V}$ be the set of integers $i \in \{0, \ldots, m - 1\}$ that are coprime with $m(m - 1)/2$. Let $p_1, \ldots, p_t$ be the distinct primes dividing m, with $p_1 < \cdots < p_t$. Then $\mathcal{V}$ consists precisely of the integers $i \in \{0, \ldots, m - 1\}$ such that $i \not\equiv 0, 2 \pmod{p_j}$ for $j = 1, \ldots, t$. Now the Chinese Remainder Theorem and a very generous estimate yield

$$|\mathcal{V}| = \left. \begin{cases} m \cdot \prod\limits_{j=1}^{t} \left(1 - 2p_j^{-1}\right) & \text{if } p_1 > 2, \\ m \cdot \frac{1}{2} \prod\limits_{j=2}^{t} \left(1 - 2p_j^{-1}\right) & \text{if } p_1 = 2 \end{cases} \right\} \geq \sqrt{m/6}. \qquad (10.3.1)$$

Given an integer k with $1 \leq k \leq m - 1$, we easily see, by repeatedly applying Lemma 10.3.3 with $g = f^{(k)}$, that a_k divides $a_{tk} - a_{(t-1)k}$ for $t = 1, 2, \ldots$. This implies that a_k divides a_{tk} for $t = 1, 2, \ldots$. Let $i \in \mathcal{V}$ with $i \geq 3$. There are $k, l \in \mathbb{Z}$ such that $ik = 1 + lm$, and so a_i divides $a_{1+lm} = 1$. Hence $a_i \in A^*$. Likewise, $a_{i-2} \in A^*$. Further, by applying Lemma 10.3.3 with $g = f^{(2)}$, $g = f^{(m-2)}$, respectively, we deduce that a_{i-2} divides $a_i - a_2$ and $a_i - a_2$ divides a_{i-2}, that is, $a_i - a_2 \in A^*$. It follows that $(a_i, a_2 - a_i)$ $(i \in \mathcal{V}, i \geq 3)$ are all solutions to the unit equation

$$x_1 + x_2 = a_2 \quad \text{in } x_1, x_2 \in A^*.$$

Hence $|\mathcal{V}| \leq N_1(A) + 2$. Together with (10.3.1) this implies $m \leq 6(N_1(A) + 2)^2$. This proves Theorem 10.3.1. $\qquad \square$

Pezda (1994) proved the following, by a totally different, local method, independent of unit equations. Let K be a field of characteristic 0 with discrete valuation v and $A = \{x \in K : v(x) \geq 0\}$ the associated discrete valuation domain. Assume that the residue class field of A is finite, say with p^f elements, where p is a prime number. Then every polynomial cycle of A has length at most

$$p^f(p^f - 1)p^{1 + \log v(p)/\log 2}.$$

By applying this with K a number field of degree d and v the discrete valuation corresponding to a prime ideal of the ring of integers O_K of K lying above 2, one obtains that every polynomial cycle of O_K has length at most

$$2^{d+1}(2^d - 1).$$

This bound is comparable with the bound of Corollary 10.3.2 in that it is exponential in rank O_K^*. In his Ph.D. thesis, Zieve (1996) proved various extensions of Pezda's result.

We now consider the finite polynomial orbits of an integral domain A. Denote by $B(A)$ the supremum of the lengths of the polynomial cycles of A. Narkiewicz and Pezda (1997), Theorem 1 proved that if $B(A)$ is finite and if moreover the number of non-degenerate solutions of $x_1 + x_2 + x_3 = 1$ in $x_1, x_2, x_3 \in A^*$ is finite, say $C(A)$, then every finite polynomial orbit of A has length at most

$$\tfrac{1}{3}B(A)(31 + C(A)) - 1.$$

We prove a variation on this result with a simpler proof. Define

$$N_2(A, b) := |\{(x_1, x_2) \in A^* \times A^* : (1 + x_1)(1 + x_2) = b\}| \quad (b \in A \setminus \{0, 1\}),$$
$$N_2(A) := \sup\{N_2(A, b) : b \in A \setminus \{0, 1\}\}.$$

Theorem 10.3.4 *Let A be an integral domain for which both $B(A)$ and $N_2(A)$ are finite. Then every finite polynomial orbit of A has length at most*

$$B(A)(2N_2(A) + 5).$$

This has the following consequence for integral domains of characteristic 0 with unit group of finite rank.

Corollary 10.3.5 *Let A be an integral domain of characteristic 0 such that A^* has finite rank r. Then every finite polynomial orbit of A has length at most* $2^{1600(r+5)}$.

Proof. The equation $(1 + x_1)(1 + x_2) = b$ in $x_1, x_2 \in A^*$ (with $b \in A$, $b \neq 0, 1$) can be rewritten as a three term unit equation

$$x_1 + x_2 + x_1 x_2 = b - 1.$$

Since $x_1, x_2 \neq -1$, there can be no solutions with $x_1 + x_1 x_2 = 0$ or $x_2 + x_1 x_2 = 0$. Further, there are at most two solutions with $x_1 + x_2 = 0$. So apart from at most two solutions, each proper subsum of the left-hand side is non-zero. Now by applying the bound (6.1.4) of Amoroso and Viada with $n = 3$, we infer

$$N_2(A) \leq 2 + 24^{324(r+4)}.$$

Together with the upper bound for $B(A)$ from Corollary 10.3.2 this implies Corollary 10.3.5. □

Proof of Theorem 10.3.4. Let

$$(a_0, \ldots, a_{k-1}, \overline{a_k, \ldots, a_{k+m-1}})$$

be a finite polynomial orbit in A, say of $f \in A[X]$, where $a_0, \ldots, a_{k+m-1}$ are distinct. Then $(a_k, \ldots, a_{k+m-1})$ is a polynomial cycle, and so $m \leq B(A)$.

We first make some reductions. Write $k = qm + r$ with $q, r \in \mathbb{Z}$ and $0 \leq r \leq m - 1$. Then $(a_r, a_{r+m}, \ldots, \overline{a_{r+qm}})$ is a finite orbit of $f^{(m)}$. Let

$$b_i := \frac{a_{r+im} - a_k}{a_r - a_k} \quad (i = 0, 1, 2, \ldots),$$

$$h(X) := (a_r - a_k)^{-1} \big(f^{(m)}((a_r - a_k)X + a_k) - a_k \big).$$

Then $h \in A[X]$, $b_0 = 1$, $b_i = 0$ for $i \geq q$, $b_0, \ldots, b_{q-1}$ are distinct, and

$$(1, b_1, \ldots, b_{q-1}, \overline{0})$$

is a finite orbit of h. We show that

$$q \leq 2N_2(A) + 3. \tag{10.3.2}$$

Then using $k + m < (q + 2)m \leq (q + 2)B(A)$ we obtain at once Theorem 10.3.4.

We use that by Lemma 10.3.3 with $g = h^{(t)}$, $b_i - b_j$ divides $b_{i+t} - b_{j+t}$ for any i, j with $0 \leq i, j \leq q$ and $i \neq j$ and any $t > 0$. Assume without loss of generality that $q \geq 5$ and let i be an index with $q/2 \leq i \leq q - 2$. Then $b_i - 1 = b_i - b_0$ divides $b_{2i} - b_i = -b_i$, hence $x_1 := b_i - 1 \in A^*$. Further, $b_{q-1} - b_i$ divides $b_{2q-2-i} - b_{q-1} = -b_{q-1}$ and hence also b_i, while $b_i = b_i - b_q$ divides $b_{q-1} - b_{2q-1-i} = b_{q-1}$, and hence also $b_{q-1} - b_i$. So $x_2 := (b_{q-1}/b_i) - 1 \in A^*$. Notice that x_1, x_2 are elements of A^* satisfying

$$(1 + x_1)(1 + x_2) = b_{q-1}$$

and that $b_{q-1} \neq 0, 1$. So the number of indices i with $\frac{1}{2}q \leq i \leq q - 2$ is at most $N_2(A)$. This implies (10.3.2) and hence Theorem 10.3.4. □

Let again A be an integral domain. We call two sequences $\{a_i\}_{i=0}^r$, $\{b_i\}_{i=0}^r$ in A (with r finite or infinite) equivalent if there are $\varepsilon \in A^*$, $a \in A$ such that $b_i = \varepsilon a_i + a$ for $i = 0, 1, 2, \ldots$. If $\{a_i\}_{i=0}^r$ is a cycle or orbit of a polynomial $f \in A[X]$, then $\{b_i\}_{i=0}^r$ is a cycle or orbit of $g(X) := \varepsilon f(\varepsilon^{-1}(X - a)) + a$.

A polynomial cycle of A is called linear if it is a cycle of a linear polynomial from $A[X]$, otherwise non-linear. A finite polynomial orbit of A is called (non-)linear if its cycle is (non-)linear.

Halter-Koch and Narkiewicz (1997, 2000) proved that if A is an integral domain of characteristic 0 that is finitely generated over $\mathbb{Z}$ and integrally closed, then it has up to equivalence only finitely many non-linear polynomial cycles and only finitely many finite non-linear polynomial orbits. The non-linearity assumption is needed here. For instance, we obtain infinitely many pairwise inequivalent linear orbits by taking $(1, \overline{0, a})$ $(a \in A \setminus \{0, 1\})$, which is a finite orbit of $f = (X - 1)(X - a)$ with linear cycle $(0, a)$ coming from $a - X$. The proof of Halter-Koch and Narkiewicz heavily uses finiteness results on unit equations. Pezda (2014) gave an effective algorithm that computes, for any given number field K, a full set of representatives for the equivalence classes of the non-linear polynomial cycles and finite orbits of O_K.

We state without proof some results on the lengths of cycles and finite orbits of rational maps on the projective line. In general, for an arbitrary field K, a rational map $\phi : \mathbb{P}^1(K) \to \mathbb{P}^1(K)$ of degree n is given by

$$\phi : (x : y) \mapsto (F(x, y) : G(x, y)), \qquad (10.3.3)$$

where $F, G \in K[X, Y]$ are two binary forms of degree n without a common factor, i.e., with resultant $R(F, G) \neq 0$. Notice that the map ϕ is unaffected if we replace F, G by $\lambda F, \lambda G$ for some $\lambda \in K^*$.

We assume henceforth that K is a number field. Let ϕ be the rational self-map of $\mathbb{P}^1(K)$ of degree n, given by (10.3.3). Let $\mathfrak{p}$ be a prime ideal of O_K. We say that ϕ has *good reduction at* $\mathfrak{p}$ if the following holds: choose F, G such that their coefficients lie in O_K but not all in $\mathfrak{p}$; then $R(F, G) \notin \mathfrak{p}$. Otherwise, we say that ϕ has *bad reduction at* $\mathfrak{p}$. It is not difficult to show that for a rational self-map of $\mathbb{P}^1(K)$ there are only finitely many prime ideals of O_K at which it has bad reduction.

This notion of good reduction has an alternative interpretation. Let $\mathbb{F}_\mathfrak{p} := O_K/\mathfrak{p}$ denote the residue class field of $\mathfrak{p}$ and denote by $F_\mathfrak{p}, G_\mathfrak{p}$ the binary forms in $\mathbb{F}_\mathfrak{p}[X, Y]$, obtained by reducing the coefficients of F, G modulo $\mathfrak{p}$. Then the reduction $\phi_\mathfrak{p}$ of ϕ at $\mathfrak{p}$ is the self-map of $\mathbb{P}^1(\mathbb{F}_\mathfrak{p})$ given by

$$(x : y) \mapsto \left(\frac{F_\mathfrak{p}(x, y)}{H(x, y)} : \frac{G_\mathfrak{p}(x, y)}{H(x, y)} \right),$$

where H is the greatest common divisor of $F_\mathfrak{p}, G_\mathfrak{p}$ in $\mathbb{F}_\mathfrak{p}[X, Y]$. Notice that $R(F, G) \notin \mathfrak{p}$ if and only if H is constant. This means that ϕ has good reduction at $\mathfrak{p}$ if and only if $\phi_\mathfrak{p}$ has the same degree as ϕ.

For more information on reduction of rational maps, see Silverman (2007), sections 2.3–2.5.

We state without proof the following result.

Theorem 10.3.6 *Let K be an algebraic number field of degree d and let ϕ be a rational self-map of $\mathbb{P}^1(K)$. Let S be the set of places of K consisting of the infinite places of K and the prime ideals at which ϕ has bad reduction. Denote by t the number of prime ideals of O_K at which ϕ has bad reduction, and let $s := |S|$.*

(i) Every cycle of ϕ has length at most

$$C_1(d, t) := \left(12(t + 2) \log \left(5(t + 2)\right)\right)^d.$$

(ii) Every finite orbit of ϕ has length at most

$$C_2(s) := \left(e^{10^{12}} (s + 1)^8 (\log 5(s + 1))^8\right)^s.$$

Part (i) has been proved by Morton and Silverman (1994), corollary B. The proof is by means of a local method, extending that of Pezda (1994). For similar and related results see Zieve (1996) and Silverman (2007), section 2.6. Part (ii) has been proved by Canci (2007), Theorem 1. His proof is an extension of that of Theorem 10.3.4. His main tools are part (i), and Theorem 6.1.3 on unit equations.

We consider the preperiodic points of rational self-maps of $\mathbb{P}^1(K)$. Let ϕ be a rational self-map of $\mathbb{P}^1(K)$. First suppose that ϕ is linear, that is, $\phi(x : y) = (ax + by : cx + dy)$ where $B := \left(\begin{smallmatrix} a & b \\ c & d \end{smallmatrix}\right) \in \mathrm{GL}(2, K)$. If B has two eigenvalues in $\overline{K}$ whose quotient is a root of unity, then there is $m > 0$ such that $\phi^{(m)}$ is the identity and every point in $\mathbb{P}^1(K)$ is a periodic point of ϕ. Otherwise, B has at most two fixed points, depending on the number of eigenvalues of B in K, and no other preperiodic points.

Assume henceforth that ϕ has degree at least 2. We denote by $\mathrm{PrePer}_K(\phi)$ the set of preperiodic points of ϕ in $\mathbb{P}^1(K)$. More generally, we may extend ϕ to a rational self-map of $\mathbb{P}^1(\overline{\mathbb{Q}})$ and consider the set $\mathrm{PrePer}_{K,D}(\phi)$ of all preperiodic points of ϕ that have degree at most D over K. Then Northcott (1950) proved that for any integer $D > 0$, the set $\mathrm{PrePer}_{K,D}(\phi)$ is finite.

We state a special case of the *Uniform Boundedness Conjecture*, which was first formulated in Morton and Silverman (1994).

Conjecture 10.3.7 *Let K be a number field of degree d, and ϕ a rational self-map of $\mathbb{P}^1(K)$ of degree $n \geq 2$. Then*

$$|\mathrm{PrePer}_K(\phi)| \leq C(d, n),$$

where $C(d, n)$ depends on d and n only.

From Theorem 10.3.6 we deduce a weaker version.

Corollary 10.3.8 *Let K, d, ϕ, n be as in Conjecture 10.3.7 and assume that ϕ has bad reduction at precisely t prime ideals of O_K. Then*

$$|\mathrm{PrePer}_K(\phi)| \leq C(d, n, t),$$

where $C(d, n, t)$ is an effectively computable number, depending on d, n and t only.

Proof. For the i-th iterate of ϕ we have $\phi^{(i)}(x : y) = (F_i(x, y) : G_i(x, y))$, where both F_i, G_i are binary forms of degree n^i with $R(F_i, G_i) \neq 0$. By Theorem 10.3.6, for every point $(x : y) \in \mathrm{PrePer}_K(\phi)$, there are k, l with $0 \leq k < l \leq C_2(s)$, such that $\phi^{(k)}(x : y) = \phi^{(l)}(x : y)$, that is, $F_k(x, y)G_l(x, y) = F_l(x, y)G_k(x, y)$. This shows that the preperiodic points of ϕ are among the zeros of the binary form

$$P := \prod_{0 \leq k < l \leq C_2(s)} (F_k G_l - F_l G_k),$$

which is not identically zero since F_i, G_i are coprime for $i \geq 0$. Now the number of preperiodic points of ϕ is at most the degree of P, which can be estimated from above effectively in terms of s and n, hence in terms of d, t and n. $\square$

10.4 Polynomials dividing many k-nomials

By a monic k-*nomial* over $\mathbb{Q}$ we will mean a polynomial of the form

$$X^{m_1} + a_2 X^{m_2} + \cdots + a_{k-1} X^{m_{k-1}} + a_k X^{m_k} \in \mathbb{Q}[X]$$

with $m_1 > \cdots > m_{k-1} > m_k = 0$.

If the polynomial is not a $(k-1)$-nomial, i.e., if all $a_i \neq 0$, we call $(m_1, \ldots, m_k)$ its *exponent k-tuple*. Put

$$PR_k := \left\{ \begin{array}{l} P \in \mathbb{Q}[X] : \exists Q \in \mathbb{Q}[X], \ r \in \mathbb{Z}_{\geq 1} \ \text{with } \deg(Q) < k \\ \text{such that } P(X) \mid Q(X^r) \ \text{over } \mathbb{Q} \end{array} \right\}.$$

Posner and Rumsey (1965) noted that $P(X) \in PR_k$ implies that $P(X)$ divides infinitely many monic k-nomials over $\mathbb{Q}$. Indeed, if $P(X)$ divides $Q(X^r)$ over $\mathbb{Q}$ for some $Q(X)$ of degree $< k$ and integer $r \geq 1$, then the vector space of polynomials in $\mathbb{Q}[X]$ modulo $Q(X)$ is at most $(k-1)$-dimensional, and hence $Q(X)$ divides infinitely many k-nomials $T(X)$ over $\mathbb{Q}$. But then $Q(X^r)$ divides $T(X^r)$ and so $P(X)$ divides $T(X^r)$ over $\mathbb{Q}$. Conversely, Posner and Rumsey conjectured that *if a polynomial $P \in \mathbb{Q}[X]$ divides infinitely many monic k-nomials over $\mathbb{Q}$ then $P \in PR_k$.*

For $k = 2$ the conjecture is obvious. For $k = 3$, Posner and Rumsey proved a weaker version of their conjecture. Later, Győry and Schinzel (1994) showed that the conjecture is true for $k = 3$ and false for $k \geq 4$. The disproof for the case $k \geq 4$ is elementary. For $k = 3$, the proof involves some deep results on S-unit equations in two unknowns.

Győry and Schinzel (1994) proved the following stronger assertion.

Theorem 10.4.1 *Let $P \in \mathbb{Q}[X]$ be a non-constant polynomial with t distinct zeros, K the splitting field of P, d the degree of K over $\mathbb{Q}$, and s the number of distinct prime ideal factors of the zeros different from 0 of P. There are effectively computable numbers C_1, C_2 depending only on d and s such that if P divides more than $C_1 \cdot C_2^t$ monic trinomials over $\mathbb{Q}$ then $P \in P\mathcal{R}_3$.*

Győry and Schinzel gave C_1 and C_2 in explicit form. It should be observed that these numbers do not depend on the size of the coefficients of P.

Proof (sketch). The proof of Theorem 10.4.1 is based on some earlier, quantitative versions of Corollary 6.1.5 and Theorem 6.1.6. We sketch the basic idea of the proof. Let P be a polynomial as in the theorem, and let $T = X^m + aX^n + b$ be a trinomial over $\mathbb{Q}$ which is divisible by P. If X divides $P(X)$ or if $ab = 0$, then $P \in P\mathcal{R}_3$ easily follows. Hence we assume that X does not divide P and $ab \neq 0$. It is easy to show that P can be written in the form $P_s P_1 P_2^2$, where P_1 and P_2 are relatively prime squarefree polynomials in $\mathbb{Q}[X]$. Denote by $\alpha_1, \ldots, \alpha_t$ the distinct zeros of $P_1 P_2$, and by S the set of prime ideal factors of these zeros in K. Then, for $i = 1, \ldots, t$, (α_i^m, α_i^n) is a solution of the S-unit equation

$$(-1/b)x_1 + (-a/b)x_2 = 1 \quad \text{in } S\text{-units } x_1, x_2. \tag{10.4.1}$$

First consider those trinomials $T = X^m + aX^n + b$ $(ab \neq 0)$ over $\mathbb{Q}$ which are divisible by P and for which the corresponding equation (10.4.1) has at most two solutions. One can show that if there are more than 15 such trinomials then $P \in P\mathcal{R}_3$.

Next consider those trinomials $T = X^m + aX^n + b$ $(ab \neq 0)$ over $\mathbb{Q}$ which are divisible by $P(X)$ and the corresponding equation (10.4.1) has more than two solutions. If $X^m + aX^n + b$ and $X^{m'} + a'X^{n'} + b'$ are such trinomials and if the corresponding equations of the form (10.4.1) are S-equivalent in the sense defined before the enunciation of Theorem 6.1.6 then $a'/a, b'/b \in O_S^* \cap \mathbb{Q}^*$, where O_S^* denotes as usual the S-unit group in K. Hence it follows from a quantitative version, due to Győry (1992b), of Theorem 6.1.6 over number fields that there is a subset $\mathcal{A}$ of $(\mathbb{Q}^*)^2$ of cardinality at most C_3 such that for each trinomial $X^m + aX^n + b$ under consideration, $a = \varepsilon a_0$, $b = \eta b_0$ with ε, $\eta \in O_S^* \cap \mathbb{Q}^*$ and some $(a_0, b_0) \in \mathcal{A}$. Here C_3 is a number depending only on d and s which can be given explicitly.

Fix such a pair $(a_0, b_0) \in \mathcal{A}$ and consider all the trinomials of the form $X^m + \varepsilon a_0 X^n + \eta b_0$ with ε, $\eta \in O_S^* \cap \mathbb{Q}^*$, which are divisible by P over $\mathbb{Q}$. If $X^m + \varepsilon a_0 X^n + \eta b_0$ and $X^m + \varepsilon' a_0 X^n + \eta' b_0$ are such trinomials then $P(X)$ divides $X^n + c$ with some $c \in \mathbb{Q}^*$ and so $P \in P\mathcal{R}_3$. Hence it suffices to deal with those trinomials for which the pairs (m, n) are pairwise distinct. We may assume that in the pairs (m, n) in question, say $m_1, \ldots, m_u$ are pairwise distinct for $u > C_4^t$ with a number C_4 specified below. Then P divides $T_i = X^{m_i} + \varepsilon_i a_0 X^{n_i} + \eta_i b_0$ over $\mathbb{Q}$ for $i = 1, \ldots, u$, and so, for each i,

$$(-1/b_0) \left(\alpha_j^{m_i} / \eta_i \right) + (-a_0/b_0) \left(\varepsilon_i \alpha_j^{n_i} / \eta_i \right) = 1$$

for $j = 1, \ldots, t$, where ε_i, $\eta_i \in O_S^* \cap \mathbb{Q}^*$ for $i = 1, \ldots, u$. By the above-mentioned version, due to Evertse (1984a), of Corollary 6.1.5, C_4 can be chosen as an explicit expression of d and s such that for each j with $1 \leq j \leq t$, $\alpha_j^{m_i}/\eta_i$ can assume at most C_4 values. Since by assumption $u > C_4^t$, there are distinct i_1 and i_2 with $1 \leq i_1, i_2 \leq u$ such that $\alpha_j^{m_{i_1}}/\eta_{i_1} = \alpha_j^{m_{i_2}}/\eta_{i_2}$ for $j = 1, \ldots, t$ and, if $m_{i_1} > m_{i_2}$, then putting $r = m_{i_1} - m_{i_2}$ and $\eta = \eta_{i_1}/\eta_{i_2}$, we get $\alpha_j^r = \eta$ for $j = 1, \ldots, t$. Consequently, $P_1 P_2$ divides $X^r - \eta$, i.e. P divides $(X^r - \eta)^2$ and so $P \in P\mathcal{R}_3$. Finally, we obtain that if P divides more than $15 + C_3 \cdot C_4^{2t}$ trinomials then $P \in P\mathcal{R}_3$. $\qquad \square$

Schlickewei and Viola (1997) improved the bound occurring in Theorem 10.4.1. They proved the theorem with a bound of the form $C_5 \cdot q^{C_6}$ where q denotes the degree of P and C_5, C_6 are explicitly given absolute constants. We note that under the above notation, $t \leq q \leq 2t$ and $q \leq d \leq q!$ hold. In their paper Schlickewei and Viola made the *conjecture* that *the bound $C_5 \cdot q^{C_6}$ may be replaced by an absolute constant* which does not involve the degree of P at all. However, as is mentioned by them, at present this seems to be out of reach.

In Győry and Schinzel (1994), the authors proposed as a problem a modified version of the conjecture of Posner and Rumsey for $k \geq 4$. Hajdu (1997) gave a negative answer to the problem and proposed a further refinement of the conjecture. For $k = 5$, this was disproved by Hajdu and Tijdeman (2003). Further, they noticed that if P divides two monic k-nomials, say T_1 and T_2, over $\mathbb{Q}$ with the same exponent k-tuple, then it divides infinitely many k-nomials, for example the k-nomials $\frac{a}{a+b} T_1 + \frac{b}{a+b} T_2$ for every pair (a, b) of positive rationals. Then in their paper (Hajdu and Tijdeman (2003)), they made the following conjecture.

Conjecture 10.4.2 *For any $k \geq 5$, a polynomial $P \in \mathbb{Q}[X]$ with $P(0) \neq 0$ divides infinitely many monic k-nomials with non-zero constant terms over $\mathbb{Q}$ if and only if either*

(i) $P \in P\mathcal{R}_k$ or
(ii) P divides over $\mathbb{Q}$ two monic k-nomials with the same exponent k-tuple.

In the same paper, the authors proved this assertion for $k = 4$ and for polynomials P with only simple zeros. Further, in Hajdu and Tijdeman (2008) they confirmed the conjecture for $k \geq 5$ in the important special case when P is irreducible over $\mathbb{Q}$ and its Galois group is $[2k/3]$-times transitive. The proof is complicated; it depends on Theorem 6.1.3 which gives an upper bound for the number of solutions of multivariate unit equations.

Finally, we note that Schlickewei and Viola (1999) described a so-called "proper" family $\mathcal{F}_k$ of monic k-nomials such that if a polynomial P having only simple zeros divides more than $C_7(k)$ elements of $\mathcal{F}_k$ with a $C_7(k)$ given explicitly in terms of k, then $P \in P\mathcal{R}_k$.

10.5 Irreducible polynomials and arithmetic graphs

Let K be an algebraic number field, S a finite set of places on K containing all infinite places, O_S the ring of S-integers, $N_S(\cdot)$ the S-norm and N a positive integer. For any finite subset $\mathcal{A} = \{\alpha_1, \ldots, \alpha_m\}$ of O_S with $m \geq 3$, we denote by $\mathcal{G}_S(\mathcal{A}) = \mathcal{G}_S(\mathcal{A}, N)$ the graph whose vertex set is $\mathcal{A}$ and whose edges are the unordered pairs $\{\alpha_i, \alpha_j\}$ with

$$N_S(\alpha_i - \alpha_j) > N.$$

When S consists of the infinite places, this graph will be denoted by $\mathcal{G}(\mathcal{A}) = \mathcal{G}(\mathcal{A}, N)$. These graphs $\mathcal{G}(\mathcal{A})$ and $\mathcal{G}_S(\mathcal{A})$ were introduced in Győry (1971, 1972, 1980c) and were studied and applied by Győry and others; see Győry (2008b), Győry, Hajdu and Tijdeman (2011) and the references given there.

Several Diophantine problems, for instance related to irreducibility of polynomials (see Theorem 10.5.3), decomposable form equations (see Subsection 9.6.2), discriminant equations (see Theorems 10.6.1–10.6.3) and resultant equations (see Theorem 10.8.1) can be reduced to the study of connectedness properties of graphs $\mathcal{G}_S(\mathcal{A}, N)$. Such properties are stated in Theorems 10.5.1 and 10.5.2 below.

In the complement of $\mathcal{G}_S(\mathcal{A}, 1)$, $\{\alpha_i, \alpha_j\}$ is an edge if and only if $\alpha_i - \alpha_j$ is an S-unit. Hence this complement is called a *difference graph of S-units*. For any finite (simple) graph $\mathcal{G}$ of order ≥ 3 there is a finite set S of places on

K containing all infinite ones such that $\mathcal{G}_S(\mathcal{A}, 1)$ is isomorphic to $\mathcal{G}$ for some subset $\mathcal{A}$ of O_S. Further, such S and $\mathcal{A}$ can be effectively determined, provided that K is effectively given; see Győry, Hajdu and Tijdeman (2014).

The subsets $\mathcal{A} = \{\alpha_1, \ldots, \alpha_m\}$, $\mathcal{A}' = \{\alpha_1', \ldots, \alpha_m'\}$ of O_S are called *S-equivalent* if, after some reordering of $\alpha_1', \ldots, \alpha_m'$,

$$\alpha_i' = \varepsilon \alpha_i + \beta, \quad i = 1, \ldots, m$$

for some $\varepsilon \in O_S^*$ and $\beta \in O_S$. In this case the graphs $\mathcal{G}_S(\mathcal{A})$ and $\mathcal{G}_S(\mathcal{A}')$ are obviously isomorphic, they have the same structure. There are infinitely many S-equivalence classes of subsets $\mathcal{A}$ of O_S with given cardinality $m \geq 3$.

We present two theorems in simplified from on the structure of graphs $\mathcal{G}_S(\mathcal{A})$. Denote by d the degree of K, and let $s := |S|$. Further, as in Chapter 4, let P denote the greatest norm and Q the product of norms of the prime ideals involved in S, and let R_S be the S-regulator of K.

The following theorem was proved by Győry (2008b) in a more precise form. Its first, weaker version can be found in Győry (1980c).

Theorem 10.5.1 *Let $m \geq 3$ be an integer, and $\mathcal{A} = \{\alpha_1, \ldots, \alpha_m\}$ a subset of O_S. Then the graph $\mathcal{G}_S(\mathcal{A}, N)$ has at most two connected components, except possibly in the case when there is an $\varepsilon \in O_S^*$ such that*

$$\max_{1 \leq i,j \leq m} h((\alpha_i - \alpha_j)/\varepsilon) \leq C_1 m^3 (C_2 s)^{2(s+2)} P R_S (\log^* R_S)(\log^* QN).$$

Here C_1, C_2 are effectively computable positive numbers such that C_1 depends only on the degree d of K and the regulator and class number of K, and C_2 only on d.

This means that the number of exceptional S-equivalence classes is finite, and a representative of each class can be, at least in principle, effectively determined.

Proof (sketch). Theorem 10.5.1 is proved by repeated application of Corollary 4.1.5. We sketch some ideas behind the proof.

For a finite graph $\mathcal{G}$ we denote by $\mathcal{G}^\triangle$ the *triangle graph* of $\mathcal{G}$, i.e. the graph whose vertices are the edges of $\mathcal{G}$, and two vertices e_1 and e_2 of $\mathcal{G}^\triangle$ are connected by an edge if and only if $\mathcal{G}$ contains a triangle having e_1 and e_2 as edges. Further, if both $\mathcal{G}$ and $\mathcal{G}^\triangle$ are connected then we say that $\mathcal{G}$ is $\triangle$-*connected*.

Consider now $\mathcal{G}_S(\mathcal{A}) = \mathcal{G}_S(\mathcal{A}, N)$ in Theorem 10.5.1 and assume that this graph has at least three connected components. It is easy to see that in this case the complement of $\mathcal{G}_S(\mathcal{A})$, for simplicity denoted by $\mathcal{G}$, is $\triangle$-connected. Let

$\{\alpha_i, \alpha_j, \alpha_k\}$ be a triangle in $\mathcal{G}$. Then we have

$$N_S(\alpha_i - \alpha_j) \leq N, \ \ N_S(\alpha_j - \alpha_k) \leq N, \ \ N_S(\alpha_k - \alpha_i) \leq N.$$

Using Proposition 4.3.12, this gives that up to unknown S-unit factors, the numbers $\alpha_i - \alpha_j, \alpha_j - \alpha_k, \alpha_k - \alpha_i$ have effectively bounded heights. But

$$(\alpha_i - \alpha_j)/(\alpha_i - \alpha_k) + (\alpha_j - \alpha_k)/(\alpha_i - \alpha_k) = 1,$$

hence Corollary 4.1.5 implies that the height of $(\alpha_i - \alpha_j)/(\alpha_j - \alpha_k)$ can be effectively bounded above. If $\{\alpha_j, \alpha_k, \alpha_l\}$ is another triangle in $\mathcal{G}$, then the heights of $(\alpha_j - \alpha_k)/(\alpha_k - \alpha_l)$ and so $(\alpha_i - \alpha_j)/(\alpha_k - \alpha_l)$ are also effectively bounded. Continuing this procedure, it follows that for any two connected vertices $\{\alpha_i, \alpha_j\}$, $\{\alpha_p, \alpha_q\}$ in $\mathcal{G}^\triangle$, the height of $(\alpha_i - \alpha_j)/(\alpha_p - \alpha_q)$ is effectively bounded. But $\mathcal{G}^\triangle$ is connected, hence for each quadruple $\{\alpha_i, \alpha_j, \alpha_p, \alpha_q\}$ for which $\{\alpha_i, \alpha_j\}$ and $\{\alpha_p, \alpha_q\}$ are edges in $\mathcal{G}$, the height of $(\alpha_i - \alpha_j)/(\alpha_p - \alpha_q)$ can be effectively bounded. Fix p and q. Since $\mathcal{G}$ is connected, each distinct α_a and α_b can be connected by a path in $\mathcal{G}$. Summing over all terms $(\alpha_i - \alpha_j)/(\alpha_p - \alpha_q)$ for the edges in this path we infer that for each pair (a, b) the height of $(\alpha_a - \alpha_b)/(\alpha_p - \alpha_q)$ can be effectively bounded. From these facts it follows easily that up to a common S-unit factor, the height of $\alpha_a - \alpha_b$ is effectively bounded for each distinct α_a, α_b, as stated in Theorem 10.5.1. $\quad\square$

The following theorem is a more precise but ineffective version of Theorem 10.5.1.

There are only finitely many pairwise non-associate $\alpha \in O_S$ with $N_S(\alpha) \leq N$. Denote by $\Psi_S(N)$ the maximal number of such α.

Theorem 10.5.2 *Let $m \geq 3$ be an integer with $m \neq 4$. Apart from at most finitely many S-equivalence classes of subsets $\mathcal{A} = \{\alpha_1, \ldots, \alpha_m\}$ of O_S,*

$$\mathcal{G}_S(\mathcal{A}) \text{ has a connected component of order at least } m - 1. \quad\quad (10.5.1)$$

Further, if

$$m > 3 \cdot 2^{16s} \Psi_S^2(N), \quad\quad\quad\quad (10.5.2)$$

then (10.5.1) holds for all subsets $\mathcal{A} = \{\alpha_1, \ldots, \alpha_m\}$ of O_S.

We note that the assumption $m \neq 4$ is necessary, and the lower bound $m - 1$ in (10.5.1) is sharp.

A more general and quantitative version of the first part of Theorem 10.5.2 is given in Győry (2008b); see also Győry (1990). The second part is a special case of Theorem 2.3 of Győry (2008b). For earlier versions of this part, see Győry (1980c, 1990).

Proof (sketch). The proof of the first part of Theorem 10.5.2 depends on Corollary 6.1.2 or, in the quantitative case, on Theorem 6.1.3 and (6.1.4) concerning S-unit equations. We now sketch the ideas behind the proof of the second statement. Let $\mathcal{A} = \{\alpha_1, \ldots, \alpha_m\}$ be a subset of O_S, and let $\mathcal{G}_1, \ldots, \mathcal{G}_l$ be the connected components of $\mathcal{G}_S(\mathcal{A})$ such that $|\mathcal{G}_1| \leq |\mathcal{G}_2| \leq \cdots \leq |\mathcal{G}_l|$. Suppose that $l \geq 3$ or $l = 2$ and $|\mathcal{G}_1| \geq 2$. If $l \geq 3$, let $\alpha_{i_1}, \alpha_{i_2}$ be vertices of $\mathcal{G}_1$ and $\mathcal{G}_2$, respectively, while if $l = 2$, let $\alpha_{i_1}, \alpha_{i_2}$ be vertices of $\mathcal{G}_1$. Then

$$\alpha_{i_2} - \alpha_{i_1} = (\alpha_{i_2} - \alpha_j) + (\alpha_j - \alpha_{i_1}) \qquad (10.5.3)$$

follows for every vertex α_j of $\mathcal{G}_3, \ldots, \mathcal{G}_l$ if $l \geq 3$, and of $\mathcal{G}_2$ if $l = 2$. Further, $\alpha_{i_2} - \alpha_j$ and $\alpha_j - \alpha_{i_1}$ have S-norms at most N for each j. There are $\Psi_S^2(N)$ pairs $(\beta_1, \beta_2) \in O_S^2$ with non-zero β_1, β_2 such that $\alpha_{i_2} - \alpha_j = \beta_1 x_1, \alpha_j - \alpha_{i_1} = \beta_2 x_2$ with S-units x_1, x_2. For fixed $\alpha_{i_1}, \alpha_{i_2}$, (10.5.3) leads to at most $\Psi_S^2(N)$ S-unit equations whose total number of solutions is by Theorem 6.1.4 at most $2^{16s} \Psi_S^2(N)$. But the number of α_j in question is at least $\frac{1}{3}m$. This shows that if (10.5.2) holds, then $l = 1$ or $l = 2$ and $|\mathcal{G}_1| = 1$, which was to be proved. $\square$

Theorems 10.5.1 and 10.5.2 have applications to irreducible polynomials. I. Schur and later A. Brauer, R. Brauer, H. Hopf and others investigated the irreducibility of polynomials of the form $g(f(X))$, where f, g are monic polynomials with integral coefficients, g is irreducible over $\mathbb{Q}$, and the zeros of f are distinct integers. For a survey of results of this type, see Győry (1972, 1982c).

These investigations were extended in Győry (1971, 1972, 1982c, 1992c) to the more general case that the zeros of f are in an arbitrary but fixed totally real number field K. Let $\mathcal{A} = \{\alpha_1, \ldots, \alpha_m\}$ be the set of zeros of such a monic polynomial $f \in \mathbb{Z}[X]$ and suppose that $g \in \mathbb{Z}[X]$ is an irreducible monic polynomial whose splitting field is a CM-field, i.e. a totally imaginary quadratic extension of a totally real number field. In this case g is called of CM-*type*. For example, cyclotomic polynomials and quadratic polynomials of negative discriminant are of CM-type. Consider the graph $\mathcal{G}(\mathcal{A}) = \mathcal{G}(\mathcal{A}, N)$ with $N = 2^d |g(0)|^{d/\deg(g)}$, where $d = [K : \mathbb{Q}]$. It was proved in Győry (1971) that if this graph $\mathcal{G}(\mathcal{A})$ has a connected component having k vertices, then the number of irreducible factors of $g(f(X))$ over $\mathbb{Q}$ is not greater than $\deg(f)/k$. This estimate is in general best possible; see Győry (1972).

For $f \in \mathbb{Z}[X]$ and $a \in \mathbb{Z}$, the polynomials $f(X)$ and $f(X + a)$ will be called *equivalent*. Then, for irreducible $g \in \mathbb{Z}[X]$, $g(f(X))$ and $g(f(X + a))$ are at the same time reducible or irreducible. Using the fact that $\Psi_{M_K^\infty}(N) \leq C_3 N$ with an effectively computable number C_3 depending only on d and the discriminant of K (see Sunley (1973)), Theorem 10.5.2 implies immediately the following theorem.

Theorem 10.5.3 *Let $g \in \mathbb{Z}[X]$ be a monic irreducible polynomial of CM-type, and K a totally real number field of degree d. There are only finitely many equivalence classes of monic polynomials $f \in \mathbb{Z}[X]$ with $\deg(f) \geq 3$, $\deg(f) \neq 4$, and with distinct zeros in K for which $g(f(X))$ is reducible over $\mathbb{Q}$. Further, if*

$$\deg(f) > C_4 |g(0)|^{2d/\deg(g)}$$

then $g(f(X))$ is irreducible over $\mathbb{Q}$. Here C_4 is an effectively computable number depending only on d and the discriminant of K.

We note that for suitable g and K, in Theorem 10.5.3 there exist infinitely many exceptional equivalence classes of quartic f for which $g(f(X))$ is reducible, and these exceptions are described in Győry (1992c). Further, Theorem 10.5.3 does not remain valid for any monic irreducible $g \in \mathbb{Z}[X]$ and for any number field K; see e.g. Győry (1992c).

In Győry, Hajdu and Tijdeman (2011), an upper bound is given for the number of exceptional equivalence classes of polynomials f. Theorem 10.5.3 is ineffective, in the sense that the method of proof does not make it possible to determine the exceptional equivalence classes. A weaker but effective version can be deduced from Theorem 10.5.1. For the first effective results of this type, see Győry (1982c).

10.6 Discriminant equations and power integral bases in number fields

Several Diophantine problems of number theory lead to discriminant equations. To illustrate applications of unit equations to such equations, we restrict ourselves here to some finiteness results in their simplest form. Many other, more general results, quantitative versions and applications are discussed in our book *Discriminant Equations in Diophantine Number Theory*.

Two important *discriminant equations* are

$$D_{K/\mathbb{Q}}(\alpha) = D \quad \text{in } \alpha \in O_K \tag{10.6.1}$$

and

$$D(f) = D \quad \text{in monic polynomials } f \in \mathbb{Z}[X], \tag{10.6.2}$$

where K is an algebraic number field, O_K its ring of integers, $D(f)$ the discriminant of f, $D_{K/\mathbb{Q}}(\alpha)$ the discriminant of the minimal polynomial, say f_α, of α over $\mathbb{Z}$, and D a non-zero rational integer. In other words, if α satisfies (10.6.1) then f_α satisfies (10.6.2). Equation (10.6.2) can have, however, other, not necessarily irreducible solutions without zeros in K. Hence equation (10.6.2) is more general than (10.6.1).

If α is a solution of (10.6.1) then so is $\alpha + a$ for all $a \in \mathbb{Z}$. Elements α, $\alpha^* \in O_K$ with $\alpha - \alpha^* \in \mathbb{Z}$ are called *equivalent*. Similarly, if f is a solution of (10.6.2), then so is $f^*(X) = f(X + a)$ for every $a \in \mathbb{Z}$. As in Section 10.5, such polynomials f, f^* are called *equivalent*. The minimal polynomials of equivalent α, α^* from O_K are obviously equivalent.

In the quadratic case, when in (10.6.1) K is a quadratic number field and in (10.6.2) the polynomials f are quadratic, the solutions of the above equations can be easily found. Delone (1930) and Nagell (1930) proved independently of each other that up to equivalence, there are only finitely many irreducible monic polynomials $f \in \mathbb{Z}[X]$ of degree 3 for which (10.6.2) holds. This implies that for a cubic number field K, equation (10.6.1) has also only finitely many equivalence classes of solutions. In the quartic case, the same assertions were obtained later by Nagell (1967, 1968a). The proofs of Delone and Nagell are ineffective. Nagell (1967) conjectured that the finiteness assertion concerning equation (10.6.1) is true for every number field K.

Let K be as above an algebraic number field, and denote by d and D_K the degree and discriminant of K. By repeatedly applying an earlier version of Theorem 4.1.1, Győry (1973) proved the following general effective result.

Theorem 10.6.1 *Every solution α of (10.6.1) is equivalent to a solution $\alpha^* \in O_K$ for which*

$$h(\alpha^*) < C_1, \tag{10.6.3}$$

where C_1 is an effectively computable number depending only on d, D_K and D.

This implies that there are only finitely many pairwise inequivalent elements in O_K with discriminant D, and a full set of representatives of such elements can be, at least in principle, effectively determined. This finiteness assertion was proved independently in an ineffective form in Birch and Merriman (1972).

In view of Minkowski's inequality (1.5.4), the degree d of K can be estimated from above in terms of $|D_K|$. Further, if (10.6.1) is solvable then D_K divides D. Hence, in (10.6.3), the dependence of the bound on d and D_K, and hence on K can be dropped; see Győry (1973).

Proof of Theorem 10.6.1 (sketch). We reduce (10.6.1) to a system of unit equations. Let G denote the normal closure of $K/\mathbb{Q}$, let g be its degree over $\mathbb{Q}$, and let $\alpha^{(1)} = \alpha, \alpha^{(2)}, \ldots, \alpha^{(d)}$ be the conjugates of α with respect to $K/\mathbb{Q}$. If $d \geq 3$ then

$$\frac{\alpha^{(1)} - \alpha^{(i)}}{\alpha^{(1)} - \alpha^{(2)}} + \frac{\alpha^{(i)} - \alpha^{(2)}}{\alpha^{(1)} - \alpha^{(2)}} = 1 \quad \text{for } i = 3, \ldots, d. \tag{10.6.4}$$

Further, the numbers $\alpha^{(1)} - \alpha^{(2)}$, $\alpha^{(1)} - \alpha^{(i)}$ and $\alpha^{(i)} - \alpha^{(2)}$ divide D in the ring of integers of G. Hence Proposition 4.3.12 implies that apart from some unknown unit factors, the heights of these differences can be effectively bounded above. Thus, equation (10.6.4) reduces indeed to finitely many unit equations in two unknowns in G. Finally, by Theorem 4.1.1 the heights of $\alpha^{(1)} - \alpha^{(i)}$, $\alpha^{(2)} - \alpha^{(i)}$ and so $\alpha^{(i)} - \alpha^{(j)}$ can be effectively estimated from above up to the common factor $\alpha^{(1)} - \alpha^{(2)}$ whose height can be effectively bounded above from (10.6.1) in terms of D, g and the class number and regulator of G. Since these parameters of G can be estimated from above in terms of d, D_K and D, Theorem 10.6.1 follows. □

Theorem 10.6.1 is in fact a consequence of Theorem 10.5.1. Let

$$\mathcal{A} = \left\{ \alpha^{(1)}, \ldots, \alpha^{(d)} \right\}, \quad N = |D|^g.$$

Then $\mathcal{A}$ is a subset of the ring of integers of G, and (10.6.1) gives

$$|N_{G/\mathbb{Q}} \left(\alpha^{(i)} - \alpha^{(j)} \right)| \leq N \quad \text{for } 1 \leq i < j \leq d.$$

Hence the graph $\mathcal{G}(\mathcal{A}, N)$ defined in Section 10.5 consists of isolated vertices. Thus Theorem 10.5.1 applies and the heights of the differences $\alpha^{(i)} - \alpha^{(j)}$ can be effectively bounded above apart from a common unit factor ε in G, while the height of ε can be estimated from above from (10.6.1). □

As was mentioned above, one may assume that in (10.6.1) D_K divides D. Let ω denote the number of distinct prime factors of the quotient D/D_K. Using Theorem 6.1.4 concerning unit equations, one can prove the following theorem, as a special case of a more general result of Evertse and Győry from their book on discriminant equations.

Theorem 10.6.2 *Equation* (10.6.1) *has at most*

$$2^{5d^2(\omega+1)}$$

equivalence classes of solutions.

The first, weaker version of this type was proved in Evertse and Győry (1985).

Concerning equation (10.6.2), Delone and Faddeev (1940) posed the problem of giving an algorithm for finding all cubic monic polynomials with integer coefficients and given non-zero discriminant. In 1973, Győry (1973) proved the following general theorem.

Theorem 10.6.3 *Every solution $f \in \mathbb{Z}[X]$ of (10.6.2) is equivalent to a solution $f^* \in \mathbb{Z}[X]$ for which*

$$\deg(f^*) \le C_2, \quad H(f^*) \le C_3, \tag{10.6.5}$$

where $H(f^*)$ denotes the maximum of the absolute values of the coefficients of f^* and C_2, C_3 are effectively computable numbers depending only on D.

This makes it possible, at least in principle, to determine all monic polynomials in $\mathbb{Z}[X]$ with given non-zero discriminant.

Later, several quantitative versions of Theorems 10.6.1 and 10.6.3, and generalizations for S-integers and for polynomials with S-integral coefficients in number fields, were established by Győry. References and the best known values for C_1 and C_3 are given in our book on discriminant equations. The best possible upper bound C_2 can be found in Győry (1974).

For irreducible polynomials $f \in \mathbb{Z}[X]$, Theorem 10.6.1 implies Theorem 10.6.3. The "reducible" case can be reduced to the "irreducible" one by means of the relation

$$D(f) = \left(\prod_{i=1}^{k} D(f_i) \right) \cdot \left(\prod_{1 \le i < j \le k} \left(R(f_i, f_j) \right)^2 \right),$$

where $f = \prod_{i=1}^{k} f_i$ is the irreducible factorization of f in $\mathbb{Z}[X]$ and $R(f_i, f_j)$ denotes the resultant of f_i and f_j. Another option is to apply Theorem 10.5.1 to equation (10.6.2) as in the proof of Theorem 10.6.1, and then estimate in the bound obtained for $H(f^*)$ the parameters involved in terms of D. An upper bound can also be derived for $\deg(f^*)$ by means of Theorem 10.5.2.

We present some consequences of Theorems 10.6.1 and 10.6.2. For other applications, for example to discriminant form and index form equations, we refer to Győry (1976, 1980b), Evertse and Győry (1988a) and our book on discriminant equations.

As is known, there exist algebraic number fields K having *power integral bases* (i.e. integral bases of the form $\{1, \alpha, \dots, \alpha^{d-1}\}$ where $d = [K : \mathbb{Q}]$), but this is not the case in general. The existence of such a basis considerably facilitates the calculations in K and the study of arithmetical properties of O_K, the ring of integers of K.

More generally, we consider *orders* in K, these are the subrings of O_K whose quotient field is K. There are infinitely many orders in K, and O_K is the maximal one among them. The order O in K is said to be *monogenic* if $O = \mathbb{Z}[\alpha]$ for some $\alpha \in O$. Equivalently, in this case $\{1, \alpha, \ldots, \alpha^{d-1}\}$ is a $\mathbb{Z}$-module basis of O, where $d = [K : \mathbb{Q}]$. In particular, the number field K is called *monogenic* if O_K is monogenic, that is, if K has a power integral basis.

It is known that $\alpha \in O$ generates O if and only if $D_{K/\mathbb{Q}}(\alpha) = D_O$, where D_O denotes the discriminant of O. If α is a generator of O then so are all $\alpha^* \in O$ which are equivalent to α. Choosing $D = D_O$, and using the fact that D_K divides D_O, Theorem 10.6.1 gives at once the following corollary, see Győry (1976):

Corollary 10.6.4 *If $O = \mathbb{Z}[\alpha]$ for some $\alpha \in O$, then there is an $\alpha^* \in O$ which is equivalent to α such that*

$$h(\alpha^*) < C_4,$$

where C_4 is an effectively computable number depending only on d and D_O.

In the special case $O = O_K$, we get immediately the following consequence, already obtained in Győry (1976).

Corollary 10.6.5 *If $\{1, \alpha, \ldots \alpha^{d-1}\}$ is an integral basis of K, then there is an $\alpha^* \in O_K$ which is equivalent to α such that*

$$h(\alpha^*) < C_5,$$

where C_5 is an effectively computable number depending only on d and D_K.

Thus, up to equivalence, there are only finitely many elements in O_K and, more generally in O, which generate a power integral basis and they can be, at least in principle, effectively determined. Combining this effective approach with some reduction procedures, all power integral bases have been determined in many number fields of relatively small degree; see e.g. Gaál (2002), Bilu, Gaál and Győry (2004) and our book on discriminant equations.

An immediate consequence of Theorem 10.6.2 is the following.

Corollary 10.6.6 *Let O be an order in K. Up to equivalence, there are at most 2^{5d^2} elements $\alpha \in O$ such that $O = \mathbb{Z}[\alpha]$.*

In particular, the same assertion is true for O_K.

An order O in K is said to be k *times monogenic* if there are at least k distinct equivalence classes of α satisfying $O = \mathbb{Z}[\alpha]$. The following result was proved by Bérczes, Evertse and Győry (2013).

Theorem 10.6.7 *Let K be an algebraic number field of degree ≥ 3. Then there are at most finitely many three times monogenic orders in K.*

The bound 3 is best possible, that is there are number fields having infinitely many two times monogenic orders. The proof of Theorem 10.6.7 depends on earlier, qualitative versions of Corollary 6.1.5 and Theorem 6.1.6 on unit equations.

A non-zero element α in an order O of an algebraic number field K is called a *basis of a canonical number system* (or CNS basis) for O if every non-zero element of O can be represented in the form

$$a_0 + a_1\alpha + \cdots + a_m\alpha^m$$

with $m \geq 0$, $a_i \in \{0, 1, \ldots, |N_{K/\mathbb{Q}}(\alpha)| - 1\}$ for $i = 0, \ldots, m$ and $a_m \neq 0$. Canonical number systems can be viewed as natural generalizations of radix representations of rational integers to algebraic integers.

If there exists a canonical number system in O, then O is called a CNS *order*. Orders of this kind have been intensively investigated; we refer to the survey paper Brunotte, Huszti and Pethő (2006) and the references given there.

It was proved in Kovács (1981) and Kovács and Pethő (1991) that O is a CNS order if and only if O is monogenic. More precisely, if α is a CNS basis in O, then it is easy to see that $O = \mathbb{Z}[\alpha]$. Conversely, $O = \mathbb{Z}[\alpha]$ does not imply in general that α is a CNS basis. However, in this case there are infinitely many α' which are equivalent to α such that α' is a CNS basis for O. A characterization of CNS bases in O is given in Kovács and Pethő (1991).

The close connection between elements α of O with $O = \mathbb{Z}[\alpha]$ and CNS bases in O enables one to apply results concerning monogenic orders to CNS orders and CNS bases. For example, it follows from Corollary 10.6.4 that up to equivalence there are only finitely many canonical number systems in O.

We say that O is a k *times* CNS order if there are at least k pairwise inequivalent CNS bases in O. Theorem 10.6.7 implies the following result, see also Bérczes, Evertse and Győry (2013).

Corollary 10.6.8 *Let K be an algebraic number field of degree ≥ 3. Then there are at most finitely many three times CNS orders in K.*

10.7 Binary forms of given discriminant

Let $F = a_0X^n + a_1X^{n-1}Y + \cdots + a_nY^n$ be a binary form of degree $n \geq 2$ with coefficients in a field K. We can factor F over an algebraic closure of K as

$$F = \prod_{i=1}^{n}(\alpha_i X - \beta_i Y); \tag{10.7.1}$$

then the discriminant of F is given by

$$D(F) = \prod_{1 \le i < j \le n} (\alpha_i \beta_j - \alpha_j \beta_i)^2.$$

We can express $D(F)$ otherwise as a homogeneous polynomial of degree $2n - 2$ in $\mathbb{Z}[a_0, \ldots, a_n]$. Define the binary form F_U by

$$F_U(X, Y) = F(aX + bY, cX + dY) \text{ for } \begin{pmatrix} a & b \\ c & d \end{pmatrix} \in GL(2, K).$$

Then we have

$$D(\lambda F_U) = \lambda^{2n-2}(\det U)^{n(n-1)} D(F) \quad \text{for } \lambda \in K^*, \ U \in GL(2, K). \quad (10.7.2)$$

Given a subring A of K, we say that two binary forms $F, G \in A[X, Y]$ are $GL(2, A)$-equivalent if there are a unit $u \in A^*$ and $U \in GL(2, A)$ such that $G = uF_U$. By (10.7.2), two $GL(2, A)$-equivalent binary forms have, up to multiplication with a unit from A, the same discriminant.

We now restrict ourselves to binary forms with coefficients in $\mathbb{Z}$. By (10.7.2),two $GL(2, \mathbb{Z})$-equivalent binary forms have the same discriminant. We have the following fundamental theorem.

Theorem 10.7.1 *Let n, D be integers with $n \ge 2$ and $D \ne 0$. Then there are only finitely many $GL(2, \mathbb{Z})$-equivalence classes of binary forms $F \in \mathbb{Z}[X, Y]$ of degree n and discriminant D.*

For $n = 2$ this is a classical theorem of Lagrange (1773) and for $n = 3$ a classical theorem of Hermite (1851). For $n \ge 4$ this was proved only in 1972 by Birch and Merriman (Birch and Merriman (1972), Theorem 2). The proofs of Lagrange and Hermite are effective, while that of Birch and Merriman is ineffective.

Proof of Birch and Merriman (sketch). We give a brief sketch of the proof of Birch and Merriman, explaining at which point it fails to be effective. Take a binary form $F \in \mathbb{Z}[X, Y]$ of degree $n \ge 4$ and discriminant $D \ne 0$. The discriminant of the splitting field of F can be estimated from above in terms of D, and by the Hermite–Minkowski Theorem, this leaves only a finite, effectively determinable collection of possible splitting fields for F. So we may restrict ourselves to binary forms F with given splitting field L, say. Let H denote the Hilbert class field of L, and let S be a finite set of places of H such that $D \in O_S^*$. Then F can be factored as in (10.7.1) with $\alpha_i, \beta_i \in O_S$ and $\alpha_i \beta_j - \alpha_j \beta_i \in O_S^*$ for $1 \le i < j \le n$. There is a matrix $U \in GL(2, O_S)$ such that

$$F_U = \varepsilon XY(X - Y)(X - \gamma_3 Y) \cdots (X - \gamma_n Y)$$

with $\varepsilon \in O_S^*$, $\gamma_3, \ldots, \gamma_n \in O_S$. Further, by (10.7.2),

$$D(F_U) = \pm \varepsilon^{2n-2} \prod_{i=3}^{n} (\gamma_i(1 - \gamma_i)) \in O_S^*,$$

which implies that γ_i, $1 - \gamma_i \in O_S^*$ for $i = 3, \ldots, n$. In this way, the problem of finding the binary forms F of given discriminant reduces to an S-unit equation in two unknowns $x + y = 1$ in x, $y \in O_S^*$.

Now by an effective finiteness result for such equations such as Theorem 4.1.3, one can show that there are only finitely many possibilities for $\gamma_3, \ldots, \gamma_n$ that can be determined effectively. This shows that the binary forms $F \in \mathbb{Z}[X, Y]$ of degree n and discriminant D lie in only finitely many $\mathrm{GL}(2, O_S)$-equivalence classes. The final step of the proof of Birch and Merriman is to show that the binary forms in $\mathbb{Z}[X, Y]$ of discriminant D in a given $\mathrm{GL}(2, O_S)$-equivalence class lie in only finitely many $\mathrm{GL}(2, \mathbb{Z})$-equivalence classes. At this point, the argument of Birch and Merriman is ineffective, since it does not give an effective procedure to check whether a given $\mathrm{GL}(2, O_S)$-equivalence class contains a binary form from $\mathbb{Z}[X, Y]$. $\square$

Evertse and Győry (1991) managed to give an effective version of the result of Birch and Merriman. The following is a less precise version of Theorem 1 from their paper. Given a binary form $F \in \mathbb{Z}[X, Y]$, denote by $H(F)$ the maximum of the absolute values of the coefficients of F.

Theorem 10.7.2 *Let n, D be integers with $n \geq 2$ and $D \neq 0$. Then there is an effectively computable number C_1, depending only on n and D, such that for every binary form $F \in \mathbb{Z}[X, Y]$ of degree n and discriminant D there is $U \in \mathrm{GL}(2, \mathbb{Z})$ such that $H(F_U) \leq C_1$.*

Proof (sketch). We give only the main idea of the proof. We may again restrict ourselves to binary forms with given splitting field L. Let $F \in \mathbb{Z}[X, Y]$ be a binary form of degree n and discriminant D with splitting field L. Take a factorization of F as in (10.7.1). After multiplying F by a small integer (effectively bounded in terms of L), we may assume that F has a factorization as in (10.7.1) with α_i, $\beta_i \in O_L$ for $i = 1, \ldots, n$. Put $\Delta_{ij} := \alpha_i \beta_j - \alpha_j \beta_i$ for $1 \leq i, j \leq n$. Then for any quadruple i, j, k, l of distinct indices we have the identity

$$\Delta_{ij} \Delta_{kl} + \Delta_{jk} \Delta_{il} + \Delta_{ki} \Delta_{jl} = 0. \tag{10.7.3}$$

Notice that all terms Δ_{ij} are in O_L and divide D; hence $|N_{L/\mathbb{Q}}(\Delta_{ij})| \leq |D|^{[L:\mathbb{Q}]}$ for all i, j. Using Proposition 4.3.12 we can express each term Δ_{ij} as a product of an element of height effectively bounded in terms of n, D, L and an element of O_L^*. By substituting this into the identities (10.7.3) we obtain homogeneous unit equations like in Theorem 4.1.1. By applying the latter, we obtain effective upper bounds for the heights of the quotients $\Delta_{ij}\Delta_{kl}/\Delta_{ik}\Delta_{jl}$. We have some freedom to choose the $\alpha_i \beta_i$ in (10.7.1). By doing this in an appropriate way, we can deduce in fact effective upper bounds for the heights of the numbers Δ_{ij} themselves. Then, with the help of an argument from the geometry of numbers, one can construct a matrix $U \in \mathrm{GL}(2, \mathbb{Z})$ as in Theorem 10.7.2. $\qquad\Box$

In our book *Discriminant Equations in Diophantine Number Theory* we give a complete proof of Theorem 10.7.2, with the explicit value

$$C_1 = \exp\left((16n^3)^{25n^2}|D|^{5n-3}\right).$$

It is possible to give a semi-effective version of Theorem 10.7.2 with for C_1 a bound with a much better dependence on D, but with an ineffective dependence on the splitting field of the binary form F. The following result is Theorem 1 of Evertse (1993).

Theorem 10.7.3 *Let $F \in \mathbb{Z}[X, Y]$ be a binary form of degree $n \geq 4$ and of discriminant $D \neq 0$. Assume that F has splitting field L. Then there is $U \in \mathrm{GL}(2, \mathbb{Z})$ such that*

$$H(F) \leq C^{\mathrm{ineff}}(n, L)|D|^{21/(n-1)}.$$

Here, $C^{\mathrm{ineff}}(n, L)$ is a number, not effectively computable from the proof, that depends only on n and L.

Proof (sketch). The proof is similar to that of Theorem 10.7.2 but one has to apply Theorem 6.1.1 with $n = 2$ to (10.7.3). Some precise combinatorics is needed to get an exponent $O(1/n)$ on $|D|$. $\qquad\Box$

It is possible to give explicit upper bounds for the number of $\mathrm{GL}(2, \mathbb{Z})$-equivalence classes of binary forms, under certain additional constraints. Although it is possible to treat reducible binary forms as well, we restrict ourselves to binary forms that are irreducible over $\mathbb{Q}$.

Let $F = a_0 X^n + a_1 X^{n-1}Y + \cdots + a_n Y^n \in \mathbb{Z}[X, Y]$ be a binary form of degree $n \geq 2$. We say that F is associated with a number field K if F is irreducible over $\mathbb{Q}$, and there is α with $F(\alpha, 1) = 0$, $K = \mathbb{Q}(\alpha)$. This being the case, we can factor F over K as

$$F = (X - \alpha Y)\left(a_0 X^{n-1} + \omega_1 X^{n-2}Y + \cdots + \omega_{n-1}Y^{n-1}\right),$$

where $\omega_1, \ldots, \omega_{n-1} \in K$. Denote by O_F the $\mathbb{Z}$-module generated by the numbers $1, \omega_1, \ldots, \omega_{n-1}$. We call O_F the *invariant order* of F. This naming is motivated by work of Simon (2001), who showed that O_F is in fact an order in K, i.e., a subring of K of rank n as a $\mathbb{Z}$-module, and that $\mathrm{GL}(2, \mathbb{Z})$-equivalent binary forms have isomorphic invariant orders. Of course O_F depends on the choice of K and α, but it is unique up to $\mathbb{Z}$-algebra isomorphism. It is not hard to show that for the discriminant of O_F, i.e., $D_{K/\mathbb{Q}}(1, \omega_1, \ldots, \omega_{n-1})$, we have

$$D(O_F) = D(F).$$

This implies

$$D(F) = c^2 D_K, \tag{10.7.4}$$

where $c = [O_K : O_F]$. The following result is a less precise form of Corollary 2.2 of Bérczes, Evertse and Győry (2004).

Theorem 10.7.4 *Let K be a number field of degree $n \geq 2$. and c a positive integer. Then for every $\epsilon > 0$, the number of $\mathrm{GL}(2, \mathbb{Z})$-equivalence classes of irreducible binary forms $F \in \mathbb{Z}[X, Y]$ are associated with K and satisfy (10.7.4) is*

$$\ll c^{(2/n(n-1))+\epsilon},$$

where the implied constant is effectively computable and depends only on n and ϵ.

It is shown in Bérczes, Evertse and Győry (2004) that the bound in Theorem 10.7.4 cannot be replaced by one of order c^α with $\alpha < \dfrac{2}{n(n-1)}$.

We subdivide the irreducible binary forms with (10.7.4) further and consider binary forms with given invariant order. By a result of Delone and Faddeev (1940), section 15, for every cubic number field K and every order O in K, there is precisely one $\mathrm{GL}(2, \mathbb{Z})$-equivalence class of cubic forms $F \in \mathbb{Z}[X, Y]$ such that $O_F \cong O$. On the other hand, in his paper referred to above, Simon proved that for every $n \geq 4$ there are number fields K of degree n such that O_K is not the invariant order of a binary form. The following result is Corollary 2.1 of Bérczes, Evertse and Győry (2004).

Theorem 10.7.5 *Let O be an order in a number field K of degree $n \geq 4$. Then there are at most 2^{24n^3} $\mathrm{GL}(2, \mathbb{Z})$-equivalence classes of irreducible binary forms $F \in \mathbb{Z}[X, Y]$ such that $O_F \cong O$.*

In our book *Discriminant Equations in Diophantine Number Theory* the bound 2^{24n^3} is improved to 2^{5n^2}. One can define more generally the invariant order of a reducible binary form of degree n, which is an order of rank n, i.e. a commutative ring which as a $\mathbb{Z}$-module is free of rank n. When F has non-zero discriminant, its invariant order has no nilpotents. In our book on discriminant equations, we have proved a generalization of Theorem 10.7.5 where O is a given nilpotent-free order of rank n.

We mention that both the proofs of Theorems 10.7.4, 10.7.5 use Theorem 6.1.4 (the result of Beukers and Schlickewei).

We finally remark that the papers Evertse and Győry (1991), Evertse (1993) and Bérczes, Evertse and Győry (2004), as well as our book on discriminant equations, contain proofs of generalizations of Theorems 10.7.2–10.7.5 for binary forms with S-integral coefficients in number fields. A further generalization of Theorem 10.7.2 is given in Evertse and Győry (1992a, 1992b) for decomposable forms of given discriminant. See also our book on discriminant equations.

10.8 Resultant equations for monic polynomials

Recall that the resultant of two monic polynomials

$$f = \prod_{i=1}^{m}(X - \alpha_i), \ g = \prod_{i=m+1}^{m+n}(X - \alpha_i)$$

is given by

$$R(f, g) = \prod_{\substack{1 \le i \le m \\ m+1 \le j \le m+n}} (\alpha_i - \alpha_j)$$

and that $R(f, g)$ is a polynomial with integer coefficients in terms of the coefficients of f and g.

Let K be an algebraic number field, and consider the *resultant equation*

$$R(f, g) = R \text{ in monic } f, g \in \mathbb{Z}[X] \text{ having all their zeros in } K, \quad (10.8.1)$$

where R is a non-zero rational integer. If f, g is a solution of (10.8.1) then so is

$$f^*(X) = f(X + a), \quad g^*(X) = g(X + a)$$

for all $a \in \mathbb{Z}$. Such pairs of polynomials f, g, and f^*, g^* are called *equivalent*. The following result was obtained in Győry (1990).

Theorem 10.8.1 *There are only finitely many equivalence classes of pairs f, g with $\deg(f) \ge 2$, $\deg(g) \ge 2$ and $\deg(f) + \deg(g) \ge 5$, without multiple zeros, such that* (10.8.1) *holds.*

We note that the assumptions concerning the degrees of f and g are necessary. Further, the condition that the zeros of f and g are contained in a fixed number field cannot be dropped. However, the restriction concerning the multiplicity of the zeros can be weakened, see Győry (1993c, 2008b).

Győry (1990, 1993c, 2008b) and Bérczes, Evertse and Győry (2007a) obtained quantitative versions of Theorem 10.8.1 which provide upper bounds for the degrees of f and g and for the number of equivalence classes of pairs f, g under consideration. For example, it is proved in Győry (1990) that, in Theorem 10.8.1,

$$\deg(f) + \deg(g) \le 12 \cdot 7^{3d+2\omega},$$

where d is the degree of K over $\mathbb{Q}$, and ω denotes the number of distinct prime factors of R.

It should be remarked that Theorem 10.8.1 is established in Győry (1990) in the more general case when the ground ring is any integrally closed integral domain of characteristic 0 which is finitely generated over $\mathbb{Z}$.

Proof of Theorem 10.8.1 (sketch). We reduce equation (10.8.1) to unit equations. The basic idea is as follows. Let f, g be a solution of (10.8.1) with

$$\deg(f) = m \ge 2, \deg(g) = n \ge 2, m + n \ge 5,$$

and let $\{\alpha_1, \ldots, \alpha_m\}$, $\{\alpha_{m+1}, \ldots, \alpha_{m+n}\}$ be the zeros of f and g in K. Since f, g are monic, these zeros are contained in O_K, the ring of integers of K, and by assumption they are distinct. Then (10.8.1) can be written in the form

$$\prod_{\substack{1 \le i \le m \\ m+1 \le j \le m+n}} (\alpha_i - \alpha_j) = R. \tag{10.8.2}$$

The differences $\alpha_i - \alpha_j$ divide R in O_K. Hence taking norms, we infer that

$$|N_{K/\mathbb{Q}}(\alpha_i - \alpha_j)| \le N \quad \text{for each } i, j, \tag{10.8.3}$$

where $N = |R|^d$. By Proposition 4.3.12, $\alpha_i - \alpha_j$ may take only finitely many values up to a unit factor from O_K. There exist several linear relations among these differences, for example

$$(\alpha_i - \alpha_j) + (\alpha_j - \alpha_k) + (\alpha_k - \alpha_l) = \alpha_i - \alpha_l,$$

with $1 \le i, k \le m$, $m + 1 \le j, l \le m + n$. This leads to inhomogeneous unit equations in three unknowns. We arrive in this way at a complicated system of unit equations. However, in contrast with the case of discriminant equations, in this situation we get unit equations in more than two unknowns. Thus one has to apply the ineffective Corollary 6.1.2 or Theorem 6.1.3. to obtain Theorem 10.8.1. Therefore, Theorem 10.8.1 is ineffective. $\qquad\square$

It is simpler to deduce Theorem 10.8.1 from Theorem 10.5.2. We recall, however, that the proof of Theorem 10.5.2 is also based on the results concerning unit equations mentioned in Chapter 6. Consider the graph $\mathcal{G}(\mathcal{A}) = \mathcal{G}(\mathcal{A}, N)$, where $\mathcal{A} = \{\alpha_1, \ldots, \alpha_m, \ldots, \alpha_{m+n}\}$. Using the above notation, it follows from (10.8.2) and (10.8.3) that $\mathcal{G}(\mathcal{A})$ has either at least three connected components or two connected components of order at least 2. Hence Theorem 10.5.2 implies that $m + n$ is bounded. Further, for fixed m and n, we have

$$\alpha_i = \varepsilon \alpha_i' + \beta, \quad i = 1, \ldots, m+n$$

with some $\varepsilon \in O_K^*, \beta \in O_K$ and with $\alpha_1', \ldots, \alpha_{m+n}' \in O_K$ which may take only finitely many values. This gives

$$\alpha_i - \alpha_j = \varepsilon(\alpha_i' - \alpha_j'), \quad 0 \leq i \leq m, \quad m+1 \leq j \leq m+n. \quad (10.8.4)$$

We see from (10.8.2) and (10.8.4) that for fixed $\alpha_1', \ldots, \alpha_{m+n}'$, ε^{mn} is also fixed, that is ε can assume only finitely many values. Finally, one can infer that $\alpha_i = \alpha_i^* + a$ with some $a \in \mathbb{Z}$ and with finitely many possible $\alpha_i^* \in O_K$, $i = 1, \ldots, m+n$, whence Theorem 10.8.1 follows. $\qquad\qquad\square$

10.9 Resultant inequalities and equations for binary forms

We keep the notation introduced in Section 10.7. Let $F = \sum_{i=1}^{m} a_i X^{m-i} Y^i$, $G = \sum_{i=1}^{n} b_i X^{n-i} Y^i$ be two binary forms with coefficients in a field K. Assume that over an algebraic closure of K, the forms F, G factor as

$$F = \prod_{i=1}^{m} (\alpha_i X - \beta_i Y), \quad G = \prod_{j=1}^{n} (\gamma_j X - \delta_j Y); \quad (10.9.1)$$

then the resultant of F, G is given by

$$R(F, G) = \prod_{i=1}^{m} \prod_{j=1}^{n} (\beta_i \gamma_j - \alpha_i \delta_j).$$

We can express $R(F, G)$ otherwise as a polynomial with integer coefficients in $a_0, \ldots, a_m, b_0, \ldots, b_n$, homogeneous of degree n in $a_0, \ldots, a_m$ and homogeneous of degree m in $b_0, \ldots, b_n$. Notice that for $\lambda, \mu \in K^*$ and $U \in \mathrm{GL}(2, K)$ we have

$$R(\lambda F_U, \mu G_U) = \lambda^n \mu^m (\det U)^{mn} R(F, G). \quad (10.9.2)$$

Let A be a subring of K. We call two pairs of binary forms $(F, G), (F', G')$ with coefficients in A $\mathrm{GL}(2, A)$-*equivalent* if $F' = u_1 F_U$, $G' = u_2 G_U$ for some $u_1, u_2 \in A^*$ and $U \in \mathrm{GL}(2, A)$. By (10.9.2), $\mathrm{GL}(2, A)$-equivalent pairs of binary forms have, up to multiplication with a unit from A^*, the same resultant.

We restrict ourselves to binary forms with coefficients in $\mathbb{Z}$. We start with formulating some results for resultant inequalities and then deduce some analogues for binary forms of some of the results from the previous section. By $C_i^{\text{ineff}}(\cdot)$ we denote positive numbers, depending on the parameters between the parentheses, that are not effectively computable by the method of proof of the theorem in which they appear. We call a binary form square-free if it is not divisible by the square of a non-constant binary form.

Our first result, which is Theorem 1 of Evertse and Győry (1993), gives a lower bound for the resultant of two binary forms in terms of their discriminants.

Theorem 10.9.1 *Let L be a finite, normal extension of $\mathbb{Q}$, and $F, G \in \mathbb{Z}[X, Y]$ binary forms such that*

$$\deg F = m \geq 3, \quad \deg G = n \geq 3, \quad FG \text{ is square-free}, \\ FG \text{ has splitting field } L. \tag{10.9.3}$$

Then

$$|R(F, G)| \geq C_1^{\text{ineff}}(m, n, L)\left(|D(F)|^{n/(m-1)}|D(G)|^{m/(n-1)}\right)^{1/18}.$$

It was shown in Evertse and Győry (1993) that the dependence on L in Theorem 10.9.1 is necessary, and that neither of the conditions $m \geq 3$, $n \geq 3$ can be removed.

Proof (sketch). Let $F, G \in \mathbb{Z}[X, Y]$ be binary forms as in the statement of Theorem 10.9.1. After multiplying F, G by small integers bounded above in terms of L which will not have an effect on our result, we may assume that F, G have factorizations as in (10.9.1) with $\alpha_i, \beta_i, \gamma_j, \delta_j \in O_L$ for $i = 1, \ldots, m$, $j = 1, \ldots, n$. Put $\Theta_{ij} := \beta_i\gamma_j - \alpha_i\delta_j$ for $i = 1, \ldots, m$, $j = 1, \ldots, n$. Then $\Theta_{ij} \in O_L$ for all i, j and $\prod_{i,j} \Theta_{ij} = R(F, G)$. Further, for all distinct $i, j, k \in \{1, \ldots, m\}$, $p, q, r \in \{1, \ldots, n\}$ we have

$$\begin{vmatrix} \Theta_{ip} & \Theta_{iq} & \Theta_{ir} \\ \Theta_{jp} & \Theta_{jq} & \Theta_{jr} \\ \Theta_{kp} & \Theta_{kq} & \Theta_{kr} \end{vmatrix} = \Theta_{ip}\Theta_{jq}\Theta_{kr} + \Theta_{iq}\Theta_{jr}\Theta_{kp} + \Theta_{ir}\Theta_{jp}\Theta_{kq}$$

$$-\Theta_{iq}\Theta_{jp}\Theta_{kr} - \Theta_{ip}\Theta_{jr}\Theta_{kq} - \Theta_{ir}\Theta_{jq}\Theta_{kp} = 0. \tag{10.9.4}$$

Similarly as in the proof of Theorem 6.1.6, we consider all possible splittings of (10.9.4) into minimal non-vanishing subsums, and then apply Theorem 6.1.1 to each of these minimal sums. This leads to lower bounds for the quantities $|N_{L/\mathbb{Q}}(\Theta_{ip}\Theta_{jp}\Theta_{kp}\Theta_{iq}\Theta_{jq}\Theta_{kq}\Theta_{ir}\Theta_{jr}\Theta_{kr})|$ for all i, j, k, p, q, r. By taking the product of these, the theorem follows. $\square$

It is also possible to give a lower bound for $|R(F, G)|$ in terms of the heights of a pair of binary forms that is $\mathrm{GL}(2, \mathbb{Z})$-equivalent to F, G. The following result is Theorem 1 of Evertse (1998).

Theorem 10.9.2 *Let* $F, G \in \mathbb{Z}[X, Y]$ *be binary forms with* (10.9.3). *Then there is* $U \in \mathrm{GL}(2, \mathbb{Z})$ *such that*

$$|R(F, G)| \geq C_2^{\mathrm{ineff}}(m, n, L)\big(H(F_U)^n H(G_U)^m\big)^{1/718}.$$

Of course, the theorem does not hold without the matrix U, since by varying the pairs (F, G) in a given $\mathrm{GL}(2, \mathbb{Z})$-equivalence class, $|R(F, G)|$ remains the same, while $H(F)$, $H(G)$ may become arbitrarily large.

Proof (sketch). Apply Theorem 10.9.1. According to Theorem 10.7.3, there is $U \in \mathrm{GL}(2, \mathbb{Z})$ such that $H(G_U)$ is bounded above in terms of $|D(G)|$, and so in terms of $|R(F, G)|$. Writing $F_U = \prod_{i=1}^{m}(\alpha_i' X - \beta_i' Y)$, we get

$$\prod_{i=1}^{m} G_U(\alpha_i', \beta_i') = \pm R(F_U, G_U) = \pm R(F, G).$$

Thus, for $i = 1, \ldots, m$, the number $G_U(\alpha_i', \beta_i')$ divides $R(F, G)$ in O_L. We may view the pairs (α_i', β_i') as solutions to a Thue equation over O_L. This leads to upper bounds for the heights of the α_i', β_i', and hence of $H(F_U)$, in terms of $H(G_U)$ and $|R(F, G)|$. Thus, both $H(F_U)$, $H(G_U)$ can be estimated from above in terms of $R(F, G)$. A precise computation gives Theorem 10.9.2. $\square$

We deduce some consequences. The first is an analogue of Theorem 10.8.1.

Corollary 10.9.3 *Let* R *be a non-zero integer. Then the pairs of binary forms* $F, G \in \mathbb{Z}[X, Y]$ *with* (10.9.3) *and with*

$$R(F, G) = R$$

lie in at most finitely many $\mathrm{GL}(2, \mathbb{Z})$-equivalence classes.

Proof. Immediate consequence of Theorem 10.9.2. $\square$

The next consequence is a special case of Theorem 1 of Evertse and Győry (1989). Given a binary form $F \in \mathbb{Z}[X, Y]$ and an integer $m > 0$, we consider the Thue inequality

$$0 < |F(x, y)| \leq m \quad \text{in } x, y \in \mathbb{Z}. \tag{10.9.5}$$

Two solutions $(x, y), (x', y')$ of (10.9.5) are called proportional if $(x', y') = a(x, y)$ for some $a \in \mathbb{Q}^*$.

Corollary 10.9.4 *Let $n \geq 3$ be an integer and L a finite normal extension of* $\mathbb{Q}$. *Then up to* GL(2, $\mathbb{Z}$)-*equivalence, there are only finitely many binary forms* $F \in \mathbb{Z}[X, Y]$ *of degree n and with splitting field L such that* (10.9.5) *has more than two pairwise non-proportional solutions.*

Proof. Let $F \in \mathbb{Z}[X, Y]$ be a binary form of degree n and splitting field L and suppose that (10.9.5) has three pairwise non-proportional solutions, say (x_1, y_1), (x_2, y_2), (x_3, y_3). Define the binary form $G := \prod_{i=1}^{3}(y_i X - x_i Y)$. Then

$$0 < |R(F, G)| = |F(x_1, y_1)F(x_2, y_2)F(x_3, y_3)| \leq m^3.$$

By applying Corollary 10.9.3 with $R = \pm 1, \ldots, \pm m^3$, we see that up to GL(2, $\mathbb{Z}$)-equivalence there are only finitely many possibilities for the pairs F, G, and so in particular only finitely many possibilities for F. $\square$

We finish with a result of LeVesque and Waldschmidt on parametrized Thue inequalities. Let $F = X^n + a_1 X^{n-1}Y + \cdots + a_n Y^n \in \mathbb{Z}[X, Y]$ be a square-free binary form of degree $n \geq 3$ and with given splitting field L. We can factor F over L as

$$F = (X - \alpha_1 Y) \cdots (X - \alpha_n Y),$$

with $\alpha_1, \ldots, \alpha_n$ distinct elements of L. Consider tuples $\boldsymbol{\varepsilon} := (\varepsilon_1, \ldots, \varepsilon_n)$ with

$$\varepsilon_1, \ldots, \varepsilon_n \in O_L^*, \quad \varepsilon_1 \alpha_1, \ldots, \varepsilon_n \alpha_n \text{ distinct}, \tag{10.9.6}$$
$$F_{\boldsymbol{\varepsilon}} := (X - \varepsilon_1 \alpha_1 Y) \cdots (X - \varepsilon_n \alpha_n) \in \mathbb{Z}[X, Y].$$

Notice that for $\boldsymbol{\varepsilon}$ with (10.9.6) we necessarily have $\varepsilon_1 \cdots \varepsilon_n = \pm 1$. Let m be an integer with $m \geq |F(0, 1)|$. Then for every ϵ with (10.9.6), the Thue inequality

$$|F_{\boldsymbol{\varepsilon}}(x, y)| \leq m \quad \text{in } x, y \in \mathbb{Z} \tag{10.9.7}$$

has solutions $(1, 0)$, $(0, 1)$. Solutions (x, y) of (10.9.7) with $xy = 0$ are called *trivial.*

The following result is a special case of Theorem 3.1 of LeVesque and Waldschmidt (2012).

Corollary 10.9.5 *There are only finitely many $\boldsymbol{\varepsilon}$ with* (10.9.6) *such that* (10.9.7) *has non-trivial solutions.*

Proof. By Corollary 10.9.4, the binary forms $F_{\boldsymbol{\varepsilon}}$ (with $\boldsymbol{\varepsilon}$ as in (10.9.6)) such that (10.9.7) has non-trivial solutions lie in only finitely many GL(2, $\mathbb{Z}$)-equivalence classes. So it suffices to show that a GL(2, $\mathbb{Z}$)-equivalence class can contain only finitely many binary forms $F_{\boldsymbol{\varepsilon}}$. Let $\boldsymbol{\varepsilon}$ be as in (10.9.6), and suppose that $F_{\boldsymbol{\varepsilon}} = \pm F_U$ for some $U = \left(\begin{smallmatrix} a & b \\ c & d \end{smallmatrix}\right) \in$ GL(2, $\mathbb{Z}$). Then

$$F(a, c) = \pm F_{\boldsymbol{\varepsilon}}(1, 0) = \pm F(1, 0) = \pm 1, \quad F(b, d) = \pm F_{\boldsymbol{\varepsilon}}(0, 1) = \pm F(0, 1).$$

Now by Thue's Theorem there are only finitely many possibilities for a, b, c, d, hence for ε. This proves Corollary 10.9.5. □

We mention that in all papers referred to above, generalizations of the theorems and corollaries stated above have been proved for binary forms with S-integral coefficients in a number field.

10.10 Lang's Conjecture for tori

Let K be an algebraically closed field of characteristic 0 and n an integer ≥ 2. Let $(K^*)^n$ denote the n-fold direct product of the multiplicative group K^* of non-zero elements of K. That is, $(K^*)^n$ is the group with coordinatewise multiplication

$$\mathbf{x} \cdot \mathbf{y} := (x_1 y_1, \ldots, x_n y_n)$$

$$\text{for } \mathbf{x} = (x_1, \ldots, x_n), \ \mathbf{y} = (y_1, \ldots, y_n) \in (K^*)^n,$$

and with unit element $\mathbf{1} = (1, \ldots, 1)$. We write polynomials $f \in K[X_1, \ldots, X_n]$ as $\sum_{\mathbf{a} \in I} c(\mathbf{a}) \mathbf{X}^{\mathbf{a}}$, where I is a finite subset of $(\mathbb{Z}_{\geq 0})^n$, $c(\mathbf{a}) \in K^*$ for $\mathbf{a} \in I$ and $\mathbf{X}^{\mathbf{a}} := X_1^{a_1} \cdots X_n^{a_n}$ if $\mathbf{a} = (a_1, \ldots, a_n)$.

A subvariety of $(K^*)^n$ is a set

$$\mathcal{X} = \{\mathbf{x} \in (K^*)^n : f_1(\mathbf{x}) = 0, \ldots, f_r(\mathbf{x}) = 0\},$$

where $f_1, \ldots, f_r \in K[X_1, \ldots, X_n]$. We do not require here that $\mathcal{X}$ is irreducible. An algebraic subgroup of $(K^*)^n$ is a subvariety of $(K^*)^n$ that is also a subgroup of $(K^*)^n$. For instance, a subvariety of $(K^*)^n$ given by equations $\mathbf{x}^{\mathbf{a}_i} = \mathbf{x}^{\mathbf{b}_i}$ $(i = 1, \ldots, r)$ with $\mathbf{a}_i, \mathbf{b}_i \in \mathbb{Z}_{\geq 0}^n$ for $i = 1, \ldots, r$ is an algebraic subgroup of $(K^*)^n$ and in fact any algebraic subgroup of $(K^*)^n$ can be expressed in this form (see, e.g., Schmidt (1996)). An *algebraic coset* is a subvariety of $(K^*)^n$ of the shape $\mathbf{u}H = \{\mathbf{u} \cdot \mathbf{x} : \mathbf{x} \in H\}$ where H is an algebraic subgroup of $(K^*)^n$ and $\mathbf{u} \in (K^*)^n$. Such a coset is more precisely called a coset of H.

The following is a more precise quantitative version of theorems of Liardet (1974, 1975) for $n = 2$, and Laurent (1984) for $n \geq 3$.

Theorem 10.10.1 *Let $\mathcal{X}$ be a subvariety of $(K^*)^n$ given by polynomials of total degree at most Δ. Let Γ be a subgroup of $(K^*)^n$ of finite rank r. Then $\mathcal{X} \cap \Gamma$ is contained in a finite union $\mathbf{u}_1 H_1 \cup \cdots \cup \mathbf{u}_t H_t$ of algebraic cosets with $\mathbf{u}_i H_i \subseteq \mathcal{X}$ for $i = 1, \ldots, t$, where*

$$t \leq C(n, \Delta)^{r+1},$$

with $C(n, \Delta)$ effectively computable in terms of n and Δ.

Below, we give a simple proof of Theorem 10.10.1 by making a reduction to unit equations

$$a_1 x_1 + \cdots + a_n x_n = 1 \quad \text{in } (x_1, \ldots, x_n) \in \Gamma, \tag{10.10.1}$$

where $a_1, \ldots, a_n \in K^*$, and using the fact that such equations have only finitely many non-degenerate solutions (see Chapter 6). But, conversely, this finiteness result for (10.10.1) is a consequence of Theorem 10.10.1. Indeed, let $\mathcal{X}$ be the subvariety of $(K^*)^n$ given by the linear equation $a_1 x_1 + \cdots + a_n x_n = 1$. Denote by $\mathcal{X}^0$ the set of points of $\mathcal{X}$ that remain if we remove all algebraic cosets of dimension > 0 that are contained in $\mathcal{X}$. By Theorem 10.10.1, the set $\mathcal{X}^0 \cap \Gamma$ is finite. It can be shown that $\mathcal{X}^0$ consists precisely of the non-degenerate points in $\mathcal{X}$, i.e., for which $\sum_{i \in J} a_i x_i \neq 0$ for each proper, non-empty subset J of $\{1, \ldots, n\}$. So Theorem 10.10.1 gives back the result that (10.10.1) has only finitely many non-degenerate solutions.

Theorem 10.10.1 may be viewed as the special case for algebraic tori of a general conjecture of Lang on semi-abelian varieties (see Lang (1960)). We do not formally define the n-dimensional algebraic torus $\mathbb{G}_{m,K}^n$ over a field K. We only need the fact that its group of K-rational points is $\mathbb{G}_{m,K}^n(K) = (K^*)^n$, endowed with coordinatewise multiplication. A *semi-abelian variety* over a field K is a commutative group variety A over K, for which there exists a short exact sequence of group varieties over K,

$$0 \to \mathbb{G}_{m,K}^n \to A \to A_0 \to 0,$$

where $n \geq 0$ and A is an abelian variety over K. If $n = 0$ then A is an abelian variety, while if $A_0 = 0$ then A is an algebraic torus. Writing $+$ for the group operation of A, we define a translate of a semi-abelian subvariety over K of A to be a subvariety of the shape $a + B := \{a + x : x \in B\}$, with B a semi-abelian subvariety of A over K and $a \in A(K)$.

Then Lang's general conjecture for semi-abelian varieties is as follows.

Conjecture *Let A be a semi-abelian variety and $\mathcal{X}$ a subvariety of A, both defined over an algebraically closed field K of characteristic 0. Let Γ be a subgroup of $A(K)$ of finite rank. Then $\mathcal{X}(K) \cap \Gamma$ is contained in a finite union of translates $(a_1 + B_1) \cup \cdots \cup (a_t + B_t)$ of semi-abelian subvarieties of A that are each contained in $\mathcal{X}$.*

Lang's Conjecture implies Mordell's Conjecture (Mordell (1922a)) that for any irreducible algebraic curve C of genus $g \geq 2$ defined over $\overline{\mathbb{Q}}$, and any number field L, the set of L-rational points $C(L)$ is finite. Indeed, we may view C as a subvariety of its Jacobian Jac_C which is a g-dimensional abelian variety over $\overline{\mathbb{Q}}$. By the Mordell–Weil Theorem, the group $\mathrm{Jac}_C(L)$ is finitely generated. One-dimensional abelian subvarieties of Jac_C are elliptic curves, and so translates of those are curves of genus 1. The curve C itself cannot be a translate of an abelian subvariety of Jac_C since it has genus at least 2. So the translates of abelian subvarieties of $\mathrm{Jac}(C)$ that are contained in C are necessarily points, and one deduces that $C(L)$ is finite.

Lang's Conjecture is now a theorem. Faltings (1983) proved Mordell's Conjecture. Laurent (1984) proved Lang's Conjecture in the case of tori. Again Faltings (1991, 1994) proved Lang's Conjecture in the case that $K = \overline{\mathbb{Q}}$, A is an abelian variety but for Γ a finitely generated subgroup of $A(\overline{\mathbb{Q}})$ instead of an arbitrary group of finite rank. Vojta (1996) proved Lang's Conjecture for arbitrary semi-abelian varieties over $\overline{\mathbb{Q}}$, but still with Γ finitely generated. Finally, McQuillan (1995), combining Vojta's arguments with Hindry's (Hindry (1988)), proved Lang's Conjecture in full generality, with K an arbitrary algebraically closed fields of characteristic 0, and Γ an arbitrary subgroup of $A(K)$ of finite rank.

We now prove Theorem 10.10.1. Our main tool is Theorem 6.1.3 on unit equations.

Proof of Theorem 10.10.1. We have

$$\mathcal{X} = \left\{ \mathbf{x} \in (K^*)^n : f_i(\mathbf{x}) = \sum_{\mathbf{a} \in I_i} c_i(\mathbf{a}) \mathbf{x}^{\mathbf{a}} = 0 \ (i = 1, \ldots, t) \right\},$$

where $I_i \subset \mathbb{Z}_{\geq 0}^n$ is finite, $c_i(\mathbf{a}) \in K^*$ for $i = 1, \ldots, t$, $\mathbf{a} \in I_i$, and the polynomials $f_1, \ldots, f_t$ have total degree at most Δ.

Let $I := \bigcup_{i=1}^t I_i$. With a point $\mathbf{x} \in \mathcal{X}$ we associate an unordered graph $\mathcal{G}_{\mathbf{x}}$ as follows. The vertices of $\mathcal{G}_{\mathbf{x}}$ are the elements of I, and a pair $\{\mathbf{p}, \mathbf{q}\}$ is an edge of $\mathcal{G}_{\mathbf{x}}$ if there are $i \in \{i, \ldots, t\}$ and a non-empty subset J of I_i such that

$$\left. \begin{array}{l} \mathbf{p}, \mathbf{q} \in I, \quad \displaystyle\sum_{\mathbf{a} \in J} c_i(\mathbf{a}) \mathbf{x}^{\mathbf{a}} = 0, \\[2mm] \displaystyle\sum_{\mathbf{a} \in J'} c_i(\mathbf{a}) \mathbf{x}^{\mathbf{a}} \neq 0 \text{ for each proper, non-empty subset } J' \text{ of } J. \end{array} \right\} \qquad (10.10.2)$$

Notice that there are at most $2^{\binom{n+\Delta}{n}^2}$ possibilities for the graph $\mathcal{G}_{\mathbf{x}}$. For a graph $\mathcal{G}$ on I, let

$$\mathcal{X}_{\mathcal{G}} = \{\mathbf{x} \in \mathcal{X} : \mathcal{G}_{\mathbf{x}} = \mathcal{G}\}.$$

We fix a graph $\mathcal{G}$ on I, and show that $\mathcal{X}_{\mathcal{G}} \cap \Gamma$ is contained in a union of at most C_1^{r+1} algebraic cosets, each of which is contained in $\mathcal{X}$, where C_1 is an effectively computable number depending only on n and Δ. This clearly suffices. In fact, these cosets will all be cosets of the algebraic group

$$H_{\mathcal{G}} := \left\{ \mathbf{x} \in (K^*)^n : \mathbf{x}^{\mathbf{p}} = \mathbf{x}^{\mathbf{q}} \text{ for each edge } \{\mathbf{p}, \mathbf{q}\} \text{ of } \mathcal{G} \right\}.$$

We first show that if $\mathbf{u} \in \mathcal{X}_{\mathcal{G}}$, then $\mathbf{u}H_{\mathcal{G}} \subset \mathcal{X}$. We can express I_1 as a union of pairwise disjoint sets $J_1 \cup \cdots \cup J_r$ such that

$$\sum_{\mathbf{a} \in J_i} c_1(\mathbf{a})\mathbf{u}^{\mathbf{a}} = 0 \quad \text{for } i = 1, \ldots, r,$$

and $\sum_{\mathbf{a} \in J'} c_1(\mathbf{a})\mathbf{u}^{\mathbf{a}} \neq 0$ if J' is a proper subset of one of the J_i. Let $\mathbf{x} \in H_{\mathcal{G}}$. Clearly, any pair $\{\mathbf{p}, \mathbf{q}\}$ contained in the same set J_i is an edge of $\mathcal{G}$, hence $\mathbf{x}^{\mathbf{p}} = \mathbf{x}^{\mathbf{q}}$. Consequently,

$$f_1(\mathbf{u} \cdot \mathbf{x}) = \sum_{i=1}^{r} \sum_{\mathbf{a} \in J_i} c_1(\mathbf{a})(\mathbf{u} \cdot \mathbf{x})^{\mathbf{a}} = 0.$$

Similarly, $f_i(\mathbf{u} \cdot \mathbf{x}) = 0$ for $i = 2, \ldots, t$. This shows that $\mathbf{u} \cdot \mathbf{x} \in \mathcal{X}$ for every $\mathbf{x} \in H_{\mathcal{G}}$.

Let $\{\mathbf{p}, \mathbf{q}\}$ be an edge of $\mathcal{G}$ and $\mathbf{x} \in \mathcal{X}_{\mathcal{G}} \cap \Gamma$. Choose a set J as in (10.10.2). Then

$$\sum_{\mathbf{a} \in J \setminus \{\mathbf{q}\}} \left(-\frac{c_i(\mathbf{a})}{c_i(\mathbf{q})} \right) \mathbf{x}^{\mathbf{a} - \mathbf{q}} = 1.$$

Notice that the tuple $(\mathbf{x}^{\mathbf{a} - \mathbf{q}} : \mathbf{a} \in J \setminus \{\mathbf{q}\})$ is a non-degenerate solution to this equation, and that it lies in a homomorphic image of the group Γ. Now Theorem 6.1.3 implies that $\mathbf{x}^{\mathbf{p} - \mathbf{q}} \in U_{\mathbf{p},\mathbf{q}}$, where $U_{\mathbf{p},\mathbf{q}}$ is a set, that may depend on $\mathbf{p}, \mathbf{q}$ but is otherwise independent of $\mathbf{x}$, of cardinality at most C_2^{r+1}, where C_2 is effectively computable and depends only on n, Δ. The values $\mathbf{x}^{\mathbf{p} - \mathbf{q}}$, taken for all edges $\{\mathbf{p}, \mathbf{q}\}$ of $\mathcal{G}$, uniquely determine the coset $\mathbf{x}H_{\mathcal{G}}$. It follows that the points $\mathbf{x} \in \mathcal{X}_{\mathcal{G}} \cap \Gamma$ lie in a union of at most C_1^{r+1} cosets of $H_{\mathcal{G}}$. This completes our proof. $\square$

We give an overview of some extensions and refinements of Theorem 10.10.1. For $\mathbf{x} = (x_1, \ldots, x_n) \in (\overline{\mathbb{Q}}^*)^n$ we define

$$\widehat{h}(\mathbf{x}) := \sum_{i=1}^{n} h(x_i),$$

where, as usual, $h(x)$ denotes the absolute logarithmic height of $x \in \overline{\mathbb{Q}}$. Let Γ be a finitely generated subgroup of $(\overline{\mathbb{Q}}^*)^n$. We denote by $\overline{\Gamma}$ the *division group* of Γ, that is the subgroup of $(\overline{\mathbb{Q}}^*)^n$ consisting of the points $\mathbf{x} \in (\overline{\mathbb{Q}}^*)^n$ for which there is $m \in \mathbb{Z}_{>0}$ such that $\mathbf{x}^m \in \Gamma$. Define the following enlargements of Γ:

$$\overline{\Gamma}_\epsilon := \left\{ \mathbf{y} \cdot \mathbf{z} : \mathbf{y} \in \overline{\Gamma}, \ \mathbf{z} \in (\overline{\mathbb{Q}}^*)^n, \ \widehat{h}(\mathbf{z}) < \epsilon \right\},$$

$$C(\overline{\Gamma}, \epsilon) := \left\{ \mathbf{y} \cdot \mathbf{z} : \mathbf{y} \in \overline{\Gamma}, \ \mathbf{z} \in (\overline{\mathbb{Q}}^*)^n, \ \widehat{h}(\mathbf{z}) < \epsilon(1 + \widehat{h}(\mathbf{y})) \right\}.$$

We may view these as a "cylinder" and "truncated cone" around $\overline{\Gamma}$. The sets $\overline{\Gamma}_\epsilon$ and $C(\overline{\Gamma}, \epsilon)$ were introduced by Poonen (1999) and Evertse (2002), respectively, in a more general context. Clearly, for $\epsilon > 0$ we have

$$\overline{\Gamma} \subset \overline{\Gamma}_\epsilon \subset C(\overline{\Gamma}, \epsilon).$$

It is important to note that $\overline{\Gamma}_\epsilon$ and $C(\overline{\Gamma}, \epsilon)$ are not groups.

Poonen (1999) formulated a "Lang–Bogomolov Conjecture" for semi-abelian varieties. In the case of algebraic tori, this states that if $\mathcal{X}$ is a subvariety of $(\overline{\mathbb{Q}}^*)^n$ and Γ is a finitely generated subgroup of $(\overline{\mathbb{Q}}^*)^n$, then there is $\epsilon > 0$ such that $\mathcal{X} \cap \overline{\Gamma}_\epsilon$ is contained in a finite union of algebraic cosets, all contained in $\mathcal{X}$. For $\epsilon = 0$ this is Lang's Conjecture for tori, and for $\overline{\Gamma} = (\overline{\mathbb{Q}}^*_{\mathrm{tors}})^n$ we get Bogomolov's Conjecture for tori. The general conjecture for semi-abelian varieties is similar, except that there one has to use a suitable canonical height on the semi-abelian variety under consideration. Poonen himself and independently S. Zhang (2000) proved the Lang–Bogomolov Conjecture for almost split semi-abelian varieties, these include algebraic tori and abelian varieties. The full conjecture was proved by Rémond (2003).

Let $\mathcal{X}$ be a subvariety of $(\overline{\mathbb{Q}}^*)^n$ and denote again by $\mathcal{X}^0$ the set that remains if we remove from $\mathcal{X}$ all algebraic cosets of positive dimension that are contained in $\mathcal{X}$. In Evertse (2002) it was stated, and sketched, that there is $\epsilon > 0$ such that $\mathcal{X}^0 \cap C(\overline{\Gamma}, \epsilon)$ is finite. Rémond (2003) proved a generalization of this for semi-abelian varieties.

Rémond (2002) obtained a quantitative version of the Lang–Bogomolov Conjecture for tori, a somewhat simplified version of which is as follows. Suppose that $\mathcal{X}$ is given by polynomials of degree at most Δ. Define the number $\epsilon := \exp(-(n^{3n+3} \log(n\Delta))$. Assume Γ has rank r. Then $\mathcal{X} \cap \overline{\Gamma}_\epsilon$ is contained in a union of at most

$$\exp\left(n^{3n^2+3} \log(n\Delta)(r+1)\right)$$

algebraic cosets, each contained in $\mathcal{X}$. In his proof, Rémond did not use the Subspace Theorem or results on unit equations, but instead the ideas introduced by Faltings (1991), which led to bounds with a better dependence on Δ. Rémond (2000a, 2000b) proved an analogous result for subvarieties of abelian varieties.

In very few cases, effective results for the above mentioned finiteness results for algebraic tori have been proved. The first case is when $\mathcal{X}$ is a curve in $\overline{\mathbb{Q}}^* \times \overline{\mathbb{Q}}^*$, i.e., $\mathcal{X}$ is given by

$$P(x_1, x_2) = 0$$

with $P \in \overline{\mathbb{Q}}[X_1, X_2]$. We assume that P is an absolutely irreducible polynomial not of the form $aX_1^m + bX_2^n$ or $aX_1^m X_2^n + b$, so that $\mathcal{X}$ is not an algebraic coset of $\overline{\mathbb{Q}}^* \times \overline{\mathbb{Q}}^*$, which means that $\mathcal{X}$ does not contain one-dimensional algebraic cosets. Let Γ be a finitely generated subgroup of $\overline{\mathbb{Q}}^* \times \overline{\mathbb{Q}}^*$. Bombieri and Gubler (2006), Theorem 5.4.5 gave an effectively computable upper bound in terms of the heights of the coefficients of P and of a set of generators for Γ for the heights of the points $\mathbf{x} \in \mathcal{X} \cap \Gamma$.

The result of Bombieri and Gubler was extended by Bérczes, Evertse, Győry and Pontreau (2009) to sets $\mathcal{X} \cap C(\overline{\Gamma}, \epsilon)$, where Γ is a finitely generated subgroup of $\overline{\mathbb{Q}}^* \times \overline{\mathbb{Q}}^*$, and $\epsilon > 0$ is an effectively computable number depending only on the coefficients of P and a generating set for Γ. Moreover, in this latter work explicit upper bounds are given both for the heights and for the degrees of the coordinates of the points of these sets, all in terms of the coefficients of P and the given generators of Γ. Further, this work contains effective versions of Lang's Conjecture for tori, with extensions to $\mathcal{X} \cap \overline{\Gamma}_\epsilon$, $\mathcal{X} \cap C(\overline{\Gamma}, \epsilon)$, for higher dimensional subvarieties $\mathcal{X}$ of $(\overline{\mathbb{Q}}^*)^n$ from a very restricted class, namely subvarieties given by polynomials with at most three non-zero terms.

Applying the specialization techniques discussed in Chapter 8, Bérczes (2015a) proved effective finiteness results for equations $P(x_1, x_2) = 0$ in $x_1, x_2 \in A^*$, where A is an integral domain that is finitely generated over $\mathbb{Z}$ and $P \in A[X_1, X_2]$. In a subsequent paper, Bérczes (2015b) he proved an effective finitenes result for $P(x_1, x_2) = 0$ in $(x_1, x_2) \in \overline{\Gamma}$, where Γ is a finitely generated subgroup of $K^* \times K^*$.

10.11 Linear recurrence sequences and exponential-polynomial equations

We give a brief overview of some results concerning zeros of linear recurrence sequences, and more generally, integer solutions of exponential-polynomial equations. Much more on these topics can be found in Schmidt (2003).

Let K be an algebraically closed field of characteristic 0. Recall that a (two-sided) linear recurrence sequence $U = \{u_m\}_{m=-\infty}^\infty$ in K is given by initial values $u_0, \ldots, u_{k-1}$ and a linear recurrence

$$u_{m+k} = c_1 u_{m+k-1} + \cdots + c_k u_m \quad \text{for } m \in \mathbb{Z},$$

where the c_i belong to K and $c_k \neq 0$. Further, assume that the length k of the recurrence has been chosen minimally. Then we call k the *order* of U, and $f_U := X^k - c_1 X^{k-1} - \cdots - c_k$ the *companion polynomial* of U. These are uniquely determined by U. Assume that

$$f_U = (X - \alpha_1)^{e_1} \cdots (X - \alpha_r)^{e_r},$$

where $\alpha_1, \ldots, \alpha_r$ are distinct elements of K and $e_1, \ldots, e_r$ positive integers. Then the terms u_m can be expressed otherwise as

$$u_m = f_1(m)\alpha_1^m + \cdots + f_r(m)\alpha_r^m \quad (m \in \mathbb{Z}),$$

where $f_i \in K[X]$ is a polynomial of degree exactly $e_i - 1$, for $i = 1, \ldots, r$.

The sequence U is called *non-degenerate* if $\alpha_1 \cdots \alpha_r \neq 0$ and α_i/α_j is not a root of unity for any two distinct indices i, j from $\{1, \ldots, r\}$.

The *zero-multiplicity* $N(U)$ of U is the number of integers m such that $u_m = 0$, that is the number of solutions of the *exponential-polynomial equation*

$$f_1(m)\alpha_1^m + \cdots + f_r(m)\alpha_r^m = 0 \quad \text{in } m \in \mathbb{Z}. \tag{10.11.1}$$

We have the following general result, due to Skolem, Mahler and Lech.

Theorem 10.11.1 *Let U be a non-degenerate linear recurrence sequence in a field K of characteristic 0. Then $N(U)$ is finite.*

Skolem (1935) proved this in the case that U has its terms in $\mathbb{Q}$ and Mahler (1935a) did so for sequences U with algebraic terms. Finally, Lech (1953) proved the general result. The proofs of Skolem, Mahler and Lech were all based on Skolem's p-adic power series method (Skolem (1933)).

We discuss a generalization of (10.11.1) to exponential-polynomial eqations in several variables. Let again K be a field of characteristic 0 and $n \geq 1$. For $\boldsymbol{\alpha} = (\alpha_1, \ldots, \alpha_n) \in (K^*)^n$ and $\mathbf{m} = (m_1, \ldots, m_n) \in \mathbb{Z}^n$, we write $\boldsymbol{\alpha}^{\mathbf{m}} := \alpha_1^{m_1} \cdots \alpha_n^{m_n}$. We consider equations

$$f_1(\mathbf{m})\boldsymbol{\alpha}_1^{\mathbf{m}} + \cdots + f_r(\mathbf{m})\boldsymbol{\alpha}_r^{\mathbf{m}} = 0 \quad \text{in } \mathbf{m} \in \mathbb{Z}^n, \tag{10.11.2}$$

where $f_i \in K[X_1, \ldots, X_n]$, $\boldsymbol{\alpha}_i \in (K^*)^n$ for $i = 1, \ldots, r$. A solution $\mathbf{m} \in \mathbb{Z}^n$ of (10.11.2) is called non-degenerate if $\sum_{i \in I} f_i(\mathbf{m})\boldsymbol{\alpha}_i^{\mathbf{m}} \neq 0$ for each proper, non-empty subset I of $\{1, \ldots, r\}$. We recall the following result, which is a special case of a more general theorem of Laurent (1984, 1989). Define the group

$$G := \{\mathbf{m} \in \mathbb{Z}^n : \boldsymbol{\alpha}_1^{\mathbf{m}} = \cdots = \boldsymbol{\alpha}_r^{\mathbf{m}}\}.$$

Theorem 10.11.2 *Assume that $G = \{\mathbf{0}\}$. Then (10.11.2) has only finitely many non-degenerate solutions.*

Proof of Theorem 10.11.2 $\Longrightarrow$ Theorem 10.11.1. We proceed by induction on r. For $r = 1$ Theorem 10.11.1 is trivial. Let $r \geq 2$ and assume that none of the quotients α_i/α_j $(1 \leq i < j \leq r)$ is a root of unity. This implies that

$$\{m \in \mathbb{Z} : \alpha_1^m = \cdots = \alpha_r^m\} = \{0\}.$$

Hence (10.11.1) has only finitely many non-degenerate solutions. By applying the induction hypothesis to any of the vanishing subsums, we infer that there are also only finitely many degenerate solutions. $\square$

Proof of Theorem 10.11.2 (sketch). The proof depends on Theorem 6.1.1. By means of a specialization argument (see for instance Schmidt (2003), section 9), Theorem 10.11.2 can be reduced to the case that the coordinates of the α_i and the coefficients of the polynomials f_i lie in an algebraic number field K. We restrict ourselves to this special case. Choose a finite set of places S of K, containing the infinite places, such that the coordinates of the α_i $(i = 1, \ldots, r)$ are all S-units. We apply Theorem 6.1.1 to (10.11.2). Pick a non-degenerate solution $\mathbf{m}$ of (10.11.2), and put $x_i := f_i(\mathbf{m})\alpha_i^{\mathbf{m}}$ for $i = 1, \ldots, r$. Then $x_1 + \cdots + x_r = 0$ and no proper subsum of the left-hand side is 0. Put $\|\mathbf{m}\| := \max(|m_1|, \ldots, |m_n|)$. Thanks to our choice of S, we have

$$N_S(x_1 \cdots x_r) = N_S(f_1(\mathbf{m}) \cdots f_r(\mathbf{m})) \leq C_1 \|\mathbf{m}\|^{C_2}, \tag{10.11.3}$$

where here and below, the C_i are constants > 1 depending on the f_i and the α_i. Further, since $\mathbf{m}$ is non-degenerate, we have $f_i(\mathbf{m}) \neq 0$ for $i = 1, \ldots, r$, which implies

$$H_S(x_1, \ldots, x_r) = \prod_{v \in S} \max(|x_1|_v, \ldots, |x_r|_v) \tag{10.11.4}$$

$$\geq C_3^{-1} \|\mathbf{m}\|^{-C_4} \cdot \prod_{v \in S} \max_{1 \leq i \leq r} |\alpha_i^{\mathbf{m}}|_v.$$

By the Product Formula and our choice for S, we have for $\mathbf{z} \in \mathbb{Z}^n$,

$$\log \prod_{v \in S} \max_{1 \leq i \leq r} |\alpha_i^{\mathbf{z}}|_v \geq \frac{1}{r^2} \sum_{1 \leq i,j \leq r} h((\alpha_i \alpha_j^{-1})^{\mathbf{z}}) =: \psi(\mathbf{z}).$$

One can easily show that ψ satisfies the triangle inequality, and $\psi(\lambda \mathbf{z}) = |\lambda| \psi(\mathbf{z})$ for $\lambda \in \mathbb{Z}$, $\mathbf{z} \in \mathbb{Z}^n$. Further, if $\psi(\mathbf{z}) = 0$, then all terms $(\alpha_i \alpha_j^{-1})^{\mathbf{z}}$ are roots of unity. By our assumption on G, this implies that $\mathbf{z} = \mathbf{0}$. Hence ψ defines a norm on $\mathbb{Z}^n$. Both ψ and the maximum norm $\| \cdot \|$ can be extended to norms on $\mathbb{R}^n$ and by a simple compactness argument one shows that there is $c > 0$ such that $\psi(\mathbf{z}) \geq c\|\mathbf{z}\|$ for $\mathbf{z} \in \mathbb{R}^n$. So

$$\prod_{v \in S} \max_{1 \leq i \leq r} |\alpha_i^{\mathbf{z}}|_v \geq \exp\left(r^{-2} c\|\mathbf{z}\|\right) \quad \text{for } \mathbf{z} \in \mathbb{Z}^n.$$

Together with (10.11.4) this implies

$$H_S(x_1, \ldots, x_r) \geq C_3^{-1} \|\mathbf{m}\|^{-C_4} C_5^{\mathbf{m}}.$$

From Theorem 6.1.1 we deduce $H_S(x_1, \ldots, x_r) \leq C_6 N_S(x_1 \cdots x_r)^2$, say. Combining this with the lower bound for $H_S(x_1, \ldots, x_r)$ just derived and the upper bound for $N_S(x_1 \cdots x_r)$ from (10.11.3), we obtain

$$C_3^{-1} \|\mathbf{m}\|^{-C_4} C_5^{\|\mathbf{m}\|} \leq C_6 (C_1 \|\mathbf{m}\|^{C_2})^2,$$

which implies that $\|\mathbf{m}\|$ is bounded. □

Below, we discuss quantitative results (upper bounds for the number of solutions) of (10.11.1) and (10.11.2), that have been obtained as consequences of the Quantitative Subspace Theorem.

It has been an open problem for a long time to obtain a uniform upper bound for the zero multiplicity $N(U)$ of a non-degenerate linear recurrence sequence U depending only on the order of U. This was finally settled by Schmidt (1999), who proved the following.

Theorem 10.11.3 *Let U be a non-degenerate linear recurrence sequence of order k in a field of characteristic 0. Then*

$$N(U) \leq \exp \exp \exp(3k \log k).$$

Schmidt's very intricate proof is based on the Quantitative Subspace Theorem from Evertse and Schlickewei (2002), but uses various other techniques. In fact, the special case where the polynomials f_i in (10.11.1) are all constants follows easily from Theorem 6.1.3, but the extension to arbitrary polynomials f_i was very difficult. Schmidt's bound has been subsequently improved by Schmidt himself (Schmidt (2000)), and Allen (2007) and Amoroso and Viada (2011). The best upper bound to date for $N(U)$, from the last mentioned paper, is $\exp \exp(70k)$. In Schmidt (2003), Schmidt worked out his method of proof in a special case, giving a flavour of the main ideas.

It is conjectured that under the assumption $G = \{0\}$, the number of solutions of (10.11.2) is bounded above by a quantity depending only on r and the total degrees of $f_1, \ldots, f_r$. Schlickewei and Schmidt (2000) proved the following weaker result for exponential-polynomial equations over number fields. We keep the notation from (10.11.2).

Theorem 10.11.4 *Assume that the coordinates of the points $\boldsymbol{\alpha}_i$ and the coefficients of the polynomials f_i lie in an algebraic number field K of degree d. Let δ_i denote the total degree of f_i for $i = 1, \ldots, r$ and put*

$$B := \max\left(n, \sum_{i=1}^{r} \binom{n + \delta_i}{n}\right).$$

Assume that $G = \{0\}$. Then equation (10.11.2) has at most $c(B, d) := 2^{35B^3} d^{6B^2}$ non-degenerate solutions.

Again the main tool in the proof is the Quantitative Subspace Theorem from Evertse and Schlickewei (2002) (which was already proved a couple of years earlier).

There are various generalizations of Theorem 10.11.4, see Schlickewei and Schmidt (2000) and Ahlgren (1999). From Schmidt (2009) and Corvaja, Schmidt and Zannier (2010) the following special case of the above conjecture can be deduced. Let K be any field of characteristic 0, $f \in K[X_1, \ldots, X_n]$ a polynomial of total degree δ, and $\boldsymbol{\alpha} = (\alpha_1, \ldots, \alpha_n)$, where $\alpha_1, \ldots, \alpha_n$ are multiplicatively independent, non-zero elements of K. Then the equation

$$\boldsymbol{\alpha}^{\mathbf{m}} = f(\mathbf{m})$$

has at most $\exp(B^{9B})$ solutions $\mathbf{m} \in \mathbb{Z}^n$, where $B := 1 + \binom{n+\delta}{\delta}$.

10.12 Algebraic independence results

A possibly infinite sequence $\alpha_1, \alpha_2, \ldots$ is called algebraically independent over a field K if there are no N and $P \in K[X_1, \ldots, X_N] - \{0\}$ such that $P(\alpha_1, \ldots, \alpha_N) = 0$. Nishioka (1986, 1987, 1989, 1994) proved various algebraic independence results for values of certain power series at algebraic arguments. All these results are applications of the semi-effective Theorem 6.1.1. Here, we prove a special case of one of Nishioka's results, and mention some of her other results. Below, by algebraic numbers we always mean complex numbers that are algebraic over $\mathbb{Q}$.

Let $K \subset \mathbb{C}$ be an algebraic number field and

$$f(z) = \sum_{k=0}^{\infty} a_k z^{e_k}$$

a power series with coefficients $a_k \in K$ and with $\{e_k\}_{k=0}^{\infty}$ a strictly increasing sequence of non-negative integers. Assume that $f(z)$ has radius of convergence $R > 0$. Further, assume that $\{e_k\}$ grows rapidly, i.e.,

$$\lim_{k \to \infty} \frac{e_k + \sum_{i=1}^{k} h(a_i)}{e_{k+1}} = 0,$$

where as usual $h(\alpha)$ denotes the absolute logarithmic height of an algebraic number α. By Cijsouw and Tijdeman (1973), the number $f(\alpha)$ is transcendental for any algebraic number α with $0 < |\alpha| < R$. Further, it was shown by Bundschuh and Wylegala (1980), that $f(\alpha_1), \ldots, f(\alpha_n)$ are algebraically independent for any algebraic numbers $\alpha_1, \ldots, \alpha_n$ with $0 < |\alpha_1| < \cdots < |\alpha_n| < R$. These results were extended by Nishioka as follows. Call non-zero algebraic numbers $\alpha_1, \ldots, \alpha_s$ $\{e_k\}$-dependent if there are γ, roots of unity $\zeta_1, \ldots, \zeta_s$, and algebraic numbers $d_1, \ldots, d_s$, not all zero, such that

$$\alpha_i = \zeta_i \gamma \quad \text{for } i = 1, \ldots, s, \qquad \sum_{i=1}^{s} d_i \zeta_i^{e_k} = 0 \quad \text{for all sufficiently large } k.$$

Denote by $f^{(l)}$ the l-th derivative of f, where $f^{(0)} = f$. The following result is Theorem 1 of Nishioka (1987).

Theorem 10.12.1 *Let $\alpha_1, \ldots, \alpha_n$ be algebraic numbers with $0 < |\alpha_i| < R$ for $i = 1, \ldots, n$. Then the following three assertions are equivalent:*

(i) $f^{(l)}(\alpha_i)$ $(i = 1, \ldots, n, l \geq 0)$ are algebraically dependent over $\mathbb{Q}$;
(ii) there are distinct $i_1, \ldots, i_s \in \{1, \ldots, n\}$ such that $\alpha_{i_1}, \ldots, \alpha_{i_s}$ are $\{e_k\}$-dependent;
(iii) $1, f(\alpha_1), \ldots, f(\alpha_n)$ are linearly dependent over the algebraic numbers.

To give a flavour of Nishioka's method of proof, we prove the following special case.

Theorem 10.12.2 *Let $f(z) = \sum_{k=0}^{\infty} z^{e_k}$, where $\{e_k\}_{k=0}^{\infty}$ is a strictly increasing sequence of non-negative integers with $\lim_{k \to \infty} e_k/e_{k+1} = 0$. Further, let $\alpha_1, \ldots, \alpha_n$ be algebraic numbers such that $|\alpha_i| < 1$ for $i = 1, \ldots, n$ and none of the quotients α_i/α_j $(1 \leq i < j \leq n)$ is a root of unity. Then the numbers $f^{(l)}(\alpha_i)$ $(i = 1, \ldots, n, l \geq 0)$ are algebraically independent over $\mathbb{Q}$.*

In fact, this was proved earlier by Nishioka (1986), with $e_k = k!$ for all k.

We first prove a crucial lemma (see Nishioka (1989), Lemma 1), which is a consequence of Theorem 6.1.1.

Lemma 10.12.3 *Let Ω be an infinite set of non-negative integers. Further, let K be a number field, $\gamma_1, \ldots, \gamma_n$ non-zero elements of K, and $\{A_i(m)\}_{m \in \Omega}$ sequences of elements of K, such that*

$$\gamma_i/\gamma_j \text{ is not a root of unity for all } i, j \text{ with } 1 \leq i < j \leq n, \text{(10.12.1)}$$

$$A_i(m) \neq 0 \quad \text{for } i = 1, \ldots, n, m \in \Omega, \tag{10.12.2}$$

$$\lim_{\substack{m \to \infty \\ m \in \Omega}} \frac{h(A_i(m))}{m} = 0 \quad \text{for } i = 1, \ldots, n. \tag{10.12.3}$$

Then for every θ with $0 < \theta < 1$ and every place v of K, there are only finitely many $m \in \Omega$ such that

$$|A_1(m)\gamma_1^m + \cdots + A_n(m)\gamma_n^m|_v \leq |\gamma_1|_v^m \theta^m. \tag{10.12.4}$$

Remark This lemma easily implies Theorem 10.11.1 (the Skolem–Mahler–Lech Theorem for linear recurrence sequences) in the case of linear recurrence sequences with terms in an algebraic number field.

Proof. The proof is by induction on n. For $n = 1$, the lemma is an easy consequence of assumptions (10.12.2), (10.12.3) and of (1.9.1), more precisely the inequality $\log |A_1(m)|_v \geq -[K : \mathbb{Q}] \cdot h(A_1(m))$. Now let $n \geq 2$ and suppose the lemma is true for sums with fewer than n terms. The induction hypothesis implies that for all but finitely many $m \in \Omega$, every *proper* subsum of $\sum_{i=1}^n A_i(m)\gamma_i^m$ is non-zero. We show that also $\sum_{i=1}^n A_i(m)\gamma_i^m$ can be 0 for at most finitely many m. Indeed, by (10.12.1) there is $v \in M_K$ such that $|\gamma_1/\gamma_n|_v \neq 1$. Assume without loss of generality that $|\gamma_1/\gamma_n|_v > 1$. Notice that by (10.12.3),

$$\frac{h(A_i(m)/A_n(m))}{m} \leq \frac{h(A_i(m)) + h(A_n(m))}{m} \to 0 \quad \text{as } m \in \Omega, \, m \to \infty.$$

Now if $\sum_{i=1}^n A_i(m)\gamma_i^m = 0$, then

$$\left| \frac{A_1(m)}{A_n(m)} \cdot \gamma_1^m + \cdots + \frac{A_{n-1}(m)}{A_n(m)} \cdot \gamma_{n-1}^m \right|_v = |\gamma_n|_v^m,$$

and by the induction hypothesis this is possible for only finitely many m.

So, after removing at most finitely many integers, we obtain an infinite set of positive integers Ω' such that for every $m \in \Omega'$, each subsum of $\sum_{i=1}^n A_i(m)\gamma_i^m$ is non-zero. By Lemma 1.9.1, for every $m \in \Omega'$ there is a positive rational integer d_m, with $\log d_m \leq [K : \mathbb{Q}] \sum_{i=1}^n h(A_i(m))$, such that $d_m A_i(m) \in O_K$ for $i = 1, \ldots, n$. Now, clearly, (10.12.3) remains valid with $d_m A_i(m)$ instead of $A_i(m)$, and $(\log d_m)/m \to 0$ as $m \to \infty$. Hence we may as well prove our lemma with $d_m A_i(m)$ instead of $A_i(m)$. So we may and will assume that all $A_i(m)$ are algebraic integers without loss of generality.

Now let S be a finite set of places of K such that $v \in S$, and $\gamma_1, \ldots, \gamma_n$ are all S-units. Put $u_i(m) := A_i(m)\gamma_i^m$ for $i = 1, \ldots, n$. We apply Proposition 6.2.1 (which is in fact equivalent to Theorem 6.1.1) with $x_i = u_i(m)$ for $i = 1, \ldots, m$, S as above, and $T = \{v\}$. Define

$$\delta_m := \frac{[K : \mathbb{Q}] \sum_{i=1}^n h(A_i(m))}{m}, \quad H := \log \prod_{w \in S} \max(|\gamma_1|_w, \ldots, |\gamma_n|_w).$$

Then $H \geq 0$. Notice that for $m \in \Omega'$ we have by (1.9.1),

$$\prod_{w \in S} \prod_{i=1}^n |u_i(m)|_w = \prod_{w \in S} \prod_{i=1}^n |A_i(m)|_w \leq \exp(\delta_m m).$$

while

$$H_S(u_1(m), \ldots, u_n(m)) \leq \exp(mH) \cdot \prod_{w \in S} \max(1, |A_1(m)|_w, \ldots, |A_n(m)_m|)$$

$$\leq \exp((H + \delta_m)m)$$

and lastly, by (1.9.1),

$$\max(|u_1(m)|_v, \ldots, |u_n(m)|_v) \geq |\gamma_1|_v^m \exp(-\delta_m m).$$

By combining these three inequalities with Proposition 6.2.1 we obtain that for every $\epsilon > 0$ there is a constant $C(\epsilon) > 0$ such that for all $m \in \Omega'$,

$$|A_1(m)\gamma_1^m + \cdots + A_n(m)\gamma_n^m|_v$$

$$\geq C(\epsilon) \max(|u_1(m)|_v, \ldots, |u_n(m)|_v)$$

$$\times \left(\prod_{w \in S} \prod_{i=1}^n |u_i(m)|_w \right)^{-1} \cdot H_S(u_1(m), \ldots, u_n(m))^{-\epsilon}$$

$$\geq C(\epsilon)|\gamma_1|_v^m \exp\left(-m((2+\epsilon)\delta_m + \epsilon H) \right).$$

Since we can choose ϵ arbitrarily small and since $\delta_m \to 0$ by (10.12.3), it follows that for every θ with $0 < \theta < 1$, inequality (10.12.4) has only finitely many solutions in $m \in \Omega$. $\square$

Proof of Theorem 10.12.2. We use the following notation. Let K be the number field $\mathbb{Q}(\sqrt{-1}, \alpha_1, \ldots, \alpha_n)$. Let v be the (necessarily complex) place of K such that $|\cdot|_v = |\cdot|^2$. For $\mathbf{x} = (x_1, \ldots, x_r) \in \mathbb{C}^r$ we put $\|\mathbf{x}\| := \max_{1 \leq i \leq r} |x_i|$. Further, for $\mathbf{x} = (x_1, \ldots, x_r) \in K^r$, $w \in M_K$ we put $\|\mathbf{x}\|_w := \max_{1 \leq i \leq r} |x_i|_w$.

Assume that Theorem 10.12.2 is false. Then there is $L \geq 0$ such that the numbers $f^{(l)}(\alpha_i)$ ($i = 1, \ldots, n, l = 1, \ldots, L$) are algebraically dependent over $\mathbb{Q}$. Assume that

$$|\alpha_1| = \max_{1 \leq i \leq n} |\alpha_i|. \tag{10.12.5}$$

For $m \geq 0$ put $f_m(z) := \sum_{k=0}^m z^{e_k}$. Define vectors $\mathbf{u} \in \mathbb{C}^{n(L+1)}$ and $\mathbf{u}_m \in K^{n(L+1)}$ ($m = 0, 1, 2, \ldots,$) by

$$\mathbf{u} := \left(\alpha_i^l f^{(l)}(\alpha_i) \right)_{\substack{i=1,\ldots,n \\ l=0,\ldots,L}}, \quad \mathbf{u}_m := \left(\alpha_i^l f_m^{(l)}(\alpha_i) \right)_{\substack{i=1,\ldots,n \\ l=0,\ldots,L}}.$$

Then $\lim_{m \to \infty} \mathbf{u}_m = \mathbf{u}$. Note that

$$\alpha_i^l f_m^{(l)}(\alpha_i) = \sum_{k=0}^m e_k(e_k - 1) \cdots (e_k - l + 1)\alpha_i^{e_k}.$$

There is a non-zero polynomial in $n(L+1)$ variables $P \in \mathbb{Q}[X_{10}, \ldots, X_{n,L+1}]$ such that

$$P(\mathbf{u}) = 0$$

(i.e., P evaluated at $X_{il} = \alpha_i^l f^{(l)}(\alpha_i)$ for $i = 1, \ldots, n$, $l = 1, \ldots, L$). We choose such P of minimal total degree. Denote the total degree of P by D.

The constants C_i introduced below will be ≥ 1 and depend on $P, \alpha_1, \ldots, \alpha_n$ only. Further, $\{C_{iw}\}_{w \in M_K}$ will be tuples of constants depending only on $P, \alpha_1, \ldots, \alpha_n$, where $C_{iw} \geq 1$ for all $w \in M_K$ and $C_{iw} = 1$ for all but finitely many w.

For $m = 0, 1, 2, \ldots$, we have by (10.12.5),

$$|P(\mathbf{u}_m)|_v = |P(\mathbf{u}_m) - P(\mathbf{u})|^2 \leq C_1 \|\mathbf{u}_m - \mathbf{u}\|^2 \leq C_2 e_{m+1}^{2L} |\alpha_1|^{2e_{m+1}}.$$

Further, for $w \in M_K \setminus \{v\}$ we have

$$|P(\mathbf{u}_m)|_w \leq C_{3w} \max(1, \|\mathbf{u}_m\|_w)^D$$
$$\leq C_{3w} \max\left(1, |(m+1)e_m^L|_w\right)^D \max(1, |\alpha_1|_w, \ldots, |\alpha_n|_w)^{De_m}$$
$$\leq C_{4w}^{e_m}. \tag{10.12.6}$$

Hence

$$\prod_{v \in M_K} |P(\mathbf{m})|_v \leq C_5^{e_m} e_{m+1}^{2L} |\alpha_1|^{2e_{m+1}},$$

which is < 1 for m sufficiently large since $|\alpha_1| < 1$ and $e_m/e_{m+1} \to 0$ as $m \to \infty$. So by the Product Formula, $P(\mathbf{u}_m) = 0$ for all sufficiently large m.

We infer that for sufficiently large m we have by Taylor's formula,

$$0 = P(\mathbf{u}_m) - P(\mathbf{u}_{m-1})$$
$$= \sum_{\mathbf{a}} P_{\mathbf{a}}(\mathbf{u}_{m-1}) \prod_{i=1}^{n} \prod_{l=0}^{L} \left(e_m(e_m - 1) \cdots (e_m - l + 1)\alpha_i^{e_m}\right)^{a_{il}},$$

where the sum is over all tuples of non-negative integers $\mathbf{a} = (a_{il})_{i=1,\ldots,n, l=0,\ldots,L}$ with $\mathbf{a} \neq \mathbf{0}$, $\sum_{i,l} a_{il} \leq D$ and where $P_{\mathbf{a}} := (\prod_{i=1}^{n} \prod_{l=0}^{L} (a_{il}!)^{-1} \partial^{a_{il}} / \partial X_{il}^{a_{il}}) P$. Estimating the terms with partial derivatives of order at least 2, we get by (10.12.5),

$$\left| \sum_{i=1}^{n} \beta_i(m)\alpha_i^{e_m} \right| \leq C_6 \max(1, \|\mathbf{u}_{m-1}\|)^D e_m^{LD} |\alpha_1|^{2e_m}$$
$$\leq C_7^{e_{m-1}} e_m^{LD} |\alpha_1|^{2e_m}, \tag{10.12.7}$$

where

$$\beta_i(m) := \sum_{l=0}^{L} e_m(e_m - 1) \cdots (e_m - l + 1) \cdot \frac{\partial P(\mathbf{u}_{m-1})}{\partial X_{il}} \quad \text{for } i = 1, \ldots, n.$$

We apply Lemma 10.12.3 with $\Omega = \{e_k\}_{k=0}^{\infty}$ minus possibly a finite subset, $\gamma_i = \alpha_i$, $A_i(e_k) = \beta_i(k)$ for $i = 1, \ldots, n$, $k \geq 0$, to show that (10.12.7) cannot hold for infinitely many m and derive a contradiction. Condition (10.12.1) is satisfied with $\gamma_i = \alpha_i$ for $i = 1, \ldots, n$ by assumption. As for condition (10.12.2), we have to show that for all sufficiently large m we have $\beta_i(m) \neq 0$ for $i = 1, \ldots, m$. Indeed, suppose that for some i we have $\beta_i(m) = 0$ for infinitely many m. Then for these m,

$$\frac{\partial P(\mathbf{u}_{m-1})}{\partial X_{iL}} = -\sum_{l=0}^{L-1} \frac{1}{(e_m - l) \cdots (e_m - L + 1)} \cdot \frac{\partial P(\mathbf{u}_{m-1})}{\partial X_{il}}$$

and by letting $m \to \infty$ we get

$$\frac{\partial P(\mathbf{u})}{\partial X_{iL}} = 0.$$

But this is impossible since we had chosen P of minimal total degree with $P(\mathbf{u}) = 0$.

To verify (10.12.3) we have to estimate the absolute logarithmic height of $\beta_i(m)$. By a similar computation as in (10.12.6), one has for all sufficiently large m,

$$|\beta_i(m)|_w \leq C_{8w}^{e_{m-1}} \quad \text{for } i = 1, \ldots, n, \ \ w \in M_K$$

and so $h(\beta_i(m)) \leq C_9 e_{m-1}$ for $i = 1, \ldots, n$. Hence

$$\frac{h(\beta_i(m))}{e_m} \leq \frac{C_9 e_{m-1}}{e_m} \to 0 \quad \text{as } m \to \infty.$$

So all conditions of Lemma 10.12.3 are satisfied. By (10.12.7) and $|\alpha_1|_v = |\alpha_1|^2 < 1$, $e_{m-1}/e_m \to 0$ as $m \to \infty$, there is θ with $0 < \theta < 1$ such that for all sufficiently large m,

$$|\beta_1(m)\alpha_1^{e_m} + \cdots + \beta_n(m)\alpha_n^{e_m}|_v \leq (|\alpha_1|_v \theta)^{e_m}.$$

But this contradicts Lemma 10.12.3. Theorem 10.12.2 follows. $\quad\square$

Nishioka (1989) proved algebraic independence results for values at algebraic points of power series

$$f_\omega(z) = \sum_{k=0}^{\infty} [k\omega] z^k,$$

where ω is a real irrational number and $[x]$ denotes the integral part of a real number x. Her method of proof is a variation on that given above, and the main tool is again Theorem 6.1.1. One of her results from that paper (i.e., Theorem 2) states that if ω has unbounded partial quotients in its continued fraction expansion, $\alpha_1, \ldots, \alpha_n$ are algebraic numbers such that $|\alpha_i| < 1$ for $i = 1, \ldots, n$ and none of the quotients α_i/α_j $(1 \leq i < j \leq n)$ is a root of unity, then the numbers $f_\omega(\alpha_1), \ldots, f_\omega(\alpha_n)$ are algebraically independent over the rationals.

In Nishioka (1994), the author applies Theorem 6.1.1 to obtain algebraic independence results for values of Mahler functions at algebraic points. For an extensive treatment of transcendence theory of Mahler functions we refer to Nishioka (1996). We state only the following special case of Nishioka (1994), Proposition: let

$$F_r(z) := \sum_{k=0}^{\infty} z^{r^k}, \quad G_r(z) := \prod_{k=0}^{\infty}(1 - z^{r^k}).$$

Then for every algebraic number α with $0 < |\alpha| < 1$, the numbers $F_r(\alpha)$ ($r = 2, 3, \ldots$), $G_r(\alpha)$ ($r = 2, 3, \ldots$) are algebraically independent over the rationals.

There are various other transcendence results that follow from the Subspace Theorem but not specifically from Theorem 6.1.1, see for instance Corvaja and Zannier (2002b), the survey Bugeaud (2011), and the book Bugeaud (2012), in particular chapters 8 and 9.

References

Adamczewski, B. and J. P. Bell (2012), On vanishing coefficients of algebraic power series over fields of positive characteristic, *Invent. Math.* **187**, 343–393.

Ahlgren, S. (1999), The set of solutions of a polynomial-exponential equation, *Acta Arith.* **87**, 189–207.

Allen, P. B. (2007), On the multiplicity of linear recurrence sequences, *J. Number Theory* **126**, 212–216.

Amoroso, F. and E. Viada (2009), Small points on subvarieties of a torus, *Duke Math. J.* **150**, 407–442.

Amoroso, F. and E. Viada (2011), On the zeros of linear recurrence sequences, *Acta Arith.* **147** (2011), 387–396.

Arenas-Carmona, L., D. Berend and V. Bergelson (2008), Ledrappier's system is almost mixing of all orders, *Ergodic Theory Dynam. Systems* **28**, 339–365.

Aschenbrenner, M. (2004), Ideal membership in polynomial rings over the integers, *J. Amer. Math. Soc.* **17**, 407–442.

Ashrafi, N. and P. Vámos (2005), On the unit sum number of some rings, *Quart. J. Math.* **56**, 1–12.

Baker, A. (1966), Linear forms in the logarithms of algebraic numbers, *Mathematika* **13**, 204–216.

Baker, A. (1967a), Linear forms in the logarithms of algebraic numbers, II, *Mathematika* **14**, 102–107.

Baker, A. (1967b), Linear forms in the logarithms of algebraic numbers, III, *Mathematika* **14**, 220–228.

Baker, A. (1968a), Linear forms in the logarithms of algebraic numbers, IV, *Mathematika* **15**, 204–216.

Baker, A. (1968b), Contributions to the theory of Diophantine equations, *Philos. Trans. Roy. Soc. London, Ser. A* **263**, 173–208.

Baker, A. (1968c), The Diophantine equation $y^2 = ax^3 + bx^2 + cx + d$, *J. London Math. Soc.* **43**, 1–9.

Baker, A. (1969), Bounds for the solutions of the hyperelliptic equation, *Proc. Camb. Philos. Soc.* **65**, 439–444.

Baker, A. (1975), *Transcendental number theory*, Cambridge University Press.

Baker, A., ed. (1988), *New Advances in Transcendence Theory*, Cambridge University Press.

Baker, A. (1998), Logarithmic forms and the abc-conjecture, in: Number Theory Diophantine, Computational and Algebraic Aspects, *Proc. Conf. Eger*, 1966, K. Győry, A. Pethő and V. T. Sós, eds., de Gruyter, 37–44.

Baker, A. (2004), Experiments on the abc-conjecture, *Publ. Math. Debrecen* **65**, 253–260.

Baker, A. and H. Davenport (1969), The equations $3x^2 - 2 = y^2$ and $8x^2 - 7 = z^2$, *Quart. J. Math. Oxford Ser.* (2) **20**, 129–137.

Baker, A. and D. W. Masser, eds. (1977), *Transcendence theory: advances and applications*, Academic Press.

Baker, A. and G. Wüstholz (2007), *Logarithmic Forms and Diophantine Geometry*, Cambridge University Press.

Barroero, F., C. Frei and R. F. Tichy (2011), Additive unit representations in rings over global fields – a survey, *Publ. Math. Debrecen* **79**, 291–307.

Belcher, P. (1974), Integers expressible as sums of distinct units, *Bull. London Math. Soc.* **6**, 66–68.

Bertók, Cs. (2013), Representing integers as sums or differences of general power products, *Acta Math. Hungar.* **141**, 291–300.

Bérczes, A. (2000), On the number of solutions of index form equations, *Publ. Math. Debrecen* **56**, 251–262.

Bérczes, A. (2015a), Effective results for unit points over finitely generated domains, *Math. Proc. Camb. Phil. Soc.* **158**, 331–353.

Bérczes, A. (2015b), Effective results for division points on curves in $\mathbb{G}_m^2$, *J. Th. Nombres Bordeaux*, to appear.

Bérczes, A., J.-H. Evertse and K. Győry (2004), On the number of equivalence classes of binary forms of given degree and given discriminant, *Acta Arith.* **113**, 363–399.

Bérczes, A., J.-H. Evertse and K. Győry (2007a), On the number of pairs of binary forms with given degree and given resultant, *Acta Arith.* **128**, 19–54.

Bérczes, A., J.-H. Evertse and K. Győry (2007b), Diophantine problems related to discriminants and resultants of binary forms, in: *Diophantine Geometry, proceedings of a trimester held from April–July 2005*, U. Zannier, ed., CRM series, Scuola Normale Superiore Pisa, pp. 45–63.

Bérczes, A., J.-H. Evertse and K. Győry (2009), Effective results for linear equations in two unknowns from a multiplicative division group, *Acta Arith.* **136**, 331–349.

Bérczes, A., J.-H. Evertse and K. Győry (2013), Multiply monogenic orders, *Ann. Sc. Norm. Super. Pisa Cl. Sci.* (5) **12**, 467–497.

Bérczes, A., J.-H. Evertse and K. Győry (2014), Effective results for Diophantine equations over finitely generated domains, *Acta Arith.* **163**, 71–100.

Bérczes, A., J.-H. Evertse, K. Győry and C. Pontreau (2009), Effective results for points on certain subvarieties of a tori, *Math. Proc. Camb. Phil. Soc.* **147**, 69–94.

Bérczes, A. and K. Győry (2002), On the number of solutions of decomposable polynomial equations, *Acta Arith.* **101**, 171–187.

Beukers, F. and H. P. Schlickewei (1996), The equation $x + y = 1$ in finitely generated groups, *Acta. Arith.* **78**, 189–199.

Beukers, F. and D. Zagier (1997), Lower bounds of heights of points on hypersurfaces, *Acta Arith.* **79**, 103–111.

Bilu, Yu. F. (1995), Effective analysis of integral points on algebraic curves, *Israel J. Math.* **90**, 235–252.

Bilu, Yu. F. (2002), Baker's method and modular curves, in: A Panorama of Number Theory, or The View from Baker's Garden, *Proc. conf. ETH Zurich, 1999*, G. Wüstholz, ed., Cambridge University Press, pp. 73–88.

Bilu, Yu. F. (2008), The many faces of the subspace theorem [after Adamczewski, Bugeaud, Corvaja, Zannier, . . .], Séminaire Bourbaki, Vol. 2006/2007, *Astérisque* 317, Exp. No. 967, vii, 1–38.

Bilu, Yu. F. and Y. Bugeaud (2000), Démonstration du théorème de Baker-Feldman via les formes linéaires en deux logarithmes, *J. Théorie des Nombres, Bordeaux*, 12, 13–23.

Bilu, Yu. F., I. Gaál and K. Győry (2004), Index form equations in sextic fields: a hard computation, *Acta Arith.* 115, 85–96.

Bilu, Yu. F. and G. Hanrot (1996), Solving Thue equations of high degree, *J. Number Theory*, 60, 373–392.

Bilu, Yu. F. and G. Hanrot (1998), Solving superelliptic Diophantine equations by Baker's method, *Compositio Math.* 112, 273–312.

Bilu, Yu. F. and G. Hanrot (1999), Thue equations with composite fields, *Acta Arith.*, 88, 311–326.

Birch, B. J. and J. R. Merriman (1972), Finiteness theorems for binary forms with given discriminant, *Proc. London Math. Soc.* 24, 385–394.

Bombieri, E. (1993), Effective diophantine approximation on $\mathbf{G}_M$, *Ann. Scuola Norm. Sup. Pisa (IV)* 20, 61–89.

Bombieri, E. (1994), On the Thue-Mahler equation (II), *Acta Arith.* 67, 69–96.

Bombieri, E. and P. B. Cohen (1997), Effective Diophantine approximation on $\mathbb{G}_m$, II, *Ann. Scuola Norm. Sup. Pisa (IV)* 24, 205–225.

Bombieri, E. and P. B. Cohen (2003), An elementary approach to effective Diophantine approximation on $\mathbb{G}_m$, in Number Theory and Algebraic Geometry, *To Peter Swinaerton Dyer on his 75th birthday, London Math. Soc.* Lecture Note Series 303, M. Reid and A. Skorobogatov, eds. Cambridge University Press, pp. 41–62.

Bombieri, E. and W. Gubler (2006), *Heights in Diophantine Geometry*, Cambridge University Press.

Bombieri, E., J. Mueller and M. Poe (1997), The unit equation and the cluster principle, *Acta Arith.* 79, 361–389.

Bombieri, E., J. Mueller and U. Zannier (2001), Equations in one variable over function fields, *Acta Arith.* 99, 27–39.

Bombieri, E. and W. M. Schmidt (1987), On Thue's equation, *Invent. Math.* 88, 69–81.

Borevich, Z. I. and I. R. Shafarevich (1967), *Number Theory*, 2nd edn., Academic Press.

Borosh, I., M. Flahive, D. Rubin and B. Treybig (1989), A sharp bound for solutions of linear Diophantine equations, *Proc. Amer. Math. Soc.* 105, 844–846.

Bosma, W., J. Cannon and C. Playoust (1997), The Magma algebra system I. The user languange, *J. Symbolic Comput*, 24, 235–265.

Brindza, B. (1984), On S-integral solutions of the equation $y^m = f(x)$, *Acta Math. Hungar.* 44, 133–139.

Brindza, B. and K. Győry (1990), On unit equations with rational coefficients, *Acta Arith.* 53, 367–388.

Broberg, N. (1999), Some examples related to the abc-conjecture for algebraic number fields, *Math. Comp.* 69, 1707–1710.

Browkin, J. (2000), The abc-conjecture, in: Number Theory, R. P. Bambah, V. C. Dumir and R. J. Hans-Gill, eds., *Birkhäuser*, pp. 75–105.

Brownawell, W. D. and D. W. Masser (1986), Vanishing sums in function fields, *Math. Soc. Camb. Phil. Soc.* **100**, 427–434.

Brunotte, H., A. Huszti and A. Pethő (2006), Bases of canonical number systems in quartic number fields, *J. Théor. Nombres Bordeaux* **18**, 537–557.

Bugeaud, Y. (1998), Bornes effectives pour les solutions des équations en S-unités et des équations de Thue-Mahler, *J. Number Theory* **71**, 227–244.

Bugeaud, Y. (2011), Quantitative versions of the subspace theorem and applications, *J. Théor. Nombres Bordeaux* **23**, 35–57.

Bugeaud, Y. (2012), Distribution Modulo One and Diophantine Approximation, *Cambridge Tracts in Mathematics* **193**, Cambridge University Press.

Bugeaud, Y. and K. Győry (1996a), Bounds for the solutions of unit equations, *Acta Arith.* **74**, 67–80.

Bugeaud, Y. and K. Győry (1996b), Bounds for the solutions of Thue-Mahler equations and norm form equations, *Acta Arith.* **74**, 273–292.

Bugeaud, Y. and F. Luca (2004), A quantitative lower bound for the greatest prime factor of $(ab + 1)(bc + a)(ca + 1)$, *Acta Arith.* **114**, 275–294.

Bundschuh, P. and F.-J. Wylegala (1980), Über algebraische Unabhängigkeit bei gewissen nichtfortsetzbaren Potenzreihen, *Arch. Math.* **34**, 32–36.

Canci, J. K. (2007), Finite orbits for rational functions, *Indag. Mathem., N.S.* **18**, 203–214.

Cassels, J. W. S. (1959), *An Introduction to the Geometry of Numbers*, Springer Verlag.

Cijsouw, P. L. and R. Tijdeman (1973), On the transcendence of certain power series of algebraic numbers, *Acta Arith.* **23**, 301–305.

Coates, J. (1969), An effective p-adic analogue of a theorem of Thue, *Acta Arith.* **15**, 279–305.

Coates, J. (1970), An effective p-adic analogue of a theorem of Thue II, The greatest prime factor of a binary form, *Acta Arith*, **16**, 392–412.

Cohen, H. (1993), *A Course in Computational Algebraic Number Theory*, Springer Verlag.

Cohen, H. (2000), *Advanced Topics in Computational Number Theory*, Springer Verlag.

Conway, J. H. and A. J. Jones (1976), Trigonometric Diophantine equations (on vanishing sums of roots of unity), *Acta Arith.* **30**, 229–240.

Corvaja, P., W. M. Schmidt and U. Zannier (2010), The Diophantine equation $\alpha_1^{x_1} \cdots \alpha_n^{x_n} = f(x_1, \ldots, x_n)$ II, *Trans. Amer. Math. Soc.* **362**, 2115–2123.

Corvaja, P. and U. Zannier (2002a), A subspace theorem approach to integral points on curves, *C.R. Math. Acad. Sci. Paris* **334**, 267–271.

Corvaja, P. and U. Zannier (2002b), Some new applications of the subspace theorem, *Compos. Math.* **131**, 319–340.

Corvaja, P. and U. Zannier (2003), On the greatest prime factor of $(ab + 1)(ac + 1)$, *Proc. Amer. Math. Soc.* **131**, 1705–1709.

Corvaja, P. and U. Zannier (2004a), On a general Thue's equation, *Amer. J. Math.* **126**, 1033–1055.

Corvaja, P. and U. Zannier (2004b), On integral points on surfaces, *Ann. Math.* **160**, 705–726.

Corvaja, P. and U. Zannier (2006), On the integral points on certain surfaces, *Int. Math. Res. Not.* **Art.ID 98623**, 20 pp.

Corvaja, P. and U. Zannier (2008), Applications of the Subspace Theorem to certain Diophantine problems: a survey of some recent results, in: *Diophantine Approximation, Festschrift for Wolfgang Schmidt*, H. P. Schlickewei, K. Schmidt and R. Tichy, eds., Springer Verlag, pp. 161–174.

Daberkow, M., C. Fieker, J. Klüners, M. Pohst, K. Roegner and K. Wildanger (1997), KANT V4, *J. Symbolic Comput.* **24**, 267–283.

David, S. and P. Philippon (1999), Minorations des hauteurs normalisées des sous-variétés des tores, *Ann. Scuola Norm. Sup. Pisa Cl. Sci.* (4) **28**, 489–543, *Errata*, **29**, 729–731.

Delone (Delaunay), B. N. (1930), Über die Darstellung der Zahlen durch die binären kubischen Formen von negativer Diskriminante, *Math. Z*, **31**, 1–26.

Delone, B. N. and D. K. Faddeev (1940), The theory of irrationalities of the third degree (Russian), *Inst. Math. Steklov* **11**, *Acad. Sci. USSR. English translation, Amer. Math. Soc.*, 1964.

Derksen, H. (2007), A Skolem-Mahler-Lech theorem in positive characteristic and finite automata, *Invent. Math.* **168**, 175–244.

Derksen, H. and D. W. Masser (2012), Linear equations over multiplicative groups, recurrences, and mixing I, *Proc. London Math. Soc.* **104**, 1045–1083.

Dombek, D., L. Hajdu and A. Pethő (2014), Representing algebraic integers as linear combinations of units, *Period. Math. Hung.* **68**, 135–142.

Dubois, E. and G. Rhin (1975) Approximation rationnelles simultanées de nombres algébriques réels et de nombres algébriques *p*-adiques, in: Journées Arithmétiques de Bordeaux (Conf. Univ. Bordeaux, 1974), W. W. Adams, ed., *Astérisque* **24/25**, *Soc. Math. France*, pp. 211–227.

Dubois, E. and G. Rhin (1976), Sur la majoration de formes linéaires à coefficients algébriques réels et *p*-adiques. Démonstration d'une conjecture de K. Mahler, *C.R. Acad. Sci. Paris Sér. A-B* **282**, A1211–A1214.

Dvornicich, R. and U. Zannier (2000), On sums of roots of unity, *Monatsh. Math.* **129**, 97–108.

Dyson, F. J. (1947), The approximation of algebraic numbers by rationals, *Acta Math.* **79**, 225–240.

Eichler, M. (1966), *Introduction to the theory of algebraic numbers and functions*, Academic Press.

Elkies, N. D. (1991), ABC implies Mordell, *Int. Math. Res. Not.* **7**, 99–109.

Erdős, P. (1976), Problems in number theory and combinatorics, *Proc. 6th Manitoba Conference on Numerical Math.* pp. 35–58.

Erdős, P., C. L. Stewart and R. Tijdeman (1988), Some Diophantine equations with many solutions, *Compos. Math.* **66**, 37–56.

Erdős, P. and P. Turán (1934), On a problem in the elementary theory of numbers, *Amer. Math. Monthly* **41**, 608–611.

Everest, G. R. and K. Győry (1997), Counting solutions of decomposable form equations, *Acta Arith.* **79**, 173–191.

Evertse, J.-H. (1983), Upper bounds for the numbers of solutions of Diophantine equations, *Ph.D. thesis, University of Leiden, Leiden*. Also published as Math. Centre Tracts No. **168**, CWI, Amsterdam.

Evertse, J.-H. (1984a), On equations in S-units and the Thue-Mahler equation, *Invent. Math.* **75**, 561–584.

Evertse, J.-H. (1984b), On sums of S-units and linear recurrences, *Compos. Math.* **53**, 225–244.

Evertse, J.-H. (1993), Estimates for reduced binary forms, *J. Reine Angew. Math.* **434**, 159–190.

Evertse, J.-H. (1995), The number of solutions of decomposable form equations, *Invent. Math.* **122**, 559–601.

Evertse, J.-H. (1996), An improvement of the quantitative subspace theorem, *Compos. Math.* **101**, 225–311.

Evertse, J.-H. (1997), The number of solutions of the Thue-Mahler equation, *J. Reine Angew. Math.* **482**, 121–149.

Evertse, J.-H. (1998), Lower bounds for resultants, II, in: Number Theory, Diophantine, Computational and Algebraic Aspects, *Proc. Conf. Eger, Hungary, 1996*, K. Győry, A. Pethö, V. T. Sós, eds., Walter de Gruyter, pp. 181–198.

Evertse, J.-H. (1999), The number of solutions of linear equations in roots of unity, *Acta Arith.* **89**, 45–51.

Evertse, J.-H. (2002), Points on subvarieties of tori, in: A Panorama of Number Theory, or the View from Baker's Garden, *Proc. conf. ETH Zürich, 1999*, G. Wüstholz, ed., Cambridge University Press, pp. 214–230.

Evertse, J.-H. (2004), Linear equations with unknowns from a multiplicative group whose solutions lie in a small number of subspaces, *Indag. Math. (N.S.)* **15**, 347–355.

Evertse, J.-H. and R. G. Ferretti (2002), Diophantine inequalities on projective varieties, *Int. Math. Res. Not.* **2002:25**, 1295–1130.

Evertse, J.-H. and R. G. Ferretti (2008), A generalization of the Subspace Theorem with polynomials of higher degree, in: *Diophantine Approximation, Festschrift for Wolfgang Schmidt*, H. P. Schlickewei, K. Schmidt and R. Tichy, eds., Springer Verlag, pp. 175–198.

Evertse, J.-H. and R. G. Ferretti (2013), A further improvement of the Quantitative Subspace Theorem, *Ann. Math.* **177**, 513–590.

Evertse, J.-H., I. Gaál and K. Győry (1989), On the numbers of solutions of decomposable polynomial equations, *Arch. Math.* **52**, 337–353.

Evertse, J.-H. and K. Győry (1985), On unit equations and decomposable form equations, *J. Reine Angew. Math.* **358**, 6–19.

Evertse, J.-H. and K. Győry (1988a), On the number of polynomials and integral elements of given discriminant, *Acta. Math. Hung.* **51**, 341–362.

Evertse, J.-H. and K. Győry (1988b), On the number of solutions of weighted unit equations, *Compos. Math.* **66**, 329–354.

Evertse, J.-H. and K. Győry (1988c), Finiteness criteria for decomposable form equations, *Acta Arith.* **50**, 357–379.

Evertse, J.-H. and K. Győry (1988d), Decomposable form equations, in: New Advances in Transcendence Theory, *Proc. conf. Durham 1986*, A. Baker, ed., pp. 175–202.

Evertse, J.-H. and K. Győry (1989), Thue-Mahler equations with a small number of solutions, *J. Reine Angew. Math.* **399**, 60–80.

Evertse, J.-H. and K. Győry (1991), Effective finiteness results for binary forms with given discriminant, *Compositio Math.*, **79**, 169–204.

Evertse, J.-H. and K. Győry (1992a), Effective finiteness theorems for decomposable forms of given discriminant, *Acta. Arith.* **60**, 233–277.

Evertse, J.-H. and K. Győry (1992b), Discriminants of decomposable forms, in: *New Trends in Probability and Statistics*, F. Schweiger and E. Manstavičius, eds., pp. 39–56.

Evertse, J.-H. and K. Győry (1993), Lower bounds for resultants, I, *Compositio Math.* **88**, 1–23.

Evertse, J.-H. and K. Győry (1997), The number of families of solutions of decomposable form equations, *Acta. Arith.* **80**, 367–394.

Evertse, J.-H. and K. Győry (2013), Effective results for unit equations over finitely generated domains, *Math. Proc. Camb. Phil. Soc.* **154**, 351–380.

Evertse, J.-H. and K. Győry (2016), *Discriminant Equations in Diophantine Number Theory*, Cambridge: Cambridge University Press, to appear.

Evertse, J.-H., K. Győry, C. L. Stewart and R. Tijdeman (1988a), On S-unit equations in two unknowns, *Invent. math.* **92**, 461–477.

Evertse, J.-H., K. Győry, C. L. Stewart and R. Tijdeman (1988b), S-unit equations and their applications, in: New Advances in Transcendence Theory, *Proc. conf. Durham 1986*, A. Baker, ed., pp. 110–174. Cambridge University Press.

Evertse, J.-H., P. Moree, C. L. Stewart and R. Tijdeman (2003), Multivariate equations with many solutions, *Acta Arith.* **107** (2003), 103–125.

Evertse, J.-H. and H. P. Schlickewei (1999), The Absolute Subspace Theorem and linear equations with unknowns from a multiplicative group, in: *Number Theory in Progress, proc. conf. Zakopane 1997 in honour of the 60th birthday of Prof. Andrzej Schinzel*, K. Győry, H. Iwaniec and J. Urbanowicz, eds., Walter de Gruyter, pp. 121–142.

Evertse, J.-H. and H. P. Schlickewei (2002), A quantitative version of the Absolute Subspace Theorem, *J. Reine Angew. Math.* **548**, 21–127.

Evertse, J.-H., H. P. Schlickewei and W. M. Schmidt (2002), Linear equations in variables which lie in a multiplicative group, *Ann. Math.* **155**, 807–836.

Evertse, J.-H. and J.-H. Silverman (1986), Uniform bounds for the number of solutions to $Y^n = f(X)$, *Math. Proc. Camb. Phil. Soc.* **100**, 237–248.

Evertse, J.-H. and U. Zannier (2008), Linear equations with unknowns from a multiplicative group in a function field, *Acta Arith.* **133**, *volume dedicated to the 75th birthday of Wolfgang Schmidt*, 157–170.

Faltings, G. (1983), Endlichkeitssätze für abelsche Varietäten über Zahlkörpern, *Invent. Math.* **73**, 349–366, *Erratum: Invent. Math.* **75** (1984), 381.

Faltings, G. (1991), Diophantine approximation on abelian varieties, *Ann. Math.* **133**, 549–576.

Faltings, G. (1994), The general case of S. Lang's conjecture, in: *Bersotti symposium in Algebraic Geometry (Abano Terme, 1991)*, 175–182, *Perspect. Math.* **15**, Academic Press.

Faltings, G. and G. Wüstholz (1994), Diophantine approximations on projective spaces, *Invent. Math.* **116**, 109–138.

Feldman, N. I. and Y. V. Nesterenko (1998), *Transcendental numbers*, Springer Verlag, Vol. 44 of Encyclopaedia of Math. Sci.

Filipin, A., R. F. Tichy and V. Ziegler (2008), The additive unit structure of pure quartic complex fields, *Funct. Approx. Comment. Math.* **39**, 113–131.

Fincke, U. and M. Pohst (1985), Improved methods for calculating vectors of short length in a lattice, including a complexity analysis, *Math. Comp.* **44**, 463–471.

Frei, C. (2012), On rings of integers generated by their units, *Bull. London Math. Soc.* **44**, 167–182.

Friedman, E. (1989), Analytic formulas for regulators of number fields, *Invent. Math.* **98**, 599–622.

Fröhlich, A. and J. C. Shepherdson (1956), Effective procedures in field theory, *Philos. Trans. Roy. Soc. London, Ser. A* **248**, 407–432.

Gaál, I. (1984), Norm form equations with several dominating variables and explicit lower bounds for inhomogeneous linear forms with algebraic coefficients, *Studia Sci. Math. Hungar* **19**, 399–411.

Gaál, I. (1985), Norm form equations with several dominating variables and explicit lower bounds for inhomogeneous linear forms with algebraic coefficients, II, *Studia Sci. Math. Hungar* **20**, 333–344.

Gaál, I. (1986), Inhomogeneous discriminant form and index form equations and their applications, *Publ. Math. Debrecen* **33**, 1–12.

Gaál, I. (1988a), Integral elements with given discriminant over function fields, *Acta Math. Hungar.* **52**, 133–146.

Gaál, I. (1988b), Inhomogeneous norm form equations over function fields, *Acta Arith.* **51**, 61–73.

Gaál, I. (2002), *Diophantine equations and power integral bases*, Birkhäuser.

Gaál, I. and M. Pohst (2002), On the resolution of relative Thue equations, *Math. Comp.* **71**, no. 237, 429–440 (electronic).

Gaál, I. and M. Pohst (2006a), Diophantine equations over global function fields I, The Thue equation, *J. Number Theory* **119**, 49–65.

Gaál, I. and M. Pohst (2006b), Diophantine equations over global function fields II, S-integral solutions of Thue equations, *Exper. Math.* **15**, 1–6.

Gaál, I. and M. Pohst (2010), Diophantine equations over global function fields IV, S-unit equations in several variables with an application to norm form equations, *J. Number Theory* **130**, 493–506.

Gebel, J., A. Pethő and H. G. Zimmer (1994), Computing integral points on elliptic curves, *Acta Arith.* **67**, 171–192.

Gelfond, A. O. (1934), Sur le septième problème de Hilbert, *Izv. Akad. Nauk SSSR* **7**, 623–630.

Gelfond, A. O. (1935), On approximating transcendental numbers by algebraic numbers, *Dokl. Akad. Nauk SSSR* **2**, 177–182.

Gelfond, A. O. (1940), Sur la divisibilité de la différence des puissances de deux nombres entiers par une puissance d'un idéal premier, *Mat. Sbornik* **7** (49), 7–26.

Gelfond, A. O. (1960), *Transcendental and algebraic numbers*, New York, Dover.

Ghioca, D. (2008), The isotrivial case in the Mordell-Lang theorem, *Trans. Amer. Math. Soc.* **360**, 3839–3856.

Grant, D. (1996), Sequences of Fields with Many Solutions to the Unit Equation, *The Rocky Mountain J. Math.* **26**, 1017–1029.

Granville, A. (1998), ABC allow us to count squarefrees, *Int. Math. Res. Not.* **19**, 991–1009.

Granville, A. and H. M. Stark (2000), *abc* implies no "Siegel zeros" for *L*-functions of characters with negative discriminant, *Invent. Math.* **139**, 509–523.

Green, B. and T. Tao (2008), The primes contain arbitrarily long arithmetic progressions, *Ann. of Math.* **167**, 481–547.

Győry, K. (1971), Sur l'irréductibilité d'une classe des polynômes, I, *Publ. Math. Debrecen* **18**, 289–307.

Győry, K. (1972), Sur l'irréductibilité d'une classe des polynômes, II, *Publ. Math. Debrecen* **19**, 293–326.

Győry, K. (1973), Sur les polynômes à coefficients entiers et de discriminant donné, *Acta Arith.* **23**, 419–426.

Győry, K. (1974), Sur les polynômes à coefficients entiers et de discriminant donné II, *Publ. Math. Debrecen* **21**, 125–144.

Győry, K. (1976), Sur les polynômes à coefficients entiers et de discriminant donné III, *Publ. Math. Debrecen* **23**, 141–165.

Győry, K. (1978a), On polynomials with integer coefficients and given discriminant IV, *Publ. Math. Debrecen* **25**, 155–167.

Győry, K. (1978b), On polynomials with integer coefficients and given discriminant V, p-adic generalizations, *Acta Math. Acad. Sci. Hung.* **32**, 175–190.

Győry, K. (1978/1979), On the greatest prime factors of decomposable forms at integer points, *Ann. Acad. Sci. Fenn., Ser. A I, Math.* **4**, 341–355.

Győry, K. (1979), On the number of solutions of linear equations in units of an algebraic number field, *Comment. Math. Helv.* **54**, 583–600.

Győry, K. (1979/1980), On the solutions of linear diophantine equations in algebraic integers of bounded norm, *Ann. Univ. Sci. Budapest. Eötvös, Sect. Math.* **22–23**, 225–233.

Győry, K. (1980a), Explicit upper bounds for the solutions of some diophantine equations, *Ann. Acad. Sci. Fenn., Ser A I, Math.* **5**, 3–12.

Győry, K. (1980b), Résultats effectifs sur la représentation des entiers par des formes désomposables, *Queen's Papers in Pure and Applied Math.*, No.56.

Győry, K. (1980c), On certain graphs composed of algebraic integers of a number field and their applications I, *Publ. Math. Debrecen* **27**, 229-242.

Győry, K. (1981a), On the representation of integers by decomposable forms in several variables, *Publ. Math. Debrecen* **28**, 89–98.

Győry, K. (1981b), On S-integral solutions of norm form, discriminant form and index form equations, *Studia Sci. Math. Hungar* **16**, 149–161.

Győry, K. (1981c), On discriminants and indices of integers of an algebraic number field, *J. Reine Angew. Math.* **324**, 114–126.

Győry, K. (1982a), Polynomials of given discriminant and integral elements of given discriminant over integral domains, *C. R. Math. Rep. Acad. Sci. Canada* **4**, 75–80.

Győry, K. (1982b), On certain graphs associated with an integral domain and their applications to Diophantine problems, *Publ. Math. Debrecen* **29**, 79–94.

Győry, K. (1982c), On the irreducibility of a class of polynomials III. *J. Number Theory* **15**, 164–181.

Győry, K. (1983), Bounds for the solutions of norm form, discriminant form and index form equations in finitely generated integral domains, *Acta Math. Hung.* **42**, 45–80.

Győry, K. (1984), Effective finiteness theorems for polynomials with given discriminant and integral elements with given discriminant over finitely generated domains, *J. Reine Angew. Math.* **346**, 54–100.

Győry, K. (1990), On arithmetic graphs associated with integral domains, in: *A Tribute to Paul Erdős*, Cambridge University Press, pp. 207–222.

Győry, K. (1992a), Some recent applications of *S*-unit equations, *Astérisque* **209**, 17–38.

Győry, K. (1992b), Upper bounds for the numbers of solutions of unit equations in two unknowns, *Lithuanian Math. J.* **32**, 40–44.

Győry, K. (1992c), On the irreducibility of a class of polynomials IV, *Acta Arith.* **62**, 399–405.

Győry, K. (1993a), On the numbers of families of solutions of systems of decomposable form equations, *Publ. Math. Debrecen* **42**, 65–101.

Győry, K. (1993b), Some applications of decomposable form equations to resultant equations, *Coll. Math.* **65**, 267–275.

Győry, K. (1993c), On the number of pairs of polynomials with given resultant or given semi-resultant, *Acta Sci. Math.* **57**, 515–529.

Győry, K. (1994), On the irreducibility of neighbouring polynomials, *Acta. Arith.* **67**, 283–294.

Győry, K. (1996), Applications of unit equations, in: *Analytic Number Theory, RIMS Kokyusoku* **958**, Kyoto, Japan, pp. 62–78.

Győry, K. (1998), Bounds for the solutions of decomposable form equations, *Publ. Math. Debrecen* **52**, 1–31.

Győry, K. (1999), On the distribution of solutions of decomposable form equations, in: *Number Theory in Progress, Proc. conf. in honour of 60th birthday of Andrzej Schinzel*, K. Győry, H. Iwaniec and J. Urbanowicz, eds., de Gruyter, pp. 237–365.

Győry, K. (2002), Solving diophantine equations by Baker's theory, in: *A Panorama of Number Theory*, Cambridge, pp. 38–72.

Győry, K. (2006), Polynomials and binary forms with given discriminant, *Publ. Math. Debrecen* **69**, 473–499.

Győry, K. (2008a), On the abc-conjecture in algebraic number fields, *Acta Arith.* **133**, 281–295.

Győry, K. (2008b), On certain arithmetic graphs and their applications to diophantine problems, *Funct. Approx. Comment. Math.*, **39**, 289–314.

Győry, K. (2010), *S*-unit equations in number fields: effective results, generalizations, ABC-conjecture, in: *Analytic number theory and related topics, RIMS Kokyusoku* **1710**, pp. 71–84.

Győry, K., L. Hajdu and R. Tijdeman (2011), Irreducibility criteria of Schur-type and Pólya-type, *Monatsh. Math.* **163**, 415–443.

Győry, K., L. Hajdu and R. Tijdeman (2014), Representation of finite graphs as difference graphs of *S*-units, I, *J. Combinatorial Theory, Ser. A*, **127**, 314–335.

Győry, K. and Z. Z. Papp (1977), On discriminant form and index form equations, *Studia Sci. Math. Hungar.* **12**, 47–60.

Győry, K. and Z. Z. Papp (1978), Effective estimates for the integer solutions of norm form and discriminant form equations, *Publ. Math. Debrecen* **25**, 311–325.

Győry, K. and A. Pethő (1980), Über die Verteilung der Lösungen von Normformen Gleichungen III, *Acta Arith.* **37**, 143–165.

Győry, K., I. Pink and Á. Pintér (2004), Power values of polynomials and binomial Thue-Mahler equations, *Publ. Math. Debrecen* **65**, 341–362.

Győry, K. and Á. Pintér (2008), Polynomial powers and a common generalization of binomial Thue-Mahler equations and *S*-unit equations, in: *Diophantine Equations, Proc. conf. in honour of Tarlok Shorey's 60th birthday*, N. Saradha, ed., New Delhi, pp. 103–119.

Győry, K. and M. Ru (1998), Integer solutions of a sequence of decomposable form inequalities, *Acta Arith.* **86**, 227–237.

Győry, K., A. Sárközy and C. L. Stewart (1996), On the number of prime factors of integers of the form $ab + 1$, *Acta Arith.* **74**, 365–385.

Győry, K. and A. Schinzel (1994), On a conjecture of Posner and Rumsey, *J. Number Theory*, **47**, 63–78.

Győry, K., C. L. Stewart and R. Tijdeman (1986), On prime factors of sums of integers I, *Compositio Math* **59**, 81–88.

Győry, K. and K. Yu (2006), Bounds for the solutions of *S*-unit equations and decomposable form equations, *Acta Arith.* **123**, 9–41.

Hajdu, L. (1993), A quantitative version of Dirichlet's *S*-unit theorem in algebraic number fields, *Publ. Math. Debrecen* **42**, 239–246.

Hajdu, L. (1997), On a problem of Győry and Schinzel concerning polynomials, *Acta Arith.* **78**, 287–295.

Hajdu, L. (2007), Arithmetic progressions in linear combinations of *S*-units, *Period. Math. Hung.* **54**, 175–181.

Hajdu, L. (2009), Optimal systems of fundamental *S*-units for LLL-reduction, *Periodica Math. Hung.* **59**, 79–105.

Hajdu, L. and F. Luca (2010), On the length of arithmetic progressions in linear combinations of *S*-units, *Archiv Math.* **94**, 357–363.

Hajdu, L. and R. Tijdeman (2003), Polynomials dividing infinitely many quadrinomials or quintinomials, *Acta Arith.* **107**, 381–404.

Hajdu, L. and R. Tijdeman (2008), A criterion for polynomials to divide infinitely many *k*-nomials, in: *Diophantine Approximation, Festschrift for Wolfgang Schmidt*, H. P. Schlickewei, K. Schmidt and R. Tichy, eds., Springer Verlag, pp. 175–198.

Halter-Koch, F. and W. Narkiewicz (1997), Polynomial cycles and dynamical units, in: *Proc. Conf. Analytic and Elementary Number Theory*, dedicated to the 80th birthday of E. Hlawka, W. G. Nowak and J. Schoißengeier, eds., Wien, 1997, 70–80.

Halter-Koch, F. and W. Narkiewicz (2000), Scarcity of finite polynomial orbits, *Publ. Math. Debrecen* **56**, 405–414.

Hardy, G. H. and E. M. Wright (1980), *An introduction to the theory of numbers*, 5th. edn., Oxford University Press.

Haristoy, J. (2003), Équations diophantiennes exponentielles, *Thèse de docteur*, Strasbourg.

Harris, J. (1992), *Algebraic Geometry, A First Course*, Springer Verlag.

Hartshorne, R. (1977), *Algebraic Geometry*, Springer Verlag.

Hermann, G. (1926), Die Frage der endlich vielen Schritte in der Theorie der Polynomideale, *Math. Ann.* **95**, 736–788.

Hermite, C. (1851), Sur l'introduction des variables continues dans la théorie des nombres, *J. Reine Angew. Math.* **41**, 191–216.

Hernández, S. and F. Luca (2003), On the largest prime factor of $(ab + 1)(ac + 1)(bc + 1)$, *Bol. Soc. Mat. Mexicana*, **9**, 235–244.

Hindry, M. (1988), Autour d'une Conjecture de Serge Lang, *Invent. Math.* **94**, 575–603.

Houriet, J. (2007), Exceptional units and Euclidean number fields, *Archiv Math.* **88**, 425–433.

Hrushovki, E. (1996), The Mordell-Lang conjecture for function fields, *J. Amer. Math. Soc.* **9**, 667–690.

Hsia, L.-C. and J. T.-Y. Wang (2004), The ABC theorem for higher-dimensional function fields, *Trans. Amer. Math. Soc.* **356**, no. 7, 2871–2887.

Jarden, M. and W. Narkiewicz (2007), On sums of units, *Monatsh. Math.* **150**, 327–332.

de Jong, R. S. (1999), *On p-adic norm form inequalities*, Master thesis, Leiden.

de Jong, R. S. and G. Rémond (2011), Conjecture de Shafarevich effective pour les revêtements cycliques, *Algebra and Number Theory* **5**, 1133–1143.

von Känel, R. (2011), An effective proof of the hyperelliptic Shafarevich conjecture and applications, *Ph.D. thesis*, ETH Zürich.

von Känel, R. (2013), On Szpiro's discriminant conjecture, *Internat. Math. Res. Notices* *1–35*. Published online: doi:10.193/imrn/vnt079.

von Känel, R. (2014a), An effective proof of the hyperelliptic Shafarevich conjecture, *J. Théorie des Nombres, Bordeaux*, **26**, 507–530.

von Känel, R. (2014b) *Modularity and integral points on moduli schemes*, arXiv:1310.7263v2 [math.NT].

Karpilovsky, G. (1988), *Unit groups of classical rings*, Oxford University Press.

Koblitz, N. (1984), *p-adic Numbers, p-adic Analysis, and Zeta-Functions*, Springer Verlag.

Konyagin, S. and K. Soundararajan (2007), Two *S*-unit equations with many solutions, *J. Number Theory* **124**, 193–199.

Kotov, S. V. (1981), Effective bound for a linear form with algebraic coefficients in the archimedean and *p*-adic metrics, *Inst. Math. Akad. Nauk BSSR*, Preprint No. 24, Minsk (Russian).

Kotov, S. V. and V. G. Sprindžuk (1973), An effective analysis of the Thue-Mahler equation in relative fields, *Dokl. Akad. Nauk BSSR* **17**, 393–395 (Russian).

Kotov, S. V. and L. Trelina (1979), *S*-ganze Punkte auf elliptischen Kurven, *J. Reine Angew. Math.* **306**, 28–41.

Kovács, B. (1981), Canonical number systems in algebraic number fields, *Acta Math. Acad. Sci. Hungar.* **37**, 405–407.

Kovács, B. and A. Pethő (1991), Number systems in integral domains, especially in orders of algebraic number fields, *Acta Sci. Math.* **55**, 287–299.

Koymans, P. (2015), *The Catalan Equation*, Master thesis, Leiden University.

Lagarias, J. C. and K. Soundararajan (2011), Smooth solutions to the abc equation: the *xyz* conjecture, *J. Théorie des Nombres de Bordeaux* **23**, 209–234.

Lagrange, J. L. (1773), Recherches d'arithmétiques, *Nouv. Mém. Acad. Berlin*, 265–312; Oeuvres III, 693–758.

Landau, E. (1918), Verallgemeinerung eines Pólyaschen Satzes auf algebraische Zahlkörper, *Nachr. Ges. Wiss.* Göttingen, 478–488.

Lang, S. (1960), Integral points on curves, *Inst. Hautes Études Sci. Publ. Math.* **6**, 27–43.

Lang, S. (1962), *Diophantine geometry*, Wiley.

Lang, S. (1970), *Algebraic Number Theory*, Addison-Wesley.

Lang, S. (1978), *Elliptic curves: Diophantine analysis*, Springer Verlag.

Lang, S. (1983), *Fundamentals of Diophantine Geometry*, Springer Verlag.

Lang, S. (1984), *Algebra*, 2nd. edn., Addison-Wesley.

Langevin, M. (1999), Liens entre le théorème de Mason et la conjecture (abc), in: *Number Theory (5th conf. of CNTA, Ottawa ON 1996)*, R. Gupta and K. S. Williams, eds. 187–213. *CRM Proc. Lecture Notes* 19, AMS, Providence RI.

Laurent, M. (1984), Équations diophantiennes exponentielles, *Invent. Math.* **78**, 299–327.

Laurent, M. (1989), Équations exponentielles polynômes et suites récurrentes linéaires, II, *J. Number Theory* **31**, 24–53.

Lech, C. (1953), A note on recurring series, *Ark. Math.* **2**, 417–421.

Lehmer, D. H. (1933), Factorization of certain cyclotomic functions, *Ann. Math.* (2) **34**, 461–479.

Leitner, D. J. (2012), Linear equations over multiplicative groups in positive characteristic, *Acta Arith.* **153**, 325–347.

Lenstra Jr., H. W. (1977), Euclidean number fields of large degree, *Inventiones Math.* **38**, 237–254.

Lenstra, A. K., H. W. Lenstra Jr. and L. Lovász (1982), Factoring polynomials with rational coefficients, *Math. Ann.* **261**, 515–534.

Leutbecher, A. (1985), Euclidean fields having a large Lenstra constant, *Ann. Inst. Fourier* **35**, 83–106.

Leutbecher, A. and J. Martinet (1982), Lenstra's constant and euclidean number fields, *Astérisque* **94**, 87–131.

Leutbecher, A. and G. Niklasch (1989), On cliques of exceptional units and Lenstra's construction of Euclidean fields, *Lecture Notes Math.* **1380**, 150–178.

LeVeque, W. J. (1964), On the equation $y^m = f(x)$, Acta Arith. **9**, 209–219.

LeVesque, C. and M. Waldschmidt (2011), Some remarks on diophantine equations and diophantine approximation, *Vietnam J. Math.* **39**, 343–368.

LeVesque, C. and M. Waldschmidt (2012), Familles d'équations de Thue-Mahler n'ayant que des solutions triviales, *Acta Arith.* **155**, 117–138.

Levin, A. (2006), One-parameter families of unit equations, *Math. Res. Lett.* **13**, 935–945.

Levin, A. (2008), The dimension of integral points and holomorphic curves on the complements of hyperplanes, *Acta Arith.* **134**, 259–270.

Levin, A. (2014), Lower bounds in logarithms and integral points on higher dimensional varieties, *Algebra Number Theory* **8**, 647–687.

Lewis, D. J. and K. Mahler (1961), Representation of integers by binary forms, *Acta Arith.* **6**, 333–363.

Liardet, P. (1974), Sur une conjecture de Serge Lang, *C.R. Acad. Sci. Paris* **279**, 435–437.

Liardet, P. (1975), Sur une conjecture de Serge Lang, *Astérisque* **24–25**, Soc. Math. France.

Liu, J. (2015), On p-adic Decomposable Form Inequalities, *Ph.D. thesis*, Leiden.

Loher, T. and D. Masser (2004), Uniformly counting points of bounded height, *Acta Arith.* **111**, 277–297.

Louboutin, S. (2000), Explicit bounds for residues of Dedekind zeta functions, values of L-functions at $s = 1$, and relative class numbers, *J. Number Theory* **85**, 263–282.

Loxton, J. H. and A. J. van der Poorten (1983), Multiplicative dependence in number fields, *Acta Arith.* **42**, 291–302.

Luca, F. (2005), On the greatest common divisor of $u - 1$ and $v - 1$ with u and v near S-units, *Monatsh. Math.* **146**, 239–256.

Mahler, K. (1933a), Zur Approximation algebraischer Zahlen I: Über den grössten Primteiler binärer Formen, *Math. Ann.* **107**, 691–730.

Mahler, K. (1933b), Zur Approximation algebraischer Zahlen III: Über die mittlere Anzahl grosser Zahlen durch binäre Formen, *Acta Math.* **62**, 91–166.

Mahler, K. (1935a), Eine arithmetische Eigenschaft der Taylor-koeffizienten rationaler Functionen, *Proc. Kon. Ned. Akad. Wetensch.* **38**, 50–60.

Mahler, K. (1935b), Über transzendente p-adische Zahlen, *Compos. Math.* **2**, 259–275.

Mann, H. B. (1965), On linear relations between roots of unity, *Mathematika* **12**, 107–117.

Mason, R. C. (1983), The hyperelliptic equation over function fields, *Math. Proc. Camb. Phil. Soc.* **93**, 219–230.

Mason, R. C. (1984), *Diophantine equations over function fields*, Cambridge University Press.

Mason, R. C. (1986a), Norm form equations I, *J. Number Theory* **22**, 190–207.

Mason, R. C. (1986b), Norm form equations III: positive characteristic, *Math. Proc. Camb. Phil. Soc.* **99**, 409–423.

Mason, R. C. (1987), Norm form equations V. Degenerate modules, *J. Number Theory* **25**, 239–248.

Mason, R. C. (1988), The study of Diophantine equations over function fields, in: *New Advances in Transcendence Theory, Proc. conf. Durham* 1986, A. Baker, ed., Cambridge University Press, pp. 229–247.

Masser, D. W. (1985), Conjecture in "Open Problems" section, in: *Proc. Symposium on Analytic Number Theory*, London, 25.

Masser, D. W. (2002), On abc and discriminants, *Proc. Amer. Math. Soc.* **130**, 3141–3150.

Masser, D. W. (2004), Mixing and linear equations over groups in positive characteristic, *Israel J. Math.* **142**, 189–204.

Masser, D. W. and G. Wüstholz (1983), Fields of large transcendence degree generated by values of elliptic functions, *Invent. Math.* **72**, 407–464.

Matveev, E. M. (2000), An explicit lower bound for a homogeneous rational linear form in logarithms of algebraic numbers, II. *Izvestiya: Mathematics* **64**, 1217–1269.

McQuillan, M. (1995), Division points on semi-abelian varieties, *Invent. Math.* **120** (1995), 143–159.

Mestre, J. F. (1981), Corps euclidiens, unités exceptionnelles et courbes elliptiques, *J. Number Theory* **13**, 123–137.

Minkowski, H. (1910), Geometrie der Zahlen, Teubner (Posthumously published; prepared by D. Hilbert and A. Speiser).

Moosa, R. and T. Scanlon (2002), The Mordell-Lang conjecture in positive characteristic revisited, In: *Model theory and applications, Quaderni di matematica 11*, L. Belair, Z. Chatzidakis, P. D'Aquino, D. Marker, M. Otero, F. Point and A. Wilkie, eds. Dipartimento di Matematica Seconda Università di Napoli. pp. 273–296.

Moosa, R. and T. Scanlon (2004), F-structures and integral points on semiabelian varieties over finite fields, *Amer. J. Math.* **126**, 473–522.

Mordell, L. J. (1922a), On the rational solutions of the indeterminate equations of the third and fourth degrees, *Proc. Cambridge Philos. Soc.* **21**, 179–192.

Mordell, L. J. (1922b), Note on the integer solutions of the equation $Ey^2 = Ax^3 + Bx^2 + Cx + D$, *Messenger Math.* **51**, 169–171.

Mordell, L. J. (1923), On the integer solutions of the equation $ey^2 = ax^3 + bx^2 + cx + d$, *Proc. London Math. Soc.* (2) **21**, 415–419.

Moree, P. and C. L. Stewart (1990), Some Ramanujan-Nagell equations with many solutions, *Indag Math. (N. S.)*, **1**, 465–472.

Morton, P. and J. H. Silverman (1994), Rational periodic points of rational functions, *Intern. Math. Res. Not.* (2), 97–110.

Mueller, J. (2000), S-unit equations in function fields via the abc-theorem, *Bull. London Math. Soc.* **32**, 163–170.

Murty, M. R. and H. Pasten (2013), Modular forms and effective Diophantine approximation, *J. Number Theory* **133**, 3739–3754.

Nagell, T. (1930), Zur Theorie der kubischen Irrationalitäten, *Acta Math.* **55**, 33–65.

Nagell, T. (1964), Sur une propriété des unités d'un corps algébrique, *Arkiv för Mat.* **5**, 343–356.

Nagell, T. (1967), Sur les discriminants des nombres algébriques, *Arkiv för Mat.* **7**, 265–282.

Nagell, T. (1968a), Quelques propriétés des nombres algébriques du quatrième degré, *Arkiv för Mat.* **7**, 517–525.

Nagell, T. (1968b), Sur les unités dans les corps biquadratiques primitifs du premier rang, *Arkiv för Mat.* **7**, 359–394.

Nagell, T. (1970), Sur un type particulier d'unités algébriques, *Arkiv för Mat.* **8**, 163–184.

Narkiewicz, W. (1989), Polynomial cycles in algebraic number fields, *Colloq. Math.* **58**, 149–153.

Narkiewicz, W. (1995), Polynomial Mappings, *Lecture Notes Math.* **1600**, Springer Verlag.

Narkiewicz, W. and T. Pezda (1997), Finite Polynomial Orbits in Finitely Generated Domains, *Monatsh. Math.* **124**, 309–316.

Neukirch, J. (1992), *Algebraische Zahlentheorie*, Springer Verlag.

Nishioka, K. (1986), Proof of Masser's Conjecture on the Algebraic Independence of Values of Liouville Series, *Proc. Japan Acad. Ser. A* **62**, 219–222.

Nishioka, K. (1987), Conditions for algebraic independence of certain power series of algebraic numbers, *Compos. Math.* **62**, 53–61.

Nishioka, K. (1989), Evertse theorem in algebraic independence, *Arch. Math.* **53**, 159–170.

Nishioka, K. (1994), Algebraic independence by Mahler's method and S-unit equations, *Compos. Math.* **92**, 87–110.

Nishioka, K. (1996), Mahler Functions and Transcendence, *Lecture Notes Math.* **1631**, Springer Verlag.

Northcott, D. G. (1950), Periodic points on an algebraic variety, *Ann. Math.* **51**, 167–177.

Parry, C. J. (1950), The p-adic generalization of the Thue-Siegel theorem, *Acta Math.* **83**, 1–100.

Pasten, H. (2014), Arithmetic problems around the abc-conjecture and connections with logic, *Ph.D. thesis*, Queen's University, Canada.

Pethő, A. and R. Schulenberg (1987), Effektives Lösen von Thue Gleichungen, *Publ. Math. Debrecen* **34**, 189–196.

Pethő, A. and B. M. M. de Weger (1986), Products of prime powers in binary recurrence sequences I. The hyperbolic case, with an application to the generalized Ramanujan-Nagell equation, *Math. Comp.* **47**, 713–727.

Pezda, T. (1994), Polynomial cycles in certain local domains, *Acta Arith.* **66**, 11–22.

Pezda, T. (2014), An algorithm determining cycles of polynomial mappings in integral domains, *Publ. Math. Debrecen* **84**, 399–414.

Poe, M. (1997), On distribution of solutions of S-unit equations, *J. Number Theory* **62**, 221–241.

Pohst, M. E. (1993), *Computational Algebraic Number Theory*, Birkhäuser Verlag.

Pohst, M. E. and H. Zassenhaus (1989), *Algorithmic algebraic number theory*, Cambridge University Press.

Poonen, R. (1999), Mordell-Lang plus Bogomolov, *Invent. Math.* **137**, 413–425.

van der Poorten, A. J. and H. P. Schlickewei (1982), The growth condition for recurrence sequences, *Macquarie University Math. Rep.* 82–0041.

van der Poorten, A. J. and H. P. Schlickewei (1991), Additive relations in fields, *J. Austral. Math. Soc.* (Ser. A) **51**, 154–170.

Posner, E. C. and H. Rumsey, Jr. (1965), Polynomials that divide infinitely many trinomials, *Michigan Math. J.*, **12**, 339–348.

Rémond, G. (2000a), Inégalité de Vojta en dimension supérieure, *Ann. Scuola Norm. Sup. Pisa Cl. Sci.* (4) **29**, 101–151.

Rémond, G. (2000b), Décompte dans une conjecture de Lang, *Invent. Math.* **142**, 513–545.

Rémond, G. (2002), Sur les sous-variétés des tores, *Compos. Math.* **134**, 337–366.

Rémond, G. (2003), Approximation diophantienne sur les variétés semi-abeliennes, *Ann. Sci. École Norm. Sup.* (4) **36**, 191–212.

Ridout, P. (1958), The p-adic generalization of the Thue-Siegel-Roth Theorem, *Mathematika* **5**, 40–48.

Robert, O., C. L. Stewart and G. Tenenbaum (2014), A refinement of the abc conjecture, *Bull. London Math. Soc.* **46**, 1156–1166.

Roquette, P. (1957), Einheiten und Divisorenklassen in endlich erzeugbaren Körpern, *Jahresber. Deutsch. Math. Verein* **60**, 1–21.

Rosser, J. B. and L. Schoenfeld (1962), Approximate formulas for some functions of prime numbers, *Illinois J. Math.* **6**, 64–94.

Roth, K. F. (1955), Rational approximations to algebraic numbers, *Mathematika* **2**, 1–20.

Ru, M. and P. Vojta (1997), Schmidt's subspace theorem with moving targets, *Invent. Math.* **127**, 51–65.

Ru, M. and P. M. Wong (1991), Integral points of $\mathbb{P}^n \setminus \{2n + 1$ hyperplanes in general position$\}$, *Invent. Math.* **106**, 195–216.

Schinzel, A. (1988), Reducibility of lacunary polynomials VIII, *Acta Arith.* **50**, 91–106.

Schlickewei, H. P. (1976a), Linearformen mit algebraischen Koeffizienten, *Manuscripta Math.* **18**, 147–185.

Schlickewei, H. P. (1976b), Die p-adische Verallgemeinerung des Satzes von Thue-Siegel-Roth-Schmidt, *J. Reine Angew. Math.* **288**, 86–105.

Schlickewei, H. P. (1976c), On products of special linear forms with algebraic coefficients, *Acta Arith.* **31**, 389–398.

Schlickewei, H. P. (1977a), Über die diophantische Gleichung $x_1 + \cdots + x_n = 0$, *Acta Arith.* **33** (1977), 183–185.

Schlickewei, H. P. (1977b), The p-adic Thue-Siegel-Roth-Schmidt theorem, *Arch. Math. (Basel)* **29**, 267–270.

Schlickewei, H. P. (1977c), On norm form equations, *J. Number Theory* **9**, 370–380.

Schlickewei, H. P. (1977d), On linear forms with algebraic coefficients and Diophantine equations, *J. Number Theory* **9**, 381–392.

Schlickewei, H. P. (1977e), Inequalities for decomposable forms, *Astérisque* **41–42**, pp. 267–271.

Schlickewei, H. P. (1990), *S*-unit equations over number fields, *Invent. Math.* **102**, 95–107.

Schlickewei, H. P. (1992), The quantitative Subspace Theorem for number fields, *Compos. Math.* **82**, 245–273.

Schlickewei, H. P. (1996a), Multiplicities of recurrence sequences, *Acta Math.* **176**, 171–243.

Schlickewei, H. P. (1996b), Equations in roots of unity, *Acta Arith.* **76**, 99–108.

Schlickewei, H. P. and W. M. Schmidt (2000), The Number of Solutions of Polynomial-Exponential Equations, *Compos. Math.* **120**, 193–225.

Schlickewei, H. P. and C. Viola (1997), Polynomials that divide many trinomials, *Acta Arith.* **78**, 267–273.

Schlickewei, H. P. and C. Viola (1999), Polynomials that divide many *k*-nomials, in: *Number Theory in Progress*, Vol. I, *Proc. conf. in honour of the 60th birthday of Andrzej Schinzel*, K. Győry, H. Iwaniec and J. Urbanowicz eds. de Gruyter, pp. 445–450.

Schlickewei, H. P. and E. Wirsing (1997), Lower bounds for the heights of solutions of linear equations, *Invent. Math.* **129**, 1–10.

Schmidt, W. M. (1971), Linearformen mit algebraischen Koeffizienten II, *Math. Ann.* **191**, 1–20.

Schmidt, W. M. (1972), Norm form equations, *Ann. Math.* **96**, 526–551.

Schmidt, W. M. (1973), Inequalities for resultants and for decomposable forms, in: *Diophantine Approximation and its Applications*, Academic Press, pp. 235–253.

Schmidt, W. M. (1975), Simultaneous approximation to algebraic numbers by elements of a number field, *Monatsh. Math.* **79**, 55–66.

Schmidt, W. M. (1978), Thue's equation over function fields, *J. Austral. Math. Soc. Ser A* **25**, 385–422.

Schmidt, W. M. (1980), Diophantine Approximation, *Lecture Notes Math.* **785**, Springer Verlag.

Schmidt, W. M. (1989), The subspace theorem in diophantine approximation, *Compos. Math.* **96**, 121–173.

Schmidt, W. M. (1990), The number of solutions of norm form equations, *Trans. Amer. Math. Soc.* **317**, 197–227.

Schmidt, W. M. (1991), Diophantine Approximations and Diophantine Equations, *Lecture Notes Math.* **1467**, Springer Verlag.

Schmidt, W. M. (1992), Integer points on curves of genus 1, *Compositio Math.* **81**, 33–59.

Schmidt, W. M. (1996), Heights of points on subvarieties of $\mathbb{G}_m^n$, In: *Number Theory 1993–94, London Math. Soc. Lecture Note Ser.* **235**, S. David, ed., 157–187. Cambridge University Press.

Schmidt, W. M. (1999), The zero multiplicity of linear recurrence sequences, *Acta Math.* **182**, 243–282.

Schmidt, W. M. (2000), Zeros of linear recurrence sequences, *Publ. Math. Debrecen* **56**, 609–630.

Schmidt, W. M. (2003), Linear recurrence sequences, in: Diophantine Approximation, C.I.M.E. Summer school, Cetraro, Italy, June 28–July 6, 2000, F. Amoroso, U. Zannier, eds., *Lecture Notes Math.* **1819**, Springer Verlag, pp. 171–247.

Schmidt, W. M. (2009), The Diophantine equation $\alpha_1^{x_1} \cdots \alpha_n^{x_n} = f(x_1, \ldots, x_n)$, in: *Analytic Number Theory. Essays in Honour of Klaus Roth*, W. W. L. Chen, W. T. Gowers, H. Halberstem and W. M. Schmidt, eds., pp. 414–420. Cambridge University Press.

Schneider, T. (1934), Transzendenzuntersuchungen periodischer Funktionen: I Transzendenz von Potenzen; II Transzendenzeigenschaften elliptischer Funktionen, *J. Reine Angew. Math.* **172**, 65–74.

Sehgal, S. (1978), *Topics in Group Rings*, Marcel Dekker.

Seidenberg, A. (1974), Constructions in algebra, *Trans. Amer. Math. Soc.* **197**, 273–313.

Serre, J.-P. (1989), Lectures on the Mordell-Weil theorem, *Aspects of Math. E15*, Vieweg.

Shorey, T. N. and R. Tijdeman (1986), *Exponential Diophantine Equations*, Cambridge University Press.

Siegel, C. L. (1921), Approximation algebraischer Zahlen, *Math. Z.* **10**, 173–213.

Siegel, C. L. (1926), The integer solutions of the equation $y^2 = ax^n + bx^{n-1} + \cdots + k$, *J. London Math. Soc.* **1**, 66–68.

Siegel, C. L. (1929), Über einige Anwendungen diophantischer Approximationen, *Abh. Preuss. Akad. Wiss., Phys. Math. Kl.*, No. 1.

Siegel, C. L. (1969), Abschätzung von Einheiten, *Nachr. Göttingen*, 71–86.

Silverman, J. H. (1984), The S-unit equation over function fields, *Math. Proc. Camb. Phil. Soc.* **95**, 3–4.

Silverman, J. H. (1995), Exceptional units and numbers of small Mahler measure, *Experiment. Math.* **4**, 70–83.

Silverman, J. H. (2007), *The arithmetic of dynamical systems*, Springer Verlag.

Simmons, H. (1970), The solution of a decision problem for several classes of rings, *Pacific J. Math.* **34**, 547–557.

Simon, D. (2001), The index of nonmonic polynomials, *Indag. Math. (N.S)* **12**, 505–517.

Skolem, Th. (1933), Einige Sätze über gewisse Reihenentwicklungen und exponentiale Beziehungen mit Anwendung auf diophantische Gleichungen, *Oslo Vid. akad. Skrifter* **6**, 1–61.

Skolem, Th. (1935), Ein Verfahren zur Behandlung gewisser exponentialer Gleichungen, *8. Skand. Mat.-Kongr. Stockholm* 163–188.

Smart, N. (1995), The solution of triangularly connected decomposable form equations, *Math. Comp.* **64**, 819–840.

Smart, N. P. (1997), S-unit equations, binary forms and curves of genus 2, *Proc. London Math. Soc.* (3) **75**, 271–307.

Smart, N. P. (1998), *The Algorithmic Resolution of Diophantine Equations*, Cambridge University Press.

Smart, N. P. (1999), Determining the small solutions to S-unit equations, *Math. Comput.* **68**, 1687–1699.

Sprindžuk, V. G. (1969), Effective estimates in "ternary" exponential diophantine equations (Russian), *Dokl. Akad. Nauk BSSR*, **13**, 777–780.

Sprindžuk, V. G. (1973), Squarefree divisors of polynomials and class numbers of algebraic number fields (Russian), *Acta Arith.* **24**, 143–149.

Sprindžuk, V. G. (1974), Representation of numbers by the norm forms with two dominating variables, *J. Number Theory*, **6**, 481–486.

Sprindžuk, V. G. (1976), A hyperelliptic diophantine equation and class numbers (Russian), *Acta Arith.* **30**, 95–108.

Sprindžuk, V. G. (1982), Classical Diophantine Equations in Two Unknowns (Russian), *Nauka*.

Sprindžuk, V. G. (1993), Classical Diophantine Equations, *Lecture Notes Math.* **1559**, Springer Verlag.

Stewart, C. L. and R. Tijdeman (1986), On the Oesterlé-Masser conjecture, *Monatsh. Math.* **102**, 251–257.

Stewart, C. L. and K. Yu (1991), On the abc conjecture, *Math. Ann.* **291**, 225–230.

Stewart, C. L. and K. Yu (2001), On the *abc* conjecture, II, *Duke Math. J.* **108**, 169–181.

Stothers, W. W. (1981), Polynomial identities and Hauptmodulen, *Quart. J. Math. Oxford Ser.* (2) **32**, 349–370.

Stroeker, R. J. and N. Tzanakis (1994), Solving elliptic Diophantine equations by estimating linear forms in elliptic logarithms, *Acta Arith.* **67**, 177–196.

Sunley, J. S. (1973), Class numbers of totally imaginary quadratic extensions of totally real fields, *Trans. Amer. Math. Soc.* **175**, 209–232.

Surroca, A. (2007), Sur l'effectivité du théorème de Siegel et la conjecture abc, *J. Number Theory*, **124**, 267-290.

Szemerédi, E. (1975), On sets of integers containing no *k* elements in arithmetic progression, *Acta Arith.* **27**, 299–345.

Taylor, R. and A. Wiles (1995), Ring-theoretic properties of certain Hecke algebras, *Ann. Math.* (2) **141**, 553–572.

Teske, E. (1998), A space efficient algorithm for group structure computation, *Math. Comp.* **67**, 1637–1663.

Thue, A. (1909), Über Annäherungswerte algebraischer Zahlen, *J. Reine Angew. Math.* **135**, 284–305.

Thunder, J. L. (2001), Decomposable Form Inequalities, *Ann. Math.* **153**, 767–804.

Thunder, J. L. (2005), Asymptotic estimates for the number of integer solutions to decomposable form inequalities, *Compos. Math.* **141** (2005), 271–292.

Tichy, R. F and V. Ziegler (2007), Units generating the ring of integers of complex cubic fields, *Colloq. Math.* **109**, 71–83.

Tzanakis, N. (2013), *Elliptic Diophantine Equations*, de Gruyter.

Tzanakis, N. and B. M. M. de Weger (1989), On the practical solution of the Thue equation, *J. Number Theory* **31**, 99–132.

Vaaler, J. (2014), Heights on groups and small multiplicative dependencies, *Trans. Amer. Math. Soc.* **366**, 3295–3323.

Vojta, P. (1983), Integral points on varieties, *Ph.D.-thesis*, Harvard University.

Vojta, P. (1987), Diophantine Approximation and Value Distribution Theory, *Lecture Notes in Math.* **1239**. Springer Verlag.

Vojta, P. (1996), Integral points on subvarieties of semiabelian varieties, I, *Invent Math.* **126**, 133–181.

Vojta, P. (2000), On the ABC-conjecture and diophantine approxination by rational points, *Amer. J. Math.* **122**, 843–872. Correction, *Amer. J. Math.* **123** (2001), 383–384.

Voloch, J. F. (1985), Diagonal equations over function fields, *Bol. Soc. Bras. Mat.* **16**, 29–39.

Voloch, J. F. (1998), The equation $ax + by = 1$ in characteristic p, *J. Number Th.* **73**, 195–200.

Voutier, P. (1996), An effective lower bound for the height of algebraic numbers, *Acta Arith.* **74**, 81–95.

Voutier, P. (2014), Modules with many non-associates and norm form equations with many families of solutions, *J. Number Theory* **138**, 20–36.

van der Waerden, B. L. (1927), Beweis einer Baudetschen Vermutung, *Nieuw. Arch. Wisk.* (2) **15**, 212–216.

Waldschmidt, M. (1973), Propriétés arithmétiques des valeurs de fonctions méromorphes algébriquement indépendantes, *Acta Arith.* **23**, 19–88.

Waldschmidt, M. (1974), *Nombres Transcendants*, Springer Verlag.

Waldschmidt, M. (2000), *Diophantine approximation on linear algebraic groups*, Springer Verlag.

Wang, J. T.-Y. (1996), The truncated second main theorem of function fields, *J. Number Theory* **58**, 139–157.

Wang, J. T.-Y. (1999), A note on Wronskians and the ABC theorem, *Manuscripta Math.* **98**, 255–264.

de Weger, B. (1987), Algorithms for Diophantine Equations, *Dissertation*, Centrum voor Wiskunde en Informatica, Amsterdam.

de Weger, B. (1989), Algorithms for Diophantine Equations, *CWI Tract* **65**, Amsterdam.

Wildanger, K. (1997), Über das Lösen von Einheiten- und Indexformgleichungen in algebraischen Zahlkörpern mit einer Anwerdung auf die Bestimmung aller ganzen Punkte einer Mordellschen Kurve, *Dissertation*, Technical University, Berlin.

Wildanger, K. (2000), Über das Lösen von Einheiten- und Indexformgleichungen in algebraischen Zahlkörpern, *J. Number Theory* **82**, 188–224.

Wiles, A. (1995), Modular elliptic curves and Fermat's Last Theorem, *Ann. Math.* (2) **141**, 443–551.

Wirsing, E. (1971), On approximation of algebraic numbers by algebraic numbers of bounded degree, in: *Proc. Sympos. Pure Math.* **20**, Amer. Math. Soc., Providence, pp. 213–247.

Wüstholz, G., ed. (2002), *A panorama of number theory or the view from Baker's garden*, Cambridge University Press.

Yu K. (2007), P-adic logarithmic forms and group varieties III, *Forum Mathematicum*, **19**, 187–280.

Zannier, U. (1993), Some remarks on the S-unit equation in function fields, *Acta Arith.* **64**, 87–98.

Zannier, U. (2003), *Some applications of diophantine approximation to diophantine equations (with special emphasis on the Schmidt subspace theorem)*, Forum.

Zannier, U. (2004), On the integer solutions of exponential equations in function fields, *Ann. Inst. Fourier (Grenoble)* **54**, 849–874.

Zannier, U. (2009), *Lecture notes on Diophantine analysis*, Edizioni della Normale.

Zannier, U. (2012), *Some Problems of Unlikely Intersections in Arithmetic and Geometry*, Princeton University Press.

Zhang, S. (2000), Distribution of almost division points, *Duke Math. J.* **103**, 39–46.

Zieve, M. E. (1996), Cycles of polynomial mappings, *Ph.D. thesis*, University of California, Berkeley.

Glossary of frequently used notation

General

$\|\mathcal{S}\|$	cardinality of a finite set $\mathcal{S}$
$\log^* x$	$\max(1, \log x)$, $\log^* 0 := 1$.
$\log_n^* x$	$\log^*$ iterated n times applied to x
$\ll, \gg$	Vinogradov symbols; $A(x) \ll B(x)$ or $B(x) \gg A(x)$ means that there is a constant $c > 0$ such that $A(x) \geq cB(x)$ for all x in the specified domain
$f(x) = O(g(x))$ as $x \to \infty$	these are constants $c_1, c_2 > 0$ such that $\|f(x)\| \in c_1 g(x)$ for all $x \geq c_2$.
$f(x) = o(g(x))$ as $x \to \infty$	$\lim\limits_{x \to \infty} f(x)/g(x) = 0$.
$\mathbb{Z}_{>0}, \mathbb{Z}_{\geq 0}$	positive integers, non-negative integers
$\mathbb{F}_p$	finite field of p elements.
$\mathbb{P}^n(K)$	n-dimensional projective space over a field K.
A, A^+, A^*	ring (always commutative with 1), additive group of A, group of units of A
$A[X_1, \ldots, X_n]$	ring of polynomials in n variables with coefficients in A
gcd	greatest common divisor
$\mathrm{GL}(n, A), \mathrm{SL}(n, A)$	multiplicative group of $n \times n$-matrices with entries in A and determinant in A^*, resp. determinant 1
L/K	field extension L/K
$\mathrm{Tr}_{L/K}(\alpha), N_{L/K}(\alpha)$	trace, norm of $\alpha \in L$ over K
$D_{L/K}(\omega_1, \ldots, \omega_n)$	discriminant of a K-basis $\{\omega_1, \ldots, \omega_n\}$ of L
$D(f), D(F)$	discriminant of a polynomial $f(X)$, binary form $F(X, Y)$
$R(f, g), R(F, G)$	resultant of polynomials $f(X), g(X)$, binary forms $F(X, Y), G(X, Y)$.

358

Number fields

$\mathrm{ord}_p(a)$	exponent of a prime number p in the unique prime factorization of $a \in \mathbb{Q}$, $\mathrm{ord}_p(0) = \infty$
$\|a\|_p$	$p^{-\mathrm{ord}_p(a)}$, p-adic absolute value of $a \in \mathbb{Q}$
$\|a\|_\infty$	$\max(a, -a)$, ordinary absolute value of $a \in \mathbb{Q}$
$\mathbb{Q}_p$	p-adic completion of $\mathbb{Q}$, $\mathbb{Q}_\infty = \mathbb{R}$
$M_\mathbb{Q}$	$\{\infty\} \cup \{\text{primes}\}$, set of places of $\mathbb{Q}$
O_K, D_K, h_K, R_K	ring of integers, discriminant, class number, regulator of a number field K
$\mathfrak{p}$, $\mathfrak{a}$	non-zero prime ideal, fractional ideal of O_K
$(\alpha) = \alpha O_K$	fractional ideal generated by α
$\mathrm{ord}_\mathfrak{p}(\mathfrak{a})$	exponent of $\mathfrak{p}$ in the unique prime ideal factorization of $\mathfrak{a}$
$\mathrm{ord}_\mathfrak{p}(\alpha)$	exponent of $\mathfrak{p}$ in the unique prime ideal factorization of (α) for $\alpha \in K$, with $\mathrm{ord}_\mathfrak{p}(0) := \infty$.
$N_K(\mathfrak{a})$	absolute norm of a fractional ideal $\mathfrak{a}$ of O_K (written as $N(\mathfrak{a})$ if it is clear which is the underlying number field)
$e(\mathfrak{P}\|\mathfrak{p})$, $f(\mathfrak{P}\|\mathfrak{p})$	ramification index, residue class degree of a prime ideal $\mathfrak{P}$ over a prime ideal $\mathfrak{p}$.
M_K	set of places of a number field K
M_K^∞	set of infinite (archimedean) places of K
M_K^0	set of finite (non-archimedean) places of K, identified with the non-zero prime ideals of O_K
$\|\cdot\|_v$ $(v \in M_K)$	normalized absolute values of K, satisfying the product formula, with $\|\alpha\|_v := N_K(\mathfrak{p})^{-\mathrm{ord}_\mathfrak{p}(\alpha)}$ if $\alpha \in K$ and $v = \mathfrak{p}$ is a prime ideal of O_K
K_v	completion of K at v
S	finite set of places of K, containing M_K^∞
O_S	$\{\alpha \in K : \|\alpha\|_v \le 1 \text{ for } v \in M_K \setminus S\}$, ring of S-integers, written as $\mathbb{Z}_S$ if $K = \mathbb{Q}$
O_S^*	$\{\alpha \in K : \|\alpha\|_v = 1 \text{ for } v \in M_K \setminus S\}$, group of S-units, written as $\mathbb{Z}_S^*$ if $K = \mathbb{Q}$
$N_S(\alpha)$	$\prod_{v \in S} \|\alpha\|_v$, S-norm of $\alpha \in K$
$\|\mathbf{x}\|_v$ $(v \in M_K)$	$\max_i \|x_i\|_v$, v-adic norm of $\mathbf{x} = (x_1, \dots, x_n) \in K^n$
$H^{\mathrm{hom}}(\mathbf{x})$	$\left(\prod_{v \in M_K} \|\mathbf{x}\|_v\right)^{1/[K:\mathbb{Q}]}$, absolute homogeneous height of $\mathbf{x} \in K^n$
$H(\mathbf{x})$	$\left(\prod_{v \in M_K} \max(1, \|\mathbf{x}\|_v)\right)^{1/[K:\mathbb{Q}]}$, absolute height of $\mathbf{x} \in K^n$

$H(\alpha)$ $(\prod_{v \in M_K} \max(1, |\alpha|_v))^{1/[K:\mathbb{Q}]}$, absolute height of $\alpha \in K$

$h^{\mathrm{hom}}(\mathbf{x}), h(\mathbf{x}), h(\alpha)$ $\log H^{\mathrm{hom}}(\mathbf{x}), \log H(\mathbf{x}), \log H(\alpha)$, absolute logarithmic heights

Function fields

$\mathbf{k}$ field of constants (always algebraically closed)

$\mathbf{k}((z))$ field of Laurent series in z

$g_{K/\mathbf{k}}$ genus of function field K with constant field $\mathbf{k}$

M_K set of (normalized discrete) valuations of K, trivial on $\mathbf{k}$

$v(\mathbf{x}) \, (v \in M_K)$ $\min_i v(x_i)$, v-adic norm of $\mathbf{x} = (x_1, \ldots, x_n) \in K^n$

$H_K^{\mathrm{hom}}(\mathbf{x})$ $-\sum_{v \in M_K} v(\mathbf{x})$, homogeneous height of $\mathbf{x} \in K^n$

$H_K(x)$ $\sum_{v \in M_K} \max(0, -v(x))$, height of $x \in K$

Index

361

Printed in the United States
By Bookmasters